Molecular Luminescence Spectroscopy

Methods and Applications: Part 3

CHEMICAL ANALYSIS

A SERIES OF MONOGRAPHS ON
ANALYTICAL CHEMISTRY AND ITS APPLICATIONS

Editor
J. D. WINEFORDNER
Editor Emeritus: **I. M. KOLTHOFF**

VOLUME 77

A WILEY-INTERSCIENCE PUBLICATION

JOHN WILEY & SONS, INC.

New York / Chichester / Brisbane / Toronto / Singapore

Vol.	29.	**The Analytical Chemistry of Sulfur and Its Compounds** (*in three parts*). By J. H. Karchmer
Vol.	30.	**Ultramicro Elemental Analysis.** By Günther Tölg
Vol.	31.	**Photometric Organic Analysis** (*in two parts*). By Eugene Sawicki
Vol.	32.	**Determination of Organic Compounds: Methods and Procedures.** By Frederick T. Weiss
Vol.	33.	**Masking and Demasking of Chemical Reactions.** By D. D. Perrin
Vol.	34.	**Neutron Activation Analysis.** By D. De Soete, R. Gijbels, and J. Hoste
Vol.	35.	**Laser Raman Spectroscopy.** By Marvin C. Tobin
Vol.	36.	**Emission Spectrochemical Analysis.** By Morris Slavin
Vol.	37.	**Analytical Chemistry of Phosphorus Compounds.** Edited by M. Halmann
Vol.	38.	**Luminescence Spectrometry in Analytical Chemistry.** By J. D. Winefordner, S. G. Schulman and T. C. O'Haver
Vol.	39.	**Activation Analysis with Neutron Generators.** By Sam S. Nargolwalla and Edwin P. Przybylowicz
Vol.	40.	**Determination of Gaseous Elements in Metals.** Edited by Lynn L. Lewis, Laben M. Melnick, and Ben D. Holt
Vol.	41.	**Analysis of Silicones.** Edited by A. Lee Smith
Vol.	42.	**Foundations of Ultracentrifugal Analysis.** By H. Fujita
Vol.	43.	**Chemical Infrared Fourier Transform Spectroscopy.** By Peter R. Griffiths
Vol.	44.	**Microscale Manipulations in Chemistry.** By T. S. Ma and V. Horak
Vol.	45.	**Thermometric Titrations.** By J. Barthel
Vol.	46.	**Trace Analysis: Spectroscopic Methods for Elements.** Edited by J. D. Winefordner
Vol.	47.	**Contamination Control in Trace Element Analysis.** By Morris Zief and James W. Mitchell
Vol.	48.	**Analytical Applications of NMR.** By D. E. Leyden and R. H. Cox
Vol.	49.	**Measurement of Dissolved Oxygen.** By Michael L. Hitchman
Vol.	50.	**Analytical Laser Spectroscopy.** Edited by Nicolo Omenetto
Vol.	51.	**Trace Element Analysis of Geological Materials.** By Roger D. Reeves and Robert R. Brooks
Vol.	52.	**Chemical Analysis by Microwave Rotational Spectroscopy.** By Ravi Varma and Lawrence W. Hrubesh
Vol.	53.	**Information Theory As Applied to Chemical Analysis.** By Karel Eckschlager and Vladimir Štěpánek
Vol.	54.	**Applied Infrared Spectroscopy: Fundamentals, Techniques, and Analytical Problem-solving.** By A. Lee Smith
Vol.	55.	**Archaeological Chemistry.** By Zvi Goffer
Vol.	56.	**Immobilized Enzymes in Analytical and Clinical Chemistry.** By P. W. Carr and L. D. Bowers
Vol.	57.	**Photoacoustics and Photoacoustic Spectroscopy.** By Allan Rosencwaig
Vol.	58.	**Analysis of Pesticide Residues.** Edited by H. Anson Moye
Vol.	59.	**Affinity Chromatography.** By William H. Scouten
Vol.	60.	**Quality Control in Analytical Chemistry.** By G. Kateman and F. W. Pijpers
Vol.	61.	**Direct Characterization of Fineparticles.** By Brian H. Kaye
Vol.	62.	**Flow Injection Analysis.** By J. Ruzicka and E. H. Hansen

(*continued on back*)

Molecular Luminescence Spectroscopy

Methods and Applications: Part 3

Edited by

STEPHEN G. SCHULMAN

College of Pharmacy
University of Florida
Gainesville, Florida

A WILEY-INTERSCIENCE PUBLICATION

JOHN WILEY & SONS, INC.

New York / Chichester / Brisbane / Toronto / Singapore

This text is printed on acid-free paper.

Copyright © 1993 by John Wiley & Sons, Inc.

All rights reserved. Published simultaneously in Canada.

Reproduction or translation of any part of this work beyond that permitted by Section 107 or 108 of the 1976 United States Copyright Act without the permission of the copyright owner is unlawful. Requests for permission or further information should be addressed to the Permissions Department, John Wiley & Sons, Inc., 605 Third Avenue, New York, NY, 10158-0012.

Library of Congress Cataloging in Publication Data:
Molecular luminescence spectroscopy.
  (Chemical analysis, ISSN 0069-2883 ; v. 77)
  "A Wiley-Interscience publication."
  Includes bibliographies and indexes.
  1. Luminescence spectroscopy. I. Schulman, Stephen G. (Stephen Gregory), 1940–

QD96.L85M65   1985   543'.085   84-21880
ISBN 0-471-51580-9 (v. 3)

Printed in the United States of America

10 9 8 7 6 5 4 3 2 1

CONTENTS

CONTRIBUTORS	vii
PREFACE	ix
CONTENTS OF PARTS 1 AND 2	xi

1. **CHEMILUMINESCENCE**   1
   By Kenichiro Nakashima and Kazuhiro Imai

2. **FLUORESCENT PROBES FOR EVALUATION OF LOCAL PHYSICAL AND STRUCTURAL PARAMETERS**   25
   By Bernard Valeur

3. **PHOTOCHEMICAL FLUOROMETRY**   85
   By Jean-Jacques Aaron

4. **APPLICATIONS OF ORGANIZED BILE SALT MEDIA FOR LUMINESCENCE ANALYSIS**   133
   By Linda B. McGown

5. **SPECTRAL HOLE-BURNING**   149
   By Keith Holliday and Urs P. Wild

6. **NEAR-INFRARED LUMINESCENCE SPECTROSCOPY**   229
   By Shuzo Akiyama

7. **MICROSPECTROFLUOROMETRY ON SUPPORTED PLANAR MEMBRANES**   253
   By Lukas K. Tamm and Edwin Kalb

8. **CLINICAL APPLICATIONS OF LUMINESCENCE SPECTROSCOPY** 307

   BY GEORGE H. SCHENK

9. **LASER-EXCITED MOLECULAR FLUORESCENCE IN ANALYTICAL SCIENCES** 323

   BY J. W. HOFSTRAAT, C. GOOIJER, AND N. H. VELTHORST

INDEX 445

CONTRIBUTORS

JEAN-JACQUES AARON
Institute of Topology and Systems
  Dynamics
University of Paris 7, associated
  with CNRS
Paris, France

SHUZO AKIYAMA
School of Pharmaceutical
  Sciences
Nagasaki University
Nagasaki, Japan

C. GOOIJER
Department of General and
  Analytical Chemistry
Free University, Amsterdam
The Netherlands

J. W. HOFSTRAAT
Department of Analytical and
  Environmental Chemistry
Akzo Research Laboratories
  Arnhem
Arnhem, The Netherlands

KEITH HOLLIDAY
Physical Chemistry Laboratory
Swiss Federal Institute of
  Technology
Zürich, Switzerland

KAZUHIRO IMAI
Branch Hospital Pharmacy
University of Tokyo
Tokyo, Japan

EDWIN KALB
Department of Biophysical
  Chemistry
Biocenter, University of Basel
Basel, Switzerland

LINDA B. MCGOWN
P. M. Gross Chemical Laboratory
Department of Chemistry
Duke University
Durham, North Carolina

KENICHIRO NAKASHIMA
School of Pharmaceutical
  Sciences
Nagasaki University
Nagasaki, Japan

GEORGE H. SCHENK
Department of Chemistry
Wayne State University
Detroit, Michigan

LUKAS K. TAMM
Department of Physiology
School of Medicine
University of Virginia
Charlottesville, Virginia

BERNARD VALEUR
Laboratoire de Chimie Générale
  (CNRS URA 1103)
Conservatoire National des
  Arts et Métiers
Paris, France

N. H. VELTHORST
Department of General and
  Analytical Chemistry
Free University, Amsterdam
The Netherlands

URS P. WILD
Physical Chemistry Laboratory
Swiss Federal Institute of
  Technology
Zürich, Switzerland

PREFACE

In Part 3 of *Molecular Luminescence Spectroscopy: Methods and Applications* is continued the coverage of fluorescence, phosphorescence, and chemiluminescence spectroscopic subjects that are current and were not included in Parts 1 and 2. Here, some of the chapters such as those on chemiluminescence by Nakashima and Imai and photochemically generated fluorophores by Aaron describe methodology already widespread in laboratory use. Valeur's chapter on fluorescent probes, McGown's on luminescence from bile salt aggregates, which is representative of luminescence from organized supramolecular assemblies, Schenk's on clinical applications of fluorescence spectroscopy, and Hofstraat, Gooijer, and Velthorst's on laser-excited fluorimetry also fall into this category.

The chapter on hole-burning spectroscopy (which complements the chapter in Part 2, on highly resolved spectra) by Holliday and Wild as well as those on laser-excited microspectrofluorometry by Tamm and Kalb and on near-infrared luminescence spectroscopy by Akiyama represent areas of activity that are as yet somewhat esoteric as far as routine analytical use is concerned but are probably only a few years from being widely accepted as common tools in the arsenal of analytical luminescence spectroscopy.

It is hoped that the present volume will contain something of interest to everyone working or contemplating working in the various areas of luminescence spectroscopy. The editor wishes to express his gratitude to the authors of the various chapters in Parts 1–3 for their excellent and timely contributions, and to Ms. Patricia Khan, Ms. Luz Jimenez, and Ms. Elaine Schulman for technical assistance with the preparation of the manuscript.

STEPHEN G. SCHULMAN

Gainesville, Florida
November 1992

CONTENTS OF PARTS 1 AND 2

PART 1

1. **LUMINESCENCE SPECTROSCOPY: AN OVERVIEW**
   By Stephen G. Schulman

2. **FLUORESCENCE AND PHOSPHORESCENCE OF PHARMACEUTICALS**
   By Willy R. G. Baeyens

3. **FLUORESCENCE OF ORGANIC NATURAL PRODUCTS**
   By O. S. Wolfbeis

4. **DETERMINATIONS OF INORGANIC SUBSTANCES BY LUMINESCENCE METHODS**
   By Alberto Fernandez-Gutierrez and Arsenio Muñoz de la Peña

5. **BIOINORGANIC LUMINESCENCE SPECTROSCOPY**
   By Harry G. Brittain

6. **EXCITED-STATE OPTICAL ACTIVITY**
   By Harry G. Brittain

7. **FLUORESCENCE DETECTION IN CHROMATOGRAPHY**
   By A. Hulshoff and H. Lingeman

8. **LUMINESCENCE IMMUNOASSAY**
   By H. Thomas Karnes, Jeffrey S. O'Neal, and Stephen G. Schulman

INDEX

PART 2

1. LUMINESCENCE FROM SOLID SURFACES

   By Robert J. Hurtubise

2. TIME-RESOLVED AND PHASE-RESOLVED EMISSION SPECTROSCOPY

   By J. N. Demas

3. FIBER OPTICAL FLUOROSENSORS IN ANALYTICAL AND CLINICAL CHEMISTRY

   By Otto S. Wolfbeis

4. HIGHLY RESOLVED MOLECULAR LUMINESCENCE SPECTROSCOPY

   By J. W. Hofstraat, C. Gooijer and N. H. Velthorst

5. APPLICATIONS OF LANTHANIDE ION LUMINESCENCE FROM INORGANIC SOLIDS

   By Harry G. Brittain

6. PROTON TRANSFER KINETICS OF ELECTRONICALLY EXCITED ACIDS AND BASES

   By Richard N. Kelly and Stephen G. Schulman

INDEX

CHAPTER

1

CHEMILUMINESCENCE

KENICHIRO NAKASHIMA

School of Pharmaceutical Sciences
Nagasaki University
Nagasaki 852, Japan

KAZUHIRO IMAI

Branch Hospital Pharmacy
University of Tokyo
Tokyo 112, Japan

1.1. Introduction
1.2. Chemiluminescence
1.3. History of Chemiluminescence
1.4. Chemiluminescent Compounds: Reaction Mechanisms and Analytical Applications
   1.4.1. Luminol Derivatives
   1.4.2. Dioxetanes
   1.4.3. Lophine
   1.4.4. Acridine Derivatives
      1.4.4.1. Lucigenin
      1.4.4.2. Acridinium Salts and Acridans
   1.4.5. Tetrakis(dimethylamino)ethylene
   1.4.6. Indole Derivatives
   1.4.7. Schiff Base
   1.4.8. Oxalates and Oxamides
   1.4.9. Diphenoylperoxide
References

Molecular Luminescence Spectroscopy, Part 3, Edited by Stephen G. Schulman. Chemical Analysis Series, Vol. 77.
ISBN 0-471-51580-9 © 1993 John Wiley & Sons, Inc.

1.1. INTRODUCTION

In nature, the phenomenon of light emission, derived from chemical reactions in luminous organisms such as certain bacteria, fungi, insects, and fishes, namely bioluminescence, has been known for a long time (1–4). In 1877, Radziszewski (5), using lophine (2,4,5-triphenylimidazole), found that synthetic compounds could also produce this phenomenon. Since then, numerous chemiluminescent compounds have been synthesized and their chemiluminescent reaction mechanisms and applications to analytical chemistry studied (2–4,6–10). Recently, chemiluminescent systems have been proven to be valuable for trace analysis. The most characteristic property of a chemiluminescent system for analytical use is its remarkable sensitivity. Certain compounds can be detected in the range from femtomole to attomole levels. Therefore, as an alternative to radiometric analysis, interest in chemiluminescent systems has focused on their utility for the determination of trace amounts of organic and inorganic compounds in biomaterials and in environmental samples (2,4,5,9,11).

1.2. CHEMILUMINESCENCE

Luminescence, or "cold light," is the emission of light by substances that have absorbed energy. Several kinds of luminescence are known, classified on the basis of the kind of energy absorbed, e.g., light, x-rays, heat, ultrasound, from chemical reactions, or from biological reactions. Photoluminescence, for example, is the emission of light that occurs when a molecule, excited to an electronically excited state by the absorption of ultraviolet (UV) or visible light, returns to the ground electronic state. In this case, the light emitted when

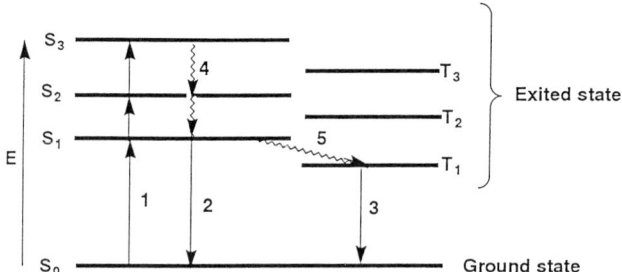

Fig. 1.1. Concept of energy level and transition: 1, absorption; 2, fluorescence; 3, phosphorescence; 4, internal conversion; 5, intersystem crossing; E, energy; S, singlet state; T, triplet state.

the molecule returns to the ground state (S_0) from the lowest singlet excited state (S_1) is called fluorescence. On the other hand, the light emitted when a molecule returns from the triplet excited state to the ground singlet state is called phosphorescence (Fig. 1.1).

Chemiluminescence is the emission of light as the result of electronic excitation of the luminescing species by a chemical reaction of a precursor of that species. The process of emission of light in chemiluminescence is the same as in photoluminescence except for its excitation process. Therefore, the resultant spectra of chemiluminescence are identical to the fluorescence spectra of the emitter. Bioluminescence is a kind of chemiluminescence produced by certain living organisms, e.g., fireflies and jellyfish. Generally, bioluminescence is derived from an oxidation reaction of a small organic molecule, catalyzed by an enzyme. Another type of bioluminescence is caused by a reaction of photoprotein with Ca^{2+} (1–4).

1.3. HISTORY OF CHEMILUMINESCENCE

The phenomenon of chemiluminescence has been known for a long time. In 1877, Radziszewski (5) found that lophine emitted green light when it reacted with oxygen in the presence of a base. This is the first example of chemiluminescence using a synthetic organic compound. Since then, a number of chemiluminescent compounds have been synthesized and studied for their chemiluminescent properties. Among them, luminol (5-amino-2,3-dihydro-1,4-phthalazinedione), whose chemiluminescence was discovered by Albrecht (16) in 1928, is one of the best known chemiluminogenic compounds. The luminol chemiluminescence reaction is catalyzed in the presence of blood (hemin) and produces a blue-white emission. Therefore, this reaction has been utilized to identify blood in many criminal investigations. In 1935, Glue and Petsch (17) demonstrated the chemiluminescence of lucigenin (N,N'-dimethyl-9,9'-biacridinium dinitrate). Like luminol, this compound produces a strong luminescence and has been utilized in analytical chemistry, e.g., for the assay of phagocytic leukocyte oxygenation activities (18). In 1963, Chandross (19) observed visible light upon reaction of hydrogen peroxide with oxalyl chloride. In 1965, Rauhut et al. (20) reviewed the oxalyl chloride chemiluminescent system and showed that oxalyl esters could also be used for this system instead of oxalyl cloride. Since then, they have synthesized numerous oxalates and oxamides and established a new strongly luminescing system, namely, the peroxyoxalate chemiluminescent system; the chemiluminescent compounds such as luminol and lucigenin produce emission of light themselves by oxidation reactions with oxygen or hydrogen peroxide. On the other hand, in peroxyoxalate chemiluminescent systems. The oxalate itself does not produce light. Instead, an acceptor of electronic excitation energy, in this case a fluorescent compound, is excited by the intermediates produced by the reaction of oxalate with hydrogen peroxide and produces emission of light. Therefore, a suitable combination of oxalates with strongly fluorescent compounds yields an intense emission. For example, Tseng et al. (21) observed in 1979, a system having a remarkable chemiluminescent quantum yield of 34%. This will be discussed later.

Generally speaking, chemiluminescent reactions are oxidation reactions. Organic chemiluminescent compounds are excited by oxidation with oxygen or hydrogen peroxide. The mechanism of chemiluminescence has been studied for some two and a half decades. In 1968, McCapra (22) explained the formation mechanism of excited carbonyl compounds by the Woodward–Hoffman rule. In 1968, Kopecky and co-workers first synthesized a simple dioxetane, trimethyl-1,2-dioxetane, and showed that excited acetone or aldehyde was formed by its decomposition (23,24). In 1977/78, Koo and Schuster (25,26) studied the chemiluminescent reaction of diphenoylperoxide and showed the mechanism to be as follows: in the presence of a strong electron donor aromatic hydrocarbon such as rubrene or 9,10-diphenylanthracene, diphenoylperoxide accepts one electron from the hydrocarbon to form an anion radical from which one electron is transferred to the aromatic hydrocarbon to produce its excited state. This mechanism is called CIEEL (chemically initiated electron exchange luminescence). The mechanism of peroxyoxalate chemiluminescence was proposed by Rauhut et al. (20) in 1965. They assumed that the intermediate, 1,2-dioxetanedione, yielded by the reaction of oxalate with hydrogen peroxide, should form a charge-transfer complex with a fluorophore and consequently the fluorophore should be excited to yield emission of light. This mechanism was further investigated by Catherall et al. (27–29) in 1984, and substituted 1,2-dioxetanedione was proposed as a key intermediate alternative to 1,2-dioxetane. This assumption was supported by the results of analysis by computer simulation by Alvarez et al. (30) in 1986.

Though the mechanism of chemiluminescence has been studied widely, the intermediates could not be isolated until now. Therefore, the mechanisms proposed are still assumptive.

1.4. CHEMILUMINESCENT COMPOUNDS: REACTION MECHANISMS AND ANALYTICAL APPLICATIONS

The chemiluminescent reaction mechanisms for organic compounds are so complex that it is difficult to understand them completely. This situation is derived from the difficulties of isolation and identification of the key intermediates in the reaction systems. Nevertheless, reasonable mechanisms have been proposed that are in accord with experimental results of kinetic studies, measurements of chemiluminescent quantum yields, identification of reaction products, and examination of reaction conditions.

Generally, the chemiluminescent processes of organic molecules include the following three steps: (1) formation of intermediate compounds by chemical reactions; (2) excited molecular formation by an intermediate through a transition state; and (3) light emission from an excited molecule or energy transfer and subsequent light emission from a fluorescent energy acceptor. Therefore, the efficiency of chemiluminescent reaction can be determined according to the equation

$$\Phi_{CL} = \Phi_C \times \Phi_E \times \Phi_F$$

where Φ_{CL} is the total chemiluminescent quantum efficiency; Φ_C, the efficiency of the chemical reaction; Φ_E, the efficiency of the energy conversion from chemical potential to electronic excitation energy; and Φ_F, the fluorescence efficiency.

Here, the proposed chemiluminescent reaction mechanisms and analytical applications of representative organic chemiluminescent compounds will be discussed.

1.4.1. Luminol Derivatives

Luminol (5-amino-2,3-dihydro-1,4-phthalazinedione, 1) is the most important and familiar compound among the chemiluminescent compounds and its chemiluminescent reaction mechanism has been extensively studied (31–35). It is well known that the luminol chemiluminescent reaction is catalyzed by many kinds of substances, e.g., ozone, halogen, Fe-complex, hemin, hemoglobin, persulfate, and oxidized transition metals. The reaction mechanism differs according to the reaction conditions used. The most acceptable scheme is shown in Fig. 1.2. Luminol forms a six-membered ring of peroxide (3) from a diazaquinone intermediate (2) and then, by the decomposition of 3, N_2 gas and the S_1 excited state of the phthalate dianion are produced, yielding emission of light. It is certain that the dianion of 2-aminophthalic acid is an emitter. White et al. (32) in 1964 confirmed this fact by the following experimental results: (1) luminol yielded 2-aminophthalic acid in 90% yield by the reaction with NaOH and O_2 in dimethylsulfoxide; (2) the fluorescence spectrum of 2-aminophthalic acid dianion agreed well with the resultant chemiluminescence spectrum; and (3) in the reaction system using $^{18}O_2$, the consumed oxygen was all found in the carbonyl group of the dianion and 1 mol N_2 gas was yielded simultaneously. Diazaquinone (2), which lumines-

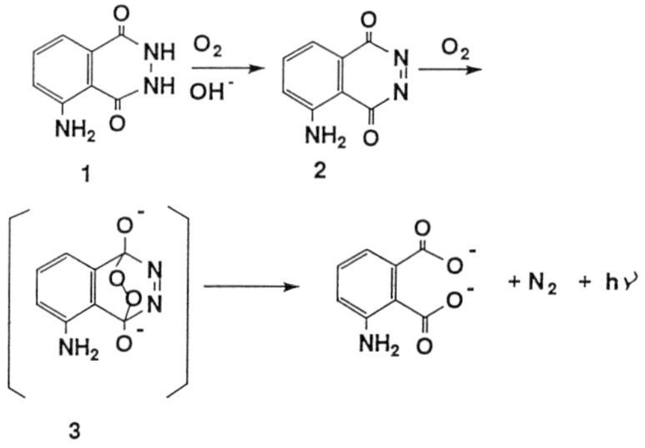

Fig. 1.2. Chemiluminescent reaction of luminol.

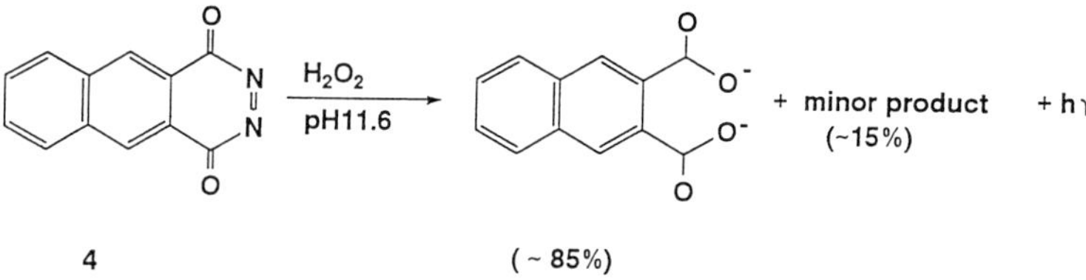

Fig. 1.3. Chemiluminescent reaction of benzo[g]phthalazine-1,4-dione.

cences with hydrogen peroxide under alkaline conditions, was first synthesized by Gundermann (36) in 1968. A chemiluminescent diazaquinone (**4**) was also synthesized by White et al. (37), who demonstrated its chemiluminescent reaction (Fig. 1.3). Several luminol derivatives have also been synthesized, as shown in Fig. 1.4. Among them, benzoperylene derivative (**11**) showed the largest chemiluminescent quantum yield ($\Phi_{CL} \simeq 0.07$) (38). Compounds (**12**) and (**13**) showed no chemiluminescence (39). The noncyclic hydrazides (**14–16**) had less efficiency of luminescence than the cyclic ones (40–42). Thankarajan et al. (43) in 1986 synthesized 2-(3-phthalhydrazidylazo)-1,3-diketones (**17**) and showed that these compounds give more intense chemiluminescence than luminol. From the structural studies of cyclic hydrazides, the following points are elucidated: (1) substitution on the hydrazide ring decreases emission of light because of steric effects; (2) substitution on the benzene ring increases the efficiency of emission; and (3) alkylation of the 3-amino substituent lowers the yield of light.

Analytical uses of luminol chemiluminescent reaction have been widely studied and include determination of trace amounts of hydrogen peroxide (44–48), measurement of oxygen in air (49), glucose assay (50–52), indicator

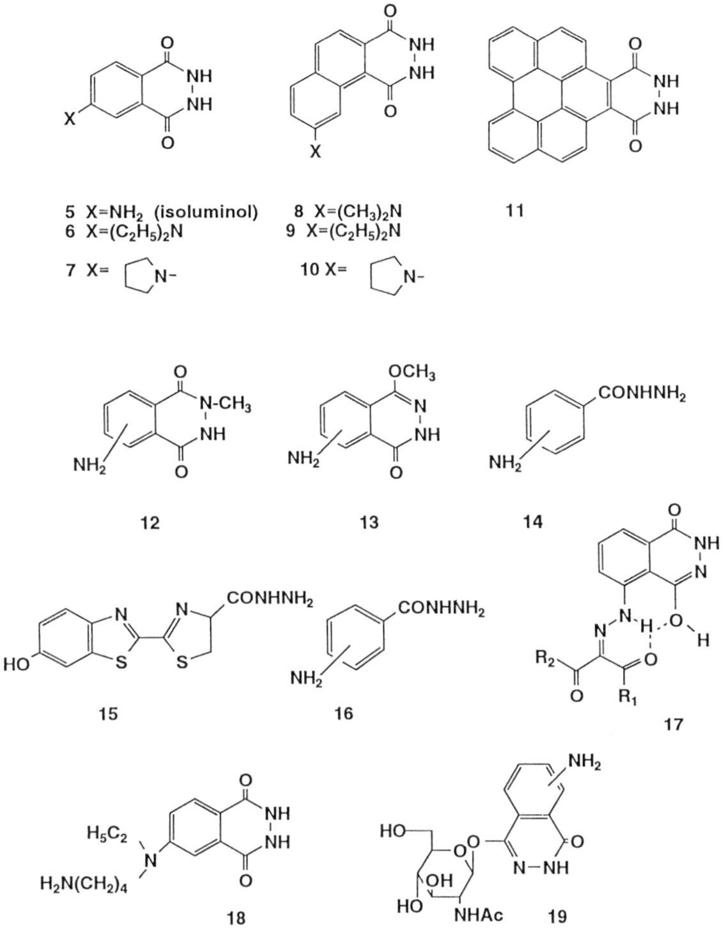

Fig. 1.4. Luminol derivatives.

for titration (53,54), immunoassay (55–57), DNA hybridization assays (58,59), and lipoperoxide assay (60). Isoluminol derivatives such as aminobutylethylisoluminol (ABEI, **18**) have been synthesized by Schroeder et al. (61) and applied to the immunoassay of biotin and thyroxine (62). Isothiocyanate derivative of isoluminol have been applied to the derivatization of amino acids (63). Patel et al. (64) in 1982, synthesized isothiocyanate derivatives of ABEI as a chemiluminescent label and utilized them for the homogeneous immunoassy of progesterone etc. (65). Chemiluminescence of ABEI was also utilized for high-performance liquid chromatographic (HPLC) determination of amines, carboxylic acids, and methamphetamine (66–68), and in flow injection analysis (69–72). Sasamoto and Ohkura developed a new chemiluminescent substrate (**19**) that was utilized for the photographic detection of N-acetyl-β-D-glucosaminidase (73).

1.4.2. Dioxetanes

1,2-Dioxetanes, having a four-membered ring, are easily converted to an excited product by heat and produce emission of light. In 1968, McCapra (22) showed that 1,2-dioxetanes are cleaved concertedly to form two carbonyl compounds and simultaneously one of them is electronically excited and produces emission of light (Fig. 1.5). This idea supported the assumption that the chemiluminescent reaction of lophine or indole includes a 1,2-dioxetane structure as an active intermediate. As already noted, a simple dioxetane, 3,3,4-trimethyl-1,2-dioxetane (**20**), was first synthesized by Kopecky and colleagues (23) in 1968. This compound was decomposed by heat to yield an excited acetone or aldehyde that produces a weak light. A strong chemiluminescence was observed by the addition of a fluorophore to this system. These facts revealed the 1,2-dioxetane itself to be the emitter, and since then many simple dioxetanes have been synthesized. On the contrary, Richardson et al. (74,75) showed in 1972 that dioxetane is decomposed by heat to an excited carbonyl compound via a biradical form (Fig. 1.5).

As the chemiluminescence of dioxetanes is generally weak, analytical use of them has been rare. But some of the dioxetanes such as **21**, **22**, and **23** are known to produce fairly strong chemiluminescence by thermal decomposition (76,77). Hummelen et al. (78,79) synthesized stable 1,2-dioxetanes such as **24** as a label for thermochemiluminescent immunoassay of protein. On the other hand, Schaap and Gagnon (80) demonstrated in 1982 that

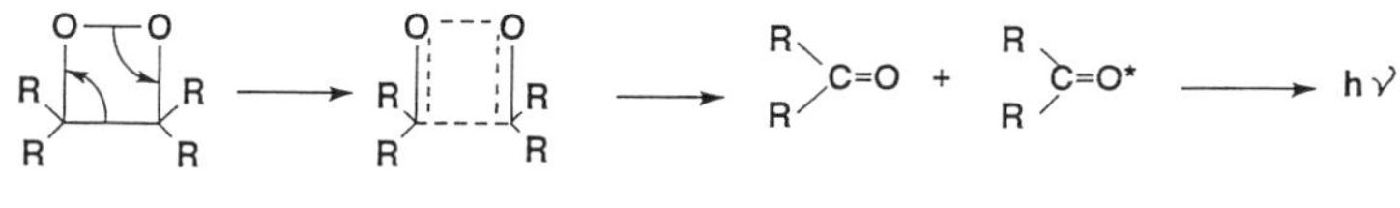

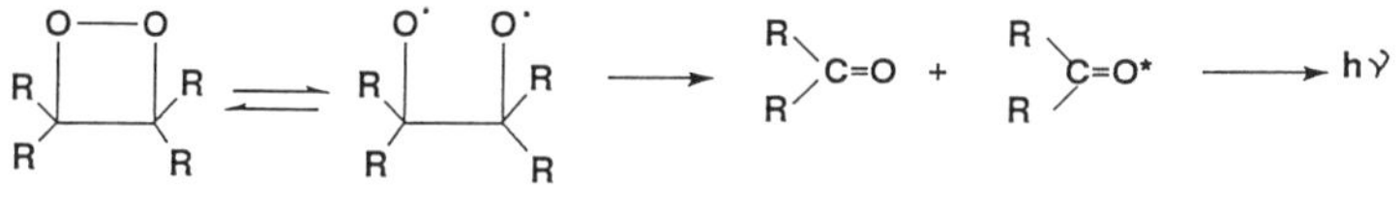

Fig. 1.5. Chemiluminescent reactions of 1,2-dioxetanes: (1) a concerted decomposition process; (2) a two-step biradical process.

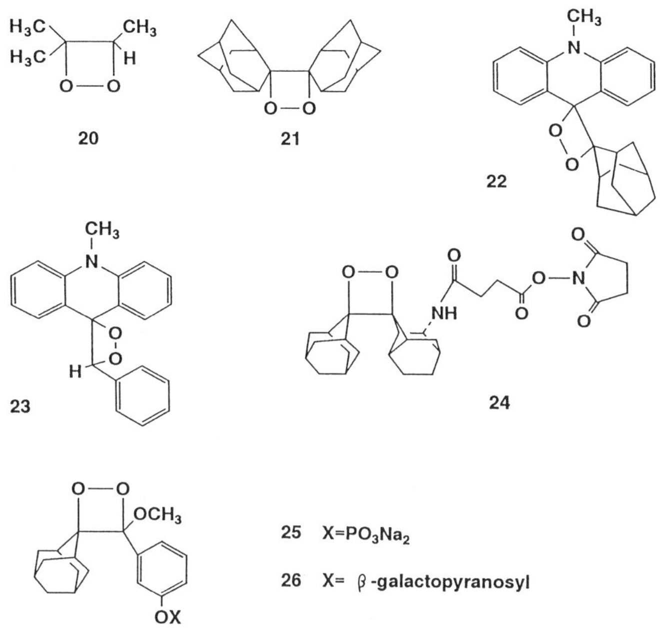

Fig. 1.6. 1,2-Dioxetane derivatives.

chemiluminescence from 1,2-dioxetane bearing phenolic substituent could be triggered by the addition of base. Since then, Schaap et al. (81–83) have synthesized many thermally stable dioxetanes and studied their chemical or enzymatically triggered chemiluminescent decompositions. Among these, **25** and **26** have been applied to enzyme immunoassays to detect alkaline phosphatase and β-D-galactosidase, respectively (Fig. 1.6) (84–86).

1.4.3. Lophine

Lophine (2,4,5-triphenylimidazole, **27**) is the classical chemiluminescent organic compound whose chemiluminescence was found by Radziszewski (5) in 1877. He observed that lophine produces a yellow light by reaction with O_2 and a strong base. In 1962, Hayashi and Maeda (87) studied the chemiluminescent reaction mechanism of lophine and showed that in the absence of O_2 lophine produces no light as a result of oxidation with Br_2 or $K_3Fe(CN)_6$ under alkline conditions, but a radical produced by the oxidation with Br_2 yields light upon treatment with O_2. In accord with these findings, they proposed the mechanism to be as follows: the anion of lophine formed under alkline conditions yields the free radical of 2,4,5-triphenylimidazole by oxidation; from this radical the

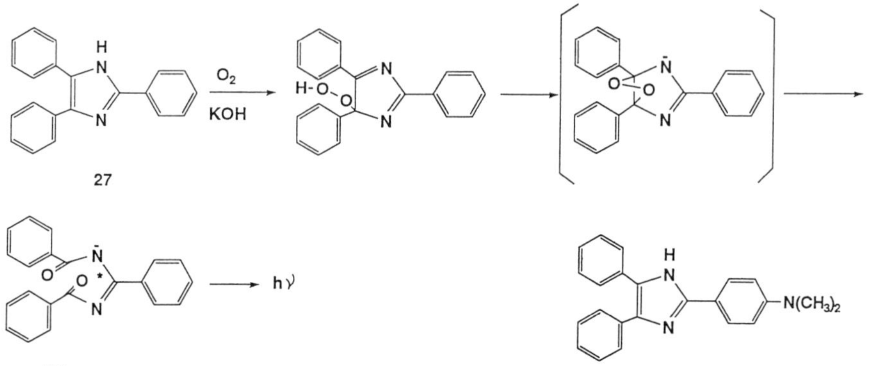

Fig. 1.7. Chemiluminescent reaction of lophine.

peroxide is produced with molecular oxygen, and finally the peroxide emits light. In 1964, Sonnenberg and White (88) and White and Harding (89) isolated the hydroperoxide of lophine, and the latter investigators subsequently synthesized many hydroperoxide derivatives (90) in order to examine their chemiluminescent reactivities. Consequently, they assumed that lophine is oxidized by O_2 and strong base to hydroperoxide, from which excited state diaroylarylamidines (**28**) should be formed via a dioxetane structure as shown in Fig. 1.7. Recently, the chemiluminescent reaction of 2-*p*-dimethylamino-lophine (DMAL, **29**) in a cationic micellar medium (cetyltrimethylammonium bromide) was demonstrated (91). Philbrook et al. (92) synthesized 16 kinds of lophine derivatives and studied their chemiluminescent characteristics. Then, they showed that the compounds bearing substituents in the meta and/or para positions of the 2-phenyl group have linear Hammett plots between $\log I/I_0$ and v. Here I and I_0 are the chemiluminescence intensities of the compound and lophine, respectively, and v is the frequency of light emitted.

Analytical applications of lophine include determination of Co^{2+} (93) and some other metal ions (94) using a lophine–hydrogen peroxide system. Lophine was also used as a chemiluminescent indicator, as were lucigenin and luminol (54).

1.4.4. Acridine Derivatives

1.4.4.1. Lucigenin

Lucigenin (*N,N'*-dimethyl-9,9'-biacridinium dinitrate, **30**) is one of the most widely studied chemiluminescent compounds, and its chemiluminescence

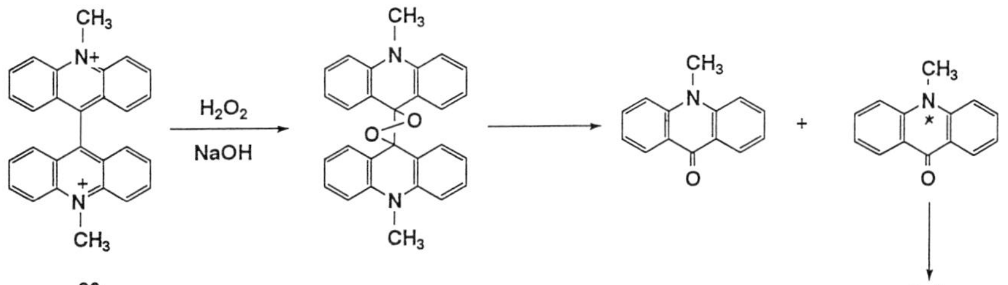

Fig. 1.8. Chemiluminescent reaction of lucigenin.

was first found by Glue and Petsch (17), in 1935. Lucigenin produces a weak light in an alkaline solution, and the intensity of its light is substantially increased by the addition of hydrogen peroxide. As to the reaction mechanism, it has been postulated (77,95) that lucigenin is oxidized to form a peroxide, which is then decomposed to yield an excited state of N-methylacridine, as shown in Fig. 1.8. Ramelow (96) in 1988 demonstrated the mechanism of enzymatically induced chemiluminescent reaction of lucigenin using vitamin K–dependent carboxylase.

Because lucigenin is soluble in aqueous ethanol, it has been utilized as an analytical reagent. For example, lucignenin was used for the determination of biological reductants (97–99), for heparin determination (100), and for immunoassay (101). A lucigenin–hydrogen peroxide chemiluminescent reaction system was used for HPLC determination of steroids and organic acids (102,103).

1.4.4.2. Acridinium Salts and Acridans

Acridinium salts such as 9-chlorocarbonyl-10-methylacridinium chloride (**31**) produce mainly peroxides by hydrogen peroxide oxidation in THF (tetrahydrofuran) with no accompanying production of gas or light. However, acridinium salts produce emission of light subsequent to dilution with water or hydrogen peroxide oxidation in alkline solution. The main reaction product is N-methylacridone (**32**). The chemiluminescence spectrum of N-methylacridone is very similar to its fluorescence spectrum. The reaction mechanism proposed by Rauhut et al. (104,105) in 1965 is shown in the upper sequence in Fig. 1.9. In this case, N-methylacridone is an active intermediate. On the other hand, McCapra and colleagues (106–108) suggested a mechanism including the dioxetane structure (**33**), as shown in the lower sequence in Fig. 1.9. Intermediates of oxidation of acridinium salts, except for dioxetanes, have been isolated and studied. Chemiluminescent quantum yields are more

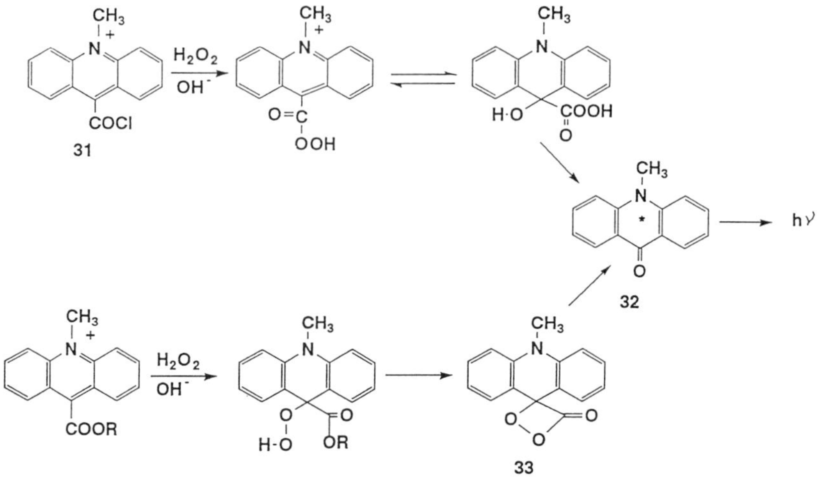

Fig. 1.9. Chemiluminescent reaction of acridinium salts.

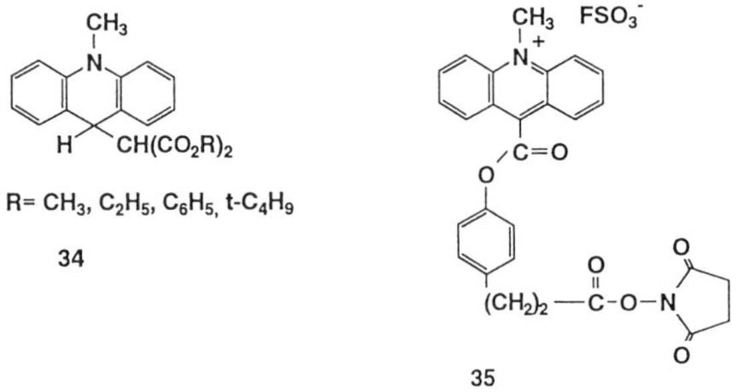

Fig. 1.10. Acridans (**34**) and acridinium salt (**35**).

than 4% and could be increased by modification of ring structure or substitution on the N atom or phenol moiety. *N*-Methyl-9-(dicarboalkoxymethyl)-acridans (**34**) were developed by Suzuki et al. (109) and demonstrated chemiluminescent reactions.

An acridinium ester, 4-(2-succinimidyloxycarbonylethyl)phenyl-10-methyl-acridinium-9-carboxylate fluorosulfonate (**35** in Fig. 1.10), was synthesized by Weeks et al. (110,111) and utilized for immunoassy of human α-fetoprotein.

1.4.5. Tetrakis(dimethylamino)ethylene

Tetrakis(dimethylamino)ethylene (TDE, **36**) was synthesized by Pruett (112) in 1950. It is a clear, slightly yellow, mobile liquid and produces chemilumines-

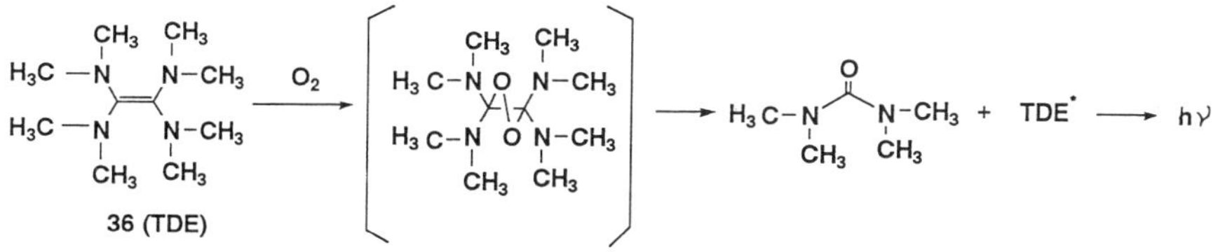

Fig. 1.11. Chemiluminescent reaction of TDE.

cence only by being exposed to oxygen or air. The chemiluminescence spectrum obtained was in good agreement with the fluorescence spectrum of TDE. The reaction mechanism is very complex. In 1967, Fletcher and Heller (113; see also Heller et al., 114) proposed the mechanism shown in Fig. 1.11, which was based on kinetic studies. Brandl (115) demonstrated chemiluminescence of TDE in 1988. Analysis of oxygen using TDE has also been reported (116,117).

1.4.6. Indole Derivatives

In 1965, Philbrook et al. (118) discovered the chemiluminescence of indole derivatives. For example, skatole (3-methylindole, **37**) produces light by reaction with oxygen under strong alkaline conditions. In 1966, McCapra and Chang (119) found that hydroperoxide (**38**) produces chemiluminescence by treatment with *t*-BuOK (potassium *tert*-butoxide) in dimethylsulfoxide. The main reaction product was amide (**39**), with 60% yield. Hydroperoxide synthesized by using $^{18}O_2$ yielded an amide that included two $^{18}O_2$ molecules. In accord with the aforementioned facts, the reaction mechanism of indole derivatives was elucidated as shown in Fig. 1.12. In 1967/1968 Sugiyama et al. (120–123) studied the substitution effects for many indole derivatives. Consequently, they supported the mechanism proposed by McCapra and Chang. A

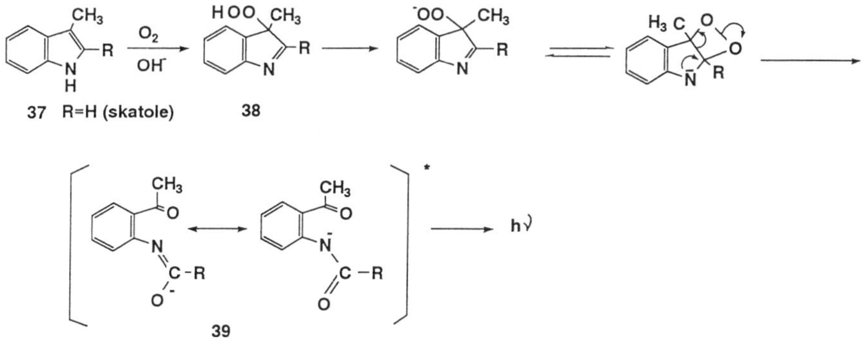

Fig. 1.12. Chemiluminescent reaction of skatole.

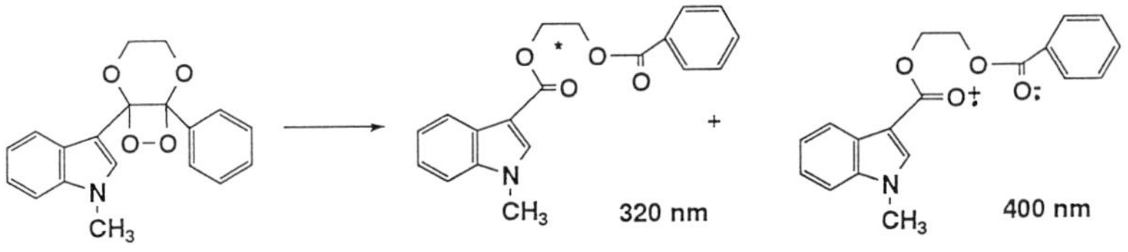

Fig. 1.13. Chemiluminescence of 1-(1-methyl-3-indolyl)-6-phenyl-2,5,7,8,-tetraoxabicyclo[4.2.0]-octane (**40**).

more complex derivative, 1-(1-methyl-3-indolyl)-6-phenyl-2,5,7,8-tetraoxabicyclo[4.2.0]octane (**40**), synthesized by Nakamura and Goto (124) in 1979, produces an emission of light at 320 nm in nonpolar solvents such as *n*-hexane and at 400 nm in polar solvents such as dichloromethane. The proposed mechanism suggests that dioxetane is decomposed by the transfer of an electron from a high-electron-density indole molecule to a dioxetane moiety as shown in Fig. 1.13.

1.4.7. Schiff Bases

Chemiluminescent reactions of Schiff bases (**41**) were studied by McCapra and Burford (125) in 1976 (Fig. 1.14). In 1977, they (126) also showed that the peroxide (trioxan, **42**), an intermediate in the oxidation of the Schiff base derived from 9-aminoanthracene and isobutylaldehyde, produces 9-formamidoanthracene accompanied with light emission by the reaction with *t*-BuOK in dimethylformamide as shown in Fig. 1.14. This mechanism is identical to

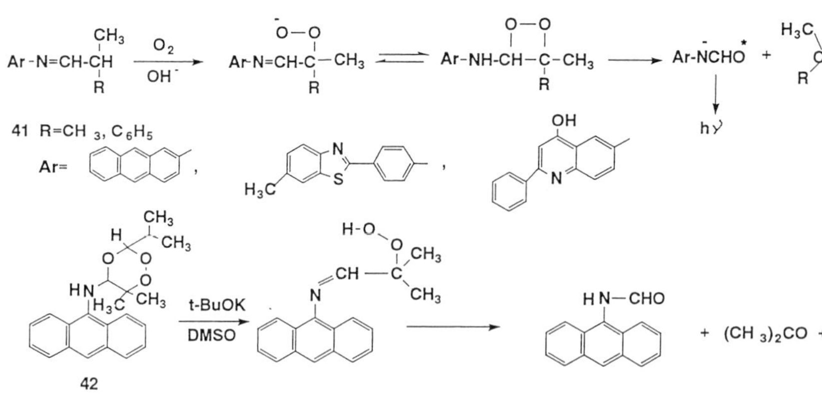

Fig. 1.14. Chemiluminescent reaction of Schiff base.

those producing the chemiluminescence of lophine and indole. The efficiency of chemiluminescence is high. However, a strong base is needed for the chemiluminescent reaction and the reaction proceeds only in aprotic or anhydrous solvents. Therefore, the chemiluminescent analytical applications of Schiff bases is restricted.

1.4.8. Oxalates and Oxamides

Oxalic acid derivatives (chloride, esters, and oxamides) react with hydrogen peroxide to yield intermediate peroxides which produce light by energy transfer to a fluorophore. This chemiluminescent reaction was first discovered by Chandross (19) in 1963. He observed that oxalyl chloride reacts with hydrogen peroxide to yield sensitized emission of light from anthracene. In 1965, Rauhut et al. (20) synthesized many kinds of derivatives and found that esters and oxamides produce chemiluminescence in the same way as does oxalyl cloride. These chemiluminescences are known as peroxyoxalate chemiluminescences. Representative aryloxalates and oxamides are shown in Fig. 1.15. The reaction mechanism is considered to be as follows (20,27–30): oxalic acid derivatives react with hydrogen peroxide to yield peroxides, and from them 1,2-dioxetanediones and/or substituted 1,2-dioxetanediones are formed as active intermediates. Then, a 1,2-dioxetanedione forms a charge-transfer complex with a fluorescer from which excited state fluorophore and

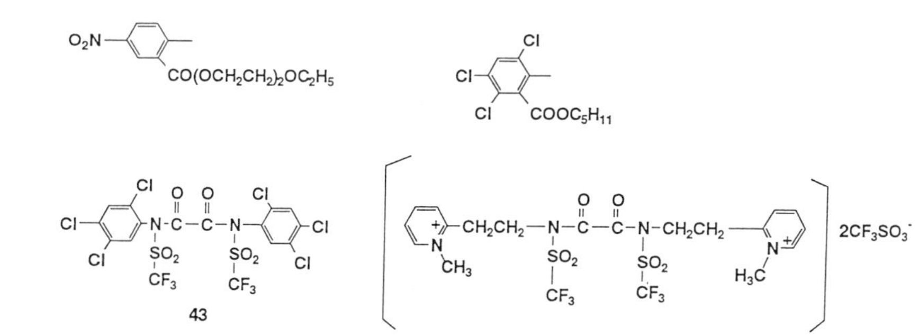

Fig. 1.15. Aryloxalates and oxamides.

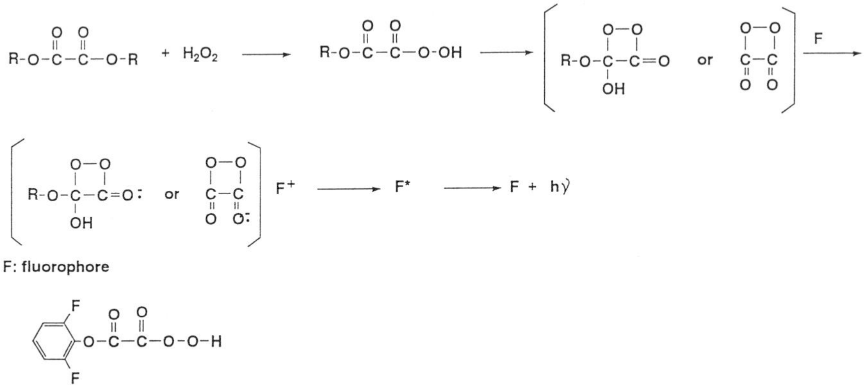

Fig. 1.16. Peroxyoxalate chemiluminescent reaction.

CO_2 are produced, as shown in Fig. 1.16. In 1990, Chokshi et al. (127) showed, by using ^{19}F-NMR (nuclear magnetic resonance), that the hydroperoxyoxalate ester (**44**) is the likely "reactive" intermediate of bis(2,6-difluorophenyl)-oxalate (DFPO)–hydrogen peroxide chemiluminescence.

Generally, the quantum yield of chemiluminescence is lower than that of bioluminescence. However the yields of some peroxyoxalate chemiluminescent systems are greater than 20%. Among them, when 9,10-bis(phenylethynyl)anthracene was used as a fluorescent compound, oxamide (**43**) showed a quantum yield of 34%. The peroxyoxalate chemiluminescent system has been used for practical purposes, e.g., as a "chemical light." This chemiluminescent system also has been widely utilized in analytical applications. For example, in the determination of hydrogen peroxide (128), biological materials such as glucose (129–131), NADH (132), aldehyde (133), cholesterol (134), and amino acids (135), and in various immunoassays (136–138). The peroxyoxalate chemiluminescent system was used as a detection device for HPLC by Kobayashi and Imai (139) in 1980. Since then, many sensitive HPLC–peroxyoxalate chemiluminescent determination methods of fluorescent compounds have been developed (140–154).

1.4.9. Diphenoylperoxide

Diphenoylperoxide (**45**) chemiluminescence was studied by Koo and Schuster (25,26) in 1977/78. Diphenoylperoxide was decomposed at 24°C for 24 h in dichloromethane in the dark to yield benzocoumarin (**46**) and polymeric peroxide. But, as benzocoumarin is nonfluorescent, no chemiluminescence is observed. However, if aromatic hydrocarbons such as diphenylanthracene

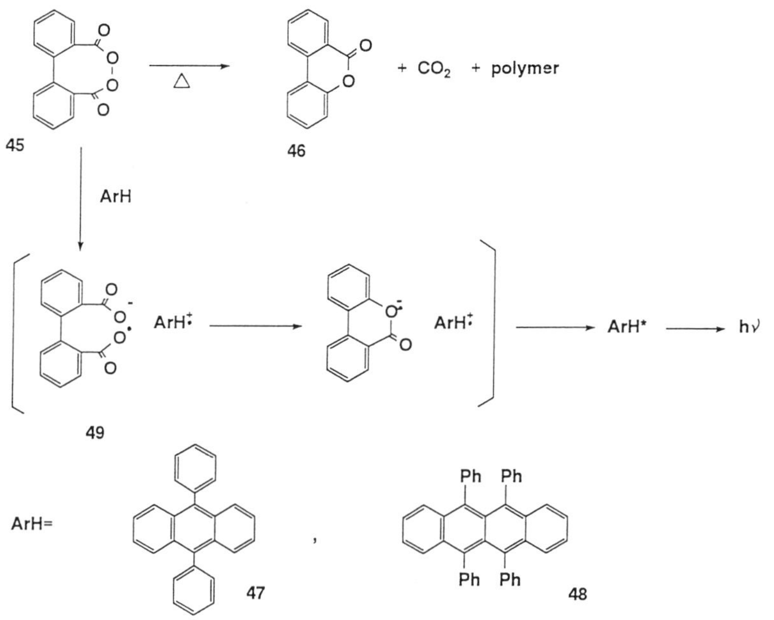

Fig. 1.17. Chemiluminescent reaction of diphenoyl peroxide.

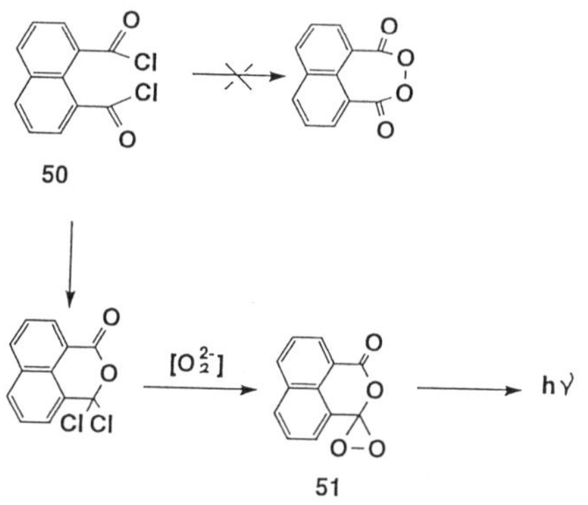

Fig. 1.18. Chemiluminescent reaction of naphthalic acid dichloride.

(47) and rubrene (48) are added to the reaction system, emission of light, corresponding to fluorescence of the hydrocarbons, is observed. It is considered that the aromatic hydrocarbons, having low oxidative electric potentials, transfer an electron to diphenoyl peroxide to form a charge-transfer complex (49) from which benzocoumarin and the excited hydrocarbon are produced.

This mechanism is known as the CIEEL mechanism (Fig. 1.17). In 1977, McCapra (155) proposed a similar mechanism. A new mechanism including a dioxiran (**51**) intermediate was proposed in 1985 for perhydrolysis of naphthalic acid dichloride (**50**, in Fig. 1.18) (156).

The phenomenon of chemiluminescence has been known for a long time. But intense chemical research in chemiluminescence started only about 25 years ago. The reason why the application of chemiluminescence to analyses has rapidly increased is its great sensitivity. In the near future, chemiluminescence will no doubt be utilized more and more in the fields of life science, environmental science, and other areas.

REFERENCES

1. E. N. Harvey, *Bioluminescence*, Academic Press, New York, 1952.
2. M. A. DeLuca, Ed., *Methods in Enzymology*, Vol. 57, Academic Press, New York, 1978.
3. M. A. DeLuca and W. D. McElroy, Eds., *Bioluminescence and Chemiluminescence*, Academic Press, New York, 1981.
4. M. A. DeLuca and W. D. McElroy, Eds., *Methods in Enzymology*, Vol. 133, Academic Press, Orlando, FL, 1986.
5. B. Radziszewski, *Chem. Ber.*, **10**, 70 (1877).
6. B. Radziszewski, *Chem. Ber.*, **10**, 321 (1877).
7. F. McCapra, *Q. Rev. Chem. Soc.*, **20**, 485 (1966).
8. G. G. Guilbault, *Practical Fluorescence: Theory, Methods, and Techniques*, Marcel Dekker, New York, 1973.
9. G. B. Schuster and S. P. Schmidt, *Adv. Phys. Org. Chem.*, **18**, 187 (1982).
10. K. Van Dyke, Ed., *Bioluminescence and Chemiluminescence: Instruments and Applications*, CRC Press, Boca Raton, 1985.
11. R. S. Givens, R. L. Schowen, J. Stobaugh, T. Kuwana, F. Alvarez, B. Matuszeuski, T. Kawasaki, O. Wong, M. Orlovic, H. Chokshi, and K. Nakashima, *ACS Symp. Ser.* **383**, 127 (1989).
12. L. J. Kricka and G. H. G. Thorpe, *Analyst*, **108**, 1274 (1983).
13. M. J. Whiting, *Clin. Biochem. Rev.*, **8**, 69 (1987).
14. G. J. Dejong and P. J. M. Kwakman, *J. Chromatogr.*, **492**, 319 (1989).
15. W. R. G. Baeyens, K. Nakashima, K. Imai, B. L. Ling, and Y. Tsukamoto, *J. Pharm. Biomed. Anal.*, **7**, 407 (1989).
16. H. D. Albrecht, *Z. Phys. Chem.*, **136**, 321 (1928).
17. K. Glue and W. Petsch. *Angew. Chem.*, **48**, 57 (1935).
18. R. C. Allen, *Methods Enzymol.*, **133**, 449 (1986).
19. E. A. Chandross, *Tetrahedron Lett.*, 1963, 761.

20. M. M. Rauhut, D. Sheehan, R. A. Clarke, and A. M. Semsel, *Photochem. Photobiol.*, **4**, 1097 (1965).
21. S. S. Tseng, A. G. Mohan, L. G. Haines, L. S. Vizcarra, and M. M. Rauhut, *J. Org. Chem.*, **44**, 4113 (1979).
22. F. McCapra, *J. Chem. Soc., Chem. Commun.*, 155 (1968).
23. K. R. Kopecky, J. H. Van De Sande, and C. Mumford, *Can. J. Chem.*, **46**, 25 (1968).
24. K. R. Kopecky and C. Mumford, *Can. J. Chem.*, **47**, 709 (1969).
25. J. Y. Koo and G. B. Schuster, *J. Am. Chem. Soc.*, **99**, 6107 (1977).
26. J. Y. Koo and G. B. Schuster, *J. Am. Chem. Soc.*, **100**, 4496 (1978).
27. C. L. R. Catherall, T. F. Palmer, and R. B. Cundall, *J. Chem. Soc., Faraday Trans. 2*, 80 (1984).
28. C. L. R. Catherall, T. F. Palmer, and R. B. Cundall, *J. Chem. Soc., Faraday Trans. 2*, 823 (1984).
29. C. L. R. Catherall, T. F. Palmer, and R. B. Cundall, *J. Chem. Soc., Faraday Trans. 2*, 837 (1984).
30. F. J. Alvarez, N. J. Parekh, B. Matusewski, R. S. Givens, T. Higuchi, and R. L. Schowen, *J. Am. Chem. Soc.*, **108**, 6435 (1986).
31. E. H. White, D. F. Roswell, and O. C. Zafiriou, *J. Org. Chem.*, **34**, 2462 (1969).
32. E. H. White, O. C. Zafiriou, H. H. Kagi, and J. H. M. Hill, *J. Am. Chem. Soc.*, **86**, 940 (1964).
33. E. H. White and M. M. Bursey, *J. Am. Chem. Soc.*, **86**, 941 (1964).
34. M. M. Rauhut, A. M. Semsel, and B. G. Roberts, *J. Org. Chem.*, **31**, 2431 (1966).
35. C. D. Kalkar and M. Arjunkwadkar, *Indian J. Pure Appl. Phys.*, 26, 433 (1988).
36. K. D. Gundermann, *Angew. Chem., Int. Ed. Engl.*, **7**, 480 (1968).
37. E. H. White, E. G. Nash, D. R. Roberts, and O. C. Zafiriou, *J. Am. Chem. Soc.*, **90**, 5932 (1968).
38. C. C. Wei and E. H. White, *Tetrahedron Lett.*, **39**, 3559 (1971).
39. E. H. White and D. F. Roswell, *Acc. Chem. Res.*, **3**, 54 (1970).
40. H. Ojima, *Naturwissenschaften*, **48**, 500 (1961).
41. E. H. White, M. M. Bursey, D. F. Roswell, and J. H. Hill, *J. Org. Chem.*, **32**, 1198 (1967).
42. E. Rapaport, M. W. Cass, and E. H. White, *J. Am. Chem. Soc.*, **94**, 3153 (1972).
43. N. Thankarajan, K. Krishnankutty, and T. K. K. Srinivasan, *J. Indian Chem. Soc.*, **63**, 977 (1986).
44. K. I. Agranov and L. V. Reiman, *Zh. Anal. Khim.*, **34**, 1533 (1979).
45. Y. Ikariyama, M. Aizawa, and S. Suzuki, *J. Solid-Phase Biochem.*, **5**, 223 (1980).
46. D. T. Bostick and D. M. Hercules, *Anal. Chem.*, **47**, 447 (1975).
47. W. R. Seitz and M. P. Neary, *Anal. Chem.*, **46**, 188A (1974).
48. D. C. Williams, G. G. Hufl, and W. R. Seitz, *Anal. Chem.*, **48**, 1003 (1976).
49. V. H. Regener, *J. Geophys. Res.*, **65**, 3975 (1960).

50. J. P. Auses, S. L. Cook, and J. T. Maloy, *Anal. Chem.*, **47**, 244 (1975).
51. M. Aizawa, Y. Ikariyama, and H. Kuno, *Anal. Lett.*, **17**, 555 (1984).
52. U. Isacsson and G. Wettermark, *Anal. Chim. Acta*, **68**, 339 (1974).
53. N. Thankarajan and K. Krishnankutty, *Talanta*, **34**, 507 (1987).
54. G. A. Shannon, *Sch. Sci. Rev.*, **68**, 81 (1986).
55. T. P. Whitehead, G. H. G. Thorpe, T. J. N. Carter, C. Groucutt, and J. Kricka, *Nature (London)*, **305**, 158 (1983).
56. M. Maeda and A. Tsuji, *Anal. Chim. Acta*, **167**, 241 (1985).
57. G. H. G. Thorpe, L. J. Kricka, S. B. Moseley, and T. P. Whitehead, *Clin. Chem. (Winston-Salem, N.C.)*, **31**, 1335 (1985).
58. J. A. Mattews, A. Batki, C. Hynds, and L. J. Kricka, *Anal. Biochem.*, **151**, 205 (1985).
59. T. Segawa, T. Kamidate, and H. Watanabe, *Anal. Sci.*, **4**, 659 (1988).
60. Y. Yamamoto, M. H. Brodsky, and J. C. Ames, *Anal. Biochem.*, **160**, 7 (1987).
61. H. R. Schroeder, R. C. Boguslaski, R. J. Carrico, and R. T. Buckler, *Methods Enzymol.*, **57**, 424 (1978).
62. H. R. Schroeder and F. M. Yeager, *Anal. Chem.*, **50**, 1114 (1978).
63. S. R. Spurlin and M. M. Cooper, *Anal. Lett.*, **19**, 2277 (1986).
64. A. Patel, M. S. Morton, J. S. Woodhead, M. E. T. Ryall, F. McCapra, and A. K. Campbell, *Biochem. Soc. Trans.*, **10**, 224 (1982).
65. A. Patel and A. K. Campbell, *Clin. Chem. (Winston-Salem, N.C.)*, **29**, 1604 (1983).
66. T. Kawasaki, M. Maeda, and A. Tsuji, *J. Chromatogr.*, **328**, 121 (1985).
67. H. Yuki, Y. Azuma, N. Maeda, and H. Kawasaki, *Chem. Pharm. Bull.*, **36**, 1905 (1988).
68. K. Nakashima, K. Suetsugu, S. Akiyama, and M. Yoshida, *J. Chromatogr.*, **530**, 154 (1990).
69. B. Olsson, *Anal. Chim. Acta*, **136**, 113 (1982).
70. T. Hara, M. Toriyama, and K. Tsukagoshi, *Bull. Chem. Soc. Jpn.*, **56**, 1382 (1983).
71. N. L. Malavolti, D. Pilosof, and T. A. Nieman, *Anal. Chem.*, **56**, 2191 (1984).
72. E. G. Sarantonis and A. Townshend, *Anal. Chim. Acta*, **184**, 311 (1986).
73. K. Sasamoto and Y. Ohkura, *Chem. Pharm. Bull.*, **38**, 1323 (1990).
74. W. H. Richardson, M. B. Yelvington, and H. E. O'Neal, *J. Am. Chem. Soc.*, **94**, 1619 (1972).
75. W. H. Richardson and H. E. O'Neal, *J. Am. Chem. Soc.*, **94**, 8665 (1972).
76. J. H. Wieringa, J. Strating, H. Wynberg, and W. Adam, *Tetrahedron Lett.*, **2**, 169 (1972).
77. K. W. Lee, L. A. Singer, and K. D. Legg, *J. Org. Chem.*, **41**, 2685 (1976).
78. J. C. Hummelen, T. M. Luider, and H. Wynberg, *Methods Enzymol.*, **133**, 531 (1986).
79. J. C. Hummelen, T. M. Luider, and H. Wynberg, *Pure Appl. Chem.*, **59**, 639 (1987).

80. A. P. Schaap and S. D. Gagnon, *J. Am. Chem. Soc.*, **104**, 3504 (1982).
81. A. P. Schaap, R. S. Handley, and B. P. Giri, *Tetrahedron Lett.*, **28**, 935 (1987).
82. A. P. Schaap, T. S. Chen, R. S. Handley, R. DeSilva, and B. P. Giri, *Tetrahedron Lett.*, **28**, 1155 (1987).
83. A. P. Schaap, M. D. Sandison, and R. S. Handley, *Tetrahedron Lett.*, **28**, 1159 (1987).
84. J. C. Voyta, B. Edward, and I. Bronstein, *Clin. Chem. (Winston-Salem, N.C.)*, 1157 (1988).
85. I. Bronstein, B. Edward, and J. C. Voyta, *J. Biolumin. Chemilumin.*, **4**, 99 (1989).
86. A. P. Schaap, H. Akhavan, and L. J. Romano, *Clin. Chem. (Winston-Salem, N.C.)*, **35**, 1863 (1989).
87. T. Hayashi and K. Maeda, *Bull. Chem. Soc. Jpn.*, **35**, 2057 (1962).
88. J. Sonnenberg and D. M. White, *J. Am. Chem. Soc.*, **86**, 5685 (1964).
89. E. H. White and M. J. C. Harding, *J. Am. Chem. Soc.*, **86**, 5686 (1964).
90. E. H. White and M. J. C. Harding, *Photochem. Photobiol.*, **4**, 1129 (1965).
91. S. Boyatzis and J. Nikokavouras, *J. Photochem. Photobiol., A* **44**, 335 (1988).
92. G. E. Philbrook, M. A. Maxwell. R. E. Taylor, and J. R. Totter, *Photochem. Photobiol.*, **4**, 1175 (1965).
93. D. F. Marino and J. D. Ingle, Jr., *Anal. Chem.*, **53**, 292 (1981).
94. A. MacDonald, K. W. Chan, and T. A. Nieman, *Anal. Chem.*, **51**, 2077 (1979).
95. R. Maskiewicz, D. Sogah, and T. Bruice, *J. Am. Chem. Soc.*, **101**, 5347 (1979).
96. U. S. Ramelow, *J. Photochem. Photobiol., B* **2**, 91 (1988).
97. R. L. Veazey and T. A. Nieman, *Anal. Chem.*, **51**, 2092 (1979).
98. R. L. Veazey and T. A. Nieman, *J. Chromatogr.*, **200**, 153 (1980).
99. L. L. Klopf and T. A. Nieman, *Anal. Chem.*, **57**, 46 (1985).
100. W. L. Hinze, T. E. Riehl, H. N. Singh, and Y. Baba, *Anal. Chem.*, **56**, 2180 (1984).
101. R. A. Steen and T. A. Nieman, *Anal. Chim. Acta*, **155**, 123 (1983).
102. R. C. Allen, *Methods Enzymol.*, **133**, 449 (1986).
103. M. Maeda and A. Tsuji, *J. Chromatogr.*, **352**, 213 (1984).
104. M. M. Rauhut, D. Sheehan, R. A. Clarke, and A. M. Semsel, *Photochem. Photobiol.*, **4**, 1097 (1965).
105. M. M. Rauhut, D. Sheehan, R. A. Clarke, B. G. Roberts, and A. M. Semsel, *J. Org. Chem.*, **30**, 3587 (1965).
106. F. McCapra, D. G. Richardson, and Y. C. Chang, *Photochem. Photobiol.*, **4**, 1111 (1965).
107. F. McCapra, D. G. Richardson, and Y. C. Chang, *Acc. Chem. Res.*, **9**, 201 (1976).
108. F. McCapra, D. G. Richardson, and Y. C. Chang, *Proc. R. Soc. London, Ser, B* **215**, 247 (1982).
109. N. Suzuki, M. Nakaminami, T. Tsukamoto, K. Iwasaki, and Y. Izawa, *Res. Rep. Fac. Eng., Mie Univ.*, **10**, 41 (1985).

110. I. Weeks, M. Sturgess, R. C. Brown, and J. S. Woodhead, *Methods Enzymol.*, **133**, 366 (1986).
111. I. Weeks, I. Beheshti, F. McCapra, A. K. Campbell, and J. S. Woodhead, *Clin. Chem. (Winston-Salem, N.C.)*, **29**, 1474 (1983).
112. R. L. Pruett, *J. Am. Chem. Soc.*, **72**, 3646 (1950).
113. A. N. Fletcher and C. A. Heller, *J. Phys. Chem.*, **71**, 1507 (1967).
114. C. A. Heller, R. A. Henry, and J. M. Fritsch, in M. J. Cormier, D. M. Hercules, and J. Lee, Eds., in *Chemiluminescence and Bioluminescence*, Plenum Press, New York, 1979, p. 249.
115. H. Brandl, *Prax. Naturwiss. Chem.*, **37**, 25 (1988).
116. T. M. Freeman and W. R. Seitz, *Anal. Chem.*, **53**, 98 (1981).
117. B. F. McDonald and W. R. Seitz, *Anal. Lett.*, **15**, 57 (1982).
118. G. F. Philbrook, J. B. Qyers, and J. R. Totter, *Photochem. Photobiol.*, **4**, 869 (1965).
119. F. McCapra and Y. C. Chang, *J. Chem. Soc., Chem. Commun.*, 522 (1966).
120. N. Sugiyama and M. Akutagawa, *Bull. Chem. Soc. Jpn.*, **40**, 240 (1967).
121. N. Sugiyama, M. Akutagawa, T. Gasha, Y. Saiga, and H. Yamamoto, *Bull. Chem. Soc. Jpn.*, **40**, 347 (1967).
122. N. Sugiyama, H. Yamamoto, and Y. Omote, *Bull. Chem. Soc. Jpn.*, **41**, 936 (1968).
123. N. Sugiyama, H. Yamamoto, and Y. Omote, *Bull. Chem. Soc. Jpn.*, **41**, 1917 (1968).
124. H. Nakamura and T. Goto, *Photochem. Photobiol.*, **30**, 27 (1979).
125. F. McCapra and A. Burford, *J. Chem. Soc., Chem. Commun.*, 607 (1976).
126. F. McCapra and A. Burford, *J. Chem. Soc., Chem. Commun.*, 874 (1977).
127. H. P. Chokshi, M. Barbush, R. G. Carlson, R. S. Givens, T. Kuwana, and R. L. Schowen, *Biomed. Chromatogr.*, **4**, 96 (1990).
128. P. Van Zoonen, D. A. Kammings, C. Gooijer, N. H. Velthorst, and R. W. Frei. *Anal. Chim. Acta.* **167**, 249 (1985).
129. D. C. Williams, III, G. F. Huff, and W. R. Seitz, *Anal. Chem.*, **48**, 1003 (1976).
130. V. I. Rigin, *J. Anal. Chem. USSR (Engl. Transl.)*, **34**, 619 (1979).
131. P. Van Zoonen, I. de Herder, C. Gooijer, N. H. Velthorst, and R. W. Frei, *Anal. Lett.*, **19**, 1949 (1986).
132. D. C. Williams, III and W. R. Seitz, *Anal. Chem.*, **48**, 1478 (1976).
133. V. I. Rigin, *J. Anal. Chem. USSR (Engl. Transl.)*, **36**, 1111 (1981).
134. V. I. Rigin, *J. Anal. Chem. USSR (Engl. Transl.)*, **38**, 1265 (1983).
135. V. I. Rigin, *J. Anal. Chem. USSR (Engl. Transl.)*, **38**, 1328 (1983).
136. H. Arakawa, M. Maeda, and A. Tsuji, *Chem. Pharm. Bull.*, **30**, 3036 (1982).
137. H. Arakawa, M. Maeda, and A. Tsuji, *Clin. Chem. (Winston-Salem, N.C.)*, **31**, 430 (1985).
138. A. Tsuji, M. Maeda, and H. Arakawa, in K. Van-Dyke, Ed., *Bioluminescence and Chemiluminescence*, Vol. 1, CRC Press, Boca Raton, FL, 1985, p. 185.

139. K. Kobayashi and K. Imai, *Anal. Chem.*, **52,** 424 (1980).
140. K. Imai, K. Miyaguchi, and K. Honda, in K. Van-Dyke, Ed., *Bioluminescence and Chemiluminescence*, Vol. 2, CRC Press, Boca Raton, FL, 1985, p. 65.
141. K. Imai, *Methods Enzymol.*, **133,** 435 (1986).
142. G. Mellbin, *J. Liq. Chromatogr.*, **6,** 1603 (1983).
143. K. Miyaguchi, H. Honda, and K. Imai, *J. Chromatogr.*, **303,** 173 (1984).
144. K. Kobayashi, K. Honda, and K. Imai, *J. Chromatogr.*, **316,** 501 (1984).
145. K. W. Sigvardson, J. M. Kennish, and J. W. Birks, *Anal. Chem.*, **56,** 1096 (1984).
146. R. Weinberger, C. A. Mannan, M. Cerchio, and M. L. Grayeski, *J. Chromatogr.*, **288,** 445 (1984).
147. K. Honda, K. Miyaguchi, and K. Imai, *Anal. Chim. Acta*, **177,** 103 (1985).
148. K. Imai, A. Nishitani, and Y. Tsukamoto, *Chromatographia*, **24,** 77 (1987).
149. K. Nakashima, C. Umekawa, S. Nakatsuji, S. Akiyama, and S. Givens, *Biomed. Chromatogr.*, **3,** 39 (1989).
150. K. Hayakawa, K. Hasegawa, N. Imaizumi, O. Wong, and M. Miyazaki, *J. Chromatogr.*, **464,** 343 (1989).
151. K. Nakashima, K. Maki, S. Akiyama, and K. Imai, *Biomed. Chromatogr.*, **4,** 105 (1990).
152. K. Imai, A. Nishitani, Y. Tsukamoto, W. H. Wang, S. Kanda, K. Hayakawa, and M. Miyazaki, *Biomed. Chromatogr.*, **4,** 100 (1990).
153. N. Imaizumi, K. Hayakawa, Y. Suzuki, and M. Miyazaki, *Biomed. Chromatogr.*, **4,** 108 (1990).
154. T. Kawasaki, K. Imai, T. Higuchi, and O. S. Wong, *Biomed. Chromatogr.*, **4,** 113 (1990).
155. F. McCapra, *J. Chem. Soc., Chem. Commun.*, 946 (1977).
156. M. F. D. Stainfatt, *J. Chem. Res., Synop.*, 140 (1985).

CHAPTER

2

FLUORESCENT PROBES FOR EVALUATION OF LOCAL PHYSICAL AND STRUCTURAL PARAMETERS

BERNARD VALEUR

Laboratoire de Chimie Générale (CNRS URA 1103)
Conservatoire National des Arts et Métiers
75003 Paris, France

2.1. **Introduction**
2.2. **Characteristics of Fluorescent Probes**
    2.2.1. Chemical Structure and Nature in Relation to the Probed Microenvironment
    2.2.2. Absorption, Excitation, and Emission Spectra: The Stokes Shift
    2.2.3. Decay of Fluorescence Intensity: Excited State Lifetime
    2.2.4. Quantum Yield
    2.2.5. Photochemical Stability
2.3. **Main Physical and Structural Parameters Measurable by Means of Fluorescent Probes**
    2.3.1. Polarity
        2.3.1.1. What Is Polarity?
        2.3.1.2. Estimation of Polarity from Spectral Shifts
        2.3.1.3. Estimation of Polarity from Changes in Vibronic Bands
    2.3.2. Fluidity: Order Parameters; Molecular Mobility
        2.3.2.1. General Considerations
        2.3.2.2. Use of Molecular Rotors
        2.3.2.3. Intermolecular Excimer Formation and Intermolecular Quenching
        2.3.2.4. Intramolecular Excimer Formation
        2.3.2.5. Fluorescence Polarization Studies
    2.3.3. Distances at a Supramolecular Scale: A "Spectroscopic Ruler"
        2.3.3.1. Mechanisms of Electronic Energy Transfer
        2.3.3.2. Determination of Transfer Efficiency
        2.3.3.3. Calculation of Distance
        2.3.3.4. Distributions of Distances

Molecular Luminescence Spectroscopy, Part 3, Edited by Stephen G. Schulman. Chemical Analysis Series, Vol. 77.
ISBN 0-471-51580-9 © 1993 John Wiley & Sons, Inc.

2.4. Concluding Remarks
Acknowledgments
References

2.1. INTRODUCTION

Fluorescent probes are extensively used in physical, chemical, biological, and medical sciences for investigating the structure and dynamics of matter or living systems at a molecular or supramolecular level. Polymeric and micellar systems, biological membranes, proteins, nucleic acids, living cells, etc., are well-known examples of systems in which estimation of local parameters such as polarity, fluidity, order, molecular mobility, pH, or electric potential are possible by means of fluorescent probes. Their success arises also from the high sensitivity of fluorimetric techniques, the specificity of changes in fluorescence characteristics of probes due to the microenvironment (Fig. 2.1), and the capability of probes to provide information on the rate of rapid phenomena and/or on the structural parameters of the system under study. In Table 2.1 are reported the various fields of application together with the relevant information that can be obtained.

Fluorescent probes are so numerous and the applications are so varied that a complete review would require several books! Therefore, the present chapter by no means intends to be exhaustive and, instead of a catalog of probes and fields of application, will rather focus on the characteristics of fluorescent probes and on the main physical and structural parameters underlying most

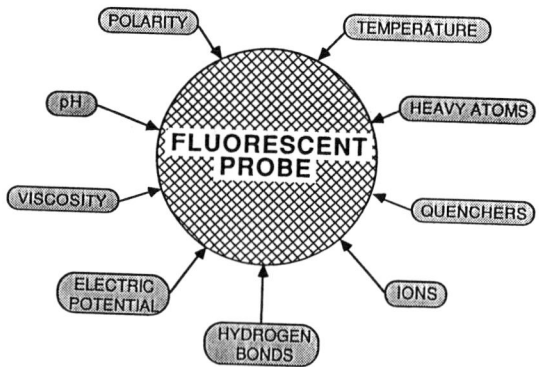

Fig. 2.1. Environmental effects on fluorescent probes.

Table 2.1. Information Provided by Fluorescent Probes in Various Fields

Field	Information	Reviews or Books
Polymers	Dynamics of polymer chains; microviscosity; free volume; orientation of chains in stretched samples; miscibility; phase separation; diffusion of species through polymer networks; end-to-end macrocyclization dynamics; monitoring of polymerization, degradation	1–11
Solid surfaces	Nature of the surface of colloid silica, clays, zeolithes, silica gels, porous Vycor glasses, alumina: rigidity, polarity and modification of surfaces	12–14
Surfactant solutions	Critical micelle concentration, distribution of reactants among particles; surfactant aggregation numbers; interface properties and polarity; dynamics of surfactant solutions; partition coefficients; phase transitions; influence of additives	12,15–17
Biological membranes	Fludity; order parameters; lipid–protein interactions; translational diffusion; site accessibility; structural changes; membrane potentials; complexes and binding; energy-linked and light-induced changes; effects of additives; location of proteins; lateral organization and dynamics	18–32
Vesicles	Characterization of the bilayer: microviscosity, order parameters; phase transition; effect of additives; internal pH, permeability	
Proteins	Binding sites; denaturation; site accessibility; dynamics; distances; conformational transition	27,29,33–35

(*continued*)

Table 2.1. (*Continued*)

Field	Information	Reviews or Books
Nucleic acids	Flexibility; torsion dynamics; helix structure; deformation due to intercalating agents; photocleavage; accessibility; carcinogenesis	36,37
Living cells	Visualization of membranes, lipids, proteins, DNA, RNA, surface antigens, surface glycoconjugates; membrane dynamics; membrane permeability; membrane potential; intracellular pH; cytoplasmic calcium, sodium, chloride, proton concentration; redox state; enzyme activities; cell–cell and cell–virus interactions; membrane fusion; endocytosis; viability, cell cycle cytotoxic activity	38–41
Fluoroimmunochemistry	Fluoroimmunoassays	42–47

of the fields of application, with special attention to some physical concepts and emphasis on the possible pitfalls. For more details on a particular field, the reader is referred to the reviews and books cited in Table 2.1 (1–47), to the papers subsequently cited in this chapter, and to other general books dealing more or less with fluorescent probes (48–55). Moreover, many references can be found in the *Handbook of Fluorescent Probes and Research Chemicals* (56).

2.2. CHARACTERISTICS OF FLUORESCENT PROBES

2.2.1. Chemical Structure and Nature in Relation to the Probed Microenvironment

Fluorescent probes can be divided into three classes: (i) intrinsic probes; (ii) extrinsic covalently bound probes; and (iii) extrinsic associating probes. Intrinsic probes are ideal, but there are only a few examples (e.g., tryptophan in proteins). The advantage of covalently bound probes over the extrinsic associating probes is that the location of the former is known. There are various examples of probes covalently attached to surfactants, polymer chains, phospholipids, proteins, polynucleotides, etc. A section of them is presented in Fig. 2.2. In particular, the anthroyloxy stearic acids with the anthracene moiety attached in various positions of the paraffinic chain allows one to probe micellar systems or bilayers at various depths.

Protein tagging can be easily achieved by means of labeling reagents having proper functional groups: covalent binding is indeed possible on amino groups (with isothiocyanates, chlorotriazinyl derivatives, or hydroxysuccinimido

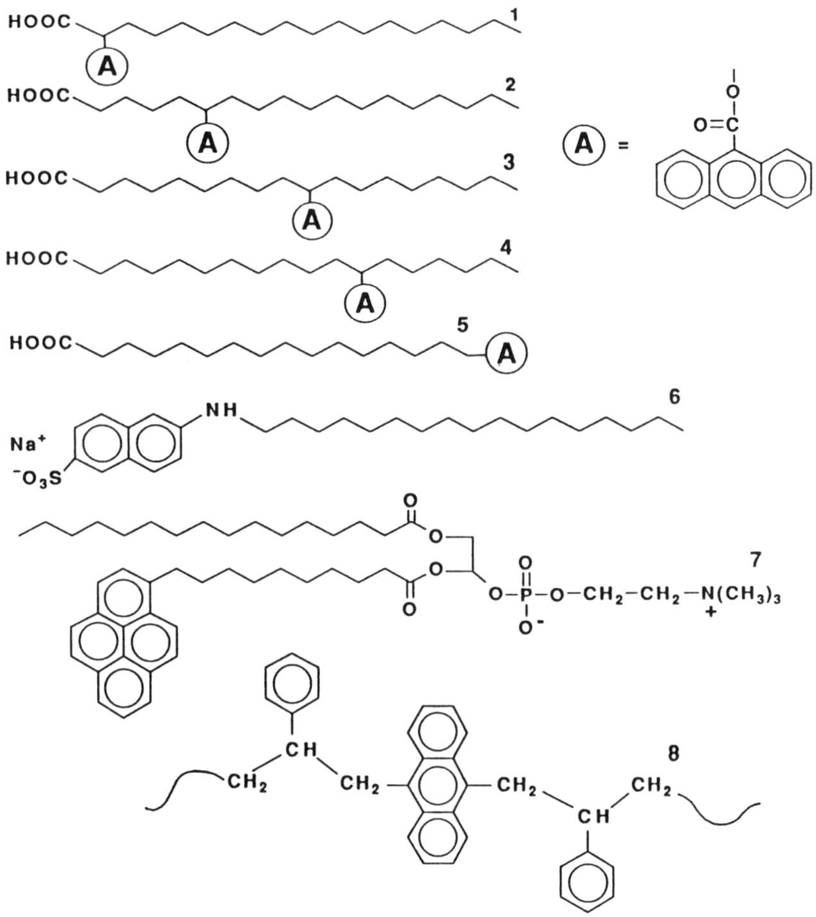

Fig. 2.2. Examples of surfactants, phospholipids, and polymers with covalently bound probes: (1) 2-(9-anthroyloxy)stearic acid; (2) 6-(9-anthroyloxy)stearic acid; (3) 10-(9-anthroyloxy)stearic acid; (4) 12-(9-anthroyloxy)stearic acid; (5) (9-anthroyloxy)palmitic acid; (6) 2-(N-octadecyl)-aminonaphthalene-6-sulfonic acid, sodium salt; (7) 3-palmitoyl-2-(1-pyrenedecanoyl)-L-α-phosphatidylcholine; (8) polystyrene labeled with anthracene.

active esters), and on sulfhydryl groups (with iodoacetamido and maleimido functional groups). Fluorescein, rhodamine, and erythrosin derivatives with these functional groups are currently used (56,57).

Owing to the difficulty of synthesis of molecules or macromolecules with covalently bound specific probes, most of the investigations are carried out with noncovalently associating probes (class iii). The sites of solubilization of extrinsic probes are governed by their chemical nature and the resulting specific interactions that can be established within the region of the system to be probed. The hydrophilic, hydrophobic, or amphiphilic character of a probe is essential. Figure 2.3 gives various examples. Pyrene is known as a probe of hydrophobic regions; furthermore its sensitivity to polarity is very useful (see Section 2.3.1.3, below). In contrast, pyranine is very hydrophilic and will be located in hydrophilic aqueous regions; moreover it is sensitive to pH. If the OH group is replaced by $O(CH_2)_n CH_3$, the resulting molecule becomes pH insensitive and amphiphilic; the fluorophore moiety plays the role of a polar head and thus is located at the surfactant–water interface of systems consisting of amphiphilic molecules (bilayers of membranes and vesicles, micellar systems, etc.). Conversely, the pyrene moiety of pyrenedodecanoic acid is deeply embedded in the hydrophobic part of an organized assembly. 1,6-Diphenyl-1,3,5-hexatriene (DPH) is located in the hydrocarbon region of bilayers of membrane and vesicles, whereas its cationic analogue TMA–DPH is anchored with its charged group at the surfactant–water interface (58). The latter is thus a probe of the upper region of bilayers. Both *cis*- and *trans*-parinaric acids (59) are good examples of probes causing minimum spatial perturbation to organized assemblies.

The aforementioned examples show that a most important criterion in the choice of a probe is its sensitivity to a particular property of the microenvironment wherein it is located (e.g., polarity or acidity). On the other hand, insensitivity to the chemical nature of the environment is preferable in some cases (e.g., fluorescence polarization or energy transfer experiments). Environment-insensitive probes are also better suited to fluorescence microscopy and flow cytometry.

A criticism often addressed to the use of extrinsic fluorescent probes is the possible local perturbation induced by the probe itself on the microenvironment to be probed. There are indeed several cases of systems perturbed by fluorescent probes [see, for instance, nuclear magnetic resonance (NMR) study of the perturbation of micelles by anthroyloxy probes (60)]. However, it should be emphasized that many examples of results consistent with those obtained by other techniques can be found in the literature (transition temperature in lipid bilayer, flexibility of polymer chains, etc.). To render the perturbation minimal, attention must be paid to the size and shape of the probe with respect to the probed region.

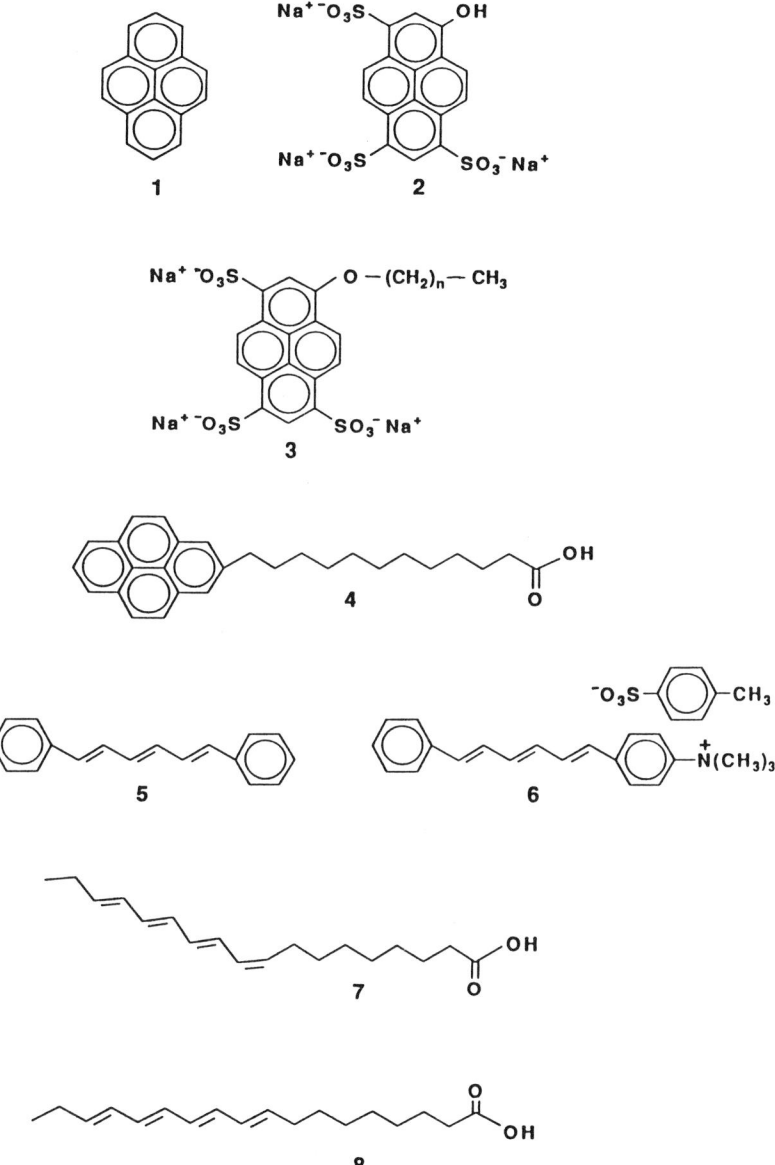

Fig. 2.3. Examples of hydrophobic, hydrophilic, and amphiphilic probes: **(1)** pyrene; **(2)** 8-hydroxypyrene-1,3,6-trisulfonic acid, trisodium salt (pyranine); **(3)** 8-alkoxypyrene-1,3,6-trisulfonic acid, trisodium salt; **(4)** 1-pyrenedodecanoic acid; **(5)** 1,6-diphenyl-1,2,5-hexatriene (DPH); **(6)** 1-(4-trimethylammoniumphenyl)-6-phenyl-1,3,5-hexatriene, p-toluene sulfonate (TMA–DPH); **(7)** cis-parinaric acid; **(8)** trans-parinaric acid.

In conclusion, the choice of a fluorescent probe is crucial for obtaining unambiguous interpretations. The design of a probe for a specific application is an art!

2.2.2. Absorption, Excitation, and Emission Spectra: The Stokes Shift

Absorption in the visible (or near-UV) region is preferable to absorption in the UV region, especially in the case of biological materials that may show a fluorescence background upon excitation in the UV. For instance excitation wavelengths beyond 450 nm are preferable for studies in cellular biology because of the fluorescence of cell components (flavins, NADH, etc.)

The absorption coefficient must be as large as possible, but it is seldom a limiting factor except in specific applications where a small number of probes must be detected in solution (e.g., in detection experiments of proteins or nucleotides) (61). The multichromophoric probes in the series of phycobiliproteins have an exceptionally high extinction coefficient ($2 \times 10^6 \, M^{-1} \cdot cm^{-1}$) (62).

Whatever the extinction coefficient, the absorbance of solutions should not be too high in order to avoid spurious effects such as inner filter effects. For instance, with 1 cm × 1 cm square cuvettes, the absorbance should generally be kept below 0.1.

Excitation to any vibrational level of an excited state is followed by very fast internal conversion to the lowest vibrational level of the lowest singlet state S_1. This process being much faster than the average residence time of the molecules in the excited state (lifetime), the emission spectrum is, in principle, independent of the excitation wavelength, but there are exceptions to this rule: emission from the second excited state S_2 may compete with internal conversion to S_1 (e.g., azulene or diphenylpolyenes).

In most cases the excitation spectrum is identical to the absorption spectrum, but this is not always true and differences between these spectra may reveal the existence of more than one species (or several forms of one species) in the ground state.

Excitation in the far-red edge of the absorption spectrum should be avoided because of possible red edge effects (63–66).

The Stokes shift, i.e., the shift of the emission maximum with respect to the absorption maximum, is an important feature. When the Stokes shift is small, (i) the detection of fluorescence without parasitic scattered light is more difficult, especially when optical filters are used, and (ii) the overlap between the absorption spectrum and the emission spectrum is large and may cause some reabsorption of the emitted fluorescence (when the probe concentration is too high, or in a scattering medium) and thus a distortion of the emission spectrum in the overlap region. Fluorescein and rhodamine derivatives are examples of dyes with a small Stokes shift, whereas aminonaphthalene derivatives exhibit a large Stokes shift that depends on the polarity of the solvent (see Section 2.3.1.2, below).

An interesting illustration of the importance of the Stokes shift is provided by two-color flow cytometry analysis using one excitation wavelength. Cell surface receptors and antigens are currently analyzed by flow cytometry using fluorescein-labeled ligands and antibodies, respectively. The concomitant use of a benzoxazinone derivative that can be excited at the excitation wavelength of fluorescein but emits at higher wavelengths allows one to detect cell subpopulations in one run with a flow cytometer equipped with a single light source (67,68); the green fluorescence from fluorescein and the red fluorescence from the benzoxazinone derivative can be detected simultaneously by appropriate optical filters.

2.2.3. Decay of Fluorescence Intensity: Excited State Lifetime

After excitation by an infinitely short pulse of light, the decay of fluorescence intensity follows, in the simplest case, a single exponential, the time constant being the fluorescence lifetime τ:

$$I(t) = I_0 \exp(-t/\tau) \tag{2.1}$$

The reciprocal of the lifetime is the sum of all the rate constants (both radiative and nonradiative) for deactivation of the excited molecules. Note that the fluorescence lifetime is one of the most important characteristics of a probe because it defines the time window of observation of dynamic phenomena. As illustrated in Fig. 2.4, no accurate information on the rate of phenomena occurring at time scales shorter than about $\tau/10$ ("private life" of the probe) or longer than about 10τ ("death" of the probe) can be obtained, whereas at intermediate times ("public life" of the probe) the time evolution of phenomena can be followed. Note also that a similar situation is found in the use of radioisotopes for dating: the period (i.e., the time constant of the exponential radioactive decay) must be of the same order of magnitude as the age of the object to be dated.

The lifetime of a homogeneous population of fluorophores is very often independent of the excitation wavelength as the emission spectrum, but there are exceptions as already mentioned. For organic probes, the lifetime ranges from tenths of nanoseconds to hundreds of nanoseconds, whereas the triplet lifetime is much longer (microseconds to seconds); therefore, observation of phosphorescence or delayed fluorescence offers the possibility of studying much slower phenomena (69–71).

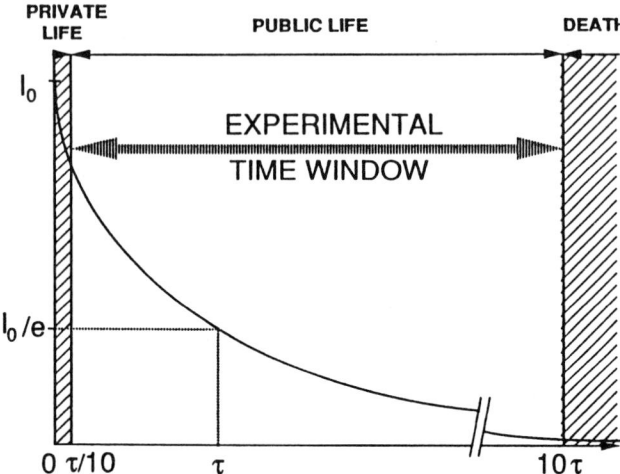

Fig. 2.4. Decay of fluorescence intensity characterized by the excited lifetime τ defining the experimental time window ($\tau/10 < t < 10\tau$).

In contrast to organic probes, lanthanide probes have long lifetimes (0.1–4 ms), which makes these probes attractive in fluoroimmunoassays. Indeed, they offer the possibility of avoiding interfering fluorescence from the biological material and from any other material (with lifetimes in the nanosecond range) by performing time-resolved experiments with a delay between the excitation pulse and the measurements of the Eu(III) luminescence (72). For a detailed review on lanthanide probes, see Bünzli (73).

In many cases, the decay of fluorescence intensity of a probe is not a single exponential:

- In a heterogeneous medium, a probe may experience different environments.
- A probe may undergo various changes after excitation: reorganization of its solvation shell, protonation or deprotonation, internal rotation, excimer or exciplex formation, energy transfer, etc. Through these excited state processes, insight into the properties of the probe microenvironment is provided: polarity, acidity, viscosity, etc.

Therefore, accurate measurements of complex fluorescence decays, either by pulse fluorometry (74–76) or multifrequency phase-modulation fluorometry (77–81), are of particular interest. Great care should be taken in the analysis of the decays [see the report of the IUPAC Commission on Photochemistry (82)] because of the considerable underlying information. This important aspect will now be discussed.

Improvement in the accuracy of the parameters involved in a physical model can be obtained by performing simultaneous analysis of different time-resolved experiments corresponding to different experimental parameters: excitation and emission wavelengths, temperature, quencher concentration, orientation of excitation and emission polarizers, pH, etc. The underlying principle of the method, called global analysis, is to utilize the relationships between decay curves by linking the common parameters in the analysis (83–90). This approach offers in principle a greater confidence in the recovered parameters and thus improves the capability of testing models. However, one has to make sure that the parameters are in fact linked as they were assumed to be.

In the absence of a model, the question may arise whether a nonexponential decay contains a sum of a limited number of exponentials or a sum of a large number of exponentials allowing one to consider a distribution of decay time constants (e.g., lifetimes). A satisfactory analysis (by a nonlinear least-squares method, for instance) with a sum of two (or three or four) exponential terms with closely spaced time constants means that the number of fitting parameters is large enough considering the limited accuracy of the experimental data, but it does not prove that the number of components is two (or three or four), unless it is already known that the system can be described by a physical model involving distinct exponentials.

However, sometimes one may expect that continuous lifetime distributions would best account for the observed phenomena (91–95). To answer the question of whether the fluorescence decay of a probe consists of a few distinct exponentials or should be interpreted in terms of a continuous distribution, it is advantageous to use an approach without a priori assumption of the shape of the distribution. In particular, the maximum entropy method has been successfully applied to the analysis of data in pulse fluorometry (96,97) and in multifrequency phase-modulation fluorometry (98). The maximum entropy method offers a new and powerful tool for the analysis of complex fluorescence decays, capable of handling both continuous and discrete lifetime distributions in a single analysis.

2.2.4. Quantum Yield

The quantum yield expresses the proportion of excited molecules that return to the ground state by emitting a fluorescence photon. In other words, the quantum yield is the ratio of the number of emitted photons to the number of absorbed photons. Hence,

$$\Phi = \frac{k_r}{k_r + k_{nr}} \qquad (2.2)$$

and the lifetime can be expressed in the same way (provided that the fluorescence decay is monoexponential):

$$\tau = \frac{1}{k_r + k_{nr}} \qquad (2.3)$$

where k_r and k_{nr} are the rate constants for radiative and nonradiative deactivations, respectively; here k_{nr} is the sum of all the rate constants for nonradiative deactivations: internal conversion and intersystem crossing. In some cases, a term $\Sigma k_Q[Q]$ (with k_Q assumed to be time independent) must be added, representing the additional routes of deactivation due to quenchers. Oxygen is a well-known quencher, but its efficiency strongly depends on the nature of the probe. Also, since the oxygen quenching is partly diffusion controlled, probes exhibiting long excited state lifetimes (e.g., naphthalene and pyrene) are more sensitive to the presence of oxygen and in viscous media the efficiency of oxygen quenching is reduced. Therefore, degassing a sample by the freeze-pump-thaw technique or by bubbling nitrogen may be required or not, depending on the system under study.

It is obvious that the larger the quantum yield of a probe, the easier the detection of fluorescence. It is worth recalling that many factors can affect the quantum yield: temperature, pH, polarity, viscosity, H-bonding ability, proximity of heavy atoms or quenchers, etc. Attention should be paid to possible misinterpretations due to concomitant phenomena affecting the quantum yield (e.g., variations of viscosity upon changing the temperature).

According to relations (2.2) and (2.3), $\Phi = k_r\tau$, but this does not mean that the quantum yield remains always proportional to the lifetime upon an external perturbation. A typical case where the quantum yield is affected without any change of the lifetime is the formation of a nonfluorescent complex in the ground state (static quenching).

2.2.5. Photochemical Stability

An obvious criterion of photostability is the absence of significant photochemical effects until the experiment is finished. The photostability of a probe should of course be very good when intense illumination is required. For instance, in fluorescence microscopy, fading of the probe (e.g., fluorescein derivatives) may be a serious problem. Conversely, good photostability is not required when photophysical studies are performed with highly sensitive instruments operating at low intensity of excitation light, especially when photon counting is employed. Photostability is also not important in flow cytofluorometry because the residence time of the probe molecules in the laser beam is very short. Conversely, in rigid matrices or very viscous media, the probe molecules cannot escape from the incident beam and good photostability is thus required.

2.3. MAIN PHYSICAL AND STRUCTURAL PARAMETERS MEASURABLE BY MEANS OF FLUORESCENT PROBES

2.3.1. Polarity

Polarity is of major importance in many physical, chemical, biochemical, and biological phenomena. The local polarity of a system can be estimated by means of a fluorescent probe by comparing its spectral properties in its environment with those in solvents of known polar characteristics. Before the methods of estimation of polarity are discussed, it is worth recalling the meaning of this term.

2.3.1.1. What Is Polarity?

The term *polarity* reflects the complex interplay of all types of solute–solvent interactions, i.e., nonspecific dielectric solute–solvent interactions and specific interactions such as hydrogen bonding. Therefore, polarity cannot be characterized by a single parameter: an oversimplification often met consists in associating the "polarity" of a solvent (or a microenvironment) with the static dielectric constant ε (macroscopic quantity) or the dipole moment μ of the solvent molecules (microscopic quantity). A molecule is said to be *polar* when it possesses a dipole moment, but the term *dipolar* would be more appropriate.

Dielectric solute–solvent interactions, in conjunction with solvatochromic shifts, are very well described in the excellent review of Suppan (99). These interactions result not only from the dipole moments but also from the polarizabilities. Let us recall that the polarizability α of a spherical molecule is defined by means of the dipole $\mu_i = \alpha E$ induced by an external electric field E in its own direction. Table 2.2 shows the four major dielectric interactions (dipole–dipole, solute dipole–solvent polarizability, solute polarizability–solvent dipole, polarizability–polarizability). Analytical expressions of the corresponding energy terms can be derived within the simple model of spherical-centered dipoles in isotropically polarizable spheres. These expres-

Table 2.2. Dielectric Solute–Solvent Interactions Resulting from the Dipole Moments and Average

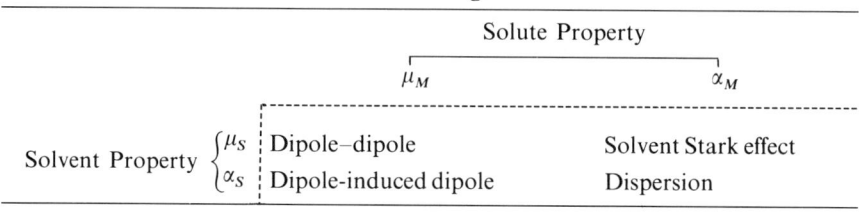

Source: From Suppan (99).

sions involve the Onsager polarity functions

$$f(\varepsilon) = \frac{\varepsilon - 1}{2\varepsilon + 1} \quad \text{and} \quad f(n^2) = \frac{n^2 - 1}{2n^2 + 1} \qquad (2.4)$$

where ε is the static dielectric constant and n is the refractive index. These two parameters are macroscopic observable quantities related to the "polarity" of a medium.

For the description of solvatochromic shifts, an additional energy term relative to the solute should be considered. This term is related to the transition dipole moment that results from the migration of electric charges during an electronic transition. Note that this transient dipole has nothing to do with the difference between the permanent dipole moment in the excited state and that in the ground state ($\mu_e - \mu_g$).

As noted by Suppan (99), each term of the solvation energy E_{solv} is a product of two factors expressing separately the "solute polarity" (P) and the "solvent polarity" (Π):

$$E_{\text{solv}} = P\Pi \qquad (2.5)$$

Therefore, Suppan suggests that this relation be used as a general quantitative definition of polarity. He emphasizes that P and Π are not simple numbers, but rather matrices describing the properties of the solute molecule (dipole moment, polarizability, transition moment, hydrogen-bonding capability) and the solvent molecule (dielectric constant, refractive index, hydrogen bonding capability).

Since there are many parameters underlying the concept of polarity, the validity of empirical scales of polarity based on a single parameter, e.g., Z (100), $E_T(30)$ (101,102) [for a discussion on the various polarity scales, see Reichardt (103) and Buncel and Rajagopal (104)], is questionable. A multiparameter approach is preferable, and the π^* scale of Kamlet and Taft (105,106) deserves special recognition because it has been established from the averaged spectral behavior of many solutes and takes into account both specific and nonspecific interactions. Any observable quantity and in particular the wavenumber $\bar{v}$ of an absorption or emission band can be written as a linear combination of individual components. For the sake of simplicity, only three main parameters will be considered. The basic relation is then

$$\bar{v} = \bar{v}_0 + S\pi^* + a\alpha + b\beta \qquad (2.6)$$

where $\bar{v}$ and $\bar{v}_0$ are the wavenumbers of the band maxima in the considered solvent and in the reference solvent (generally cyclohexane), respectively; π^* is a measure of the polarity/polarizability effects of the solvent; the α scale is an index of hydrogen bond donor capability of the solvent, and β is the relevant quantity for hydrogen bond acceptor capability; and the S, a, and b coefficients describe the sensitivity of a process to each of the individual contributions.

2.3.1.2. Estimation of Polarity from Spectral Shifts

The shifts of absorption and emission spectra when the nature of the solvent (solvatochromic shifts) is changing result from changes in energy of the ground state and the excited state. It should be emphasized that the energy of the emitting state is often different from that of the Franck–Condon (FC) state; one of the reasons pertaining to the polarity effect is the process called *solvent relaxation*, the origin of which will now be explained. In most cases the dipole moment of an aromatic molecule in the excited state μ_e differs from that in the ground state μ_g. As a matter of fact, absorption of a photon by a fluorophore occurs in a very short time ($\sim 10^{-15}$ s) with respect to the displacement of nuclei (FC principle) but in time enough to allow for a redistribution of electrons, which results in an almost instantaneous change of the dipole moment. Polarity probes are chosen so that $\mu_e > \mu_g$. Therefore, subsequent to excitation, the solvent cage undergoes a relaxation, i.e., a reorganization leading to a relaxed state of minimum free energy (Fig. 2.5). The larger the polarity of the solvent, the lower the energy of the relaxed state and the larger the red shift of the emission spectrum. It is important to note that the rate of solvent relaxation depends on the viscosity of the solvent. In viscous media, this relaxation may not be complete within the lifetime of the excited state of the probe, which may lead to misinterpretation of the shift in terms of polarity. At the limit, in a very viscous or rigid polar medium, emission arises from a state close to the FC state, as in the case of a nonpolar medium.

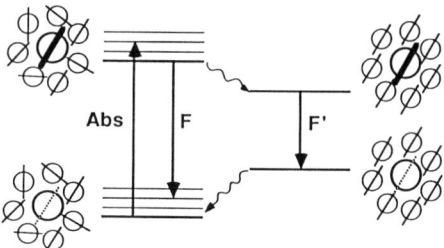

Fig. 2.5. Illustration of solvent relaxation around a probe that has a weak dipole moment in the ground state and a large dipole moment in the excited state.

If solvent (or environment) relaxation is complete, equations for the dipole–dipole interaction solvatochromic shifts can be derived within the simple model of spherical-centered dipoles in isotropically polarizable spheres and within the assumption of equal dipole moments in FC and relaxed states. The solvatochromic shifts (expressed in wavenumbers) are then given by Eqs. (2.7) and (2.8) for absorption and emission, respectively:

$$\bar{v}_a = -\frac{2}{hc}\boldsymbol{\mu}_g \cdot (\boldsymbol{\mu}_e - \boldsymbol{\mu}_g)a^{-3}\Delta f + \text{constant} \tag{2.7}$$

$$\bar{v}_f = -\frac{2}{hc}\boldsymbol{\mu}_e \cdot (\boldsymbol{\mu}_g - \boldsymbol{\mu}_e)a^{-3}\Delta f + \text{constant} \tag{2.8}$$

where h is Planck's constant; c, the velocity of light; a, the radius of the cavity in which the solute resides; and Δf, the orientation polarizability:

$$\Delta f = f(\varepsilon) - f(n^2) = \frac{\varepsilon - 1}{2\varepsilon + 1} - \frac{n^2 - 1}{2n^2 + 1} \tag{2.9}$$

Subtraction of Eqs. (2.7) and (2.8) leads to Lippert's equation (107):

$$\bar{v}_a - \bar{v}_f = \frac{2}{hc}(\mu_e - \mu_g)^2 a^{-3}\Delta f + \text{constant} \tag{2.10}$$

This expression of the Stokes shift depends only on the absolute magnitude of the charge transfer dipole moment $\Delta\boldsymbol{\mu}_{ge} = \boldsymbol{\mu}_e - \boldsymbol{\mu}_g$ and not on the angle between the dipoles. The validity of Eq. (2.10) can be checked by using various solvents and by plotting $\bar{v}_a - \bar{v}_f$ as a function of Δf. A linear variation is not always observed because only the dipole–dipole interaction has been taken into account. By choosing solvents without H-bonding donor or acceptor ability, a linear behavior is often observed, which allows one to determine the increase in dipole moment $\Delta\mu_{ge}$ upon excitation, provided that a correct estimation of the cavity radius is possible.

Furthermore, Suppan (108) derived another useful equation from Eqs. (2.7) and (2.8) under the assumption that $\mathbf{\mu}_g$ and $\mathbf{\mu}_e$ are colinear but without any assumption about the cavity radius a or the form of the solvent polarity function. This equation involves the differences in absorption and emission solvatochromic shifts between two solvents 1 and 2:

$$\frac{(\bar{v}_f)_1 - (\bar{v}_f)_2}{(\bar{v}_a)_1 - (\bar{v}_a)_2} = \frac{\mu_e}{\mu_g} \qquad (2.11)$$

Equations (2.10) and (2.11) provide a means of determining excited dipole moments together with dipole vector angles, but they are valid only if (i) the dipole moments in the FC and relaxed states are identical, (ii) the cavity radius remains unchanged upon excitation, and (iii) the solvent shifts are measured in solvents of the same refractive index but of different dielectric constants.

An ideal polarity probe (i) should undergo a large change in dipole moment upon excitation but without change in direction, (ii) should bear no permanent charge in order to avoid contributions due to ionic interactions, and (iii) should be soluble in solvents of various polarity, from the apolar solvents to the most polar ones. Among the polarity probes given in Fig. 2.6, PRODAN [6-propionyl-2-(dimethylamino)naphthalene] (109) fulfills these requirements. Figure 2.7 shows the fluorescence spectra of this probe in various solvents: the very large increase in the Stokes shift from 4300 cm^{-1} in cyclohexane to 8640 cm^{-1} in water is due to an increase in the dipole moment upon excitation from about 2 to 18–20 D (debyes, calculated from the Lippert equation). The structural, conformational, and electronic density distribution changes in the lowest singlet excited state have been studied by Ilich and Prendergast (110). A chemical variant, DANCA [4,2'-(dimethylamino)-6'-naphthoylcyclohexanecarboxylic acid], was used by MacGregor and Weber (111) to determine the polarity of the myoglobin hemepocket. Prendergast et al. (112) introduced another chemical variant, ACRYLODAN [6-acryloyl-2-(dimethylamino)naphthalene], which covalently binds to protein —SH groups.

In contrast to PRODAN, 1-anilino-8-naphthalene sulfonate (ANS) bears a charge. This probe, which was discovered by Weber and Lawrence a long time ago (113), exhibits the interesting feature of being nonfluorescent in aqueous solutions and highly fluorescent in solvents of low polarity. This feature allows one to visualize only hydrophobic regions of biological systems without interference of the probe molecules remaining in the surrounding aqueous

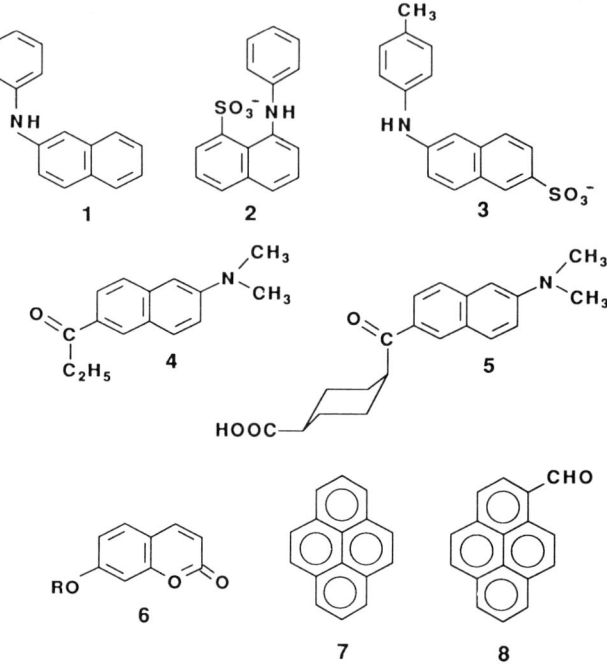

Fig. 2.6. Examples of polarity-sensitive probes: **(1)** 2-anilinonaphthalene; **(2)** 1-anilino-8-naphthalene sulfonate (ANS); **(3)** 2-*p*-toluidinyl-6-naphthalene sulfonate (TNS); **(4)** 6-propionyl-2-(dimethylamino)naphthalene (PRODAN); **(5)** 4-2′-(dimethylamino)-6′-naphthoylcyclohexane-carboxylic acid (DANCA); **(6)** 7-alkoxycoumarin; **(7)** pyrene; **(8)** pyrene-1-carboxaldehyde.

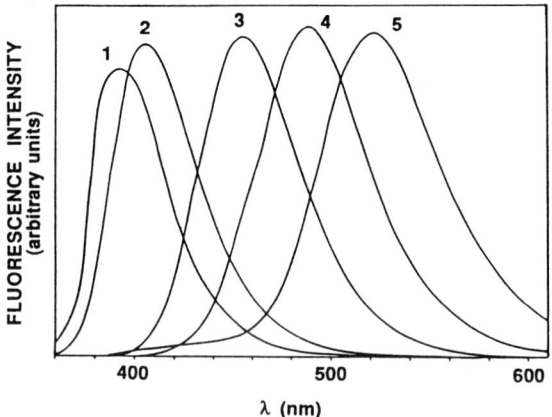

Fig. 2.7. Uncorrected fluorescence spectra of PRODAN in cyclohexane **(1)**, chlorobenzene **(2)**, dimethylformamide **(3)**, ethanol **(4)**, and water **(5)**. [From Weber and Farris (109).]

environment. ANS and its derivative TNS (2-p-toluidinyl-6-naphthalene sulfonate) have been extensively used for probing proteins, biological membranes, and micellar systems [see the references cited in Haugland (56)]. The reason why these probes undergo large variations in fluorescence with solvent polarity has been the object of numerous investigations focusing on: (a) intersystem crossing to the triplet state (114); (b) specific solvent–solute interactions (115,116); (c) intramolecular charge transfer (117); (d) change in molecular conformation (118); and (e) monophotonic photoionization (119,120).

Owing to solvent (or environment) relaxation, the position and shape of the emission spectrum of polar probes change as a function of time after excitation. Microscopic solvation dynamics is of fundamental interest and can be studied by time-resolved fluorescence spectroscopy. This subject, which is beyond the scope of this chapter, has been described in several reviews (121–124). It should be emphasized again that solvent (or environment) relaxation must be complete during the lifetime of the excited state for a correct interpretation of the shift of the fluorescence spectrum in terms of polarity.

2.3.1.3. Estimation of Polarity from Changes in Vibronic Bands

In aromatic molecules with a minimum D_{2h} symmetry (e.g., benzene and pyrene), the first singlet absorption ($S_1 \leftarrow S_0$) is symmetry forbidden and leads to a weak oscillator strength. The relevant forbidden vibronic bands are sensitive to solvent polarity (the Ham effect). This effect, illustrated in Fig. 2.8, provides a simple way to estimate the polarity of an environment by measuring

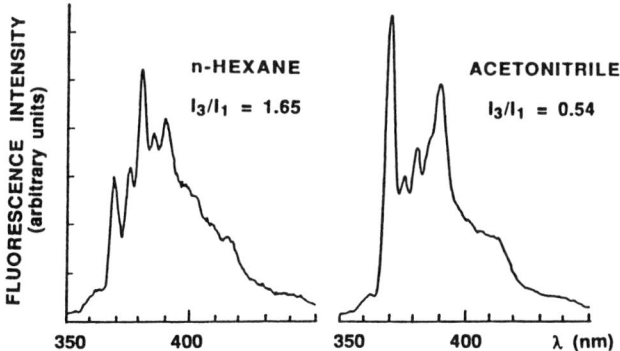

Fig. 2.8. Fluorescence spectra of pyrene in n-hexane and acetonitrile showing the polarity dependence of vibronic band intensities. [From Kalyanasundaram and Thomas (125).]

the ratio of the fluorescence intensities of the third and first vibronic bands (125–130). For pyrene (Py scale), this ratio I_3/I_1 ranges from ~1.7 in hydrocarbon media to ~0.6 in water. When solvents are divided by classes (aprotic aliphatics, protic aliphatics, aprotic aromatics), each class gives excellent correlation between the Py scale and the π^* scale (130). The Py scale appears to be relatively insensitive to the hydrogen capability of protic solvents.

Pyrene has been extensively used for probing polarity in micellar systems [see the reviews of Zana (15) and Kalyanasundaram (12)].

2.3.2. Fluidity: Order Parameters; Molecular Mobility

Since viscosity is a macroscopic parameter that loses its physical meaning at a molecular scale, the term *fluidity* will be used in this section to characterize, in a very general way, the effects of viscous drag and cohesion of the probed microenvironment.

Only the methods providing information at a molecular or supramolecular scale will be discussed. Fluorescence correlation spectroscopy (131) and photobleaching recovery (132) will not be considered here.

2.3.2.1. General Considerations

The main fluorescence techniques permitting evaluation of the fluidity of a microenvironment by means of a fluorescent probe are reported in Table 2.3. The underlying physical quantity in all these techniques is a diffusion constant (either rotational or translational) expressing the viscous drag of the surrounding molecules. The major problem is then to relate this diffusion constant to the viscosity η. For a spherical particle, the well-known Stokes–Einstein relation is often used

$$D = \frac{kT}{\xi} \tag{2.12}$$

where k is the Boltzmann constant; T, the absolute temperature; ξ, the frictional coefficient. For translational and rotational diffusion, the diffusion constants are respectively

$$D_t = \frac{kT}{6\pi\eta r} \tag{2.13}$$

and

$$D_r = \frac{kT}{8\pi\eta r^3} \tag{2.14}$$

Table 2.3. Main Fluorescence Techniques for the Determination of Fluidity

Technique	Measured Fluorescence Characteristics	Phenomenon	Comments
Molecular rotors	Fluorescence quantum yield and/or lifetime	Internal torsional motion	Very sensitive to free volume; fast experiment
Excimer formation	Fluorescence spectra		Fast experiment
(1) Intermolecular	Comparison between monomer and excimer bands	Translational diffusion	Diffusion perturbed by microheterogeneities
(2) Intramolecular	Same as above	Internal rotational diffusion	More reliable than intermolecular formation
Fluorescence quenching	Fluorescence quantum yield and/or lifetime	Translational diffusion	Addition of two probes, same drawback as intermolecular excimer formation
Fluorescence polarization	Emission anisotropy	Rotational diffusion of the whole probe	
(1) Steady-state			Simple technique, but Perrin's law often not valid
(2) Time-resolved			Sophisticated technique but very powerful; also provides order parameters

where r is the hydrodynamic radius of the sphere. However, these relations are valid only for a rigid sphere that is large compared to the molecular dimensions, moving in a homogeneous Newtonian fluid, and obeying the Stokes hydrodynamic law. Many other relations have been proposed. They are gathered in a review by André et al. (30) in which this problem is thoroughly discussed. These authors point out the absence of a satisfactory relationship between diffusion and bulk viscosity for probes. The main reason is that the size of probes is comparable to that of the surrounding molecules forming the microenvironment to be probed. Moreover, this microenvironment may not be isotropic, as in the case of organized assemblies such as micellar systems and biological membranes.

In other words, viscosity is a macroscopic parameter and any attempt to get absolute values of the viscosity of a medium from measurements using a fluorescent probe is hopeless. The term *microviscosity* is often employed, but again no absolute values can be given, and the best we can do is to speak of "equivalent viscosity" (133,134), i.e., the viscosity of a homogeneous medium in which the response of the probe is the same. But a difficulty arises as to the choice of the reference solvent because the rotational relaxation rate of a probe in various solvents of the same macroscopic viscosity depends on the internal order of the solvent (133).

A microscopic approach to the viscosity problem was developed by Gierer and Wirtz (135), and it is worthwhile giving the main aspects of this theory, which is of interest here in that it accounts for the finite thickness of the solvent layers and for the existence of holes in the solvent (free volume). This theory with further developments is well described in the review by Alwattar et al. (136). Stokes's law is modified by using a microscopic frictional coefficient ξ_{micro},

$$D = \frac{kT}{\xi_{micro}} \quad (2.15)$$

and by introducing a microfriction factor $f(<1)$ defined as

$$\xi_{micro} = \xi_{Stokes} f \quad (2.16)$$

A solute molecule moves according to two diffusional processes: a viscous process with displacement of solvent molecules (Stokes diffusion), and a process associated with migration into holes of the solvent (free volume diffusion) as illustrated in Fig. 2.9. The total velocity of the solute molecule is given by $v = v_1 + v_2$. Since the frictional coefficient is the ratio of the viscous force to the velocity ($\xi = F/v$), the microscopic frictional coefficient ξ_{micro} consists of two parts:

$$\frac{1}{\xi_{micro}} = \frac{1}{\xi_1} + \frac{1}{\xi_2} \quad (2.17)$$

Here ξ_1, corresponding to the Stokes diffusional process, can be written as the

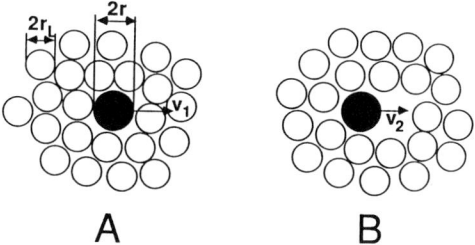

Fig. 2.9. Translational diffusion processes: (A) Stokes diffusion; (B) free volume diffusion. *Key:* ●, solute molecules; ○, solvent molecules.

product of the Stokes frictional coefficient multiplied by a correcting factor f'_t taking into account the finite thickness of the solvent layers (135):

$$\xi_1 = \xi_{\text{Stokes}} f'_t \tag{2.18}$$

with

$$f'_t = \left[2\frac{r_L}{r} + \frac{1}{1 + r_L/r} \right]^{-1} \tag{2.19}$$

where r and r_L are the radii of the solute and the solvent molecules, respectively, assuming that they are spherical.

Using Eqs. (2.15)–(2.18), we obtain the microfriction factor f_t for translation:

$$f_t = \frac{\xi_{\text{micro}}}{\xi_{\text{Stokes}}} = \frac{1}{1/f'_t + \xi_{\text{Stokes}}/\xi_2} = \frac{f'_t}{1 + v_2/v_1} \tag{2.20}$$

The translational diffusion coefficient on the molecular scale is then

$$D_t = \frac{kT}{6\pi\eta r f_t} \tag{2.21}$$

For rotational diffusion, the correcting factor is (135)

$$f'_r = \left[6\frac{r_L}{r} + \frac{1}{(1 + r_L/r)^3} \right]^{-1} \tag{2.22}$$

The importance of free volume effects in diffusional processes at a molecular level should be further emphasized. An empirical relation between viscosity and free volume was proposed by Doolittle (137)

$$\eta = \eta_0 \exp\left(\frac{V_0}{V_f}\right) \qquad (2.23)$$

where η_0 is a constant; V_0 and V_f are the van der Waals volume and the free volume of the solvent, respectively. From this relation, a very important empirical relation was derived by Williams, Landel, and Ferry (WLF) (138) for polymers and liquids at temperatures ranging from the glass transition temperature T_g to roughly $T_g + 100\,^\circ\text{C}$:

$$\ln \frac{\eta}{\eta_g} = -\frac{C_1(T-T_g)}{C_2 + (T-T_g)} \qquad (2.24)$$

where C_1 and C_2 are constants. The WLF prediction for the temperature dependence of η for dimethylphthalate and glycerol was tested by Loufty and Arnold (139) with fluorescent probes in the series of aminobenzylidenemalononitriles. The agreement with WLF equation was found to be excellent.

These preliminary considerations should be borne in mind while considering the following discussion on the various methods of characterization of "microviscosity."

2.3.2.2. Use of Molecular Rotors

A molecular rotor as a fluorescent probe is a molecule exhibiting emissive properties that depend on internal rotation. Various examples are given in Fig. 2.10. Oster and Nishijima (140) reported that the quantum yield of the diphenylmethane dye auramine O is strongly dependent on the medium viscosity. They concluded that internal rotation of the molecule is the main route for nonradiative deactivation and that an increase in viscosity reduces this process by frictional resistance, thus resulting in an increased quantum yield. This interpretation is consistent with the fact that the reciprocal of the quantum yield was found to vary linearly with η/T. These findings show that the energy resulting from Brownian motion is predominant with respect to the potential energy change involved in the internal rotation of the phenyl rings. Auramine O was thus used by Oster and Nishijima (141) to probe the viscosity of polymers in the solid state.

Subsequently, many investigations were devoted to the viscosity dependence of the quantum yield and lifetime of various dyes (142–151). Förster and Hoffmann (142) found that the quantum yield of crystal violet, a triphenylmethane dye, was proportional to $\eta^{2/3}$. Tredwell and Osborne (144) observed

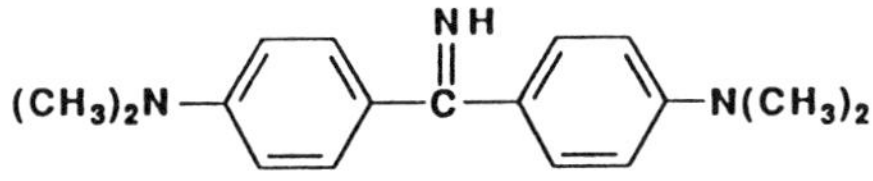

Fig. 2.10. Examples of molecular rotors: (1) Auramine O; (2) crystal violet; (3) p-N,N-dimethylaminobenzylidenemalononitrile; (4) julolidinebenzylidenemalononitrile; (5) p-dimethylaminobenzonitrile; (6) p-dimethylaminobenzoate.

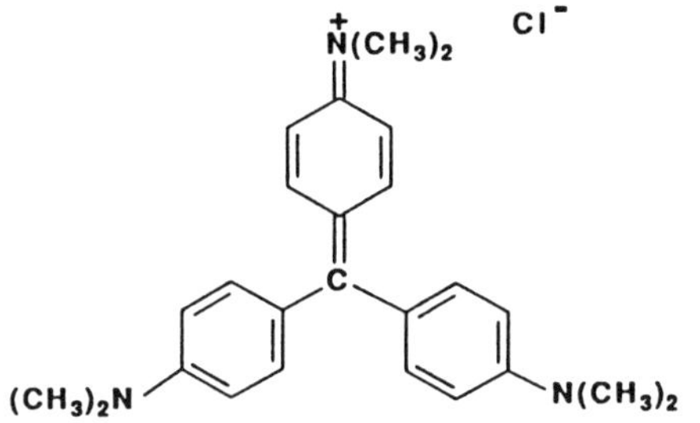

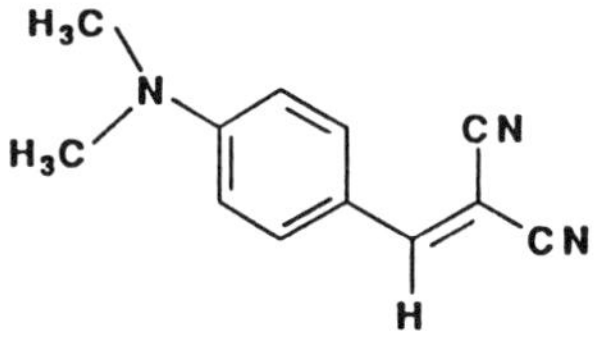

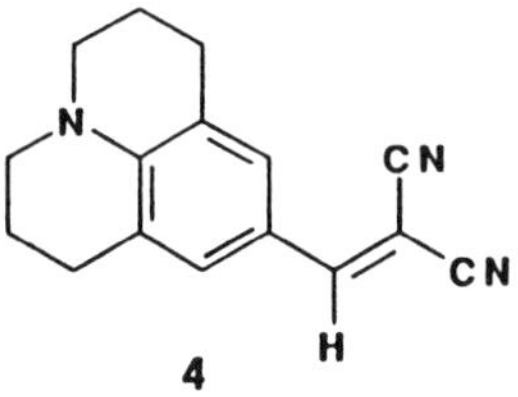

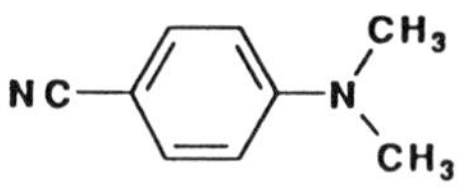

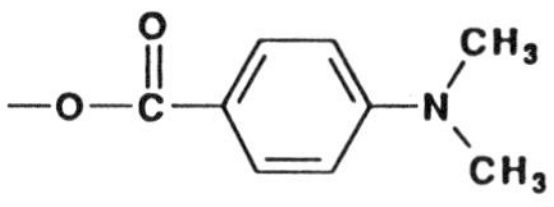

the same viscosity dependence for the lifetime of the xanthene dye fast acid violet 2R. For seven cyanine dyes, Sundström and Gillbro (146) found that the lifetime obeyed the simple empirical relation $\tau = a\eta^\alpha$, the exponent α being dependent on the solvent nature: α is equal to ·1 in n-alcohols and 0.5 in glycerol/methanol and glycerol/water mixtures. The fact that intermediate values of α can be found indicates that the 2/3 power law found by Förster and Hoffmann has no particular significance.

Various triphenylmethane dyes with large and small, rigid and flexible rotors groups were examined by Vogel and Rettig (150). These authors found that they observed viscosity dependence of the rotational relaxation rate k_{rot}

of these dyes do not obey the Stokes–Einstein relationship but can be fitted within the microviscosity theory of Gierer and Wirtz (*vide supra*), from which the authors derived the following relation:

$$k_{rot}\eta = A + B\eta^x \qquad (2.25)$$

The quantity $k_{rot}\eta$ is constant within Stokes–Einstein theory but depends on η^x for the Gierer–Wirtz microviscosity theory. This equation was obeyed with an exponent $x = 0.7$ for crystal violet, which means that the activation energy for diffusion into solvent holes represents 30% of the activation energy of the viscous flow. It is important to note that *the effective viscosity is less than the bulk viscosity η owing to free volume effects*.

Molecular rotors in the series of aminobenzylidenemalononitriles also offer interesting properties as regards viscosity dependence (139,152–158). The free volume concept was used by Loufty and Arnold (139) for the interpretation of the behavior of *p-N,N*-dialkylaminobenzylidenemalononitrile. In solvents of medium and high viscosity, these authors assumed that the nonradiative rate constant for deactivation is linked to the ratio of the van der Waals volume to the free volume according to

$$k_{nr} = k_{nr}^{o} \exp\left(-x\frac{V_0}{V_f}\right) \qquad (2.26)$$

where k_{nr}^{o} is the free-rotor reorientation rate and x is a constant for a particular probe. By combining this equation, the relation linking k_{nr}, k_r, and the quantum yield [Eq. (2.2) in Section 2.2.4, above] and the Doolittle equation [Eq. (2.23) in Section 2.3.2.1, above], one obtains

$$\Phi = a\left(\frac{\eta}{T}\right)^x \qquad (2.27)$$

The variations of $\ln \Phi$ versus $\ln(\eta/T)$ was indeed found to be linear for *p-N,N*-dimethylaminobenzylidenemalononitrile in glycerol ($x = 0.69$) and in dimethylphtalate ($x = 0.43$), which indicates that the viscosity dependence of the quantum yield arises from the dependence of the viscosity on the free volume. However, this is true only for solvents of medium and high viscosity. At high temperatures, i.e., much above the solvent glass transition temperature, and in solvents of low viscosity, the free volume concept does not apply. In these cases, Loufty and Arnold (139) found that the simple Stokes–Einstein relation gives a good description of the dynamics of torsional motion of the probes.

Among the various possible internal rotations in molecular rotors of the donor–acceptor type, those which lead to the formation of a twisted intramolecular charge transfer (TICT) state were extensively studied (159–161) with special attention to *p*-dimethylaminobenzonitrile, which exhibits dual fluorescence: one of the bands is the normal fluorescence, whereas the other arises from the TICT state resulting from the rotation of the dimethylamino group. This type of molecule is well suited for probing free volume, segment mobility, microheterogeneity in solid polymers (162–165).

2.3.2.3. Intermolecular Excimer Formation and Intermolecular Quenching

Excimers (excited dimers) can be formed by collision of an excited fluorophore with an identical unexcited one:

$$A + A^* \underset{k_{\bar{1}}}{\overset{k_1}{\rightleftharpoons}} (AA)^*$$

Excimer formation (pyrene is a good example) is revealed by a new unstructured emission band at lower energy (see Fig. 2.11). This translational diffusion process is viscosity dependent and is thus expected to provide information on the fluidity of a microenvironment, provided that excimer formation occurs on a time scale comparable to the monomer excited lifetime (experimental time window). The rate constant k_1 that characterizes this process is often assumed to be independent of time and expressed by the familiar form of the

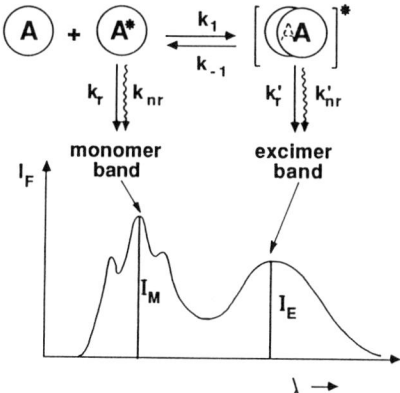

Fig. 2.11. Kinetic scheme for excimer formation.

Einstein–Smoluchowski expression:

$$k_1 = \frac{4\pi N_A RD}{1000} \qquad (2.28)$$

where R is the encounter radius; D, the mutual diffusion coefficient of A and A* ($D_A + D_{A^*}$); and N_A is Avogadro's number. However, k_1 is, under general conditions, time dependent; for noninteracting spherical particles diffusing in a continuous three-dimensional medium, k_1 is described by the well-known Smoluchowski expression:

$$k_1 = \frac{4\pi N_A RD}{1000}\left[1 + \frac{R}{(\pi Dt)^{1/2}}\right] \qquad (2.29)$$

which asymptotically approaches the time-independent expression (2.28) with increasing time. More general models involving diffusion effects have been developed (166,167). In media of low viscosity, the transient term is significant only at short times (less than 100 ps for viscosities comparable to that of water) and may be neglected, whereas in viscous media the transient term may not be safely ignored.

The excimer-to-monomer fluorescence intensity ratio I_F/I_M is often used to characterize the efficiency of excimer formation. By comparing the value of this ratio to that obtained in a reference solvent, one can determined the equivalent viscosity of an unknown medium. If we assume that the time dependence of k_1 can be ignored, analysis of the kinetic scheme (Fig. 2.11) leads to

$$\frac{I_E}{I_M} = \frac{k'_r}{k_r}\frac{k_1}{k_{-1} + k'_r + k'_{nr}} \qquad (2.30)$$

Under conditions where the excimer dissociation rate becomes very slow with respect to deactivation, the preceding equation reduces to

$$\frac{I_E}{I_M} = \frac{k'_r}{k_r}\tau_E k_1 \qquad (2.31)$$

where $\tau_E\ [= 1/(k'_r + k'_{nr})]$ is the lifetime of the excimer. These equations show that the ratio I_E/I_M is proportional to the rate constant k_1 for excimer formation. However, it should be emphasized that erroneous conclusions may be drawn as to the changes in fluidity of the host matrix as a function of temperature if the excimer lifetime τ_E is not constant over the range of temperature investigated.

Intermolecular quenching is also a translational diffusion-controlled process involving the encounter of an excited fluorophore and a quencher molecule:

$$A^* + Q \xrightarrow{k_Q} A$$

The above considerations on time dependence of k_1 also apply to k_Q. When the transient term is negligible, k_Q can easily be determined by measuring the fluorescence quantum yield or lifetime as a function of the quencher concentration; the results are analyzed using the Stern–Volmer relation:

$$\frac{\Phi_0}{\Phi} = \frac{\tau_0}{\tau} = 1 + k_Q \tau_0 [Q] \qquad (2.32)$$

where Φ_0 and Φ are the emission quantum yields of the probe in the absence and presence of quencher, respectively; τ_0 and τ are the fluorescence lifetimes of the probe in the absence and presence of quencher, respectively.

The theoretical approaches of intermolecular quenching and excimer formation have been reviewed by Alwattar et al. (136). The experimental observations can be interpreted in terms of the Gierer–Wirtz theory, using the modified Stokes coefficient of viscous friction ξ_{micro} ($= \xi_{Stokes} f_t$) (vide supra). It is thus possible to account for the experimentally found viscosity and temperature dependences of the rate constants:

$$\frac{k\eta}{T} = A + B\eta^x \qquad (2.33)$$

The serious drawback of the methods of determination of fluidity based on intermolecular excimer formation and quenching is that the translational diffusion can be perturbed in constrained media. André et al. (30) emphasized that, in the case of biological membranes, problems in the estimation of the fluidity arise from the presence of proteins and possible additives (e.g., cholesterol). Nevertheless, excimer formation with pyrene or pyrene-labeled phospholipids can provide interesting information on phospholipid bilayers and in particular on phase transition, lateral diffusion, and lateral distribution (168–178).

In the case of surfactants solution, the fluidity cannot be easily determined by intermolecular excimer formation and quenching because of the distribution of the probe (and quencher) molecules. However, these methods are very useful for the determination of aggregation numbers and for the study of dynamics of surfactant solutions (12,15).

Molecular mobility in polymer films and bulk polymers can also be probed by excimer formation of pyrene (179).

2.3.2.4. Intramolecular Excimer Formation

Bichromophoric molecules consisting of two identical chromophores linked by a short flexible chain may form an excimer (excited dimer). Examples of such "supramolecular rotors" are given in Fig. 2.12 [see the reviews on this subject (180–183)]. In contrast to intermolecular excimer formation, this process is not translational but requires close approach of the two moieties through internal rotations during the lifetime of the excited state. Information on viscosity is thus obtained without the difficulty of possible perturbation of the diffusion process by microheterogeneity of the medium, as mentioned in the previous subsection for intermolecular excimer formation. This advantage explains the extensive use of intramolecular excimers. Owing to the long excited state lifetime of pyrene derivatives (hundreds of nanoseconds in nonpolar environments), dipyrenylalkanes (134,184–189) and dipyrenylmethylether (185,190) can be used to estimate "microviscosities" ranging from 10 to 200 cP and is thus well suited for probing the hydrocarbon region of micelles, vesicles, biological membranes, etc.

Equations (2.30) and (2.31), which express the excimer-to-monomer fluorescence intensity ratio, are still valid provided that the time dependence of k_1 can be ignored. It is worth mentioning again that difficulties may arise from the possible temperature dependence of the excimer lifetime, when effects of temperature on fluidity are investigated. Moreover, Eq. (2.30) only applies to the case where one excited monomer leads to a single excimer through a single determining step. For all these reasons, it is recommended to perform time-resolved fluorescence experiments.

The viscosity dependence of intramolecular excimer formation is complex (134,191–196). As in the case of molecular rotors already discussed (Section 2.3.2.2), most of the experimental observations can be interpreted in terms of

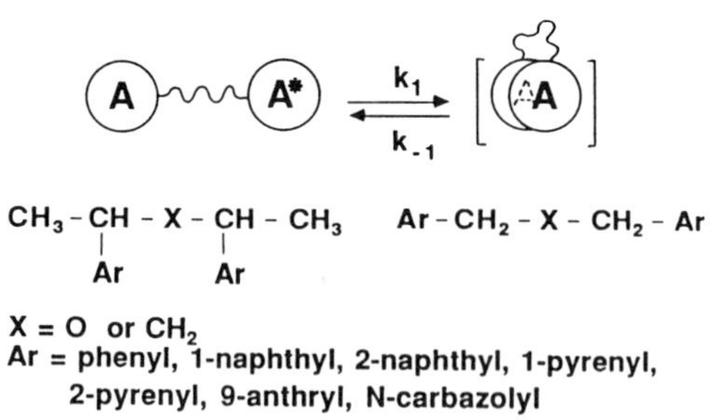

$$\text{(A)} \sim \text{(A*)} \underset{k_{-1}}{\overset{k_1}{\rightleftarrows}} [\text{(A A)}]$$

$$\text{CH}_3 - \text{CH} - \text{X} - \text{CH} - \text{CH}_3 \quad \quad \text{Ar} - \text{CH}_2 - \text{X} - \text{CH}_2 - \text{Ar}$$
$$\quad \quad \quad | \quad \quad \quad \quad |$$
$$\quad \quad \quad \text{Ar} \quad \quad \text{Ar}$$

X = O or CH$_2$
Ar = phenyl, 1-naphthyl, 2-naphthyl, 1-pyrenyl, 2-pyrenyl, 9-anthryl, N-carbazolyl

Fig. 2.12. Examples of excimer forming bichromophoric molecules.

free volume. Again using the Doolittle equation and assuming that the rate constant for excimer formation is given by

$$k_1 = k_1^o \exp\left(-x\frac{V_0}{V_f}\right) \quad (2.34)$$

[note the analogy with Eq. (2.26)], we obtain

$$k_1 = A\eta^{-x} \quad (2.35)$$

Experiments performed with 1,3-di-(N-carbazolyl)propane (191) and 2,4-diphenylpentanes (194). Viriot et al. (134) measured the variations of I_E/I_M for α,ω-di-(1-pyrenyl)propane in ethanol/glycerol mixtures and in hexadecane/paraffin mixtures at 20 °C (either degassed or undegassed). The results show significant differences between I_E/I_M values for the same viscosity but with different solvent mixtures. Moreover the comparison between degassed and undegassed solutions reveals the strong effect of oxygen at low viscosities. A dependence on the chain length was also found (134, 188).

The effects of chain length and solvent can be explained in terms of internal rotations involved in excimer formation. These rotations depend on the torsional potential of the various bonds and on the solvent nature; the solvent intervenes not only by the viscous drag but also by its microstructure and possible interactions with the probe. The distribution of interchromophoric distances at the instant of excitation evolves during the excited state lifetime of the monomer (134). Some of the bichromophoric molecules with one excited chromophore form an excimer because of favorable initial interchromophoric distance and favorable time evolution. Once the two chromophores are close to each other, the formation of the excited dimer (excimer) is very rapid, so that the overall process is diffusion dependent. The initial distribution and its further evolution depends on chain length, solvent nature, viscosity, and temperature (134). The equilibrium distribution of interchromophoric distances in bichromophoric molecules can be predicted from conformational analysis (197), and a molecular dynamics simulation of the diffusion process would be of great interest for a quantitative approach to excimer formation.

Note that in organized assemblies the local order may affect the internal rotations and the distribution of interchromophoric distances. Therefore, it is not surprising that the values of the equivalent viscosity may depend on the probe and, in particular, on the chain length. The results relative to SDS (sodium dodecyl sulfate) micelles reported in Table 2.4 illustrate the effects of chain length, oxygen, and choice of the reference solvents (134).

The effect of oxygen leading to smaller values of the ratio I_E/I_M is simple to understand on the basis of Eq. (2.31): the ratio I_E/I_M is proportional to the

Table 2.4. Values of Equivalent Viscosities (cP) in SDS Micelles by Using Dipyrenylpropanes as Probes and Two Different Reference Solvents

Probe	Reference Solvent			
	Ethanol/Glycerol Mixture		Hexadecane/Paraffin Mixture	
	Aerated	Deaerated	Aerated	Deaerated
$Py(CH_2)_3Py$	19	20	6	21
$Py(CH_2)_6Py$	17	30	17	13
$Py(CH_2)_9Py$	64	65	1	6

Source: From Viriot et al. (134).

excited state lifetime of the excimer, which is reduced by oxygen quenching. The rate of oxygen quenching decreases with increasing viscosity, and its effect on excimer formation in dipyrenylalkanes becomes negligible at viscosities higher than 100 cP (134).

In conclusion, the method of intramolecular excimer formation is rapid and convenient, but the foregoing discussion has shown that great care should be taken for a reliable interpretation of the experimental results. In some cases it has been demonstrated that the results in terms of equivalent microviscosity are consistent with those obtained by fluorescence polarization (134), but this is not a general rule. Nevertheless, the relative changes in fluidity and local dynamics upon an external perturbation are less dependent on the probe, and useful applications to the study of temperature effects (183–185) or pressure effects (198–201) have been described.

2.3.2.5. *Fluorescence Polarization Studies*

Excitation with vertically polarized light sets up a population of excited molecules with their transition moments oriented preferentially in the direction of the electric vector of the exciting light. There are various causes of depolarization of the emitted fluorescence, and the extent of depolarization is a very useful piece of information, as shown hereafter. The state of polarization of the fluorescence is best characterized by the emission anisotropy defined as

$$r = \frac{I_\| - I_\perp}{I_\| + 2I_\perp} \quad (2.36)$$

where $I_\|$ and $I_\perp$ are the intensity components respectively parallel and perpendicular to the electric vector of the polarized incident light. The quantity $I_\parallel + 2I_\perp$ is proportional to the total fluorescence intensity I. Emission anisotropy is preferable to the polarization ratio, defined as $(I_\parallel - I_\perp)/(I_\parallel + I_\perp)$, because in contrast to the polarization ratio, emission anisotropy is an additive quantity with weighting factors equal to the corresponding fractional intensities; that is, the overall contribution of n components r_i corresponding to fractional intensities f_i is given by

$$r = \sum_{i=1}^{n} f_i r_i \quad \text{with} \quad \sum_{i=1}^{n} f_i = 1 \qquad (2.37)$$

In the absence of external causes of depolarization, the emission anisotropy is called fundamental emission anisotropy r_0, which is a function of the angle α between the absorption and emission transition moments:

$$r_0 = \tfrac{1}{5}(3\cos^2\alpha - 1) \qquad (2.38)$$

Here r_0 can take values ranging from -0.2 to 0.4, the latter value corresponding to parallel absorption and emission transition moments. The wavelength dependence of r_0, which is due to different values of α according to the involved excited state, allows one to distinguish electronic and even vibronic bands. The case of indole and tryptophan offers a good illustration (202). The experimental values of r_0 can differ significantly from the theoretical values given by Eq. (2.38) and, in particular, will seldom reach the highest value of 0.4. In fact, the fundamental anisotropy is not strictly a molecular constant because it depends slightly but sometimes significantly on temperature and solvent, which affect the torsional vibrations and librational motions of the probe (203–205). Changes in molecular geometry between the ground and excited states might lead also to an actual value of r_0 lower than the theoretical one (206).

When a probe undergoes movements that are slow compared to the lifetime ($r \cong r_0$), or fast ($r \cong 0$: unpolarized fluorescence), no information is available because the time scale of the movements is outside of the experimental time window (discussed in Section 2.2.3 above). On the other hand, when rotations occur on a time scale comparable to the excited lifetime, information on the rotational rate of a probe (Fig. 2.13), and thus on the fluidity of its microenvironment can be obtained.

After excitation by an infinitely short pulse of light, the emission anisotropy generally decays from r_0 to zero at times that are long compared to the reciprocal of the rate of rotation, i.e., when orientational randomization of the excited molecules is complete. The time dependence of the emission anisotropy is directly related to the rotational autocorrelation function (which here is the

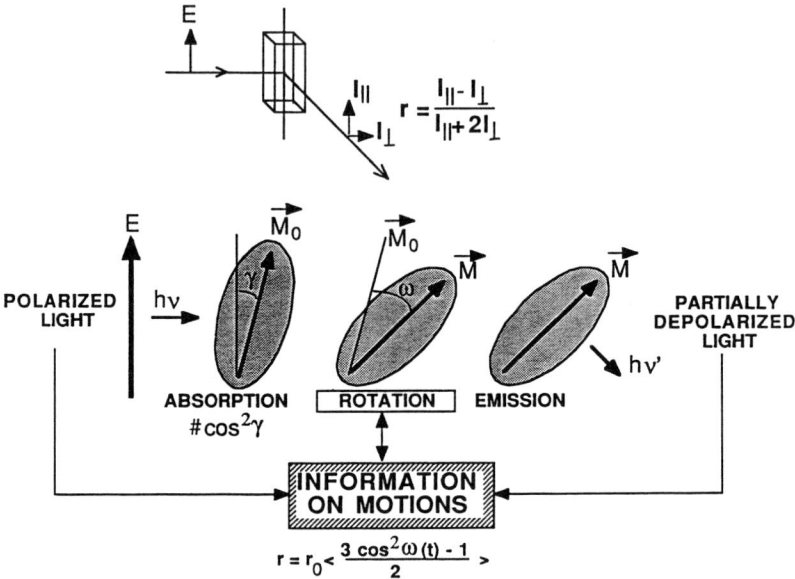

Fig. 2.13. Principles of the fluorescence polarization method for the study of rotational motions. The absorption and emission moments are assumed to be parallel (M_0).

Legendre polynomial of order 2, P_2):

$$r(t) = r_0 \langle P_2[\cos \omega(t)] \rangle = r_0 \left\langle \frac{3\cos^2 \omega(t) - 1}{2} \right\rangle \quad (2.39)$$

where $\omega(t)$ is the angle through which a molecule has rotated between time $t = 0$ and time t; the angular brackets $\langle \ \rangle$ denote an ensemble average over the whole population of excited probe molecules. In the case of *isotropic rotations*, the decay is exponential according to

$$r(t) = r_0 \exp(-6D_r t) \quad (2.40)$$

where D_r is the rotational diffusion constant, which is related to the viscosity as discussed above. Here D_r can be determined by time-resolved techniques, i.e., from the decay of the $I_\parallel$ and $I_\perp$ components in pulse fluorometry, or the phase shift between the $I_\parallel$ and $I_\perp$ components versus frequency in phase fluorometry.

Under continuous illumination, the measured steady-state anisotropy r is

$$\bar{r} = \frac{\int_0^\infty r(t) I(t) \, dt}{\int_0^\infty I(t) \, dt} \tag{2.41}$$

Hence, in the case of unique lifetime τ

$$\bar{r} = \frac{1}{\tau} \int_0^\infty r(t) \exp(-t/\tau) \, dt \tag{2.42}$$

Eqs. (2.40) and (2.42) lead to Perrin's relationship:

$$\frac{1}{\bar{r}} = \frac{1}{r_0}(1 + 6D_r \tau) \tag{2.43}$$

which allows one to determine D_r. Assuming that the Stokes–Einstein relation is valid, we can then calculate a value of η ($= kT/6VD_r$) provided that the hydrodynamic volume is known. In addition to the problem of validity of the Stokes–Einstein relation [Eq. (2.12) Section 2.3.2.1, above], the hydrodynamic volume cannot be calculated on a simple geometrical basis but must also take into account the solvation shell. Moreover, Perrin's relation is not valid for anisotropic rotations and hindered rotations.

Note that the steady-state technique (continuous illumination) is by far simpler than the time-resolved technique, but knowledge of the probe lifetime is required. As a matter of fact, the variations of the steady-state anisotropy with an external perturbation (e.g., temperature) may not be due only to changes in rotational rate, because this perturbation may also affect the lifetime.

In most cases, fluorescent probes undergo *anisotropic rotations* because of the asymmetry of their shape (204,207–209). A totally asymmetric rotor has three different rotational diffusion coefficients; and in the case where the absorption and emission transition moments are not directed along one of the principal diffusion axes, the decay of $r(t)$ is a sum of five exponentials (210–213). The use of different excitation wavelengths corresponding to different values of r_0 is helpful (214,215). In particular, for disk-like probes, only the out-of-plane rotations are detected at the excitation wavelength corresponding to $r_0 = 0.1$, whereas the contribution of both in-plane and out-of-plane rotations with equal weight are observed at a wavelength corresponding to r_0 close to 0.4, i.e., for parallel absorption and emission transition moments (214). As an example of application, these two types of rotation in an anthroyloxy probe were taken into consideration by Vincent et al. (216) in their investigation of phospholipid vesicles.

The case of anisotropic media such as lipid bilayers and liquid crystals requires further attention. If the rotational motions of the probe are hindered, the emission anisotropy does not decay to zero but to a value r_∞, called non-zero-limiting anisotropy (Fig. 2.14), which implicitly contains information on the order of the medium. Let us first consider the model in which the rotations of a rod-like probe (with the direction of its absorption and emission transition moments coinciding with the long molecular axis) are restricted within a cone ("wobble-in-cone" model) (217,218). The rotational motions are described by the rotational diffusion coefficient $D_\perp$ around an axis perpendicular to the long molecular axis (the rotations around this axis having no effect on the emission anisotropy) and an order parameter (half angle of the cone θ_c) reflecting the degree of orientational constraint due to the

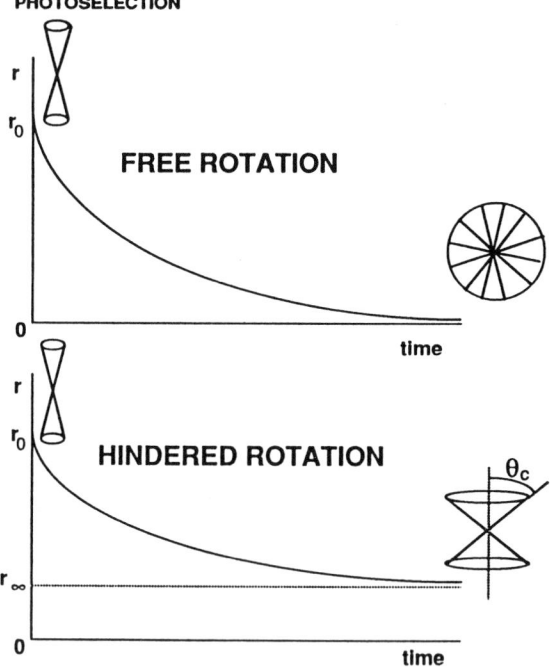

Fig. 2.14. Illustration of the difference in the decay of emission anisotropy for free and hindered rotations.

MAIN PHYSICAL AND STRUCTURAL PARAMETERS 61

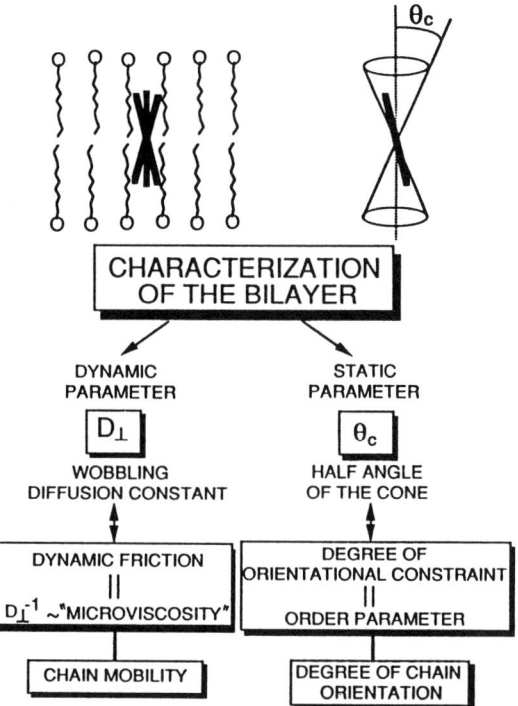

Fig. 2.15. "Wobble-in-cone" model for the characterization of bilayers. The absorption and emission moments are assumed to coincide with the long molecular axis.

surrounding paraffinic chains (Fig. 2.15): θ_c can be determined from the ratio r_0/r_∞:

$$r_0/r_\infty = [\tfrac{1}{2}\cos\theta_c(1+\cos\theta_c)]^2 \qquad (2.44)$$

An approximate expression of the anisotropy decay is (217)

$$r(t) = r_\infty + (r_0 - r_\infty)\exp(-6D_\perp t/\langle\sigma\rangle) \qquad (2.45)$$

The quantity $\langle\sigma\rangle/D_\perp$ is the effective relaxation time of $r(t)$, that is, the time with which the initially photoselected distribution of orientations approaches the stationary distribution. This time depends on θ_c. An analytical expression of $\langle\sigma\rangle$ as a function of θ_c was derived by Lipari and Szabo (218). Thus, if θ_c is determined from r_0/r_∞, the wobbling diffusion constant $D_\perp$ can be calculated from the effective relaxation time of $r(t)$.

More general theories have been developed especially for probes embedded in bilayers or liquid crystals (219–224). A general expression for $r(t)$ can be written in the following form:

$$r(t) = r_0[G_0(t) + 2G_1(t) + 2G_2(t)] \quad (2.46)$$

involving three autocorrelation functions, $G_0(t)$, $G_1(t)$, and $G_2(t)$. The values of these functions at times $t=0$ and $t=\infty$ are related to the two order parameters $\langle P_2 \rangle$ and $\langle P_4 \rangle$, which are orientational averages of the second and fourth rank Legendre polynomial $P_2(\cos\beta)$ and $P_4(\cos\beta)$, respectively, relative to the orientation β of the probe axis with respect to the normal to the local bilayer surface or with respect to the liquid crystal direction. The order parameters are defined as

$$\langle P_2 \rangle = \langle 3\cos^2\beta - 1 \rangle/2 \quad (2.47)$$

and

$$\langle P_4 \rangle = \langle 35\cos^4\beta - 30\cos^2\beta + 3 \rangle/8 \quad (2.48)$$

and the autocorrelation functions are given by

$$G_0(0) = \tfrac{1}{5} + \tfrac{2}{7}\langle P_2 \rangle + \tfrac{18}{35}\langle P_4 \rangle \quad (2.49a)$$

$$G_1(0) = \tfrac{1}{5} + \tfrac{1}{7}\langle P_2 \rangle - \tfrac{12}{35}\langle P_4 \rangle \quad (2.49b)$$

$$G_2(0) = \tfrac{1}{5} - \tfrac{2}{7}\langle P_2 \rangle + \tfrac{3}{35}\langle P_4 \rangle \quad (2.49c)$$

$$G_0(\infty) = \langle P_2 \rangle^2 \quad (2.49d)$$

$$G_1(\infty) = G_2(\infty) = 0 \quad (2.49e)$$

It is assumed that the probe molecules undergo Brownian rotational motions with an angle-dependent ordering potential $U(\beta)$ (221)

$$U(\beta) = kT[\lambda_2 P_2(\cos\beta) + \lambda_4 P_4(\cos\beta)] \quad (2.50)$$

For a rodlike probe with its absorption transition moment direction coinciding with the long molecular axis, the rotational motions in this potential well is described by the diffusion coefficient $D_\perp$ (as defined above). The decay of the autocorrelation functions is then shown to be an infinite sum of exponential terms (221):

$$G_k(t) = \sum_{m=0}^{\infty} b_{km} \exp(-a_{km} D_\perp t) \qquad k = 0, 1, 2 \quad (2.51)$$

The coefficients a_{km} and b_{km} are complex functions of the parameters λ_2 and λ_4 that describe the ordering potential. In many practical situations, $G_k(t)$ is essentially monoexponential (221):

$$G_k(t) = [G_k(0) - G_k(\infty)] \exp(-a_{k1} D_\perp t) + G_k(\infty) \qquad (2.52)$$

The diffusion constant $D_\perp$ with the underlying "microviscosity," and the two order parameters $\langle P_2 \rangle$ and $\langle P_4 \rangle$ reflecting the degree of orientational constraint have been determined from the fluorescence anisotropy decay in vesicles and liquid crystals. Examples can be found in van Langen et al. (225, 226), Ruggiero and Hudson (227), and Arcioni et al. (228,229).

In systems with lifetime and dynamics heterogeneity, i.e., when a probe exists in distinct environments having different fluorescence decays and rotational behaviors (230), Eq. (2.37) should be rewritten as

$$r = \sum_{i=1}^{n} f_i(t) r_i(t) \qquad (2.53)$$

where

$$f_i(t) = \frac{a_i \exp(-t/\tau_i)}{I(t)} = a_i \exp(-t/\tau_i) \bigg/ \sum_{i=1}^{n} a_j \exp(-t/\tau_j) \quad \text{with} \quad \sum_{i=1}^{n} a_i = 1$$

The fluorescence polarization technique is a very powerful tool for studies of the fluidity and orientational order of organized assemblies: aqueous micelles (231–235), reverse micelles and microemulsions (236–241), lipid bilayers (216,225–227,242–248), synthetic nonionic vesicles (249), liquid crystals (228, 229,250,251). This technique is also highly useful for probing the segmental mobility of polymers (252–256) and antibody molecules (257,258). Information on orientation of chains in solid polymers can also be obtained (9,259–262).

Finally, it is worth pointing out that many artifacts can alter the measurements of emission anisotropy. It is necessary to control the instrument with a scattering nonfluorescent solution (r close to 1) and with a solution of a fluorophore with a long lifetime in a solvent of low viscosity ($r \approx 0$). It is also recommended that the probe concentration be kept low enough to avoid interaction between probes. Otherwise, depolarization would also be due to energy transfer between probe molecules. Since the transition moments of two interacting probes are unlikely to be parallel, this effect is indeed formally equivalent to a rotation. Moreover, artifacts may arise from scattering media.

2.3.3. Distances at Supramolecular Scale: A "Spectroscopic Ruler"

Electronic excitation energy can be transferred from a donor to a suitable acceptor provided that the emission spectrum of the donor overlaps the absorption spectrum of the acceptor. Such a transfer can occur over distances as large as 80 Å. According to Förster's theory, the rate of transfer depends on the inverse sixth power of the distance between the donor and acceptor. Stryer and Haugland (263) suggested that energy transfer from a donor probe to an acceptor probe could be used as a spectroscopic ruler for the determination of distances in the 10–60 Å range. Several reviews deal with this method (264–267).

2.3.3.1. Mechanisms of Electronic Energy Transfer

Only the mechanisms of nonradiative transfer will be discussed here. The "trivial" radiative transfer in which fluorescence photons emitted by the donor are absorbed by the acceptor (268) can, in principle, be avoided by using samples of low optical density, or very thin samples.

In nonradiative transfer of electronic energy, deactivation of the donor and excitation of the acceptor occur simultaneously without emission of photons during this process. Some interaction between the donor and the acceptor is required, but the interaction energy can be very small so that long-range energy transfer is possible. The resonance condition to be fulfilled implies that the energy of some emission transitions of the donor corresponds to the energy of absorption transitions of the acceptor (Fig. 2.16). This condition is fulfilled in the region where the donor emission spectrum overlaps the acceptor absorption spectrum.

Two kinds of interaction are involved in nonradiative transfer. The total interaction energy can be indeed divided into two terms, a coulombic term and an exchange term. The first is often approximated by a dipole–dipole interaction between the transition moments of the donor and the acceptor.

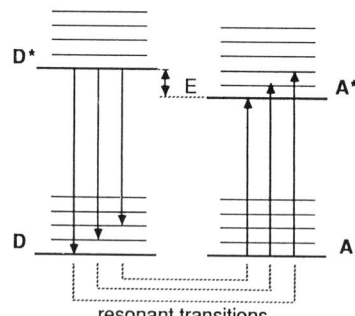

Fig. 2.16. Energy level diagram for a donor D and an acceptor A (ΔE represents the difference in electronic energy).

This is a long-range interaction that can extend up to 60–80 Å. In contrast, the exchange interaction, representing the electrostatic interaction between the charge clouds, occurs only at short distances ($\lesssim 8$ Å), allowing the orbitals to overlap.

By a quantum mechanical treatment of resonance transfer via dipole–dipole interaction between two separated molecules, Förster (269,270) obtained the following expression for the transfer rate constant:

$$k_T = k_D \left[\frac{R_0}{R}\right]^6 = \frac{1}{\tau_D^0} \left[\frac{R_0}{R}\right]^6 \tag{2.54}$$

where k_D is the emission rate constant of the donor and τ_D^0 its lifetime in the absence of transfer; R is the distance between the donor and the acceptor (which is assumed to remain unchanged during the lifetime of the donor); and R_0 is the critical distance at which transfer and spontaneous decay of the excited donor are equally probable ($k_T = k_D$). Here R_0, which can be determined from spectroscopic data (see tables in reference 270a), is given by

$$R_0^6 = \frac{9000(\ln 10)}{128 \pi^5 N_A} n^{-4} \kappa^2 \Phi_D^0 J \tag{2.55}$$

where n is the refractive index of the medium; N_A, Avogadro's number; Φ_D^0, the donor quantum yield in the absence of transfer; and J, the overlap integral:

$$J = \int_0^\infty F_D(\lambda) \varepsilon_A(\lambda) \lambda^4 \, d\lambda \tag{2.56}$$

in which the extinction coefficient, ε_A, of the acceptor is expressed in units of L·mol·cm^{-1}, whereas the spectral distribution of the donor fluorescence $F_D(\lambda)$ is normalized so that

$$\int_0^\infty F_D(\lambda) \, d\lambda = 1 \tag{2.57}$$

In Eq. (2.55) κ^2 is an orientation factor for the chromophores, given by

$$\kappa = \cos \theta_{DA} - 3 \cos \theta_A \cos \theta_D \tag{2.58}$$

where θ_{DA} is the angle between the donor and acceptor transition moments, and θ_D and θ_A are the angles between these, respectively, and the separation vector. When the molecules are free to rotate at a rate that is much larger than the deexcitation rate of the donor (dynamic averaging), the average value of κ^2 is 2/3. In a rigid medium, the square of the average of κ is 0.476 for an ensemble of acceptors that are statistically randomly distributed about the donor with respect to both distance and orientation (this case is often called "static isotropic average"). The case of donor–acceptor pairs at fixed separations has been investigated by Dale (271).

The effect of limited rotational mobility of the donor and/or acceptor on κ^2 has been studied in detail by Dale, Eisinger, and coworkers (272–275). These authors suggest that the range of possible κ^2 values be evaluated by means of fluorescence polarization experiments.

The transfer efficiency is defined as

$$\Phi_T = \frac{k_T}{k_D + k_T} = \frac{k_T}{1/\tau_D^0 + k_T} \quad (2.59)$$

According to Eq. (2.54), the transfer efficiency is related to the ratio R/R_0:

$$\Phi_T = \frac{1}{1 + (R/R_0)^6} \quad (2.60)$$

for a given interchromophoric distance.

It should be emphasized that the inverse sixth power dependence on distance is characteristic of the dipole–dipole mechanism of energy transfer. In contrast, an exponential dependence is to be expected from the exchange mechanism according to Dexter's theory (276); the rate constant for transfer can be written in the form

$$k_T^{ex} = \frac{2\pi}{h} KJ' \exp(-2R/L) \quad (2.61)$$

where

$$J' = \int_0^\infty F_D(\lambda)\varepsilon_A(\lambda)\,d\lambda \quad (2.62)$$

with the normalization condition

$$\int_0^\infty F_D(\lambda)\,d\lambda = \int_0^\infty \varepsilon_A(\lambda)\,d\lambda = 1 \quad (2.63)$$

Here L is the average Bohr radius. Since K is a constant that is not related to any spectroscopic data, it is difficult to characterize the exchange mechanism by experiments.

From the above description of mechanisms, it can be concluded that Förster nonradiative energy transfer can be used as a spectroscopic ruler if the distance between donor and acceptor is longer than ~ 8 Å and constant during the lifetime of the donor. The validity of such a spectroscopic ruler has been confirmed by studies on model systems in which the donor and acceptor are separated by well-defined rigid spacers (263,277,278).

2.3.3.2. Determination of Energy Transfer Efficiency

Three steady-state methods can be used for the determination of the energy transfer efficiency (265,279). In the following description of these methods, the subscripts D and A refer to the donor and acceptor, respectively, and the superscript 0 refers to the properties of the donor or acceptor alone in the system to be studied; F will represent the fluorescence intensities, with the excitation wavelength λ indicated in parentheses; Φ, the quantum yields; and $\mathscr{A}$, the absorbances. Since the characteristics of the donor and/or acceptor are measured in the presence and absence of transfer, the concentrations of donor and acceptor and their microenvironments must be the same in both these conditions.

Method 1. *Decrease in donor fluorescence.* Transfer from donor to acceptor causes the quantum yield of the donor to decrease. The measurement of this quantum yield in the presence (Φ_D) and in the absence of transfer (Φ_D^0) permits determination of the transfer efficiency

$$\Phi_T = 1 - \frac{\Phi_D}{\Phi_D^0} \tag{2.64}$$

A good separation of the donor and acceptor emission spectra is required. Subtraction of the acceptor fluorescence is possible but would decrease the accuracy if it is too large. Since only the relative quantum yields are to be determined, a single observation wavelength is sufficient. If there is no contribution of the acceptor fluorescence at this wavelength, Eq. (2.64) can be rewritten in terms of absorbances at the excitation wavelength λ_D and fluorescence intensities of the donor:

$$\Phi_T = 1 - \frac{\mathscr{A}(\lambda_D) \, F_D}{\mathscr{A}_D(\lambda_D) \, F_D^0} \tag{2.65}$$

The factor $\mathscr{A}/\mathscr{A}_D$ arises from the contribution of the acceptor moiety to the overall absorption at the excitation wavelength λ_D.

Method 2. *Comparison between the absorption spectrum and the excitation spectrum (through observation of the acceptor fluorescence).* The corrected excitation spectrum is represented by

$$F_A(\lambda) = K\Phi_A[\mathscr{A}_A(\lambda) + \mathscr{A}_D(\lambda)\Phi_T] \tag{2.66}$$

where K is a constant depending on the gain of the spectrofluorometer. The absorption spectrum is defined as

$$\mathscr{A}(\lambda) = \mathscr{A}_A(\lambda) + \mathscr{A}_D(\lambda) \tag{2.67}$$

In the case of total transfer ($\Phi_T = 1$), these two spectra are similar. But for any value of Φ_T lower than 1, the excitation band corresponding to the donor is relatively lower than the absorption band. The comparison of the absorption and excitation spectra can be done at two wavelengths λ_D and λ_A corresponding to the absorption maxima of the donor and the acceptor, respectively. If there is no absorption of the donor at λ_A, one obtains

$$\frac{F_A(\lambda_A)}{\mathscr{A}_A(\lambda_A)} = K\Phi_A \tag{2.68}$$

$$\frac{F_A(\lambda_D)}{\mathscr{A}_A(\lambda_D)} = K\frac{\Phi_A[\mathscr{A}_A(\lambda_D) + \mathscr{A}_D(\lambda_D)\Phi_T]}{\mathscr{A}_A(\lambda_D)} \tag{2.69}$$

The ratio of Eqs. (2.68) and (2.69) yield

$$\Phi_T = \frac{\mathscr{A}_A(\lambda_A)}{\mathscr{A}_D(\lambda_D)}\left[\frac{F_A(\lambda_D)}{F_A(\lambda_A)} - \frac{\mathscr{A}_A(\lambda_D)}{\mathscr{A}_A(\lambda_A)}\right] \tag{2.70}$$

Method 3. *Enhancement of acceptor fluorescence.* The fluorescence intensity of the acceptor is enhanced in the presence of transfer. Comparison with the intensity in the absence of transfer provides the transfer efficiency

$$\Phi_T = \frac{\mathscr{A}_A(\lambda_D)}{\mathscr{A}_D(\lambda_D)}\left[\frac{F_A(\lambda_D)}{F_A^0(\lambda_D)} - 1\right] \tag{2.71}$$

Note that $\mathscr{A}_A(\lambda_D)$ is often small and thus difficult to measure accurately, which may lead to large errors in Φ_T.

Method 1 appears to be more straightforward than methods 2 and 3. However it cannot be used in the case of very low donor quantum yields. Note also that quenching of the donor by the acceptor may occur. This point can be checked by a complementary study based on observation of the acceptor fluorescence (method 2 or 3).

Time-resolved fluorescence experiments also provide information on transfer efficiency and rate and are free from inner filter effects. The measurement of the donor lifetime in the presence and absence of transfer is a straightforward way to determine the transfer efficiency:

$$\Phi_T = 1 - \frac{\tau_D}{\tau_D^0} \qquad (2.72)$$

Information on transfer rate can also be obtained from the rise of the acceptor fluorescence following pulse excitation of the donor (280), from the phase shift between the fluorescence of the acceptor excited directly and via donor excitation (281), or from ground state recovery measurements of the donor chromophore (282).

2.3.3.3. Calculation of Distance

According to Eq. (2.60), the distance between the donor and the acceptor can be calculated from the values of the transfer efficiency and the critical distance R_0:

$$R = \left(\frac{1}{\Phi_T} - 1\right)^{1/6} R_0 \qquad (2.73)$$

Several precautions must be taken for a correct use of the spectroscopic ruler:

a. The critical distance R_0 should be determined under the same experimental conditions as those of the investigated system because R_0 [see Eq. (2.55)] involves the quantum yield of the donor and the overlap integral, which both depend on the nature of the microenvironment.

b. As regards the orientation factor, attention should be paid to possible preferred mutual orientations between the donor and the acceptor. Complementary fluorescence polarization experiments are helpful in this respect.

c. The distance to be measured should not be too different from R_0 because, according to Eq. (2.60), the transfer efficiency varies very rapidly as a function of the distance around R_0. At $R < R_0/2$, the transfer efficiency is close to 1, and at $R > 2R_0$, it approaches 0.

The foregoing considerations are of course valid only for a unique distance between the donor and the acceptor. The case of distributions of distances will now be examined.

2.3.3.4. Distributions of Distances

In three-dimensional homogeneous mixtures of donor and acceptor molecules, theories have been developed to take into account the distribution of donor–acceptor distances. When the mean translational diffusion length l of the fluorophores during the donor lifetime is large with respect to the critical distance R_0 (that is, $l = (D_t\tau)^{1/2} \gg R_0$, with D_t being the mutual diffusion coefficient) the fluorescence decay of the donor is monoexponential (Stern–Volmer kinetics). In contrast, when the donor–acceptor distances do not change significantly during the donor lifetime ($l = (2D_t\tau)^{1/2} \ll R_0$), the fluorescence decay of the donor is not strictly monoexponential, because the transfer rate constant is time dependent [Förster kinetics; see Förster (269,270)]. Intermediate cases where diffusion effects play a role have been considered (283–285).

The donor–acceptor distance is also not unique in the following interesting cases: (i) bichromophoric molecules in which the donor is linked by a short flexible chain (e.g. oligomers of polymethylene or polyoxyethylene, oligopeptides, or oligonucleotides) [197,279,281,286–292; for a review, see Valeur (293)]; (ii) end-labeled polymer chains (294–298); (iii) supramolecular assemblies (micelles, vesicles, etc.) where donor and acceptor molecules are distributed [299–302; for a review, see Yamazaki et al. (303)]; and (iv) biosystems (304,305).

Distance distributions can be recovered by the following means:

Steady-State Methods
a. Variations of the relative quantum yield of the donor (Φ_D/Φ_D^0) as a function of acceptor concentration (if the latter can be varied) (299)
b. Measurements of donor emission at various levels of an external quencher which causes R_0 to vary through changes in donor quantum yield (306); however, because of the sixth root dependence of R_0 on the donor quantum yield [Eq. (2.55)], a variation of this parameter by a factor of 10 leads to a variation of only 30% in R_0; moreover, static quenching and transient diffusion phenomena may be sources of difficulty.

Time-Resolved Methods
a. Time-resolved emission of the donor fluorescence measured by time domain (289,307) or frequency domain fluorometry (290); the impulse

response of the donor emission can be written as

$$F_D(t) = K \int_0^\infty f(R) \exp\left[-\frac{t}{\tau_D^0} - \frac{t}{\tau_D^0}\left(\frac{R_0}{R}\right)^6 \right] dr \qquad (2.74)$$

where K is a constant and $f(R)$ is the distribution function of donor–acceptor distances

b. Excite-and-probe spectroscopy, which consists of measuring the change in optical density of the acceptor after pumping the donor by a very short laser pulse; this technique has been used for checking the validity of calculated distance distributions in end-labeled oligomers (292) and can also be used for the recovery of unknown Gaussian distance distributions (308). Alternatively, excite-and-probe spectroscopy can be used for ground state recovery measurements of the donor chromophore (282).

c. Time-resolved emission anisotropy measurements performed with identical fluorophores (296).

The case of chromophores linked by a flexible chain has received special attention because they can serve as models. The distance distribution is generally assumed to be Gaussian, but this assumption may not be valid in the case of short chains [see, for instance, Valeur et al. (197); Ikeda et al. (309)]. Effort should be made to develop non a priori methods of data analysis. The maximum entropy method (see Section 2.2.3) is of great interest in this respect. Distance distributions can be determined when the interchromophoric distance does not change significantly during the donor lifetime. Otherwise, energy transfer is enhanced by translational diffusion of the donor and acceptor moieties toward each other. Then, information can be obtained on relative diffusion coefficients of chain ends in oligopeptides (310) and polymers (311,312).

2.4. CONCLUDING REMARKS

Fluorescent probes are most useful for estimation of local physical and structural characteristics of an unknown environment. However, one should not attempt to obtain absolute values of these parameters. For instance, viscosity and dielectric constant are macroscopic parameters that lose their meaning at a microscopic scale. Nevertheless, in many cases the measurement of relative changes of environment upon an external perturbation has been found to be relatively independent of the probe and the technique.

Great care should be taken in the use of fluorescent probes and relevant techniques. Erroneous interpretation may arise from misunderstanding of basic phenomena. For example, a polarity-dependent probe does not unequivocally indicate a hydrophobic environment whenever a blue shift of the fluorescence spectrum is observed. The demonstration of complete environmental relaxation is required before such an interpretation can be accepted.

As regards estimation of fluidity, note that (i) only an equivalent viscosity can be determined, (ii) the response of a probe may be different in solvents of the same viscosity but of different chemical nature and structure, and (iii) the measured equivalent viscosity often depends on the probe and on the fluorescence technique. Nevertheless, the relative variations of the equivalent viscosity resulting from an external perturbation are generally much less dependent on the technique and on the nature of the probe. Therefore, the fluorescence techniques are very valuable for monitoring changes in fluidity upon an external perturbation such as temperature, addition of compounds (e.g., cholesterol added to lipid vesicles, or addition of alcohols and oil to micellar systems), and pressure. The choice of the method depends on the system to be investigated. The methods of intramolecular excimer formation and molecular rotors are convenient and rapid, but the time-resolved fluorescence polarization technique provides much more detailed information, in particular about the order of an anisotropic medium. The methods of intermolecular quenching and intermolecular excimer formation are not recommended for probing fluidity of microheterogeneous media because of possible perturbation of the translational diffusion process.

Distances at a supramolecular level can be estimated by means of suitable donor-acceptor probes. It is important to determine the Förster critical radius under the same experimental conditions as those of the system under study because this radius depends on the environment through the donor quantum yield and the integral overlap between donor emission and acceptor absorption. Attention should also be paid to the orientation factor.

It is to be hoped that this chapter will have given readers a comprehensive overview of the physical and structural parameters measurable by fluorescent probes, and helpful advice for proper use of them. Applications of fluorescent probes to the estimation of chemical parameters (pH, ion recognition) have not been reviewed in this chapter because of space limitations but will be the object of a separate review.

N.B. *Many fluorescent probes are commercially available from Molecular Probes, Inc. (USA) and from Lambda Probes & Diagnostics (Austria).*

ACKNOWLEDGMENTS

The author is specially indebted to Drs. J. C. André, E. Bardez, and M. N. Berberan-Santos for critical reading of the manuscript. Thanks are also due to Professors L. Monnerie and M. Monsigny and Drs. J. C. Brochon, L. Bokobza, N. Cittanova, and Th. Montenay-Garestier for helpful advice on specific points.

REFERENCES

1. J. Guillet, *Polymer Photophysics and Photochemistry*, Cambridge University Press, London and New York, 1985.
2. D. Phillips, Ed., *Polymer Photophysics: Luminescence, Energy Migration and Molecular Motion in Synthetic Polymers*, Chapman & Hall, London, 1985.
3. M. A. Winnik, Ed., *Photophysical and Photochemical Tools in Polymer Science*, Reidel, Dordrecht, The Netherlands, 1986.
4. L. Zlatkevitch, Ed., *Luminescence Techniques in Solid-State Polymer Research*, Marcel Dekker, New York, 1989.
5. Y. Nishijima, *Prog. Polym. Sci. Jpn.*, **6**, 199 (1973).
6. Y. Nishijima, *J. Macromol. Sci., Phys.*, **8**, 407 (1973).
7. S. W. Beavan, J. S. Hargreaves, and D. Phillips, *Adv. Photochem.*, **11**, 207 (1979).
8. H. Morawetz, *Science*, **203**, 4379 (1979).
9. L. Monnerie, in R. A. Pethrick and R. W. Richards, Eds., *Static and Dynamic Properties of the Polymeric State*, Reidel, Dordrecht, The Netherlands, 1982, p. 383.
10. M. A. Winnik, *Acc. Chem. Res.*, **18**, 73 (1985).
11. H. Morawetz, *J. Lumin.*, **43**, 59 (1989).
12. K. Kalyanasundaram, *Photochemistry in Microheterogeneous Systems*, Academic Press, Orlando, FL, 1987.
13. J. K. Thomas, *J. Phys. Chem.*, **91**, 267 (1987).
14. J. K. Thomas, *Acc. Chem. Res.*, **21**, 275 (1988).
15. R. Zana, in R. Zana, Ed., *Surfactants Solutions: New Methods of Investigation*, Marcel Dekker, New York, 1987, p. 241.
16. M. Grätzel and J. K. Thomas, in E. L. Wehry, Ed., *Modern Fluorescence Spectroscopy*, Vol. 2, Plenum Press, New York, 1976, p. 169.
17. F. Grieser and C. J. Drummond, *J. Phys. Chem.*, **92**, 5580 (1988).
18. G. Weber, *Adv. Protein Chem.*, **8**, 415 (1953).
19. G. K. Radda and J. Vanderkooi, *Biochim. Biophys. Acta*, **265**, 509 (1972).
20. L. Brand and J. R. Gohlke, *Annu. Rev. Biochem.*, **41**, 843 (1972).
21. M. Edidin, *Annu. Rev. Biophys. Bioenerg.*, **3**, 179 (1974).
22. A. Azzi, *Q. Rev. Biophys.*, **8**, 237 (1975).

23. R. A. Badley, in E. L. Wehry, Ed., *Modern Fluorescence Spectroscopy*, Vol. 2, Plenum Press, New York, 1976, p. 91.
24. W. W. Mantulin and H. J. Pownall, *Photochem. Photobiol*, **26**, 69 (1977).
25. M. Donner, M. Bouchy, M. L. Viriot and J. C. André, *Biochimie*, **63**, 961 (1981).
26. M. Shinitzky and I. Yuli, *Chem. Phys. Lipids*, **30**, 261 (1982).
27. J. R. Lakowicz, *Principles of Fluorescence Spectroscopy*, Plenum Press, New York, 1983.
28. K. Kinosita, Jr., S. Kawato, and A. Ikegami, *Adv. Biophys.*, **17**, 147 (1984).
29. B. Hudson, D. L. Harris, R. D. Ludescher, A. Ruggiero, A. L. Cooney-Freed, and S. A. Cavalier, in D. L. Taylor, A. S. Waggoner, F. Lanni, R. F. Murphy, and R. R. Birge, Eds., *Applications of Fluorescence in the Biomedical Sciences*, Alan R. Liss, New York, 1986, p. 159.
30. J. C. André, M. Bouchy, and M. Donner, *Biorheology*, **24**, 237 (1987)
31. L. M. Loew, *Spectroscopic Membrane Probes*, CRC Press, Deerfield Beach, FL, 1988.
32. H. J. Hilderson, Ed., *Subcellular Biochemistry*, Vol. 13, Plenum Press, New York, 1988.
33. J. E. Churchich, in E. L. Wehry, Ed., *Modern Fluorescence Spectroscopy*, Vol. 2, Plenum Press, New York, 1976, p. 217.
34. J. M. Beechem and L. Brand, *Annu. Rev. Biochem.*, **54**, 43 (1985).
35. A. G. Szabo, in A. Cooper, J. L. Houben, and L. C. Chien, Eds., *The Enzyme Catalysis Process*, Plenum Press, New York, 1989, p. 123.
36. J. B. Le Pecq, in R. F. Chen and H. Edelhoch, Eds., *Biochemical Fluorescence: Concepts*, Marcel Dekker, New York, 1975, p. 711.
37. R. F. Steiner and Y. Kubota, in R. F. Steiner and Y. Kubota, Eds., *Excited States of Biopolymers*, Plenum Press, New York, 1983, p. 203.
38. A. S. Waggoner, in D. L. Taylor, A. S. Waggoner, F. Lanni, R. F. Murphy, and R. R. Birge, Eds., *Applications of Fluorescence in the Biomedical Sciences*, Alan R. Liss, New York, 1986, p. 3.
39. H. M. Shapiro, *Practical Flow Cytometry*, Alan R. Liss, New York, 1988.
40. Y. L. Wang and D. L. Taylor, Eds., *Methods in Cell Biology*, Vol. 29, Part A, Academic Press, San Diego, CA, 1989.
41. Y. L. Wang and D. L. Taylor, Eds., *Methods in Cell Biology*, Vol. 30, Part B, Academic Press, San Diego, CA, 1989.
42. R. M. Nakamura, W. R. Dita, and E. S. Tucker, III, Eds., *Immunoassays: Clinical Laboratory Techniques for the 1980s*, Alan R. Liss, New York, 1980.
43. D. S. Smith, M. H. H. Al-Hakiem, and J. Landon, *Ann. Clin. Biochem.*, **18**, 253 (1981).
44. I. Hemmila, *Clin. Chem. (Winston-Salem, N.C.)*, **31**, 359 (1985).
45. H. T. Karnes, J. S. O'Neal, and S. G. Schulman, in S. G. Schulman, Ed., *Molecular Luminescence Spectroscopy*, Part 1, Wiley, New York, 1985, p. 717.

46. P. Urios, *Immunoanal. Biol. Spectrosc.*, **18**, 23 (1989).
47. R. S. Davidson and M. M. Hilchenbach, *Photochem. Photobiol.*, **52**, 431 (1990).
48. R. F. Steiner and I. Weinryb, Eds., *Excited States of Proteins and Nucleic Acids*, Plenum Press, New York, 1971.
49. R. F. Chen and H. Edelhoch, Eds., *Biochemical Fluorescence: Concepts*, Marcel Dekker, New York, 1975.
50. G. S. Beddard and M. A. West, Eds., *Fluorescent Probes*, Academic Press, London, 1981.
51. R. B. Cundall and R. E. Dale, Eds., *Time-Resolved Fluorescence Spectroscopy in Biochemistry and Biology*, Plenum Press, New York, 1983.
52. S. G. Schulman, Ed., *Molecular Luminescence Spectroscopy*, Wiley, New York, 1985.
53. D. L. Taylor, A. S. Waggoner, F. Lanni, R. F. Murphy, and R. R. Birge, Eds., *Applications of Fluorescence in the Biomedical Sciences*, Alan R. Liss, New York, 1986.
54. D. M. Jameson and G. D. Reinhart, Eds., *Fluorescent Biomolecules, Methodologies and Applications*, Plenum Press, New York, 1989.
55. W. R. G. Baeyens, D. De Keukeleire, and K. Kordidis, Eds., *Luminescence Techniques in Chemical and Biochemical Analysis (Practical Spectroscopy Series)*, Marcel Dekker, New York, 1990.
56. R. P. Haugland, *Handbook of Fluorescent Probes and Research Chemicals*, Molecular Probes, Inc., Eugene, OR, 1989.
57. E. Köller, *Appl. Fluorescence Technol*, **1**(2), 1 (1989).
58. F. G. Prendergast, R. P. Haugland, and P. J. Callahan, *Biochemistry*, **20**, 7333 (1984).
59. L. A. Sklar, B. Hudson, and R. D. Simoni, *Proc. Natl. Acad. Sci. U.S.A.*, **72**, 1649 (1975).
60. D. Fornasiero, F. Grieser, and W. H. Sawyer, *J. Phys. Chem.*, **92**, 2301 (1988).
61. R. A. Mathies and L. Stryer, in D. L. Taylor, A. S. Waggoner, F. Lanni, R. F. Murphy, and R. R. Birge, Eds., *Applications of Fluorescence in the Biomedical Sciences*, Alan R. Liss, New York, 1986, p. 129.
62. A. N. Glazer and L. Stryer, *Trends Biochem. Sci.*, **9**, 425 (1984).
63. G. Weber and M. Shinitzky, *Proc. Natl. Acad. Sci. U.S.A.*, **65**, 878 (1970).
64. W. C. Galley and R. M. Purkey, *Proc. Natl. Acad. Sci. U.S.A.*, **67**, 1116 (1970).
65. B. Valeur and G. Weber, *Chem. Phys. Lett.*, **45**, 140 (1977).
66. B. Valeur and G. Weber, *J. Chem. Phys.*, **69**, 2393 (1978).
67. M. Monsigny, P. Midoux, M. T. Le Bris, A.-C. Roche, and B. Valeur, *Biol. Cell*, **67**, 193 (1989).
68. C. Depierreux, M. T. Le Bris, M. F. Michel, B. Valeur, M. Monsigny, and F. Delmotte, *FEMS Microbiol. Lett.*, **67**, 237 (1990).
69. T. M. Jovin, M. Bartholdi, W. L. C. Vaz, and R. G. Austin, *Ann. N.Y. Acad. Sci.*, **366**, 176 (1981).

70. B. Garland and J. J. Birmingham, in D. L. Taylor, A. S. Waggoner, F. Lanni, R. F. Murphy, and R. R. Birge, Eds., *Applications of Fluorescence in the Biomedical Sciences*, Alan R. Liss, New York, 1986, p. 245.
71. N. E. Geacintov and H. C. Brenner, *Photochem. Photobiol.*, **50**, 841 (1989).
72. E. Soini and T. Lögren, *CRC Crit. Rev. Anal. Chem.*, **18**, 105 (1987).
73. J.-C. Bünzli, in J.-C. Bünzli and G. R. Choppin, Eds., *Lanthanides Probes in Life, Chemical and Earth Sciences*, Elsevier, Amsterdam, 1989.
74. J. N. Demas, *Excited State Lifetime Measurements*, Academic Press, New York, 1983.
75. D. O'Connor and D. Phillips, *Time-Correlated Single Photon Counting*, Academic Press, New York, 1983.
76. G. R. Fleming, *Chemical Applications of Ultrafast Spectroscopy*, Oxford University Press, London and New York, 1986.
77. E. Gratton and Limkeman, *Biophys. J.*, **44**, 315 (1983).
78. E. Gratton, D. M. Jameson, and R. D. Hall, *Annu. Rev. Biophys. Bioeng.*, **13**, 105 (1984).
79. J. R. Lakowicz and B. P. Maliwal, *Biophys. Chem.*, **21**, 61 (1985).
80. J. Pouget, J. Mugnier, and B. Valeur, *J. Phys. E*, **22**, 855 (1989).
81. G. Laczko, I. Gryczynski, Z. Gryczynski, W. Wiczk, H. Malak, and J. R. Lakowicz, *Rev. Sci. Instrum.*, **61**, 2331 (1990).
82. D. Eaton, *Pure Appl. Chem.*, **62**, 1631 (1990).
83. J. M. Beechem, J. R. Knutson, J. B. A. Ross, B. W. Turner, and L. Brand, *Biochemistry*, **12**, 6054 (1983).
84. J. R. Knutson, J. M. Beechem, and L. Brand, *Chem. Phys. Lett.*, **102**, 501 (1983).
85. J. M. Beechem, M. Ameloot, and L. Brand, *Chem. Phys. Lett*, **120**, 466 (1985).
86. J. M. Beechem, M. Ameloot, and L. Brand, *Anal. Instrum*, **14**, 379 (1985).
87. J. M. Beechem and L. Brand, *Photochem. Photobiol.*, **44**, 323 (1986).
88. M. Ameloot, J. M. Beechem, and L. Brand, *Biophys. Chem.*, **23**, 155 (1986).
89. S. R. Flom and J. H. Fendler, *J. Phys. Chem.*, **92**, 5908 (1988).
90. L. D. Janssens, N. Boens, M. Ameloot, and F. C. De Schryver, *J. Phys. Chem.*, **94**, 3564 (1990).
91. D. R. James and W. R. Ware, *Chem. Phys. Lett.*, **120**, 455 (1985).
92. J. R. Alcala, E. Gratton, and G. Prendergast, *Biophys. J.*, **51**, 587, 597, 925 (1987).
93. J. R. Lakowicz, H. Cherek, I. Gryczynski, and M. L. Johnson, *Biophys. Chem.*, **28**, 35 (1987).
94. D. R. James, J. R. Turnbull, B. D. Wagner, W. R. Ware, and N. O. Petersen, *Biochemistry*, **26**, 6272 (1987).
95. A. Siemiarczuk and W. R. Ware, *Chem. Phys. Lett.*, **167**, 263 (1990).
96. A. K. Livesey and J. C. Brochon, *Biophys. J.*, **52**, 693 (1987).
97. A. Siemiarczuk, B. D. Wagner, and W. R. Ware, *J. Phys. Chem.*, **94**, 1661 (1990).

98. J.-C. Brochon, A. K. Livesey, J. Pouget, and B. Valeur, *Chem. Phys. Lett.*, **174**, 517 (1990).
99. P. Suppan, *J. Photochem. Photobiol.*, **A50**, 293 (1990).
100. E. M. Kosower, *J. Am. Chem. Soc.*, **80**, 3253 (1958).
101. K. Dimroth, C. Reichardt, T. Siepmann, and F. Bohlmann, *Justus Liebigs Ann. Chem.*, **661**, 1 (1963).
102. C. Reichardt, *Angew. Chem.*, **91**, 119 (1979).
103. C. Reichardt, *Solvent Effects in Organic Chemistry*, Verlag Chemie, Weinheim, Germany, 1988.
104. E. Buncel and S. Rajagopal, *Acc. Chem. Res.*, **23**, 226 (1990).
105. M. J. Kamlet, J.-L. M. Abboud, and R. W. Taft, *J. Am. Chem. Soc.*, **99**, 6027 (1977).
106. M. J. Kamlet, J.-L. M. Abboud, and R. W. Taft, *Prog. Phys. Org. Chem.*, **13**, 485 (1981).
107. E. Lippert, *Z. Elektrochem.*, **61**, 962 (1957).
108. P. Suppan, *Chem. Phys. Lett.*, **94**, 272 (1983).
109. G. Weber and F. J. Farris, *Biochemistry*, **18**, 3075 (1979).
110. P. Ilich and F. G. Prendergast, *J. Phys. Chem.*, **93**, 4441 (1989).
111. R. B. MacGregor and G. Weber, *Nature (London)*, **319**, 70 (1986).
112. F. G. Prendergast, M. Meyers, G. L. Carlson, S. Fida, and J. D. Potter, *J. Biol. Chem.*, **258**, 7541 (1983).
113. G. Weber and D. J. R. Lawrence, *Biochem. J.*, **56**, 31 (1954).
114. C. J. Seliskar and L. Brand, *J. Am. Chem. Soc.*, **93**, 5405, 5414 (1971).
115. R. P. Detoma, J. H. Easter, and L. Brand, *J. Am. Chem. Soc.*, **98**, 5001 (1976).
116. T. W. Ebbensen and C. A. Ghiron, *J. Phys. Chem.*, **93**, 7139 (1989).
117. E. M. Kosower and H. J. Kanety, *J. Am. Chem. Soc.*, **105**, 6236 (1983).
118. G. R. Penzer, *Eur. J. Biochem.*, **25**, 218 (1972).
119. G. R. Fleming, G. Porter, R. J. Robbins and J. A. Synowiec, *Chem. Phys. Lett.*, **52**, 228 (1977).
120. P. J. Sadkowski and G. R. Fleming, *Chem. Phys.*, **54**, 79 (1980).
121. J. D. Simon, *Acc. Chem. Res.*, **15**, 128 (1988).
122. P. F. Barbara and W. Jarzeba, *Acc. Chem. Res.*, **15**, 195 (1990).
123. P. F. Barbara and W. Jarzeba, *Adv. Photochem.*, **15**, 1 (1990).
124. M. Maroncelli, J. MacInnis, and G. R. Fleming, *Science*, **243**, 1674 (1989).
125. K. Kalyanasundaran and J. K. Thomas, *J. Am. Chem. Soc.*, **99**, 2039 (1977).
126. A. Nakajima, *J. Lumin.*, **11**, 429 (1976); *J. Mol. Spectrosc.*, **61**, 467 (1976).
127. A. Nakajima, *Bull. Chem. Soc. Jpn.*, **44**, 3272 (1977).
128. A. Nakajima, *Spectrochim. Acta.*, **A39**, 913 (1983).
129. D. C. Dong and M. A. Winnik, *Photochem. Photobiol.*, **35**, 17 (1982).
130. D. C. Dong and M. A. Winnik, *Can. J. Chem.*, **62**, 2560 (1984).

131. W. Webb, *Q. Rev. Biophys.*, **9**, 49 (1976).
132. D. Axelrod, D. E. Koppel, J. Schlessinger, E. Elson, and W. W. Webb, *Biophys. J.*, **16**, 1055 (1976).
133. F. Hare and C. Lussan, *Biochim. Biophys. Acta*, **467**, 262 (1977).
134. M. L. Viriot, M. Bouchy, M. Donner, and J. C. André, *Photobiochem. Photobiophys.*, **5**, 293 (1983).
135. A. Gierer and K. Wirtz, *Z. Naturforsch.*, **8A**, 535 (1953).
136. A. H. Alwattar, M. D. Lumb, and J. B. Birks, in J. B. Birks, Ed., *Organic Molecular Photophysics*, Vol. 1, Wiley, New York, 1973, Chap. 8.
137. A. K. Doolittle, *J. Appl. Phys.*, **22**, 1471 (1951); **23**, 236 (1952).
138. M. L. Williams, R. F. Landel, and J. D. Ferry, *J. Am. Chem. Soc.*, **77**, 3701 (1955).
139. R. O. Loutfy and B. A. Arnold, *J. Phys. Chem.*, **86**, 4205 (1982).
140. G. Oster and Y. Nishijima, *J. Am. Chem. Soc.*, **78**, 1581 (1956).
141. G. Oster and Y. Nishijima, *Fortschr. Hochpolym.-Forsch.*, **3**, 313 (1964).
142. Th. Förster and G. Hoffmann, *Z. Phys. Chem.* [N.S.] **75**, 63 (1971).
143. J. S. McCaskill and R. G. Gilbert, *Chem. Phys.*, **44**, 389 (1979).
144. C. D. Tredwell and A. D. Osborne, *J. Chem. Soc., Faraday Trans. 2*, **76**, 1627 (1980).
145. B. Bagchi, G. R. Fleming, and W. Oxtoby, *J. Chem. Phys.*, **78**, 7375 (1983).
146. V. Sundström and T. Gillbro, *J. Chem. Phys.*, **61**, 257 (1981).
147. V. Sundström and T. Gillbro, *J. Chem. Phys.*, **81**, 3463 (1984).
148. A. D. Osborne, *J. Chem. Soc., Faraday Trans. 2*, **76**, 1638 (1980).
149. M. Vogel and W. Rettig, *Ber. Bunsenges. Phys. Chem.*, **89**, 962 (1985).
150. M. Vogel and W. Rettig, *Ber. Bunsenges. Phys. Chem.*, **91**, 1241 (1987).
151. B. Bagchi and G. R. Fleming, *J. Phys. Chem.*, **94**, 9 (1990).
152. K. Y. Law, *Chem. Phys. Lett.*, **75**, 545 (1980).
153. R. O. Loutfy and K. Y. Law, *J. Phys. Chem.*, **84**, 2803 (1980).
154. R. O. Loutfy, *Pure Appl. Chem.*, **58**, 1239 (1986).
155. C. E. Kung and K. Reed, *Biochemistry*, **28**, 6678 (1989).
156. M. S. A. Abdel-Mottaleb, R. O. Loutfy, and R. Lapouyade, *J. Photochem. Photobiol.*, **A48**, 87 (1989).
157. S. Mqadmi and A. Pollet, *J. Photochem. Photobiol.*, **A53**, 275 (1990).
158. Z. Bentikouk, M. L. Viriot, and J. C. André, to be published.
159. Z. R. Grabowski, K. Rotkiewicz, A. Siemiarczuk, D. J. Cowley, and W. Baumann, *Nouv. J. Chim.* **3**, 443 (1979).
160. W. Rettig, *Angew. Chem.*, **98**, 969 (1986).
161. E. Lippert, W. Rettig, V. Bonacic-Koutecky, F. Heisel, and J. A. Miehé, *Adv. Chem. Phys.*, **68**, 1 (1987).
162. C. Cazeau-Dubroca, A. Peirigua, M. Ben Brahim, G. Nouchi, and P. Cazeau, *Chem. Phys. Lett.*, **157**, 393 (1989).

REFERENCES

163. R. K. Guo and S. Tazuke, *Macromolecules*, **22**, 3286 (1989).
164. S. Tazuke and R. K. Guo, *Macromolecules*, **23**, 719 (1990).
165. S. Tazuke, R. K. Guo, T. Ikeda, and T. Ikeda, *Macromolecules*, **23**, 1208 (1990).
166. J. C. André, F. Baros, and M. A. Winnik, *J. Phys. Chem.*, **94**, 2942 (1990).
167. M. N. Berberan-Santos and J. M. G. Martinho, *Chem. Phys. Lett.*, **178**, 1 (1991).
168. J. M. Vanderkooi and J. B. Callis, *Biochemistry*, **13**, 4000 (1974).
169. H. J. Galla and W. Hartmann, *Chem. Phys. Lipids*, **27**, 199 (1980).
170. M. E. Jones and B. R. Lentz, *Biochemistry*, **25**, 567 (1986).
171. J. Eisinger, J. Flores, and W. P. Petersen, *Biophys. J.*, **49**, 987 (1986).
172. P. J. Somerharju, J. A. Virtanen, K. K. Eklund, P. Vainio, and P. K. J. Kinnuanen, *Biochemistry*, **24**, 2773 (1985).
173. J. R. Wiener, R. Pal, Y. Barenholz, and R. R. Wagner, *Biochemistry*, **24**, 7651 (1985).
174. R. C. Hresko, I. P. Sugar, Y. Barenholz, and T. E. Thompson, *Biochemistry*, **25**, 3813 (1986).
175. R. C. Hresko, I. P. Sugar, Y. Barenholz, and T. E. Thompson, *Biophys. J.*, **51**, 725 (1987).
176. H. Lemmetyinen, M. Yliperttula, J. Mikkola, J. A. Virtanen, and P. K. J. Kinnunen, *J. Phys. Chem.*, **93**, 7170 (1989).
177. M. Van den Zegel, N. Boens, and F. C. De Schryver, *Biophys. Chem.*, **20**, 333 (1984).
178. D. Daems, M. Van den Zegel, N. Boens, and F. C. De Schryver, *Eur. Biophys. J.*, **12**, 97 (1985).
179. D. Y. Chu and J. K. Thomas, *Macromolecules*, **23**, 2223 (1990).
180. W. Klöpffer, in J. B. Birks, Ed., *Organic Molecular Photophysics*, Vol. 1, Wiley, New York, 1973, p. 357.
181. L. Bokobza and L. Monnerie, in M. A. Winnik, Ed., *Photophysical and Photochemical Tools in Polymer Science*, Reidel, Dordrecht, The Netherlands, 1986, pp. 449–466.
182. F. C. De Schryver, P. Collart, J. Vandendriessche, R. Goedeweeck, A. M. Swinnen, and M. Van der Auweraer, *Acc. Chem. Res.*, **20**, 159 (1987).
183. L. Bokobza, *Prog. Polym. Sci.*, **15**, 337 (1990).
184. K. A. Zachariasse, W. Kühnle, and A. Weller, *Chem. Phys. Lett.*, **73**, 6 (1980).
185. K. A. Zachariasse, W. L. C. Vaz, and C. Sotomayor, *Biochim. Biophys. Acta*, **688**, 323 (1982).
186. K. A. Zachariasse, B. Kozankiewicz, and W. Kühnle, in A. H. Zewail, Ed., *Photochemistry and Photobiology*, Vol. 2, Harwood, London, 1983, p. 941.
187. K. A. Zachariasse, G. Duveneck, and R. Busse, *J. Am. Chem. Soc.*, **106**, 1045 (1984).
188. C. N. Henderson, B. K. Selinger, and A. R. Watkins, *J. Photochem.*, **16**, 215 (1981).

189. M. J. Snare, P. J. Thistlethwaite, and K. P. Ghiggino, *J. Am. Chem. Soc.*, **105**, 3328 (1983).
190. D. Georgescauld, J. P. Desmasez, R. Lapouyade, A. Babeau, H. Richard, and M. A. Winnik, *Photochem. Photobiol.*, **31**, 539 (1980).
191. G. E. Johnson, *J. Chem. Phys.*, **63**, 4047 (1975).
192. Y. C. Wang and H. Morawetz, *J. Am. Chem. Soc.*, **98**, 3611 (1976).
193. M. Goldenberg, J. Emert, and H. Morawetz, *J. Am. Chem. Soc.*, **100**, 7171 (1978).
194. L. Bokobza, B. Jasse, and L. Monnerie, *Eur. Polym. J.*, **16**, 715 (1980).
195. P. D. Fitzgibbon and C. W. Franck, *Macromolecules*, **14**, 1650 (1981).
196. J. Vandendriessche, M. Van der Auweraer, and F. C. De Schryver, *Bull. Soc. Chim. Belg.*, **94**, 991 (1985).
197. B. Valeur, J. Mugnier, J. Pouget, J. Bourson, and F. Santi, *J. Phys. Chem.*, **93**, 6073 (1989).
198. M. L. Viriot, R. Guillard, I. Kauffmann, J. C. André, and G. Siest, *Biochim. Biophys. Acta*, **733**, 34 (1983).
199. W. D. Turley and H. W. Offen, *J. Phys. Chem.*, **89**, 3962 (1985); **90**, 1967 (1986).
200. B. D. Freeman, L. Bokobza, P. Sergot, L. Monnerie, and F. C. De Schryver, *Macromolecules*, **23**, 2566 (1990).
201. K. Hara and H. Suzuki, *J. Phys. Chem.*, **94**, 1079 (1990).
202. B. Valeur and G. Weber, *Photochem. Photobiol.*, **25**, 441 (1975).
203. A. Jablonski, *Acta Phys. Pol.*, **10**, 193 (1950); **38**, 717 (1965).
204. P. E. Zinsli, *Chem. Phys.*, **20**, 299 (1977).
205. V. Veissier, J. L. Viovy, and L. Monnerie, *J. Phys. Chem.*, **93**, 1709 (1989).
206. L. B-Å. Johansson, *J. Chem. Soc., Faraday Trans.*, **86**, 2103 (1990).
207. W. W. Mantulin and G. Weber, *J. Chem. Phys.*, **66**, 4092 (1977).
208. J. L. Viovy, *J. Phys. Chem.*, **89**, 5465 (1985).
209. R. L. Christensen, R. C. Drake, and D. Phillips, *J. Phys. Chem.*, **90**, 5960 (1986).
210. T. Tao, *Biopolymers*, **8**, 609 (1969).
211. T. J. Chuang and K. B. Eisenthal, *J. Chem. Phys.*, **57**, 5094 (1972).
212. M. Ehrenberg and R. Rigler, *Chem. Phys. Lett.*, **14**, 539 (1972).
213. G. G. Belford, R. L. Belford, and G. Weber, *Proc. Natl. Acad. Sci. U.S.A.*, **69**, 1392 (1972).
214. G. Weber, in A. A. Thaer and M. Sternetz, Eds., *Fluorescence Techniques in Cell Biology*, Springer-Verlag, Berlin, 1973, p. 5.
215. M. D. Barkley, A. A. Kowalczyk, and L. Brand, *J. Chem. Phys.*, **75**, 3581 (1981).
216. M. Vincent, B. de Foresta, J. Gallay, and A. Alfsen, *Biochemistry*, **21**, 708 (1982).
217. K. Kinoshita, S. Kawato, and A. Ikegami, *Biophys. J.*, **20**, 289 (1977).
218. G. Lipari and A. Szabo, *Biophys. J.*, **30**, 489 (1980).
219. W. Van der Meer, R. P. H. Kooyman, and Y. K. Levine, *Chem. Phys.*, **66**, 39 (1982).

220. W. Van der Meer, H. Pottel, W. Herreman, M. Ameloot, H. Hendrickx, and H. Schröder, *Biophys. J.*, **46**, 515 (1984).
221. C. Zannoni, A. Arcioni, and P. Cavatorta, *Chem. Phys. Lipids*, **32**, 179 (1983).
222. A. Szabo, *J. Chem. Phys.*, **81**, 150 (1984).
223. J. J. Fisz, *Chem. Phys.*, **99**, 177 (1985); **132**, 303, 315 (1989).
224. H. Pottel, W. Herreman, B. W. Van der Meer, and M. Ameloot, *Chem. Phys.*, **102**, 37 (1986).
225. H. van Langen, Y. K. Levine, M. Ameloot, and H. Pottel, *Chem. Phys. Lett.*, **140**, 394 (1987).
226. H. van Langen, G. Van Ginkel, D. Shaw, and Y. K. Levine, *Eur. Biophys. J.*, **17**, 37 (1989).
227. A. Ruggiero and B. Hudson, *Biophys. J.*, **55**, 1125 (1989).
228. A. Arcioni, F. Bertinelli, R. Tarroni, and C. Zannoni, *Mol. Phys.*, **61**, 1161 (1987).
229. A. Arcioni, F. Bertinelli, R. Tarroni, and C. Zannoni, *Chem. Phys.*, **143**, 259 (1990).
230. R. D. Ludescher, L. Petting, S. Hudson, and B. Hudson, *Biophys. Chem.*, **28**, 59 (1987).
231. M. Shinitzky, A. C. Dianoux, C. Gitler, and G. Weber, *Biochemistry*, **10**, 2106 (1971).
232. M. Shinitzky, *Isr. J. Chem.*, **12**, 879 (1974).
233. R. C. Dorrance and T. F. Hunter, *J. Chem. Soc., Faraday Trans. 2*, **73**, 89 (1977).
234. U. K. A. Klein and H.-P. Haar, *Chem. Phys. Lett.*, **58**, 531 (1978).
235. E. Blatt, K. P. Ghiggino, and W. H. Sawyer, *J. Phys. Chem.*, **86**, 4461 (1982).
236. B. Valeur and E. Keh, *J. Phys. Chem.*, **83**, 3305 (1979).
237. E. Keh and B. Valeur, *J. Colloid Interface Sci.*, **79**, 465 (1981).
238. P. E. Zinsli, *J. Phys. Chem.*, **83**, 3223 (1979).
239. E. Bardez, E. Monnier, and B. Valeur, *J. Colloid Interface Sci.*, **112**, 200 (1986).
240. A. J. W. G. Visser, K. Vos, A. van Hoek, and G. Santema, *J. Phys. Chem.*, **92**, 759 (1988).
241. V. Chen, G. G. Warr, D. F. Evans, and F. G. Prendergast, *J. Phys. Chem.*, **92**, 768 (1988).
242. M. Vincent and J. Gallay, *Biochem. Biophys. Res. Commun.*, **113**, 799 (1983).
243. S. Kawato, K. Kinosita, and A. Ikegami, *Biochemistry*, **16**, 2319 (1977); **17**, 5026 (1978).
244. J. R. Lakowicz and F. G. Prendergast, *Science*, **200**, 1399 (1978).
245. J. R. Lakowicz, F. G. Prendergast, and D. Hogen, *Biochemistry*, **18**, 508 (1979).
246. L. A. Sklar, G. P. Miljanich, and E. A. Dratz, *Biochemistry*, **18**, 1707 (1979).
247. P. L. Pugh, M. Kates, and A. G. Szabo, *Chem. Phys. Lipids*, **30**, 55 (1982).
248. P. A. Van Paridon, J. K. Shute, K. W. A. Wirtz, and A. J. W. G. Visser, *Eur. Biophys. J.*, **16**, 53 (1988).

249. A. Ribier, R. M. Handjani-Vila, E. Bardez, and B. Valeur, *Colloids Surf.*, **10**, 155 (1984).
250. E. V. Gordeev, V. K. Dolganov, and V. V. Korshunov, *JETP Lett. (Engl. Transl.)*, **43**, 766 (1986).
251. B. Kalman, N. Clarke, and L. B.-A. Johansson, *Chem. Phys.*, **57**, 5094 (1989).
252. B. Valeur and L. Monnerie, *J. Polym. Sci., Polym. Phys. Ed.*, **14**, 11 (1976).
253. B. Valeur and L. Monnerie, *J. Polym. Sci., Polym. Phys. Ed.*, **14**, 29 (1976).
254. J. P. Jarry and L. Monnerie, *Macromolecules*, **12**, 927 (1979).
255. J. L. Viovy, L. Monnerie, and J. C. Brochon, *Macromolecules*, **16**, 1845 (1983).
256. J. L. Viovy, L. Monnerie, and F. Merola, *Macromolecules*, **18**, 1130 (1985).
257. J. Yguerabide, H. F. Epstein, and L. Stryer, *J. Mol. Biol.*, **51**, 573 (1970).
258. D. C. Hanson, J. Yguerabide, and V. N. Schumaker, *Biochemistry*, **20**, 6842 (1981).
259. Y. Nishijima, *J. Polym. Sci., Part C*, **31**, 353 (1970).
260. J. P. Jarry and L. Monnerie, *Ann. N.Y. Acad. Sci.*, **366**, 328 (1981).
261. L. Monnerie, in M. A. Winnik, Ed., *Photophysical and Photochemical Tools in Polymer Science*, Reidel, Dordrecht, The Netherlands, 1986, p. 371.
262. J. H. Nobbs and I. M. Ward, in D. Phillips, Ed., *Polymer Photophysics: Luminescence, Energy Migration and Molecular Motion in Synthetic Polymers*, Chapman & Hall, London, 1985, p. 159.
263. L. Stryer and R. P. Haugland, *Proc. Natl. Acad. Sci. U.S.A.*, **58**, 719 (1967).
264. I. Z. Steinberg, *Annu. Rev. Biochem.*, **40**, 83 (1971).
265. P. W. Schiller, in R. F. Chen and H. Edelhoch, Eds., *Biochemical Fluorescence: Concepts*, Vol. 1, Marcel Dekker, New York, 1975, p. 285.
266. L. Stryer, *Annu. Rev. Biochem.*, **47**, 819 (1978).
267. R. H. Fairclough and C. R. Cantor, *Methods Enzymol*, **48**, 347 (1978).
268. J. C. Conte and J. M. G. Martinho, *J. Lumin.*, **22**, 273 (1981).
269. Th. Förster, *Ann. Phys. (Leipzig)* [6], **2**, 55 (1948).
270. Th. Förster, *Z. Naturforsch.*, **4A**, 321 (1949).
270a. I. Berlman, *Energy Transfer Parameters of Aromatic Molecules*, Academic Press, New York, 1973.
271. R. E. Dale, *Acta Phys. Pol.*, **A54**, 743 (1978).
272. R. E. Dale and J. Eisinger, *Biopolymers*, **13**, 1573 (1974).
273. R. E. Dale and J. Eisinger, in R. F. Chen and H. Edelhoch, Eds., *Biochemical Fluorescence: Concepts*, Vol. 1, Marcel Dekker, New York, 1975, p. 115.
274. W. E. Blumberg, R. E. Dale, J. Eisinger, and D. M. Zuckerman, *Biopolymers*, **13**, 1607 (1974).
275. J. Eisinger, W. E. Blumberg, and R. E. Dale, *Ann. N. Y. Acad. Sci.*, **366**, 155 (1981).
276. D. L. Dexter, *J. Chem. Phys.*, **21**, 836 (1964).

277. S. A. Latt, H. T. Cheung, and E. R. Blout, *J. Am. Chem. Soc.*, **87**, 995 (1965).
278. H. C. Chiu and Bersohn, *Biopolymers*, **16**, 277 (1977).
279. J. Mugnier, J. Pouget, J. Bourson, and B. Valeur, *J. Lumin.*, **33**, 273 (1985).
280. B. Kopainsky, W. Kaiser, and F. P. Schäfer, *Chem. Phys. Lett.*, **56**, 458 (1978).
281. J. Mugnier, B. Valeur, and E. Gratton, *Chem. Phys. Lett.*, **119**, 217 (1985).
282. N. P. Ernsting, M. Kaschke, J. Kleinschmidt, K. H. Drexhage, and V. Huth, *Chem. Phys.*, **122**, 431 (1988).
283. U. K. A. Klein, R. Frey, M. Hauser, and U. Gösele, *Chem. Phys. Lett.*, **41**, 139 (1976).
284. P. R. Butler and M. J. Pilling, *Chem. Phys.*, **41**, 239 (1979).
285. D. P. Millar, R. J. Robbins, and A. H. Zewail, *J. Chem. Phys.*, **75**, 3649 (1981).
286. R. H. Conrad and L. Brand, *Biochemistry*, **7**, 777 (1968).
287. C. R. Cantor and P. Pechukas, *Proc. Natl. Acad. Sci. U.S.A.*, **68**, 2099 (1971).
288. A. Grinvald, E. Haas, and I. Z. Steinberg, *Proc. Natl. Acad. Sci. U.S.A.*, **69**, 2273 (1972).
289. E. Haas, M. Wilchek, E. Katchalski-Katzir, and I. Z. Steinberg, *Proc. Nat. Acad. Sci. U.S.A.*, **72**, 1807 (1975).
290. J. R. Lakowicz, M. L. Johnson, W. Wiczk, A. Bhat, and R. F. Steiner, *Chem. Phys. Lett.*, **138**, 587 (1987).
291. W. M. Wiczk, I. Gryczynski, H. Szmacinski, M. L. Johnson, M. Kruszynski, and J. Zboinska, *Biophys. Chem.*, **32**, 43 (1988).
292. M. Kaschke, N. P. Ernsting, B. Valeur, and J. Bourson, *J. Phys. Chem.*, **94**, 5757 (1990).
293. B. Valeur, in D. M. Jameson and G. D. Reinhart, Eds., *Fluorescent Biomolecules, Methodologies and Applications*, Plenum Press, New York, 1989, p. 269.
294. G. H. Fredrickson, H. C. Andersen, and C. W. Frank, *J. Chem. Phys.*, **79**, 3572 (1983).
295. K. A. Peterson and M. D. Fayer, *J. Chem. Phys.*, **85**, 4702 (1986).
296. M. B. Zimmt, K. A. Peterson, and F. D. Fayer, *Macromolecules*, **21**, 1145 (1988).
297. G. Liu and J. E. Guillet, *Macromolecules*, **23**, 1388 (1990).
298. G. Liu, J. E. Guillet, E. T. B. Al-Takrity, A. D. Jenkins, and D. R. M. Walton, *Macromolecules*, **23**, 1393 (1990).
299. M. N. Berberan-Santos and M. J. E. Prieto, *J. Chem. Soc., Faraday Trans. 2*, **83**, 1391 (1987).
300. N. Tamai, T. Yamazaki, I. Yamazaki, A. Mizuma, and N. Mataga, *J. Phys. Chem.*, **91**, 3503 (1987).
301. M. Kaschke, O. Kittelmann, K. Vogler, and A. Graness, *J. Phys. Chem.*, **92**, 5998 (1988).
302. K.-J. Choi, L. A. Turkevich, and R. Loza, *J. Phys. Chem.*, **92**, 2248 (1988).
303. I. Yamazaki, N. Tamai, and T. Yamazaki, *J. Phys. Chem.*, **94**, 516 (1990).
304. D. Amir and E. Haas, *Biochemistry*, **26**, 2162 (1987).

305. R. Lakowicz, I. Gryczynski, H. C. Cheung, C. K. Wang, M. L. Johnson, and N. Joshi, *Biochemistry*, **27**, 9149 (1988).
306. I. Gryczynski, W. Wiczk, M. L. Johnson, and R. Lakowicz, *Chem. Phys. Lett.*, **145**, 439 (1988).
307. S. Albaugh and R. F. Steiner, *J. Phys. Chem.*, **93**, 8013 (1989).
308. M. Kaschke, B. Valeur, J. Bourson, and N. P. Ernsting, *Chem. Phys. Lett.*, **179**, 544 (1991).
309. T. Ikeda, B. Lee, S. Kurihara, S. Tazuke, S. Ito, and M. Yamamoto, *J. Am. Chem. Soc.*, **110**, 8299 (1988).
310. E. Haas, E. Katchalski-Katzir, and I. Z. Steinberg, *Biopolymers*, **17**, 11 (1978).
311. G. Liu and J. E. Guillet, *Macromolecules*, **23**, 2969,2973 (1990).
312. G. Liu, J. E. Guillet, E. T. B. Al-Takrity, A. D. Jenkins, and D. R. M. Walton, *Macromolecules*, **23**, 4164 (1990).

CHAPTER

3

PHOTOCHEMICAL FLUOROMETRY

JEAN-JACQUES AARON

Institute of Topology and Systems Dynamics
University of Paris 7, associated with CNRS
75005 Paris, France

3.1. Introduction
3.2. Theoretical Considerations
    3.2.1. Theoretical Intensity Expression for the Decrease of the Fluorescence Signal of the Analyte
    3.2.2. Theoretical Intensity Expression for the Increase of the Fluorescence Signal of the Photoproduct
3.3. Instrumentation
    3.3.1. Photochemical Fluorescence Equipment for Liquid Solution Studies
    3.3.2. Photochemical Fluorescence Equipment for TLC and Solid Surface Studies
    3.3.3. Photochemical Fluorescence Equipment for HPLC Studies
        3.3.3.1. Photochemical Reactors
        3.3.3.2. Fluorometers and Spectrophotofluorometers
    3.3.4. Photochemical Fluorescence Equipment for Flow Analysis Studies
3.4. Photoreactions in Photochemical Fluorescence Analysis
    3.4.1. Photocyclization Reactions
    3.4.2. Photoisomerization Reactions
    3.4.3. Photolysis Reactions
    3.4.4. Sensitized Photoreduction Reactions
    3.4.5. Photooxidation Reactions
3.5. Applications
    3.5.1. General Analytical Performances
    3.5.2. Pharmaceutical Analysis
    3.5.3. Clinical Analysis Applications
    3.5.4. Pesticides and Environmental Analysis
3.6. Future Trends
References

Molecular Luminescence Spectroscopy, Part 3, Edited by Stephen G. Schulman. Chemical Analysis Series, Vol. 77.
ISBN 0-471-51580-9 © 1993 John Wiley & Sons, Inc.

3.1. INTRODUCTION

Molecular fluorescence spectrometery is a very powerful method of analysis for organic compounds because high sensitivity and high selectivity can frequently be achieved. However, its usefulness is impeded in several cases. Photolysis of the analyte can occur and compete with fluorescence when a fluorescent compound is excited with intense ultraviolet (UV) radiation. It constitutes a serious reproducibility problem since the fluorescence intensity of the analyte then changes during the time of measurement (1–3). Also, a number of compounds exist that, because of their particular molecular structure, present very low fluorescence quantum yields and therefore cannot be determined fluorometrically with good sensitivity.

In order to eliminate these analytical drawbacks, a new approach has recently been proposed using photochemical reactions for improving the selectivity, sensitivity, and reproducibility fluorometric detection. This method is generally called photochemical fluorescence analysis (or photochemical fluorometry). It consists of the initiation of a photoreaction of a photochemically unstable analyte by means of UV radiation and the fluorometric observation of the formation of the photoproducts. When the photoproduct is less fluorescent than the analyte, the decay of the fluorescence signal is measured (4,5). However, in the majority of compounds, the photochemical reaction yields a photoproduct with an enhanced molecular absorption coefficient and a higher fluorescence quantum yield relative to those of the analyte (6–8). Therefore, the kinetics of the increase of the photoproduct fluorescence signal can be evaluated, resulting in an improvement of the sensitivity of fluorometric detection (6–8). As will be seen, specific photochemical reagents, such as photosensitizers, have been used to absorb UV light and initiate a photoreaction. It has helped the development of the method by enabling investigators fluorometrically to detect compounds with only very weak chromophores (9–11) or without any native fluorescence (12).

Photochemical fluorometry can be utilized in a variety of experimental setups, in solution as well as on solid surfaces. Also, it has been combined with dynamic analytical techniques such as high-performance liquid chromatography (HPLC) (8), thin-layer chromatography (TLC) (13–15), and flow injection analysis (FIA) (16). In the case of HPLC, photochemical fluorescence has been used as a postcolumn detection system instead of postcolumn chemical derivatization (8).

Laboratory-constructed instruments and modifications to commercial instruments have appeared in the last 15 years for measuring photochemical fluorescence in various media. A variety of new photochemical reactors have been constructed. A relatively large number of applications have been described in the literature in areas such as pharmaceutical analysis, pesticide analysis, clinical chemistry, biochemistry, and trace metals analysis. In this chapter, we will present the theory, instrumental aspects, and applications of photochemical fluorometry.

3.2. THEORETICAL CONSIDERATIONS

The theoretical aspects of photochemical fluorescence analysis have been studied by very few authors (3,4,6; J.-J. Aaron, unpublished results, 1982). They have concerned themselves mainly with the relations between the initial concentration of the analyte and the fluorescence signal of the photoproduct, and the laws of photochemical kinetics involved. Generally, photochemical kinetics are relatively rapid, and the fluorescence of the analyte or of the photoproduct is a convenient method for following the rate of the photoreaction.

In dilute solution, assuming that absorbance is $< 10^{-2}$, the rate of photoreaction of an analyte A is given by the following equation (3,4):

$$v = -\frac{dC_A}{dt} = \frac{dC_B}{dt} = \sum_{\lambda_p} \Phi_{\lambda_p} \cdot I_{0,\lambda_p} \cdot a_{\lambda_p} \cdot b \cdot C_A \quad (3.1)$$

where the summation is over those wavelengths incident on the sample which produce the photoreaction; Φ_{λ_p} is the quantum yield for the photoreaction at each wavelength; I_{0,λ_p} is the incident intensity of each wavelength; a_{λ_p} is equal to $2.303 \times \varepsilon_A$ (ε_A = molecular absorption coefficient of A at each wavelength); b is the optical path length along the excitation radiation axis; C_A is the concentration of the analyte; and C_B is the concentration of the photoproduct.

In admitting that I_{0,λ_p} is constant and is large relative to C_A, it is possible to transform relation (3.1) into a conventional first-order rate equation (4):

$$(C_A)_t = (C_A)_0 \exp\left(-\sum_{\lambda_p} \Phi_{\lambda_p} \cdot I_{0,\lambda_p} \cdot a_\lambda \cdot b \cdot t\right) \quad (3.2)$$

where $(C_A)_t$ is the concentration of analyte at time t and $(C_A)_0$ is the initial concentration of analyte.

Now, two cases must be considered:

- The analyte is fluorescent, and its decay is followed fluorometrically.
- The analyte is not (or weakly) fluorescent, but the photoproduct is strongly fluorescent. The increase of the concentration of photoproduct is then followed fluorometrically.

3.2.1. Theoretical Intensity Expression for the Decrease of the Fluorescence Signal of the Analyte

The fluorescence intensity of the analyte (I_{FA}) is described by the equation:

$$I_{FA} = f(\theta) g(\lambda') \sum_{\lambda_{FA}} \Phi_{\lambda_{FA}} \cdot I_{0,\lambda_{FA}} \cdot a_{\lambda_{FA}} \cdot b \cdot C_A \tag{3.3}$$

where $f(\theta)$ is the geometry factor, and $g(\lambda')$ is the detector response at the analytical wavelength (4). The remaining symbols are the same as in Eq. (3.1), except that they now concern fluorescence excitation wavelengths of the analyte (λ_{FA}).

It can be seen that Eqs. (3.1) and (3.3) have a similar form, which is expected since they are relative to two competing processes of energy dissipation from the same singlet excited state obtained from a single absorption process (4).

If we assume that the incident intensity is constant, the decrease of analyte fluorescence intensity with time depends only on C_A [Eq. (3.1)], and Eq. (3.2) shows that the photoreaction is pseudo-first-order in $(C_A)_t$. Therefore, three procedures were proposed for evaluating the initial fluorophor concentration.

The initial-rate method is based on extrapolation at $t = 0$ of the fluorescence intensity versus time curve, since the initial photoreaction rate is a measure of $(C_A)_0$.

Single-point direct fluorescence analysis consists of measuring at a constant time t the analyte fluorescence intensity, which is proportional to $(C_A)_0$ for a given reaction time.

The digital integration of the analyte fluorescence intensity over a given time interval was found to give an integrated signal that is proportional to initial fluorophor concentration. Taking into account Eqs. (3.2) and (3.3), we can define the following parameters:

$$k = \sum_{\lambda_p} \Phi_{\lambda_p} \cdot I_{0,\lambda_p} \cdot a_{\lambda_p} \cdot b \tag{3.4}$$

$$\chi = f(\theta) g(\lambda') \sum_{\lambda_{FA}} \Phi_{\lambda_{FA}} \cdot I_{0,\lambda_{FA}} \cdot b \tag{3.5}$$

Therefore, the analyte fluorescence signal at any time is

$$(I_{FA})_t = \chi (C_A)_t = \chi (C_A)_0 \exp(-kt) \tag{3.6}$$

The integration of Eq. (3.6) over a given time interval τ leads to the integrated fluorescence intensity $(I_{FA})_{int}$:

$$(I_{FA})_{int} = \int_0^\tau (I_{FA})_t \, dt = (C_A)_0 \chi \int_0^\tau \exp(-kt) \, dt \qquad (3.7)$$

$$(I_{FA})_{int} = -\frac{\chi}{k}(C_A)_0 [\exp(-k\tau) - 1] \qquad (3.8)$$

Using the same τ value for all integrations, when we plot integrated fluorescence intensity versus $(C_A)_0$ we obtain a linear calibration curve.

3.2.2. Theoretical Intensity Expression for the Increase of the Fluorescence Signal of the Photoproduct

Let us now consider a simple photochemical reaction (J.-J. Aaron, unpublished results, 1982): the concentration of the photoproduct $(C_B)_t$ at any time t is

$$(C_B)_t = (C_A)_0 - (C_A)_t \qquad (3.9)$$

Considering Eq. (3.2), we can transform Eq. (3.9) into the expression

$$(C_B)_t = (C_A)_0 \left[1 - \exp\left(-\sum_{\lambda_p} \Phi_{\lambda_p} \cdot I_{0,\lambda_p} \alpha_{\lambda_p} \cdot b \cdot t \right) \right] \qquad (3.10)$$

The fluorescence intensity of the photoproduct (I_{FB}) is

$$I_{FB} = f(\theta) g(\lambda') \sum_\lambda \Phi_{\lambda_{FB}} \cdot I_{0,\lambda_{FB}} \cdot a_{\lambda_{FB}} \cdot b \cdot C_B \qquad (3.11)$$

where the meaning of the symbols is the same as in Eq. (3.3) except that the parameters refer now to the photoproduct B.

If we assume that the incident intensity is constant, the change of photoproduct fluorescence signal with time depends only on C_B [Eq. (3.10)].

Using definitions identical to those given in Eqs. (3.4) and (3.5), except that the χ value refers to the photoproduct, we obtain the photoproduct fluorescence signal at any time:

$$(I_{FB})_t = \chi(C_B)_t = \chi(C_A)_0 [1 - \exp(-kt)] \qquad (3.12)$$

When the same time (t) value is used for each fluorescent measurement, a plot of photoproduct fluorescence intensity $(I_{FB})_t$ versus $(C_A)_0$ should yield a straight line calibration curve. However, nonlinearity could occur if the photoreaction is more complex than expected and leads to a variety of photoproducts.

This method allows one to perform photochemical fluorescence analysis of a nonfluorescent analyte when the photoproduct is characterized by a high fluorescence quantum yield ($\Phi_{\lambda_{FB}}$) value.

Aaron et al. (6) proposed an alternative method using the digital integration of the photoproduct fluorescence signal over a given time interval. This integration procedure is similar to that performed in the case of the decay of a fluorophor fluorescence signal.

3.3. INSTRUMENTATION

Presently, there exists practically no commercial instrument available for photochemical fluorescence analysis. In most cases, laboratory-constructed equipment as well as modifications to commercial instruments have been designed by researchers. Birks and Frei (8) have discussed various photochemical reactor designs used for photochemical fluorescence detection in HPLC. Krull and La Course (17) have described postcolumn photochemical reaction systems for fluorescence, UV, and electrochemical detection. This section will summarize some of the laboratory-constructed instruments, photochemical reactors designs, and innovative modifications to commercial instruments. Equipment used for photochemical fluorescence studies in bulk liquid solution, on solid surfaces, in TLC, in HPLC, and in flow analysis will be successively considered.

3.3.1. Photochemical Fluorescence Equipment for Liquid Solution Studies

Lukasiewicz and Fitzgerald (4) were probably the first researchers to design and construct an apparatus especially for photochemical fluorescence studies in liquid solution (Fig. 3.1). Several excitation sources were tested for producing fluorescence and initiating photolysis. A Hanovia medium-pressure mercury arc was found to be the most stable and effective source (Fig. 3.1), and it was used for most studies. An electrically driven shutter for the excitation source and the readout system were both activated by switching them from a common time control system (T). The sample cell was positioned rigidly inside a covered optical bench. For compounds emitting in the UV region, a UV-transmitting, visible-absorbing filter was placed between the cell and the entrance slit of the emission monochromator in order to reduce the background signal. The authors used two readout devices, a time-drive recorder and a

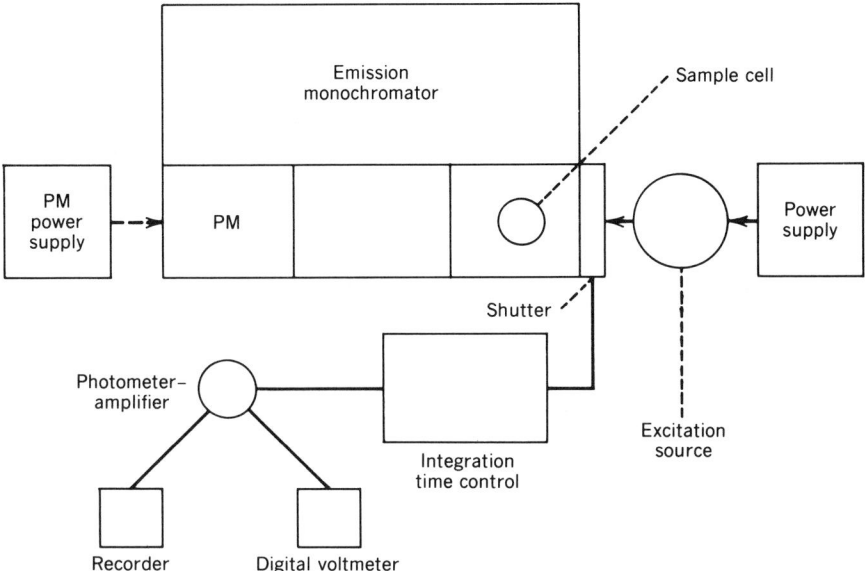

Fig. 3.1. Block diagram of a photochemical fluorometric system for liquid solution studies. [From Aaron (7), with permission.]

fixed-time-interval summing-digital voltmeter. Digital integration of the signal from the photolysed sample was performed with a Heath universal digital instrument, using a fixed integration time of about 30 s. Integration of the background signal was accomplished under the same conditions. This digital integration procedure is convenient for measurement of a small signal on top of a large background, because a small difference in relatively large numerical values can be evaluated with good precision. The relative standard deviation of integrated signals was found to be about 0.3%. The limits of detection were in the nanogram range.

Arakawa et al. (18) described an apparatus for photoreaction and fluorescence measurements in liquid solution. The equipment is shown in Fig. 3.2. A quartz reaction cuvette (Fig. 3.2C) containing the sample was surrounded by a cuvette holder (Fig. 3.2D). After being heated at 70 °C, the solution was irradiated by two low-pressure mercury lamps (Fig. 3.2B). The fluorescence intensity of the sample was measured at fixed irradiation times ranging between 0 and 40 min, with a Shimadzu spectrofluorometer.

In our laboratory, we investigated, in several studies (12,19,20), the use of a photochemical fluorescence method for liquid solutions. The sample was placed in a cylindrical quartz tube and irradiated for a fixed time on an optical

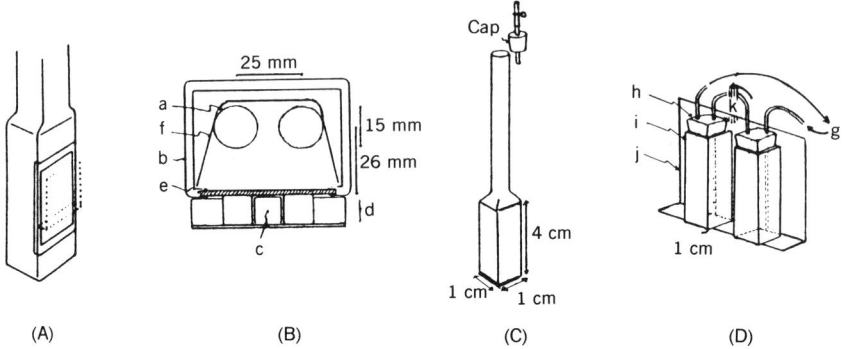

Fig. 3.2. Apparatus for photochemical fluorometric studies in liquid solution. (A) Lamp case; dashed lines show the position of the cuvette holder. (B) Cross section of the irradiation apparatus from above: (a) lamp; (b) lamp case; (c) reaction cuvette; (d) cuvette holder; (e) UV transparent filter; (f) stainless steel reflecting plate. (C) Quartz reaction cuvette. (D) Controlled-temperature cuvette holder: (g) water circulator; (h) silicone rubber; (i) spectrofluorometer quartz cuvettes; (j) aluminum plate; (k) position for reaction cuvette. [From Arakawa et al. (18), with permission.]

bench at about 60 cm from a 200-W Osram mercury arc lamp. The fluorescence intensity of the sample was determined with a Turner filter fluorometer.

Pastore et al. (21) studied a photochemical fluorometric method of analysis. UV irradiation of liquid solutions was performed either in the fluorescence attachment of a Zeiss spectrophotometer or in a Rayonet photochemical reactor operating with 12 24-W lamps (λ_{max} at 350 nm), using quartz test tubes. As far as the sensitivity is concerned, the best performances were obtained with 4- to 30-min irradiation times. Limits of detection were found to be in the 2–10 ng range for most compounds.

Procopio et al. (22) also reported a photochemical fluorometric technique for liquid solutions. The photochemical reaction was performed by irradiation for 20 min with a 300-W Osram high-pressure mercury arc lamp. Fluorescence measurements were done with a Shimadzu spectrofluorometer. A limit of detection of $0.3 \, \text{ng} \cdot \text{mL}^{-1}$ was found by the authors.

3.3.2. Photochemical Fluorescence Equipment for TLC and Solid Surfaces Studies

Photochemical fluorescence analysis has also been applied to TLC and to various solid surfaces during the last decade, using a combination of commercial instruments and newly designed sample holders.

Aaron and Fidanza (13,14) proposed a technique for evaluating the fluorescence enhancement of samples that were photolysed on a silica-gel TLC plate. Following TLC separation, the silica-gel plates were placed on a Camag TLC scanner and irradiated for a fixed time, using a 200-W Osram mercury arc lamp. The image of the mercury arc was focused on the sample spots. The fluorescence intensities of the spots were then recorded as a function of time by automatic scanning of the TLC plates, using a Turner filter fluorometer. The entire procedure including development of the TLC plates required about 2 h, but the photochemical fluorometric measurement itself took less than 10 min. Fives samples could be examined successively on the same TLC plate. Detection limits ranged between 4 and 50 ng.

The same photochemical fluorescence method was applied by Fidanza and Aaron (23,24) to solid surfaces such as filter paper, silica gel, and aluminum oxide plates. No TLC separation was required in this procedure. The authors designed and constructed a new aluminum single-sample holder (Fig. 3.3), which was inserted in the standard fluorescence sample compartment of a Turner 111 fluorometer (23). The holder was used to hold 0.6-cm-diameter filter paper disks or 0.5-cm^2 squares of plastic-backed silica gel and aluminum oxide thin layers. The samples were placed under the cover plate of the sample holder and held in place by two screws. Aaron and Fidanza (24) developed also a multiple-sample holder (Fig. 3.4). The new device was made of aluminum and was blackened to avoid scattering of radiations. It could hold

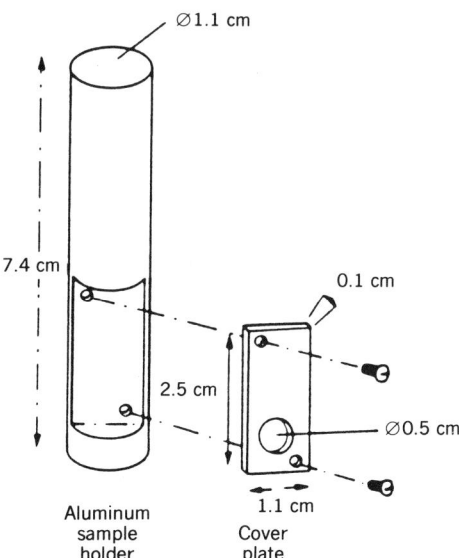

Fig. 3.3. Sample holder for photochemical fluorometric studies on solid surfaces. [From Fidanza and Aaron (23), with permission.]

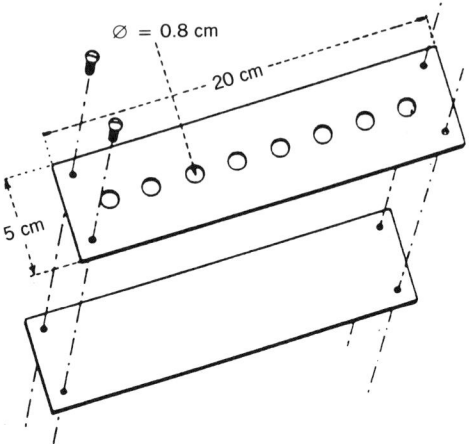

Fig. 3.4. Multiple-sample holder for photochemical fluorometric studies on solid surfaces. [From Aaron and Fidanza (24), with permission.]

simultaneously eight samples spotted onto 40-cm^2 filter paper bands or 30-cm^2 plastic-backed silica gel and aluminum oxide thin layers. The paper or plastic-backed bands were placed under the cover plate of the multiple-sample bar and held in place by four screws. The multiple-sample bar was positioned on the Camag model 110–710 automatic TLC scanner of a Turner 111 filter fluorometer. It was moved automatically in front of the Turner optical window, and the fluorescence intensities of the eight samples were successively measured. The authors found that the total sampling time was significantly shorter using the multiple-sample holder (about 20 min for eight samples) than using the single-sample holder (about 48 min for eight samples) (24).

3.3.3. Photochemical Fluorescence Equipment for HPLC Studies

Photochemical reactions have been employed extensively as postcolumn derivatization devices in HPLC. In most cases, dramatic improvements in both sensitivity and selectivity of the fluorescence detection have been observed. The HPLC-photochemical fluorescence systems are constituted generally by three separate functional units—an HPLC column, a photochemical reactor, and a fluorometer—which are connected in series. A general schematic diagram is shown in Fig. 3.5. In this section, HPLC instrumentation and chromatographic conditions will not be discussed because they would be beyond the scope of this chapter. Several recent reviews detailed the various HPLC equipment and conditions utilized for postcolumn photoderivatization

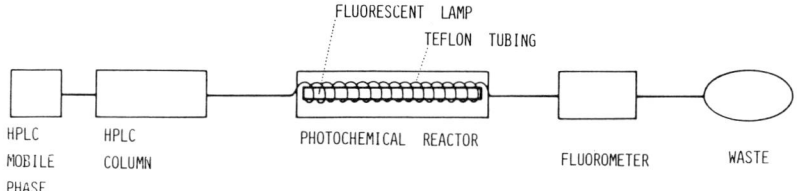

Fig. 3.5. Block diagram of an HPLC photochemical fluorescence system.

(8,17,25–28). These papers are recommended to interested readers. The various photochemical reactors described recently in the literature will be discussed in the following subsection. A survey will also be given of the main fluorometers and spectrophotofluorometers used in combination with these photochemical reactors.

3.3.3.1. Photochemical Reactors

The photochemical reactors constitute an essential part of the photochemical fluorescence detection systems. A broad variety of designs were constructed.

In the first photochemical reactors to be proposed, medium- or high-pressure arc lamps, such as a mercury source with water cooling (29) and xenon and xenon–mercury sources with different types of air cooling (30–33), were used. In all cases, quartz capillaries were utilized as reaction coils to produce good transparency to the incident UV light. The use of quartz capillaries seemed to be required since most of the analytes investigated so far presented absorption maxima well below 300 nm. An example of photochemical reactor is shown in Fig. 3.6 (32). Scholten et al. (32) used a 200-W Xe–Hg lamp as the light source. A vacuum cleaner sucked air through the reactor, which provided reproducible cooling, and an aluminum reflector shield was placed around the lamp and capillaries in order to increase the intensity of the reflected radiations of the source. Brown et al. (33) proposed a photochemical reactor made of a quartz capillary coil interposed between two mercury UV lamps, which were placed in an air-cooled housing. Each mercury lamp could be separately powered, enabling the experimenter to vary illumination intensity.

However, quartz capillaries are known to be fragile and expensive, and they are not readily available in various geometries. Therefore, the use of poly(tetrafluoroethylene) (PTFE, or Teflon) coils was proposed as a substitute for quartz capillaries by Scholten et al. (34). PTFE capillaries were found to transmit light by an internal multiple reflectance mechanism in which the radiations pass through the pores of the polymer. The surprisingly good transmission of

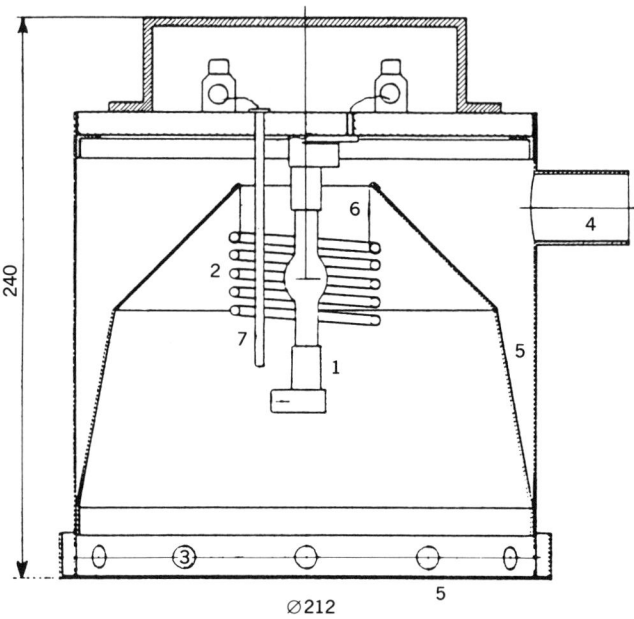

Fig. 3.6. Photochemical reactor for HPLC photochemical fluorometric studies: (1) Hanovia 200-W Xe–Hg high-pressure lamp; (2) quartz capillary, 0.5 mm i.d.; (3) inlet for air cooling; (4) outlet for air cooling; (5) reflector shield; (6) clip; (7) glass insulator (dimensions in millimeters). [From Scholten et al. (32), with permission.]

light below 300 nm made the PTFE capillaries an excellent material for photochemical reactors. Moreover, PTFE tubing is inexpensive and readily available in a variety of internal diameters (between 0.3 and 1.0 mm i.d.) Also, it was found that improved chromatographic peaks with better symmetry and less tailing were obtained with PTFE coils than with quartz capillaries, and the intensities of fluorescence signals were comparable with both materials (34). Lefevere et al. (35) used reaction coils made of PTFE tubing (0.16 cm o.d. and 0.42 and 1.07 mm i.d.) in another photochemical reactor design, this one equipped with a 200-W Xe–Hg high-pressure lamp (Fig. 3.7). The PTFE reactors coils had a varying length to adjust the reaction time and were cooled by a continuous flow of water (Fig. 3.7: 2,7) through the jackets formed by two quartz (Fig. 3.7: 6) and one Pyrex cylinders. Lang et al. (36) proposed a photochemical reactor in which PTFE tubing used as an irradiation coil was wrapped around the quartz well of a UV light source assembly (Fig. 3.8); the irradiation system was maintained at 25 °C with circulating water. Shih and Carr (37) studied the performances of a postcolumn photochemical reactor made of a medium-pressure 175-W mercury lamp placed close to coiled Teflon

INSTRUMENTATION 97

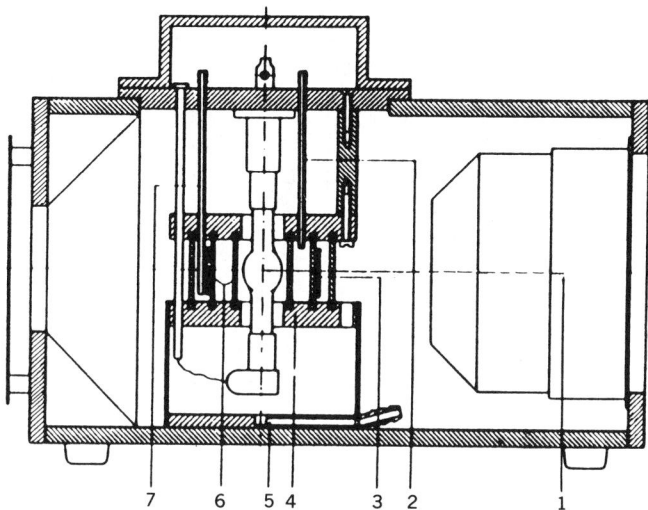

Fig. 3.7. Photochemical reactor: (1) fan; (2) outlet for circulating water; (3) PTFE capillary; (4) filter compartment; (5) additional cooling via pressurized air; (6) quartz cylinders; (7) liquid inlet. The light source is a 200-W Hg-Xe high-pressure lamp. [From Lefevere et al. (35), with permission.]

tubing and cooled with a fan (Fig. 3.9). This instrument was applied to the determination of traces of metals ions as organic complexes at the 10^{-8} M level. De Ruiter et al. (38) used postcolumn photolysis in order to decompose (to dansyl-OH) the dansyl derivatives of several phenolic derivatives. These authors utilized a fan-cooled 90-W mercury lamp and a PTFE reaction coil.

However, the use of Teflon components in photochemical reactors has been cautioned against because significant amounts of fluoride and hydrogen ions are released in irradiated samples when the temperature of the solution reaches about 50 °C (39). Fluoride release is a function of solution residence time in the coil. It can increase the background signal and/or quench the fluorescence of the photoproduct under study.

In order to minimize the photoreaction temperature effect and to simplify the construction of photochemical reactors, several authors have suggested the use of low-pressure mercury lamps and low-intensity fluorescent black lights (9–11,40–47). Uihlein and Schwab (40) built a simple and versatile photochemical reactor made of knitted PTFE capillaries placed around bars close to a low-pressure 15-W bar-shaped mercury lamp (Fig. 3.10). The type of capillary arrangement was investigated to minimize peak broadening. No cooling device was necessary for this reactor. Since the fluorescence yield of the photoproduct as a function of the total irradiation energy goes through a

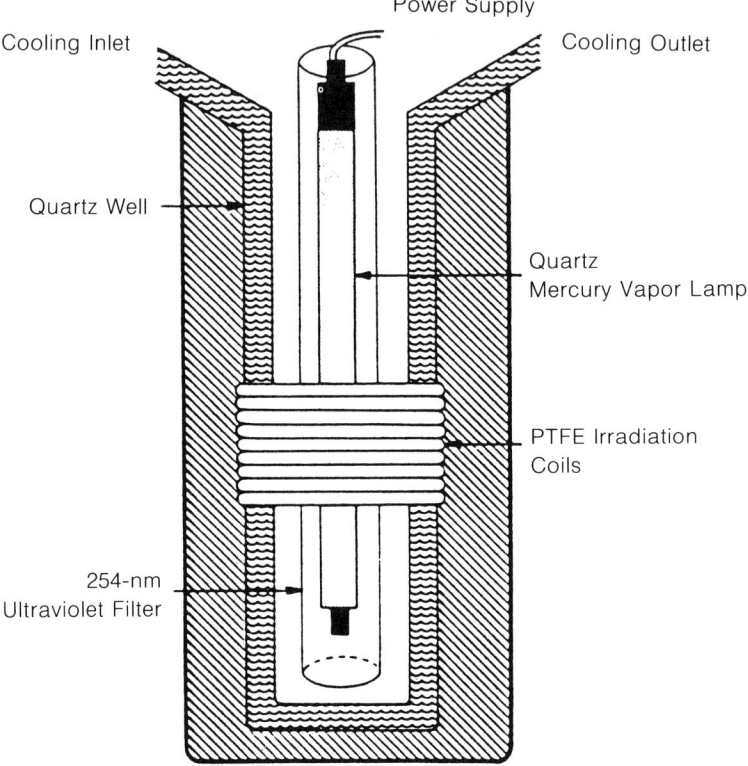

Fig. 3.8. Photochemical reactor with PTFE irradiation coils. [From Lang et al. (36), with permission.]

maximum that is substance dependent, the system was optimized by changing the intensity of UV irradiation according to the compound under investigation. The simplest way to control the intensity of irradiation was to adjust with a sledge the length of the UV source descending into the irradiation chamber (Fig. 3.10). Improved specificity and sensitivity were observed with this design.

Gandelman et al. (9–11) developed a photoreduction fluorescence (PRF) detector system based on the generation of fluorescent products from alcohols, aldehydes, ethers, amines, glycosides, and saccharides via a photoreduction reaction. The authors investigated several types of photochemical reactors. In all designs, PTFE tubing was knitted into a cylindrical shape and slid over a Pyrex (or quartz) sleeve which was then placed over a 8-W fluorescent black lamp (Fig. 3.11). In most cases, the photoreactors were placed within a nitrogen-flushed plexiglass housing to avoid any oxygen diffusion through the

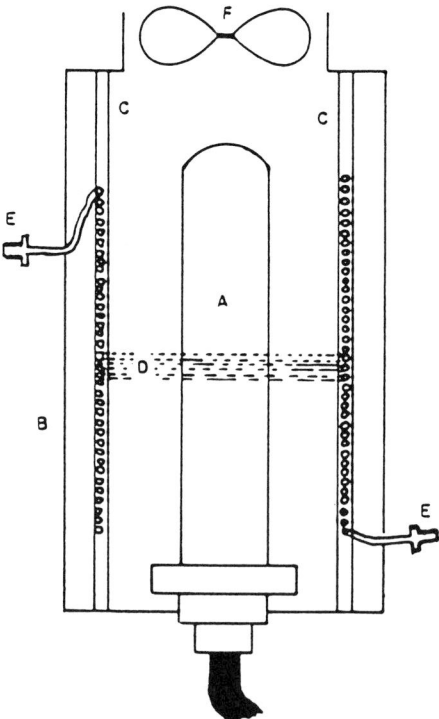

Fig. 3.9. Photochemical reactor: (A) medium-pressure mercury lamp; (B) lamp housing; (C) Teflon rod (coil holder); (D) coiled Teflon tubing; (E) Altex to Swagelok connectors; (F) cooling fan. [From Shih and Carr (37), with permission.]

PTFE capillary. The PFR technique was found to provide low detection limits (in the nanogram range) for compounds exhibiting extremely low UV–visible absorptivities and no native fluorescence.

Arakawa et al. (41) designed a photochemical reactor consisting of a temperature-controlled box and a Teflon reaction tube coiled around a 4-W low-pressure mercury lamp covered with a quartz cylindrical tube (Fig. 3.12).

Poulsen et al. (42) described the construction of crocheted PTFE cylindrical photochemical reactors that were suitable for use with low-intensity fluorescent (or "Pen Ray") lamps. No solid support for mounting on the lamp was needed, and the crocheted reactor was merely slipped over the light source. Band broadening of chromatographic peaks was reduced by the use of tightly crocheted narrow tubing and fittings modified to reduce dead volume.

Poulsen and Birks (43) designed a photochemical reactor prepared from PFTE tubing, plumbed and crocheted by the aforementioned method of

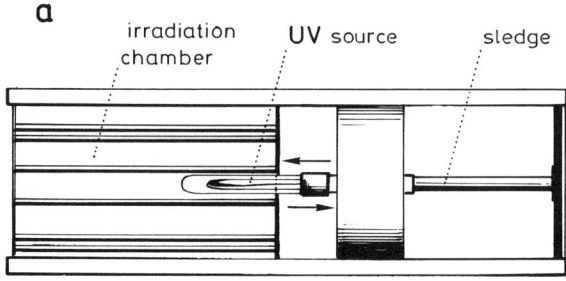

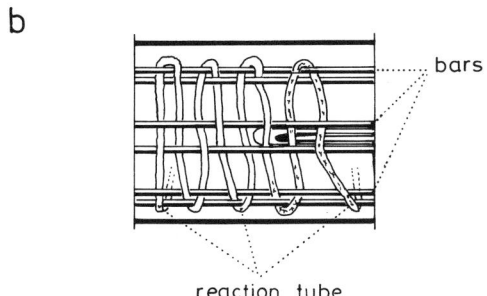

Fig. 3.10. Photochemical reactor. (a) Reactor without reaction tube; from left, irradiation chamber, source housing with sledge, and UV source. (b) Irradiation chamber with reaction tube woven around the bars (detail). [Redrawn from Uihlein and Schwab (40).]

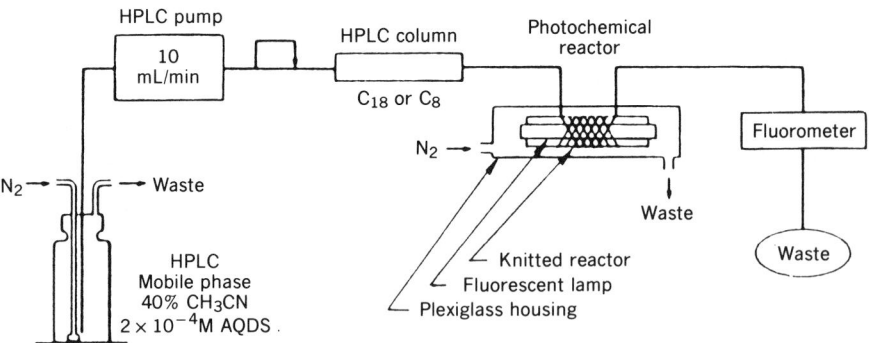

Fig. 3.11. Diagram of the photoreduction fluorescence detection system. [From Gandelman et al. (10), with permission.]

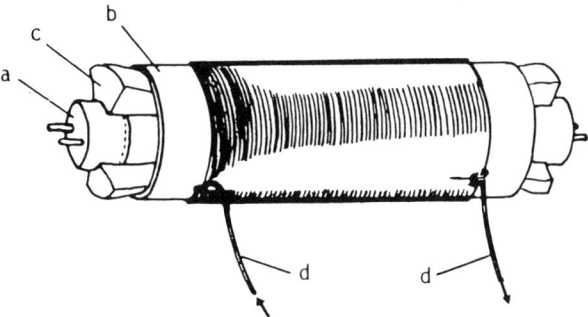

Fig. 3.12. Teflon reaction tube coil with an irradiation lamp: (a) 100-V 4-W low-pressure mercury lamp coated with inorganic fluorescent compound (emission maximum, 312 nm) inside; (b) clear quartz tube (26-mm i.d., 30-mm o.d., and 100-mm length); (c) silicone rubbers for fixing a lamp in the center of the quartz tube; (d) Teflon reaction tube (0.5-mm i.d., 1.0-mm o.d., and 20-m length). [From Arakawa et al. (41), with permission.]

Poulsen et al. (42). It was applied to the photoreduction fluorescence detection of quinones. A water-filled housing allowed the temperature of the photoreactor to be regulated by an external circulation pump, maintaining an anaerobic environment.

A 254-nm pencil lamp was supported by a cylindrical quartz tube (Fig. 3.13). Circulating water was flushed by nitrogen inside the reactor housing. Band broadening and analysis time were minimized with this photoreactor. Detection limits were in the low-picogram range, with good selectivity.

Lutchefeld (44) used postcolumn photolysis of phenylurea herbicides to generate methylamine, which reacted with o-phthalaldehyde–mercaptoethanol to produce a fluorophore. The photochemical reactor consisted of a knitted coil of PTFE capillary tubing fitted over a UV lamp. The selectivity of fluorescence detection was markedly improved relative to UV detection.

Miles and Moye (45) constructed a postcolumn photolytic reactor similar to that described by Lutchefeld (44) and used it for the fluorometric determination of several classes of pesticides and herbicides. A jacketed UV lamp was inserted in the center of a cylindrical, woven Teflon coil. The UV lamp–Teflon tube assembly was located in an electrical PVC pipe, and holes were drilled in the bottom of the PVC pipe to allow cool air to enter the photoreactor. Aluminum foil was placed inside the PVC pipe to increase photolysis efficiency.

Jansen et al. (46) studied a photochemical reaction detection system coupled with a 320-μm-i.d. packed fused-silica HPLC analytical column. The authors compared the efficiency, on fluorescence detection, of in-column and

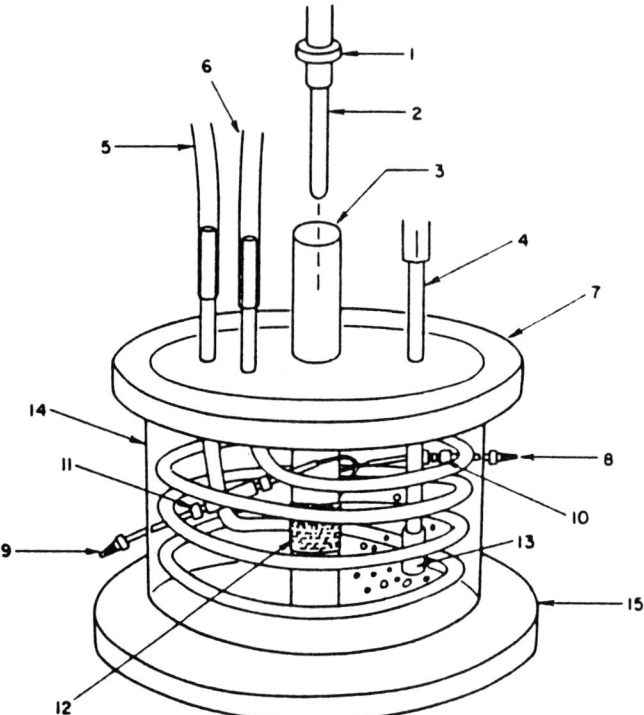

Fig. 3.13. Diagram of the photoreduction fluorescence photoreactor housing: (1) rubber ring for positioning the lamp within the quartz tube; (2) 254-nm pencil lamp; (3) quartz tube; (4) nitrogen inlet; (5) coolant inlet; (6) coolant outlet; (7) top of housing; (8) photoreactor inlet from HPLC column; (9) photoreactor outlet to fluorometer; (10) adapter stainless steel to PTFE capillary tubing; (11) adapter PTFE to stainless steel capillary; (12) crocheted PTFE photoreactor; (13) gas dispersion tube (Pyrex); (14) body of the (water-filled) photoreactor housing; (15) bottom of housing. [From Poulsen and Birks (43), with permission.]

postcolumn UV irradiation that was carried out through windows made on the column, using a Philips mercury lamp. The in-column photochemical reaction was found to be more rapid and to give a 10-fold increase in signal-to-noise ratio, as compared to the postcolumn photochemical reaction.

Kikuta and Schmid (47) developed a new on-line photochemical reactor, consisting of a 30-cm-long, 8-W low-pressure mercury lamp built in a small plastic container. The reaction capillary was made of Teflon tubing crocheted according to the procedure of Poulsen et al. (42) and fixed on a stainless steel net arranged very close in a half circle around the light tube.

3.3.3.2. *Fluorometers and Spectrophotofluorometers*

Various kinds of fluorometers and spectrophotofluorometers have been utilized in combination with photochemical reactors to detect the fluorescence of photoproducts and/or record their spectra. In most cases, sinple models such as Schoeffel & Kratos FS 970 (9–11,36,37,43,45), Kontron 100 (10), Varian Fluorichrom (33), Merck-Hitachi (47), and Perkin-Elmer 650-10 LC (37) fluorometers were used. Generally, the excitation wavelength was set at a fixed value and a filter was placed on the emission side.

Some authors have determined the fluorescence spectra of photoproducts or carried out quantitative measurements with more sophisticated instruments, such as the Perkin-Elmer 204A, 3000 (32,34,35,40,48–50), or MPF44B (45) spectrophotofluorometers. Arakawa et al. (41) utilized a Shimadzu RF500LCA spectrophotofluorometer.

3.3.4. Photochemical Fluorescence Equipment for Flow Analysis Studies

Photochemical fluorescence analysis was applied also to a flow of solvent. Tsuchiya et al. (16,51) designed and constructed a photochemical fluorometric system for flow analysis studies (Fig. 3.14). It consisted of a separatory funnel reservoir (containing the solvent or sample solution) connected via transparent PTFE tubing to a fluorescence flow cell. Between the separatory funnel and the flow cell, the PTFE (about 1 m) tubing was coiled around inside a housing containing a 200-W xenon–mercury arc lamp. A fan blower assembly was used to cool the PTFE tubing, and aluminum foil reflectors were wrapped

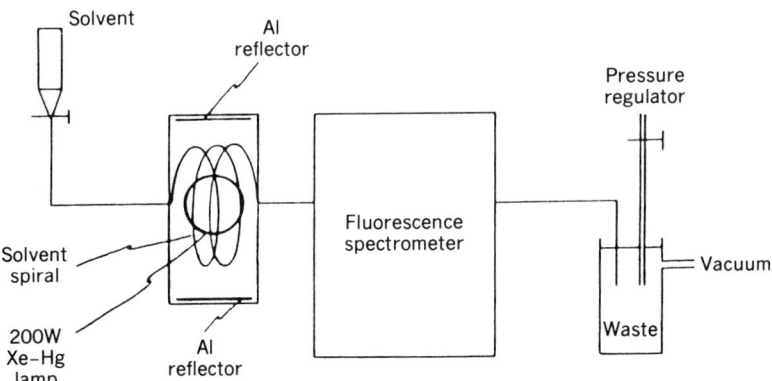

Fig. 3.14. Schematic diagram of photochemical fluorometric flow cell system. [From Tsuchiya et al. (16), with permission.]

inside the arc lamp housing in order to increase the intensity of the reflected radiation of the lamp. The fluorescence flow cell was placed in an Amino-Bowman spectrophotofluorometer. This method was found to be rapid (about 5 min per measurement), and particularly suitable for the determination of primaquine, an antimalarial drug. The photochemical fluorescence detector proposed by Tsuchiya et al. (16) should also be directly interfaceable to FIA systems.

An FIA device was recently developed by Chen et al. (52), using photochemical spectrofluorometric determination of phenothiazine compounds by two unsegmented-flow methods. The first system consisted of a Gibson peristaltic pump, a Rheodyne rotary injection valve, a reaction coil (254-nm UV lamp, 50 Hz, 200 V) and a fluorescence flow cell of 1.1 mm path length; fluorescence measurements were performed with a Kontron spectrofluorometer after irradiation of the injected sample in the reaction coil. In the second system, a laboratory-built timer was used to control the pump and the injection valve in the stopped-flow mode; the injected sample was propelled along the photochemical reactor, then halted and irradiated directly in the flow cell with the excitation radiation.

3.4. PHOTOREACTIONS IN PHOTOCHEMICAL FLUORESCENCE ANALYSIS

Photochemistry provides many types of photoreactions that might be used for photochemical derivatization and enhancement of fluorescence detection. However, to be of interest from an analytical standpoint, the photoreactions must fulfill several important requirements:

- Reagents (analyte molecules) should strongly absorb radiations in the UV–visible region to initiate the photochemical reaction.
- These radiations should be of wavelengths not significantly absorbed by photoproducts.
- Photoproducts (derivatives formed) should present increased rigidity and/or aromaticity, resulting in an enhanced absorption coefficient and a higher fluorescence quantum yield relative to those of the analyte.
- Photoproducts should be chemically and thermally stable on the time scale needed for analytical measurements.
- Photoconversion process should be very efficient, with. high photochemical quantum yields.

In this section, we shall examine the characteristics of the main photochemical processes that have been used in photochemical fluorometry. Photocyclization, photoisomerization, photolysis, photooxidation, and photosensitized photoreduction appear to be the most promising candidates in this respect.

3.4.1. Photocyclization Reactions

Photocyclization reactions have been shown to be useful for converting stilbene-based molecules into highly fluorescent phenanthrene derivatives by UV irradiation (8,33,47,50,51). Harman et al. (50) found that clomiphene (Scheme I), which was previously not fluorescent, produced upon UV irradiation a strongly fluorescent photoproduct with an emission maximum at 367 nm when excited at 257 nm. The authors postulated a ring closure with formation of a phenanthrene derivative. This is in agreement with the conversion of stilbene and triphenylethylene to phenanthrene derivatives by UV irradiation (53).

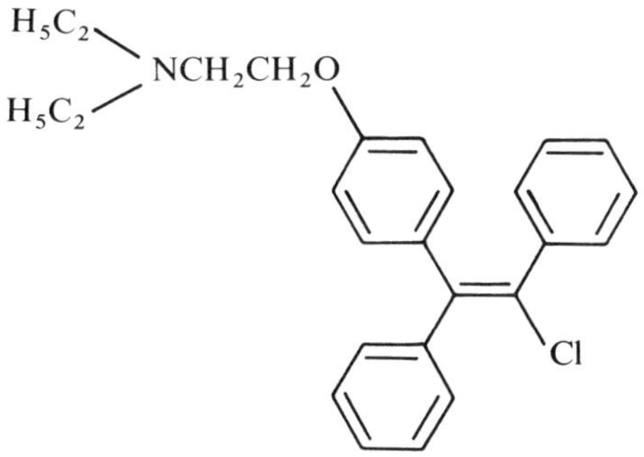

Scheme I

Several researchers have reported HPLC methods based upon the property of tamoxifen and some of its metabolites to react under UV irradiation to form fluorescent phenanthrene derivatives (Scheme II) (33,47,54–57). Kikuta and Schmid (47) found that the fluorescent photoproduct of tamoxifen formed upon irradiation was characterized by excitation and emission maxima at 256 and 380 nm, respectively. Irradiation times for on-line photochemical reactions were short, ranging from 12 s for tamoxifen and 4-hydroxytamoxifen to about 23 s for N-desmethyltamoxifen (47).

Rhys-Williams et al. (49) proposed an HPLC method based on the photocyclization of diethylstilbestrol into a fluorescent phenanthrene compound (Scheme III), with excitation and emission maxima at 280 and 390 nm, respectively.

Another photocyclization reaction consisted of conversion of nonfluorescent demoxepam (1) into a highly fluorescent photoproduct, which presented

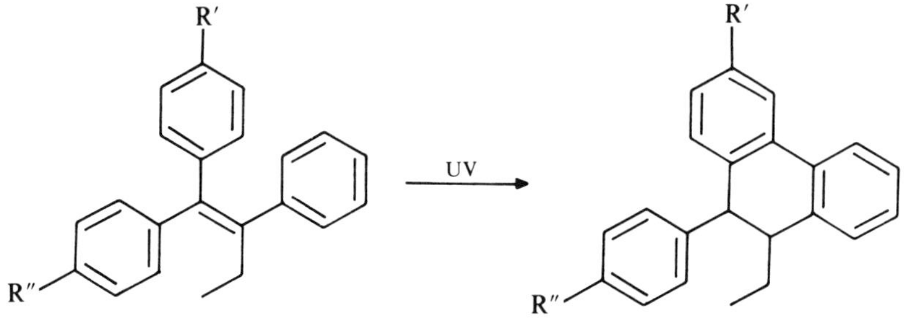

Scheme II

Tamoxifen: R' = OCH$_2$CH$_2$N(CH$_3$)$_2$, R'' = H
N-Desmethyltamoxifen: R' = OCH$_2$CH$_2$NHCH$_3$, R'' = H
4-Hydroxytamoxifen: R' = OCH$_2$CH$_2$N(CH$_3$)$_2$, R'' = OH

Scheme II

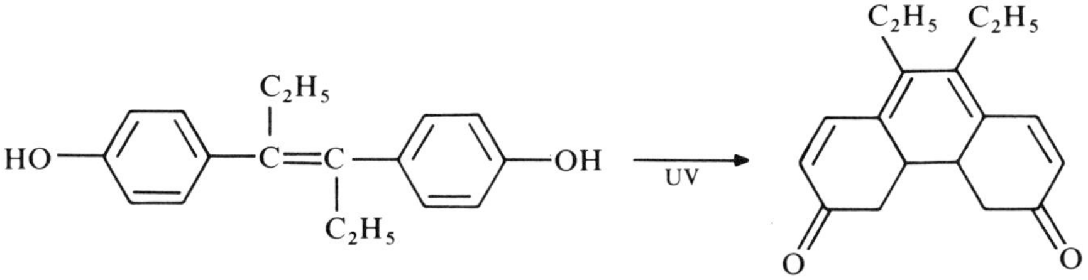

Scheme III

structure (**2**) given in Scheme IV, according to Strojny and de Silva (58). Moreover, Brinkman et al. (48) found that the quinazolone (**2**) in 0.1 M sodium hydroxide solution presented excitation and emission maxima at 380 and 460 nm, respectively, whereas the photoproduct obtained after a 1-min irradiation of compound **1** in a pH 8 buffer solution exhibited excitation and emission maxima at 340 and 410 nm, respectively.

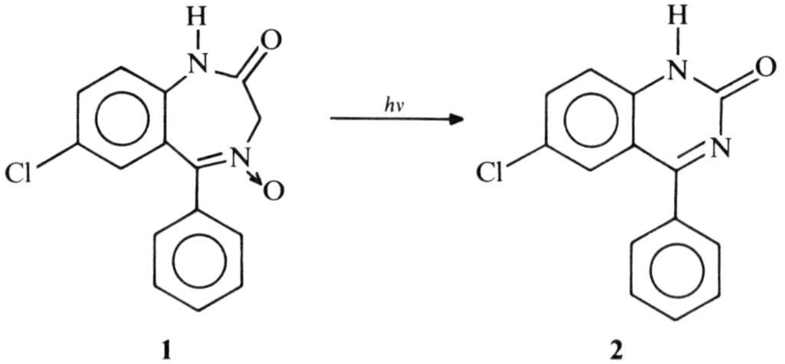

Scheme IV

This difference in fluorescence spectra indicates that the photoproduct has a structure not identical with that of compound **2**. Brinkman et al. (48) also obtained results from gas chromatographic–mass spectrometric (GC–MS) measurements that showed the absence of a chlorine atom in the main photoproduct and suggested the presence of a hydroxy group in the 2 position.

The photocyclization reaction of diphenylamine to fluorescent carbazole was found to be analytically useful in photochemical fluorometry by Aaron et al. (59). The proposed reaction mechanism is given in Scheme V.

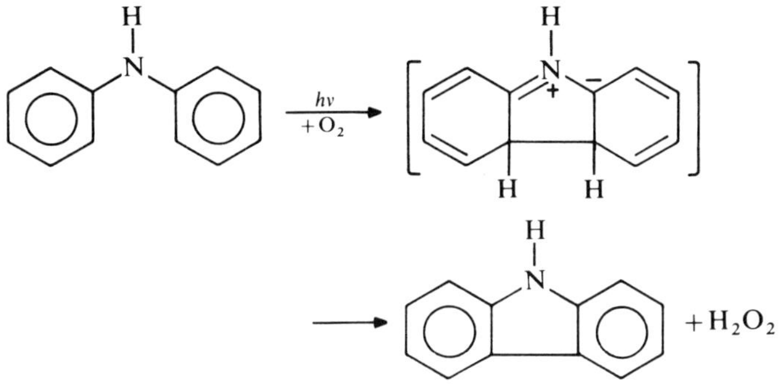

Scheme V

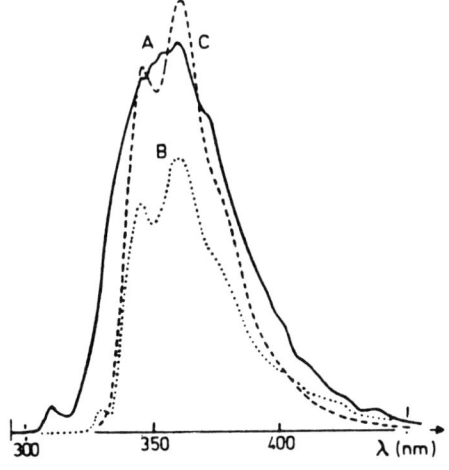

Fig. 3.15. Fluorescence emission spectra of 5×10^{-5} M nonirradiated diphenylamine (curve A, $\lambda_{ex} = 306$ nm), irradiated diphenylamine (curve B, $t_{irr} = 5$ min; $\lambda_{ex} = 337$ nm) and 5×10^{-5} M carbazole (curve C, $\lambda_{ex} = 338$ nm) in acetonitrile. [From Fall et al. (59), with permission.]

The formation of carbazole after an irradiation time of 10–15 min was confirmed by the fact that fluorescence excitation and emission spectra of irradiated diphenylamine solution were strictly identical to those of an authentic sample of carbazole (see Fig. 3.15).

3.4.2. Photoisomerization Reactions

A few photoisomerization reactions have been found to lead to fluorescent photoproducts (20,22). Aaron and Diop (20) suggested a photochemical fluorometric method based on the photoisomerization of purpurogallin into a naphthalene derivative (according to Scheme VI, given for the photoisomerization of tetra-O-methylpurpurogallin).

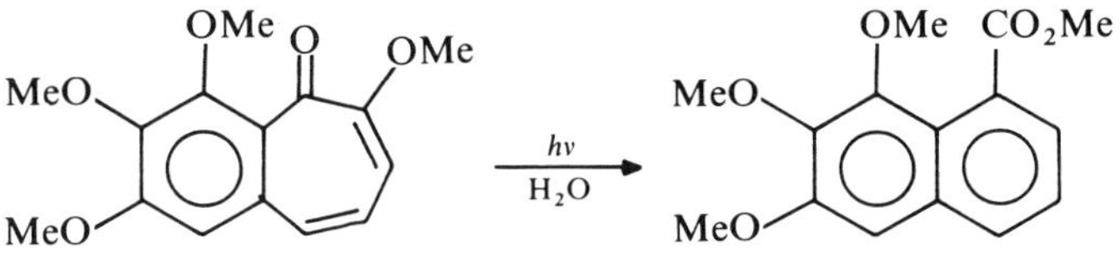

Scheme VI

Upon irradiation for 65 min of an aqueous solution of purpurogallin, the authors observed the appearance of a broad fluorescence emission band at 445 nm, with excitation at 338 nm, which was attributed to the formation of a naphthalene derivative (20).

Procopio et al. (22) found that UV irradiation of lorazepam (3) in 0.1 M sodium hydroxide solution led to the formation of a fluorescent photoproduct that they identified as the photoisomer 4 (Scheme VII). The structure of photoproduct 4 was elucidated by infrared (IR), nuclear magnetic resonance (NMR), and MS. Compound 4 exhibited a broad fluorescence emission band

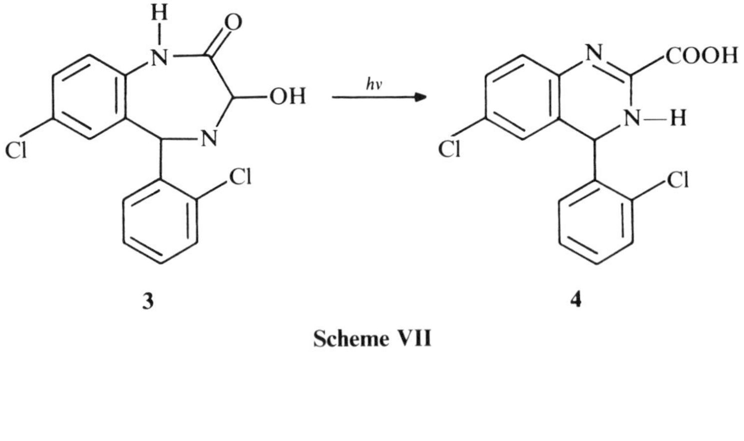

Scheme VII

with a maximum at 435 nm and an excitation spectrum with two maxima at 256 and 360 nm. During the irradiation, which lasted between 5 and 30 min, the absorption band of lorazepam occurring at 345 nm decreased while a new absorption band appeared at 368 nm.

3.4.3. Photolysis Reactions

The usefulness of various photolysis reactions in photochemical fluorometry has been studied by several researchers (18,37,38,45,60).

Arakawa et al. (18) showed that the photoinduced fluorogenic reaction for the fluorometric detection of catecholamine-O-sulfates in aqueous solution proceeded by the photolysis of the O-sulfate bond in a first step (Scheme VIII).

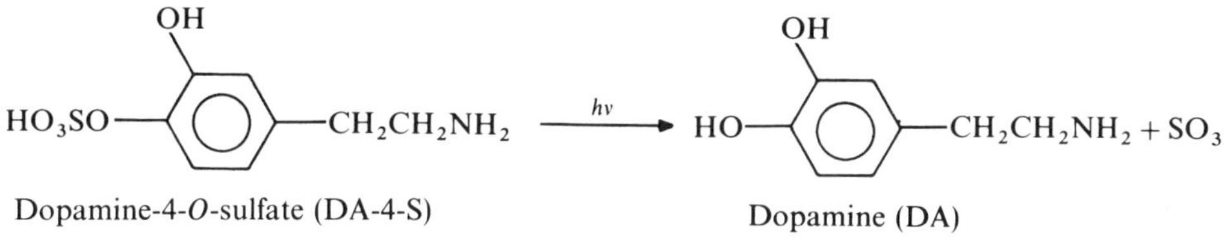

$$HO_3SO\text{—}\langle\text{ring}\rangle(OH)\text{—}CH_2CH_2NH_2 \xrightarrow{h\nu} HO\text{—}\langle\text{ring}\rangle(OH)\text{—}CH_2CH_2NH_2 + SO_3$$

Dopamine-4-O-sulfate (DA-4-S)   Dopamine (DA)

Scheme VIII

An alternative possibility would be the formation of sulfonated DA, resulting from the occurrence of the photoinduced Fries rearrangement. The second step of the reaction is the condensation of DA or sulfonated DA with 1,2-diaminoethane under aerobic conditions. Indeed, the fluorescence excitation and emission spectra of the photoreaction product were similar to those of the classic fluorogenic condensation product of the corresponding free catecholamines with 1,2-diaminoethane (18).

Shih and Carr (37) found that the photolysis of n-butyl-2-naphthylmethyl-dithiocarbamate (BNMDTC) complexes with transition metal ions [M = Zn(II), Ni(II), Cu(II), Hg(II)] led to a fluorescent photoproduct. Since the BNMDTC complexes photolysis products presented the same fluorescence spectra (excitation and emission maxima wavelengths at 221 and 330 nm, respectively) as the BNM amine, the authors postulated that the mechanism of photodecomposition was as shown in Scheme IX. UV excitation energy would be absorbed by the NCS_2 group, and would break the N—C bond to produce the fluorescent BNM amine (37). Formation of BNM amine during photolysis was confirmed by UV absorption spectra and HPLC results.

Miles and Moye (45) examined the effect of UV irradiation on the fluorescence of several classes of aromatic pesticides. Most of the compounds exhibited a fluorescence signal only when irradiated, indicating that the

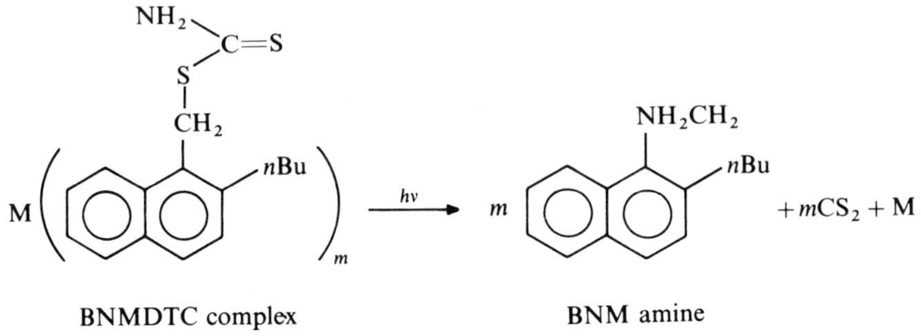

Scheme IX

(BNMDTC complex) $\xrightarrow{h\nu}$ m (BNM amine) + mCS_2 + M

fluorescence was due to the photolysis products. The phototransformation of the aromatic moieties of herbicides was probably responsible for the fluorescence observed. Aniline and substituted anilines presented fluorescence spectra that were similar to those of the photolysis products of the phenylene herbicides, suggesting that these photoproducts are amino-substituted aromatic compounds (45). However, HPLC analysis of photolysed phenylurea herbicide solutions indicated the presence of several peaks that showed fluorescence. Therefore, a mixture of photolysis products was formed (45).

Using fluorescence detection, Werkhoven-Goewie et al. (60) investigated the HPLC photochemical reaction of chlorophenols. The authors found that UV irradiation of mono- and dichlorophenols produced strong fluorescence signals with excitation and emission maxima of 271 and 290 nm, respectively, corresponding to those of phenol. Therefore, the authors concluded that photolytic dechlorination of mono- and dichlorophenols took place. In contrast, irradiation of more highly chlorinated phenols such as tri-, tetra-, and pentachlorophenols gave very little fluorescence. The latter behavior probably results from the fact that photolysis of more highly chlorinated phenols gives rise to many other photoproducts besides phenol, with a higher chlorine content, which may cause fluorescence quenching. The presence of lower chlorinated phenols in the photochemical reactor effluent was confirmed by LC analysis of the photoreaction mixture formed (60).

3.4.4. Photoreduction Reactions

Anaerobic photoreduction reactions of nonfluorescent quinones, which produce highly fluorescent dihydroquinones, can be applied either to the detection of hydrogen atom donating compounds (9–12) or to that of quinones themselves (6,35,43).

$$AQ + h\nu \longrightarrow AQ^*(\text{singlet}, S_1)$$

$$AQ^*(S_1) \longrightarrow AQ^*(\text{triplet}, T_1)$$

$$AQ^*(T_1) + (CH_3)_2CHOH \longrightarrow AQH\cdot + (CH_3)_2\dot{C}OH$$

$$(CH_3)_2\dot{C}OH + AQ \longrightarrow AQH\cdot + (CH_3)_2C{=}O$$

$$2AQH\cdot \longrightarrow AQ + AQH_2$$

Scheme X

In the case of the detection of hydrogen atom donors, Gandelman and Birks (9) used the photoreduction of anthraquinone-2,6-disulfonate (AQ). These authors suggested the reaction mechanism shown in Scheme X for the sensitized photoreduction of AQ by 2-propanol. In the first step, the lowest excited AQ singlet state (S_1) is produced by absorption of UV light. In the second step, the molecule undergoes intersystem crossing from S_1 (n, π^*) state to the lowest triplet state (T_1, n, π^*). The photochemical reaction itself begins with the abstraction of an hydrogen atom of 2-propanol by the AQ $T_1(n, \pi^*)$ state (step 3). This hydrogen abstraction is the rate-limiting step for the photoreduction of AQ. A subsequent thermal disproportionation reaction between the semiquinone radicals (AQH·) leads to the highly fluorescent 9,10-dihydroxyanthracene (AQH$_2$). Those compounds with C—H bond dissociation energies less than $400 \, kJ \cdot mol^{-1}$ can photoreduce AQ and thus be detected. This photoreduction reaction that is photosensitized by AQ is particularly useful in photochemical fluorometry, since it allows one to detect hydrogen-atom-donating compounds having extremely low UV–visible absorption coefficients. Gandelman and Birks (9) applied this photosensitized photoreduction of AQ to the determination of aliphatic alcohols, ethers, and aldehydes in 80:20 v/v acetonitrile/water mixtures. Cardiac glycosides, saccharides, and hydrocortisone were also analyzed by this HPLC-photoreduction-fluorescence method, using AQ (10) or 2-t-Bu-AQ (11) in the same photoreduction pathway. Traore and Aaron (12) proposed a similar mechanism for the photoreduction of AQ by nonfluorescent dinitroaniline derivative herbicides (Scheme XI). The photosensitized fluorescence signal was found to be proportional to the herbicide concentration (12).

Nonfluorescent quinone derivatives were also determined using this photoreduction reaction. Aaron et al. (6) found that the UV irradiation of 2-methyl-3-phytyl-1,4-naphthoquinone (vitamin K_1) in several bulk organic solutions (dioxane, 2-propanol, ethanol, cyclohexane) yielded fluorescent photoproducts. By applying the same photoreduction reaction to HPLC, Lefevere et al. (35) quantitated vitamin K_1 via the formation of the corresponding naphthohydroquinone. Photoreduction reaction conditions were optimized by adding small

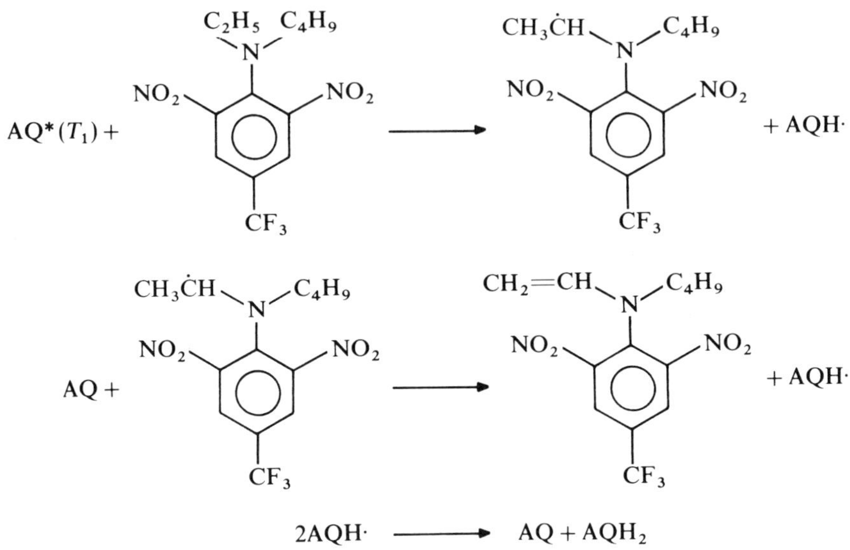

Scheme XI

AQ*(T₁) + [C₂H₅,C₄H₉-N-substituted 2,6-dinitro-4-trifluoromethyl benzene] ⟶ [CH₃ĊH,C₄H₉-N-substituted 2,6-dinitro-4-trifluoromethyl benzene] + AQH·

AQ + [CH₃ĊH,C₄H₉-N-substituted 2,6-dinitro-4-trifluoromethyl benzene] ⟶ [CH₂=CH,C₄H₉-N-substituted 2,6-dinitro-4-trifluoromethyl benzene] + AQH·

2AQH· ⟶ AQ + AQH₂

amounts of ascorbic acid to the methanolic solution used as eluant, which produced intense fluorescence (35). This increase of fluorescence could be explained by the fact that the formation of only two highly fluorescent photoproducts (Fig. 3.16, compounds I and II) occurred in the presence of ascorbic acid among the variety of reaction products and intermediates that may be formed in this complex photochemical reaction scheme (Fig. 3.16).

Poulsen and Birks (43) utilized strictly anaerobic conditions in order to photoreduce a large number of quinones into corresponding dihydroquinones via hydrogen abstraction.

3.4.5. Photooxidation Reactions

White et al. (61) examined the use of a photochemical oxidation reaction for enhancing the fluorescence intensities of several phenothiazines (in aqueous solutions adjusted to different pH'_R). In all cases, decay of native fluorescence began immediately upon UV irradiation whereas fluorescence ingrowth of photoproducts occurred within 30–60 s. But the fluorescent properties of the photoproducts depended largely on the nature of the substituent at the 2-position of the phenothiazine ring (61). The fluorescence emission wavelength maxima of the photooxidation products of the trifluoromethyl-substituted compounds (at 402 and 408 nm, respectively, for fluphenazine and trifluoperazine) were similar to those measured for the chemically oxidized compounds.

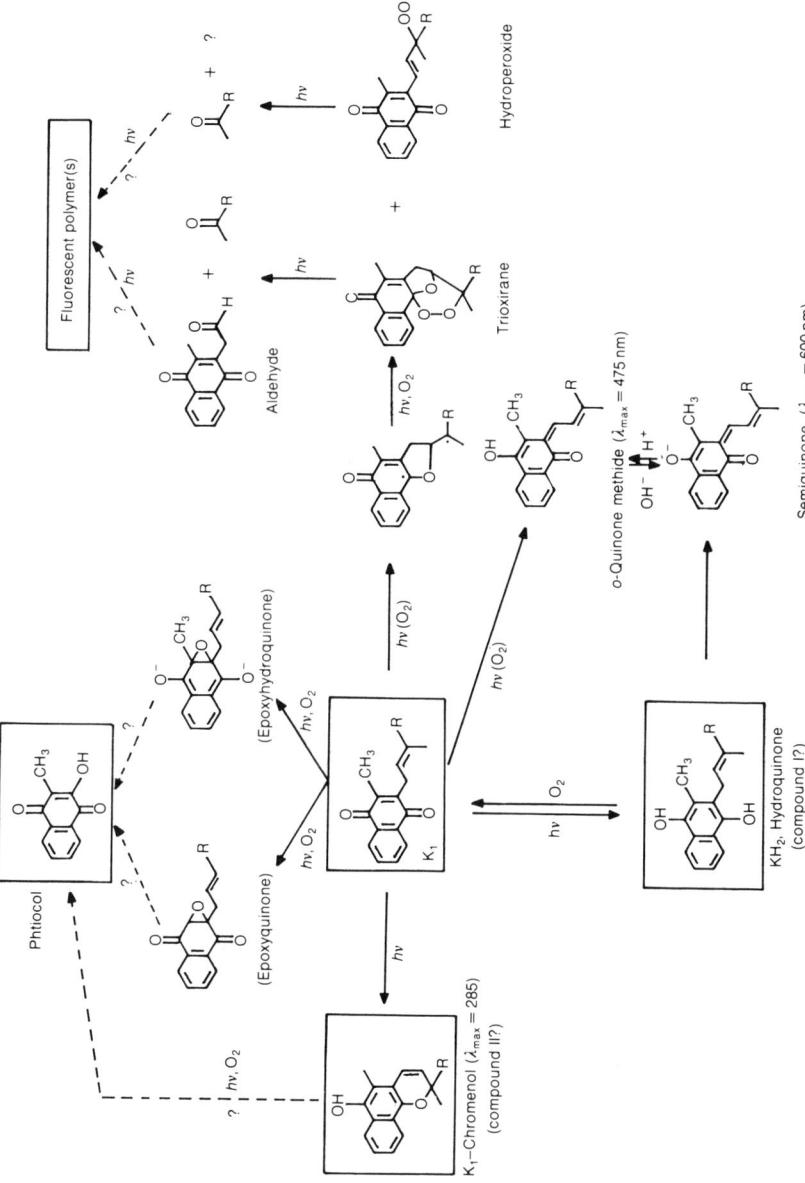

Fig. 3.16. Proposed photochemical reactions scheme of vitamin K_1. [From Lefevere et al. (35), with permission.]

In contrast, 2-chloro-substituted phenothiazines such as perphenazine and chlorpromazine yielded upon UV irradiation a fluorescence emission band at 469 nm, not previously reported, when peroxide or permanganate oxidation was performed. Therefore, apparently photooxidation led to photoproducts different from those obtained by chemical oxidation of the latter compounds.

Gandelman and Birks (62) also utilized a sensitized photooxygenation reaction for detecting hydrogen-atom-donating compounds, such as alcohols, aldehydes, and ethers, that do not absorb UV–visible radiation. In that photoreaction scheme, the authors added an anthraquinone to the HPLC solvent saturated with O_2. The photooxygenation reaction produced hydrogen peroxide, which was subsequently detected by its chemiluminescence with luminol.

3.5. APPLICATIONS

Several applications have been reported for photochemical fluorometry in various areas such as pharmaceutical analysis, clinical chemistry, pesticide residue analysis, and environmental chemistry. In this section, we shall discuss the main results and performance of several photochemical fluorescence analysis methods described over the past decade.

3.5.1. General Analytical Performances

Pastore et al. (21) investigated the effect of UV irradiation on the product of the fluorogenic reaction of several indolic acids with o-phthalaldehyde in sulfuric acid medium. The fluorescence signal was markedly increased for most compounds. Limits of detection (LOD) ranged between 2 and 110 ng·mL^{-1} (Table 3.1).

Fall et al. (59) found that the fluorescence intensity of diphenylamine was significantly enhanced upon UV irradiation, reaching a maximum within 10–15 min in several organic solvents (Fig. 3.17).

The photochemical fluorescence enhancement factors (PFEF), defined as the ratio of the relative fluorescence signal for the irradiated sample (maximum intensity at the optimal irradiation time) and of the relative fluorescence signal for the nonirradiated sample, varied between 3 and 35 according to the solvent system (Table 3.1).

Photochemically induced fluorescence of BNMDTC complexes with various metals was studied by Shih and Carr (37) in bulk methanol/water 95:5 solutions and in a flow system. Optimal irradiation times were in the 5–10 min range for bulk solutions, and in the 50–70 s range in the flow system. LOD values ranged between 0.2 and 1.2 ng·mL^{-1} in the flow system (Table 3.1).

Table 3.1. Analytical Figures of Merit of Selected Compounds, Obtained by Photochemical Fluorometry in Liquid Solution

Compound	Solvent System	Irradiation Time[a] (min)	PFEF[b]	LDR[c]	LOD[d] (ng·mL^{-1})	Absolute[e] LOD (ng)	Ref.
Diphenylamine	Acetonitrile	15	13	10^2	30	60	59
Diphenylamine	Cyclohexane	15	35	10^2	150	300	59
Diphenylamine	Ethanol	15	27	10^3	40	80	59
Diphenylamine	1-Propanol	13	7.3	10^2	100	200	59
Diphenylamine	Water/acetonitrile (50:50 v/v)	10	3.1	50	20	40	59
Indole acetic acid	H_2SO_4 (14–16 N)	30	0.44	4.3×10^2	2.3	6.9	21
Indole propionic acid	H_2SO_4 (13.8–14.3 N)	10	10.0	4.8×10^2	5.2	15.6	21
Indole pyruvic acid	H_2SO_4 (13–14 N)	4	1.4	9.1	110	330	21
Indole lactic acid	H_2SO_4 (15 N)	10	7.0	4.2×10^2	6.0	18	21
Indole butyric acid	H_2SO_4 (16 N)	30	5.0	3.1×10^2	8.0	24	21
Hydroxyindole acetic acid	H_2SO_4 (12.7–14.0 N)	30	1.0	4.3×10^2	2.3	6.9	21
Methoxyindole acetic acid	H_2SO_4 (11.4 N)	10	2.3	5×10^2	2.0	6.0	21
Indole acetic acid ethyl ester	H_2SO_4 (12.8–13.3 N)	60	2.0	42	60	180	21
Purpurogallin	Water	60–140	—	2×10^2	30	60	20
Fe(BNMDTC)$_3$[g]	Methanol/water (95:5)	5	—	2×10^2	0.2[f]	—	37
Ni(BNMDTC)$_2$[g]	Methanol/water (95:5)	5	—	2×10^2	1.2[f]	—	37
Hg(BNMDTC)$_2$[g]	Methanol/water (95:5)	5	—	2×10^2	1.2[f]	—	37
Co(BNMDTC)$_2$[g]	Methanol/water (95:5)	10	—	2×10^2	0.5[f]	—	37

[a] Irradiation time corresponding to the maximum fluorescence intensity value.
[b] PFEF = photochemical fluorescence enhancement factor (see text for definition).
[c] LDR = linear dynamic range, corresponding to the ratio of upper concentration of linearity (within 5%) and the detection limit.
[d] LOD = limit of detection, defined as the concentration of the solution giving a signal-to-noise ratio of 3.
[e] Absolute LOD, calculated for a 2-mL (21,59) or a 3-mL (21) sample volume.
[f] LOD evaluated using a HPLC postcolumn photochemical detector.
[g] BNMDTC = n-butyl-2-naphthylmethyldithiocarbamate.

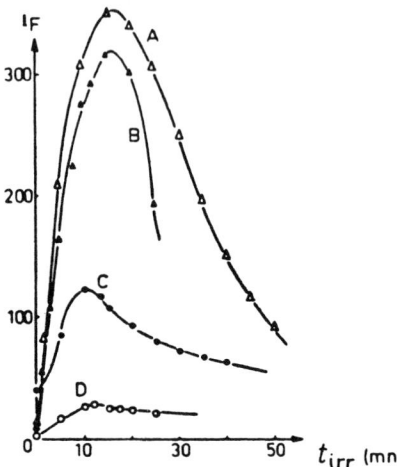

Fig. 3.17. Effect of irradiation time on the fluorescence intensity of diphenylamine (10^{-4} M) in several solvents. Curves: (A) in ethanol; (B) in propan-1-ol; (C) in water/acetonitrile (50:50); (D) in cyclohexane. [From Fall et al. (59), with permission.]

Table 3.2. Absolute Limits of Detection (LOD) of Several Aliphatic Organic Compounds Determined by Photoreduction Fluorescence in HPLC[a]

Compound	Absolute LOD[b] (ng)
Ethanol	4
2-Propanol	4
1-Butanol	5
2-Methyl-1-propanol	5
1-Hexanol	5
1-Octanol	7
1,3-Dioxane	12
Tetrahydrofuran	3
Diethylether	7
1,5-Pentanedial	31
Propanal	4
3-Methylbutanal	7

Source: From Gandelman and Birks (9).

[a] HPLC mobile phase: acetonitrile/water (80:20).
[b] The absolute LOD was determined from the equation $m_L = 3\sigma/(S/m_i)$, where 3σ is half the peak-to-peak noise; S is the signal peak height obtained from an injection of a 0.1% solution of the analyte in 80% acetonitrile; and m_i is the mass of the analyte in that sample.

Table 3.3. Absolute Limits of Detection (LOD) of Quinones and nitro-PAHs Compounds Determined by Photoreduction Fluorescence in HPLC[a]

Compound	t_r^b (min)	Absolute LOD[c] (pmol)	
		$\lambda_{ex} =$ 243.5 nm	$\lambda_{ex} =$ 257.5 nm
1,4-Naphthoquinone 2,3-epoxide	3.1	0.20	1.2
2-Methoxy-1,4-naphthoquinone	3.4	0.17	1.5
1,4-Naphthoquinone	3.6	0.20	2.9
Menadione (vitamin K_3)	4.3	0.29	2.5
2,7-Dimethyl-1,4-naphthoquinone	5.2	0.26	1.6
2,6-Dimethyl-1,4-naphthoquinone	5.2	0.25	1.5
2,3-Dimethyl-1,4-naphthoquinone	5.5	0.22	1.7
9,10-Phenanthrequinone	3.9	0.19	0.21
9,10-Anthraquinone-2-sulfonate	1.6	0.21	0.050
2-Hydroxymethyl-9,10-anthraquinone	3.9	0.32	0.077
2-Methyl-1-nitro-9,10-anthraquinone	4.9	5.9	2.1
2-Amino-9,10-anthraquinone	3.7	0.058	0.038
1-Amino-9,10-anthraquinone	5.1	1.7	0.82
9,10-Anthraquinone	6.6	0.078	0.026
1,5-Dichloro-9,10-anthraquinone	7.8	3.4	1.4
1-(3-Methoxyphenoxy)-9,10-anthraquinone	8.1	0.18	0.055
2-Methyl-9,10-anthraquinone	8.7	0.078	0.021
2-Ethyl-9,10-anthraquinone	10.6	0.010	0.027
2-Chloro-9,10-anthraquinone	11.1	0.55	0.10
1,4-Dimethyl-9,10-anthraquinone	13.8	0.77	0.19
2-tert-Butyl-9,10-anthraquinone	14.6	0.12	0.037
9-Anthrone	5.4	0.078	0.042
1-Nitronaphthalene	4.7	0.69	2.8
9-Nitroanthracene	7.6	0.16	0.077
1-Nitropyrene	12.2	0.15	0.22

Source: From Poulsen and Birks (43).

[a] HPLC mobile phase: methanol/water (85:15).

[b] t_r = uncorrected retention time.

[c] Defined for a signal-to-noise ratio of 3. LOD were evaluated at two different excitation wavelengths ($\lambda_{ex} = 243.5$ and 257.5 nm) to compare the way the method performed in each case.

Gandelman and Birks (9) detected very low amounts, in the nanogram range, of aliphatic oxygen-containing compounds, using the HPLC photosensitized photoreduction fluorescence (PRF) method (Table 3.2). The PRF detector was believed by the authors to be both the least complex and most sensitive detector for these types of compounds. In particular, it was found to be much more sensitive than the detector based on quinone-sensitized photooxidation (62). This HPLC photoreduction fluorescence method was also applied by Poulsen and Birks (43) to the detection of a large number of quinone derivatives (Table 3.3). The best performances, in terms of signal-to-noise ratio, were found at excitation wavelength 243.5 nm for naphthoquinones and 257.5 nm for anthraquinones. LOD values were in the low-picogram range, depending on the substituent functional groups and the positions of these functionalities (Table 3.3). The selectivity of PRF of quinones was limited by the fact that naturally fluorescent compounds, at least those that were not destroyed in the photochemical reactor, produced a response. PRF detection was found to be more selective than UV detection. Biogenic quinones such as the K vitamins could also be detected by this method at low-picogram levels. In all cases, the optimal photoreduction reaction times were extremaly short, between 5 and 15 s, which allowed relatively high flow rates and easy subsequent fluorometric measurement of the dihydroquinones within the fluorometer detector flow cell (43).

3.5.2. Pharmaceutical Analysis

Photochemical fluorometry was extensively applied to the determination of pharmaceuticals in the pure state and in preparations.

Biological compounds of pharmaceutical interest, such as DL-tryptophan, DL-5-hydroxytryptophan, tryptamine, 5-hydroxytryptamine, DL-dopa, dopamine, quinine sulfate, and norepinephrine, could be measured in liquid aqueous solutions of various pH's at the nanogram per milliliter level by Lukasiewicz and Fitzgerald (4) (Table 3.4). Initial fluorophor concentrations were evaluated by digital integration of the decay of fluorescence signal for a fixed time of 0.5 min. Calibration curves were linear over 1–2 decades, and the uncertainty in the slopes and the scatter of points about the straight lines were better by about an order of magnitude than those obtained using conventional fluorescence measurements. The high precision and the increased sensitivity of photochemistry-fluorescence measurements make it possible to distinguish between small differences in concentration of pharmaceutical at the $ng \cdot mL^{-1}$ level. Generally, the measurements of fluorophores reported in this work were found to be more sensitive and precise than measurements by previously reported fluorometric methods (4).

The same photochemical fluorometric technique was used by White et al. (61) for determining in liquid solution between 10 and 100 $ng \cdot mL^{-1}$ of several phenothiazines (perphenazine, chlorpromazine, fluphenazine, and trifluoperazine) (Table 3.4). With the exception of thioridazine, linear calibration curves were obtained with correlation coefficients greater than 0.999. Optimal values of pH, fluorescence emission wavelength, and photooxidation time were reported for each drug (Table 3.4).

Antimalarial compounds such as chloroquine, primaquine, plasmocid, and mefloquine were quantitated by photochemical fluorescence on solid surfaces or in liquid solutions (5,13,14,19,63,64). Optimal irradiation times were in the 0.5–6 min range for solid surfaces studies and between 0.5 and 43 min in liquid solution. Absolute LOD values were generally in the low-nanogram range (Table 3.4).

Fricoteaux et al. (65) reported on photochemical fluorescence determination of phenylbutazone (a nonsteroidal, anti-inflammatory drug), and its degradation products in liquid solution. Optimal irradiation times ranged between 4 and 15 min. Photochemical fluorescence enhancement factors were between about 2 and 7. Linear log-log calibration plots were obtained over a 50- to 1000-fold range of concentration, and LOD values were between $1\,\text{ng}\cdot\text{mL}^{-1}$ and $1.2\,\mu\text{g}\cdot\text{mL}^{-1}$, which showed photochemical fluorometry to be a simple, sensitive, and precise method for these compounds.

Fidanza and Aaron (63,64) determined quantitatively by photochemical fluorometry (in liquid solution) acetylsalicylic acid, chloroquine, mefloquine, and theophylline in various pharmaceutical preparations (tablets, powders, and syrups). The authors obtained $> 100\%$ overall recoveries from pharmaceutical tablets, with relative standard deviations of about 4%.

Procopio et al. (22) evaluated lorazepam in tablets by the same method. The mean overall recovery from pharmaceutical tablets was 99.2%, with a relative standard deviation of 0.6%.

HPLC photochemical-fluorescence analysis was also applied by several authors (10,11,32–36,40,47,50) to pharmaceuticals. Analytical performances are summarized in Table 3.5.

Scholten et al. (32,34) and Uihlein and Schwab (40) determined clobazam and its metabolites at the nanogram level, using short residence times in the photochemical reactor (between 15 and 54 s) (Table 3.5). The sensitivity of the HPLC photochemical fluorescence detector was significantly better than that of the UV absorption detector (40). Also, clobazam and desmethylclobazam could be detected in serum and urine samples, with recoveries of 95% and 70%, respectively (32).

Clomiphene, which is a nonsteroidal triphenylethylene compound used as an ovulatory stimulant, was detected at the picogram level by Harman et al. (50) (Table 3.5). The authors separated and quantitated the cis and trans isomers of clomiphene in plasma (Fig. 3.18), with a minimal detectable level of $350\,\text{pg}\cdot\text{mol}^{-1}$. HPLC photochemical fluorometry gives a 20-fold improvement in sensitivity over UV detection.

Brinkman et al. (48) reported on the HPLC photochemical fluorometric detection of the photoproducts of demoxepam, an important tranquilizer, and of several phenothiazines [Fenergan (promethazine hydrochloride), Largactil (chlorpromazine hydrochloride), levopromazine, nedaltran (10-(3-dimethyl-

Table 3.4. Photochemical Fluorometric Analytical Characteristics of Pharmaceuticals in Liquid Solution and on Solid Surfaces

Compound	Liquid solution or solid surface	λ_F^a (nm)	t_{irr}^b (min)	PFEF^c	LDR^d	LOD^e (ng·mL^{-1})	Absolute LOD (ng)	Ref.
Acetylsalicylic acid	Filter paper	408^g	5	—	10	—	—	63
Chloroquine	Borate buffer (pH 11)	480	0.5	∫	6 × 10^3	0.1	2.5	5
Chloroquine	Silica gel	470	0.5	—	260	200	4	13,14
Chloroquine	Filter paper	375^g	5	∫	50	—	—	63
Chlopromazine	HCl 0.1 M	469	1	—	100	10	—	61
3,4-Dihydroxy-phenylalanine	Methanol	329^g	0.5	∫	32	25	625	4
Fluphenazine	Water	402	2	—	80	10	—	62
5-Hydroxytryptamine	Water	355^g	0.5	—	100	0.7	17.5	4
5-Hydroxytryptophan	Borate buffer (pH 10)	355^g	0.5	∫	100	1.0	25	4
3-Hydroxytyramine	95% Ethanol	345^g	0.5	∫	60	10	250	4
Lorazepam	NaOH 0.2 M	435	20	—	100	0.3	—	22
Mefloquine	Methanol	370	5	157	1 × 10^3	50	150	64
Norphenylephrine	HCl 0.1 M	318^g	0.5	∫	32	25	625	4
Perphenazine	Aqueous NH$_3$ (pH 10)	469	1	—	100	10	—	61
Phenylbutazone	Ethanol	340	30	7.3	200	1	3	65
Phenylbutazone	Chloroform	390	4	5.8	50	1.2 × 10^3	3.6 × 10^3	65
CHB^h	Ethanol	360	15	18	1 × 10^3	24	72	65

Analyte	Solvent	λ						
MADH[i]	Ethanol	370	4	4.6	200	100	300	65
TADH[j]	Ethanol	365	15	3.7	400	8	24	65
Plasmocid	H_3PO_4 0.66 M	477	43	9.2	2×10^3	50	150	19
Plasmocid	Silica gel	—	6	12	200	500	10	19
Primaquine	Silica gel	335	0.5	—[f]	200	45	0.9	15
Quinine	H_2SO_4 0.1 M	450[g]	0.5	—[f]	200	0.1	2.5	4
Theophylline	Filter paper	315[g]	5	—	10	—	—	63
Thioridazine	Water	421	2	—[f]	27	10	—	61
Trifluoperazine	Water	408	2	—	80	10	—	61
Tryptamine	Borate buffer (pH 10)	355[g]	0.5	—[f]	100	10	—	4
Tryptophan	Aqueous NaOH (pH 10.5)	345[g]	0.5	∞^k	10	20	—	4
Vitamin K_1	Dioxane	431	1	—	40	5	125	6

[a] Fluorescence emission maximum wavelength of the photoproduct, unless otherwise noted.
[b] Optical irradiation time corresponding to the maximum fluorescence signal value or to the integrated fluorescence signal value.
[c] PFEF = photochemical fluorescence enhancement factor (see text for definition).
[d] LDR = linear dynamic range (see footnote c of Table 3.1 for definition).
[e] LOD = limit of detection (see footnote d of Table 3.1 for definition).
[f] No PFEF value available because a decrease of fluorescence was observed upon UV irradiation.
[g] Fluorescence emission maximum wavelength of the analyte.
[h] CHB = N-caproylhydrazobenzene.
[i] MADH = butylmalonic acid mono-(N,N'-diphenyl)hydrazide.
[j] TADH = butyltartronic acid mono-(N,N'-diphenyl)hydrazide.
[k] The PFEF value is infinite because the nonirradiated analyte is nonfluorescent.

Table 3.5. HPLC Photochemical Fluorescence Analysis of Pharmaceuticals

Compound	Eluant	Residence Time $(s)^a$	LDR^b	Absolute LOD $(ng)^c$	Ref.
Clobazam and desmethylclobazam	Methanol–0.01 M sodium acetate (6:4)	15	36	0.02^d	32, 34
Clobazam and metabolites	0.1 M $(NEt)_4$ phosphate buffer pH 7.0–methanol (16:13)	54	186	0.13	40
Clomiphene	Methanol	10	45	0.06	50
Denoxepam	Methanol–0.1 M phosphate buffer pH 8 (3:2)	110	10^3	0.10	48
Diethylstilbestrol	Acetonitrile–water pH 3.5 (1:1)	2	4	0.2	49
Diginatin	Acetonitrile–water (60:40)	42–81	10	20	10, 11
Digoxin	Acetonitrile–water (60:40)	42–81	10	20	10, 11
Fenbendazole and metabolites	Methanol–0.033 M H_3PO_4 (1:1)	78	100	1.2	40
Hydrocortisone	Acetonitrile–water (60:40)	21–42	10	20	11
Phenothiazines: Mesoridazine Sulforidazine Thioridazine	Methanol–7.6 M sodium acetate (4:1), 0.01% NH_4 peroxodisulfate	25	77	0.5	34
Phenothiazines: Fenergan Largactil Levopromazine Nedaltran Thiodiphenylamine	Methanol–2.6 M sodium acetate (4:1)	35	50	0.04–0.10	48
Reserpine	Acetonitrile–0.05 M phosphate buffer pH 6 (70:30)	—	10	0.08	36
Tamoxifen and metabolites	Methanol–0.04% diethylamine acetate^e	—	100	0.2	33

Table 3.5. (*Continued*)

Compound	Eluant	Residence Time (s)[a]	LDR[b]	Absolute LOD (ng)[c]	Ref.
Tamoxifen and desmethyltamoxifen	Methanol–5×10^{-3} M NH$_4$ acetate buffer (9:1)	18	—	0.2	47
Tamoxifen and 4-hydroxytamoxifen	Methanol–water–acetic acid–N,N'-dimethyl-hexylamine (73:27:0.5:0.2)	12	—	0.02	47
Vitamin K$_1$	Methanol[f]	87	—	0.15	35

[a] Optical residence time of the analyte in the photochemical reactor.
[b] LDR = linear dynamic range (see footnote c of Table 3.1 for definition).
[c] LOD = limit of detection (see footnote d of Table 3.1 for definition).
[d] 0.05 ng for desmethylclobazam.
[e] Other eluants used: methanol–water–triethylamine–acetic acid (98:2:0.03:0.3) and hexane-2-propanol (98.7:1.3).
[f] Other eluants used: methanol–acetate buffer (98:2) + 0.25 mg·mL^{-1} ascorbic acid. Postcolumn addition of a reagent consisting of methanol–acetate buffer (85:15) + 1 mg·mL^{-1} ascorbic acid was also performed for increasing sensitivity.

amino-2-methylpropyl)phenothiazine)]. The calibration graphs showed good linearity. Limits of detection were 100 pg for demoxepam, Fenergan and Largactil, 10 pg for levopromazine, and 50 pg for nedaltran. Also, human serum was spiked with 8 μg·L^{-1} of demoxepam, and the recovey of demoxepam was 95%. A detection limit of about 10 ng·mL^{-1} in serum was also found for phenothiazines (48).

Diethylstilbestrol, which is a very controversial estrogen because of its carcinogenicity, could be determined at low-ng·mL^{-1} levels in biological samples, using HPLC photochemical fluorometry (49). Calibration curves were found to be linear in the low-nanogram range for pure diethylstilbestrol in ethanol solution as well as for diethylstilbestrol extracted from human urine. The authors concluded that this method was highly specific, rapid, and simple, as compared to the lengthy extraction and derivatization procedures characterizing other analytical techniques (49).

Gandelman et al. (10,11) Applied the PRF detector to the quantitation of two cardiac glycosides (diginatin and digoxin) and of hydrocortisone. Residence times were relatively short (21–81 s, according to chromatographic conditions). Lower detection limits were found for the cardiac glycosides

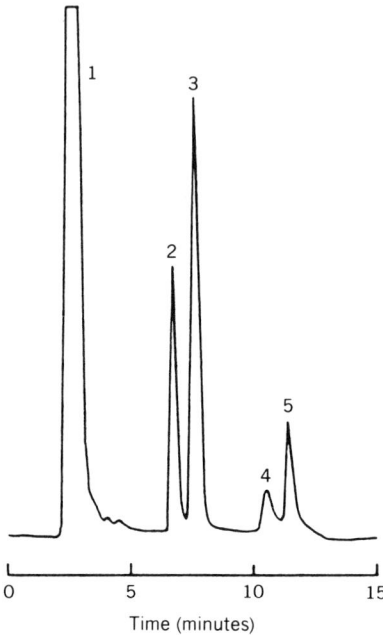

Fig. 3.18. HPLC chromatogram of a plasma sample obtained using photochemical fluorescence detection: (1) solvent peak; (2) *cis*-clomiphene; (3) *trans*-clomiphene; (4 & 5) clomiphene metabolites. [From Harman et al. (50), with permission.]

(2 ng), compared to those obtained with the UV–visible detector, while no significant improvement in sensitivity was found for hydrocortisone. Unfortunately, when the PRF detector was used, the calibration curves were only linear over 1 order of magnitude. However, the authors pointed out that this nonlinearity does not prevent its use in quantitative analysis. Two pharmaceutical products, a digoxin tablet and an hydrocortisone cream, were also determined using this method (10,11).

The HPLC photochemical fluorescence method was shown by Uihlein and Schwab (40) to be highly specific for the evaluation of fenbendazole and its metabolites in serum extracts. Calibration curves were linear over 2 orders of magnitude for fenbendazole. The sensitivity of the photochemical fluorescence detector was found by the authors to be better than that of the UV detector (40).

Brown et al. (33) and Kikuta and Schmid (47) investigated tamoxifen (TAM), an antiestrogenic drug, and its metabolites in serum and plasma samples by HPLC photochemical fluorometry. Calibration curves of TAM and two of its metabolites, recovered from sera extracts, were found to be linear over 2 orders of magnitude. The limit of detection of each compound was about 0.2 ng (33). Because of the high specificity of the method, small volumes (between 20 and 40 µL) of deproteinized plasma samples containing TAM and metabolites could be injected directly without further sample preparation. Accuracy of the overall HPLC photochemical fluorometric procedure was very acceptable, with recoveries of TAM and desmethyl-TAM ranging between 97% and 104% (47).

Lefevere et al. (35) applied HPLC photochemical fluorometry to the detection of vitamin K_1. A limit of detection of 150 pg was found, representing a 3.7-fold increase in sensitivity as compared to UV detection. Vitamin K_1 was also determined in serum extracts. UV detection failed to find physiological concentrations of vitamin K_1, whereas photochemical-fluorescence detection resulted in a clear chromatogram with a well-separated vitamin K_1 peak.

Chen et al. (52) determined three phenothiazine compounds (chlorpromazine, promethazine, and perphenazine) by normal and stopped-flow FIA photochemical spectrofluorometric methods. The detection limits were between 20 and 50 $ng \cdot mL^{-1}$, with relative standard deviations of less than 2%. The effect of various foreign species on the photochemical fluorometric determination of these phenothiazines was studied, and the application of the method to authentic pharmaceutical preparations was investigated and found to be satisfactory.

Mahedero and Aaron (66) quantitatively determined several sulfonamides by photochemical fluorometry in aqueous solution. The fluorescence intensity of heterocyclic sulfonamides was increased upon UV irradiation, whereas a decrease occurred for nonheterocyclic sulfonamides. Linear log-log calibration plots were obtained over a 15- to 480-fold range of concentration, and limits of detection ranged between 1 and 97 $ng \cdot mL^{-1}$.

3.5.3. Clinical Analysis Applications

Several of the applications of photochemical fluorometry described in the previous section (3.5.2) can also be considered as belonging to clinical analysis, since they are related to the separation and detection of pharmaceuticals in biological samples (32,33,35,40,48,50).

Other investigators have used photochemical fluorescence in clinical analysis (15,41,51,52,67).

Arakawa et al. (41) determined six catecholamine sulfoconjugate (CAS) isomers in human urine, using HPLC with photochemical fluorescence detection. The six CAS isomers (dopamine 3-sulfate, dopamine 4-sulfate, norepinephrine 3-sulfate, norepinephrine 4-sulfate, epinephrine 3-sulfate, and epinephrine 4-sulfate) were separated satisfactorily by HPLC, and peaks corresponding to the CAS isomers in urine were identified quantitatively by two different separation methods. Linear calibration curves were found between the CAS amounts and the fluorescence intensities of the peaks over LDR values of 20. The limits of detection ranged from 1 to 2 pmol for the CAS isomers. The excretion rates of all CAS isomers were evaluated in urine. The authors concluded that photochemical fluorometry was a highly sensitive and selective method of detection of physiological levels of CAS isomers in human urine (41).

Tsuchiya et al. (16,51) were able to apply photochemical fluorescence in a flowing solvent to the determination of primaquine and chloroquine in blood serum, with detection limits of 100 and 0.25 ng·mL^{-1}, respectively. Photochemical fluorescence enhancement factors were between 3 and 4 for primaquine, but only 1.6 for chloroquine. The recoveries were better than 96% for primaquine concentrations >500 ng·mL^{-1}, and better than 94% for chloroquine concentrations >5.5 ng·mL^{-1}. The authors pointed out that the photochemical-fluorometric flow cell method is sensitive, simple, and fairly rapid. Its successful application to the detection of primaquine and chloroquine in blood serum suggests its use in routine clinical analysis.

Schmid et al. (68) demonstrated the value of applying HPLC photochemical fluorometry to the detection of several drugs, in therapeutic concentrations, in biological samples. Enhanced fluorescence signals were reported for ambroxol, acetylsalicylic acid, loperamid, and 1,4-benzodiazepine, with irradiation times ranging between 120 and 160 s, according to the compound. The interest of this method for clinical applications results from its high sensitivity and specificity, and from the small volume of sample required.

Wolf and Schmid (67) applied HPLC with photochemical fluorescence detection to the separation and the analysis of mianserin, an antidepressant drug, in blood plasma. As compared to UV absorbance detection, this method is particularly interesting because it lowers detection limits of mianserin to low-nanogram levels in blood plasma and it simplifies the preparation of plasma samples without sacrificing the specificity of the analysis. Therefore, HPLC photochemical fluorometry should be a valuable method for routine clinical analysis of mianserin (67).

3.5.4. Pesticides and Environmental Analysis

Photochemical fluorometry was also developed for the quantitative analysis of pesticides and other environmental pollutants. Analytical performances are presented in Table 3.6.

Werkhoven-Goewie et al. (60) applied this method to the trace analysis of chlorophenols in effluent water samples and in biological samples. The fluorescence signals of photolysed chlorophenols showed a linear response to the amounts of solutes over 2–3 orders of magnitude. For the mono- and di-

Table 3.6. Photochemical Fluorometric Limits of Detection (LOD) of Pesticides and Other Environmental Pollutants

Compound	Conditions	Absolute LOD^a (ng)	Ref.
2-Chlorophenol	HPLC	4	60
3-Chlorophenol	HPLC	1	60
4-Chlorophenol	HPLC	1	60
2,4-Dichlorophenol	HPLC	50	60
Dinitroanilines:			
Benfluralin	Liquid solution	300	12
Isopropalin	Liquid solution	3.3×10^3	12
Oryzalin	Liquid solution	4.3×10^3	12
Trifluralin	Liquid solution	300	12
Phenylcarbamates:			
Karbutilate	HPLC	10	45
Propham	HPLC	5	45
Phenmedipham	HPLC	87	45
Chlorpropham	HPLC	5	45
Barban	HPLC	4.9	45
Phenylamides:			
Pyracarbolid	HPLC	4.5	45
Propachlor	HPLC	88	45
Propanil	HPLC	0.7	45
Monalide	HPLC	6.3	45
Phenylurea herbicides:			
Monuron	HPLC	24	60
Fluometuron	HPLC	1.4	60
Diuron	HPLC	13	60
Siduron	HPLC	3.3	60
Linuron	HPLC	4.7	60
Chlorbromuron	HPLC	12	60
Neburon	HPLC	11	60

^{a} See footnote d of Table 3.1 for definition.

chlorophenols, the detection limits were in the low-nanogram range (Table 3.6). Good repeatability (relative standard deviation = 7%) and recoveries of $100 \pm 4\%$ were found for river water samples spiked with amounts of 35, 5, and 70 ng·mL^{-1} of 2-chloro-, 4-chloro-, and 2,4-dichlorophenol, respectively (60). Excellent selectivity was also noted for this HPLC photochemical fluorescence detection system after on-line preconcentration of the river water samples.

Traore and Aaron (12) analyzed (in liquid solution) traces of several nonfluorescent dinitroaniline derivative herbicides by a photoreduction fluorescence method. Optimal irradiation times were found to be between 2 and 4 min, and LDR values ranged from 10 to 10^3 concentration units. LOD values were between 0.3 and 4.3 μg, according to the molecular structure of the herbicide (Table 3.6). The authors concluded that photoreduction fluorescence detection was valuable for the sensitive, rapid, and precise analysis of dinitroaniline herbicides. This method should enable investigators to evaluate these herbicide residues in plants or soils (12).

Miles and Moye (45) applied HPLC photochemical fluorometry to the fluorescence detection of several classes of pesticides such as phenylcarbamates, phenylamides, and phenylurea herbicides. Most of the compounds gave linear calibration curves over about 2 orders of magnitude, while limits of detection were in the low-nanogram range (Table 3.6). The trace analysis of selected pesticides in ground water was also performed using HPLC postcolumn photolysis and fluorogenic labeling with o-phthalaldehyde–2-mercaptoethanol (OPA–MERC) (45). The latter method led to recoveries of five pesticides (aldicarb sulfoxide, aldicarb, propoxur, thiram, and neburon) in three fortified groundwater samples, ranging from 82.7% to 113%, at 10- and 50-ng·mL^{-1} level (45). As mentioned in Section 3.3.3.1, Luchtefeld (44) demonstrated previously that phenylurea herbicides could be analyzed by HPLC photolysis and labeling with OPA–MERC to form fluorescent products.

Patel et al. (69,70) built a postcolumn photolysis fluorescence detector and studied several classes of nitrogenous pesticides to determine their fluorescence signal photoinduced by UV photolysis in HPLC and FIA. Several solvent systems were evaluated as typical reversed-phase mobile phases, and the presence of photosensitizers increased the fluorescence response of most pesticides. Analytical figures of merit were given for the determination of several pesticides in groundwater. LOD ranged between 5 and 50 ng/g (70).

3.6. FUTURE TRENDS

Photochemical fluorometry has demonstrated itself to be a very selective and sensitive method of analysis of a variety of organic compounds. It has developed into a versatile technique, adaptable to various media such as HPLC, solid surfaces, and liquid solutions. Recently, important improvements have been made in photochemical reactor designs. The use of miniaturized excitation sources and of UV-transmitting knitted Teflon tubing has allowed researchers to minimize undesirable temperature effects and to reduce reaction times in the photoreactors when HPLC and flowing solvent systems are utilized.

Nevertheless, more advances might well be made in the construction of new, more efficient photochemical reactors. There has been practically no use of laser sources in photochemical fluorometry. Laser sources could enhance selectivity and sensitivity of photochemical fluorescence detection (58); they could also improve positioning of the adsorbed compound in the path of irradiation when solid surfaces are utilized. Since the precision of irradiation times is critical for obtaining reproducible fluorescence intensity values of photoproducts, it would be advisable to use computers and digital electronics in order to precisely monitor illumination time and to integrate time-dependent fluorescence signals.

Several photoreactions have already been shown to produce fluorescent photoproducts and therefore to be analytically useful (see Section 3.4). It seems likely that many more photoreactions and photoreactive analytes will be discovered, leading to enhanced fluorescence sensitivities. Sensitized photoreactions should be good candidates from this standpoint. Also, the use of solid-state photochemical reactions has been very limited until now, and it should be developed in the future.

Besides fluorescence, other means of detection of the photoproducts might be proposed, such as phosphorescence, delayed fluorescence, and chemiluminescence.

Because of the simplicity, sensitivity, and selectivity of photochemical fluorescence analysis, applications will no doubt continue to proliferate in the future, especially in the pharmaceutical and clinical fields.

REFERENCES

1. S. Udenfriend, *Fluorescence Assay in Biology and Medicine*, Vol. 2, Academic Press, New York, 1969.
2. W. Fink and W. R. Koehler, *Anal. Chem.*, **42**, 990, (1970).
3. J. M. Fitzgerald, in J. M. Fitzgerald, Ed., *Analytical Photochemistry and Photochemical Analysis*, Marcel Dekker, New York, 1971, p. 145.
4. R. J. Lukasiewicz and J. M. Fitzgerald, *Anal. Chem.*, **45**, 511 (1973).
5. R. J. Lukasiewicz and J. M. Fitzgerald, *Appl. Spectrosc.*, **28**, 151 (1974).
6. J. J. Aaron, J. E. Villafranca, V. R. White, and J. M. Fitzgerald, *Appl. Spectrosc.*, **30**, 159, (1976).
7. J. J. Aaron, *Methods Enzymol.*, **67F**, 140 (1980).
8. J. W. Birks and R. W. Frei, *Trends Anal. Chem.*, **1**, 361 (1982).
9. M. S. Gandelman and J. W. Birks, *Anal. Chem.*, **54**, 2131 (1982).
10. M. S. Gandelman, J. W. Birks, U. A. Th. Brinkman, and R. W. Frei, *J. Chromatogr.*, **282**, 193 (1983).
11. M. S. Gandelman and J. W. Birks, *Anal. Chim. Acta*, **155**, 159 (1983).

12. S. Traore and J. J. Aaron, *Anal. Lett.*, **20**, 1995 (1987).
13. J. Fidanza and J. J. Aaron, *Analusis*, **9**, 118 (1981).
14. J. J. Aaron and J. Fidanza, *Talanta*, **29**, 383 (1982).
15. J. J. Aaron, S. A. Ndiaye, and J. Fidanza, *Analusis*, **10**, 433 (1982).
16. M. Tsuchiya, E. Torres, J. J. Aaron, and J. D. Winefordner, *Anal. Lett.* **17** (B16), 1831 (1984).
17. I. S. Krull and W. R. La Course, in I. S. Krull, Ed., *Reaction Detection in Liquid Chromatography*, Marcel Dekker, New York, 1986, p. 303.
18. Y. Arakawa, K. Imai, and Z. Tamura, *Anal. Chim. Acta*, **147**, 325 (1983).
19. J. J. Aaron, J. Fidanza, and M. D. Gaye, *Talanta*, **30**, 649 (1983).
20. J. J. Aaron and A. Diop, *Analusis*, **13**, 40 (1985).
21. T. C. M. Pastore, E. M. de M. Nicola, and C. G. de Lima, *Analyst*, **109**, 243 (1984).
22. J. R. Procopio, P. H. Hernandez, and L. H. Hernandez, *Analyst*, **112**, 79 (1987).
23. J. Fidanza and J. J. Aaron, *Talanta*, **33**, 215 (1986).
24. J. J. Aaron and J. Fidanza, *Analusis*, **16**, 353 (1988).
25. W. R. La Course and I. S. Krull, *TrAC, Trends Anal. Chem. (Pers. Ed.)* **4**, 118, (1985).
26. J. T. Stewart and W. J. Bachman, *TrAC, Trends Anal. Chem. (Pers. Ed.)*, **7**, 106 (1988).
27. W. J. Bachman and J. T. Stewart, *LC-GC*, **7**, 38 (1989).
28. I. S. Krull, C. M. Selavka, M. Lookabaugh, and W. R. Childress, *LC-GC*, **7**, 758 (1989).
29. P. J. Twitchett, P. L. Williams, and A. C. Moffat, *J. Chromatogr.*, **149**, 683 (1978).
30. W. Iwaoka and S. R. Tannenbaum, *IARC Sci. Publ.*, **14**, 51 (1979).
31. A. H. M. T. Scholten and R. W. Frei, *J. Chromatogr.*, **176**, 349 (1979).
32. A. H. M. T. Scholten, U. A. Th. Brinkman, and R. W. Frei, *Anal. Chim. Acta*, **114**, 137 (1980).
33. R. R. Brown, R. Bain, and V. Craig Jordan, *J. Chromatogr., Biomed. Appl.*, **272**, 351 (1983).
34. A. H. M. T. Scholten, P. L. M. Welling, U. A. Th. Brinkman, and R. W. Frei, *J. Chromatogr.*, **199**, 239 (1980).
35. M. F. Lefevere, R. W. Frei, A. H. M. T. Scholten and U. A. Th. Brinkman, *Chromatographia*, **15**, 459 (1982).
36. J. R. Lang, J. T. Stewart, and I. L. Honigberg, *J. Chromatogr.*, **264**, 144 (1983).
37. Y. T. Shih and P. W. Carr, *Anal. Chim. Acta*, **159**, 211 (1984).
38. C. de Ruiter, J. F. Bohle, G. J. Dejong, U. A. Th. Brinkman, and R. W. Frei, *Anal. Chem.*, **60**, 666 (1988).
39. G. E. Batley, *Anal. Chem.*, **56**, 2262 (1984).
40. M. Uihlein and E. Schwab, *Chromatographia*, **15**, 140 (1982).
41. Y. Arakawa, K. Imai, and Z. Tamura, *Anal. Biochem.*, **132**, 389 (1983).

42. J. R. Poulsen, K. S. Birks, M. S. Gandelmans, and J. W. Birks, *Chromatographia*, **22**, 231 (1986).
43. J. R. Poulsen and J. W. Birks, *Anal. Chem.*, **61**, 2267 (1989).
44. R. G. Luchtefeld, *J. Chromatogr. Sci.*, **23**, 516 (1985).
45. C. J. Miles and H. A. Moye, *Anal. Chem.*, **60**, 220 (1988).
46. H. Jansen, J. J. Vreuls, T. A. J. Van Der Heide, C. J. De Jong, U. A. Th. Brinkman, and R. W. Frei, *J. Liq. Chromatogr.*, **11**, 1855 (1988).
47. C. Kikuta and R. Schmid, *J. Pharmacol. Biomed. Anal.* **7**, 329 (1989).
48. U. A. Th. Brinkman, P. L. M. Welling, G. De Vries, A. H. M. T. Scholten, and R. W. Frei, *J. Chromatogr.*, **217**, 463 (1981).
49. A. T. Rhys-Williams, S. A. Winfield, and R. C. Belloli, *J. Chromatogr.*, **235**, 461 (1982).
50. P. J. Harman, G. L. Blackman, and G. Philipou, *J. Chromatogr.*, **225**, 131 (1981).
51. M. Tsuchiya, J. J. Aaron, E. Torres, and J. D. Winefordner, *Anal. Lett.*, **18** (B13), 1647 (1985).
52. D. Chen, A. Rios, M. D. Luque del Castro, and M. Valcarcel, *Analyst*, **116**, 171 (1991).
53. R. O. Kan, *Organic Photochemistry*, McGraw-Hill, New York, 1966, p. 219.
54. B. J. Wilbur, Ch. C. Benz, and M. W. Degregorio, *Anal. Lett.*, **18**, 1915 (1985).
55. D. W. Mendenhall, M. Kubayashi, F. M. L. Shih, L. A. Sternson, T. Higuchi, and C. Fabian, *Clin. Chem. (Winsten-Salem, N.C.)*, **24**, 1518 (1978).
56. Y. Golander and L. A. Sternson, *J. Chromatogr.*, **181**, 41 (1980).
57. M. Nieder and H. Jager, *J. Chromatogr.*, **413**, 207 (1987).
58. N. Strojny and J. A. F. de Silva, *Anal. Chem.*, **52**, 1554 (1980).
59. C. A. T. Fall, A. Diop, and J. J. Aaron, *Bull. Soc. Chim. Belg.*, **95**, 631 (1986).
60. C. E. Werkhoven-Goewie, W. M. Boon, A. J. J. Praat, R. W. Frei, U. A. Th. Brinkman, and C. J. Little, *Chromatographia*, **16**, 53 (1982).
61. V. R. White, C. S. Frings, J. E. Villafranca, and J. M. Fitzgerald, *Anal. Chem.*, **48**, 1314 (1976).
62. M. S. Gandelman and J. W. Birks, *J. Chromatogr.*, **242**, 21 (1982).
63. J. Fidanza and J. J. Aaron, *J. Pharm. Biomed. Anal.*, **5**, 619 (1987).
64. J. Fidanza and J. J. Aaron, *Anal. Chim. Acta*, **227**, 325 (1989).
65. R. Fricoteaux, M. Quaglia, and J. J. Aaron, *J. Pharm. Biomed. Anal.*, **7**, 1585 (1989).
66. M. C. Mahedero and J. J. Aaron, 4th International Symposium on Quantitative Luminescence Spectrometry in Biomedical Sciences, Ghent, Belgium, May, 1991.
67. C. Wolf and R. W. Schmid, 2nd International Symposium on Pharmaceutical and Biomedical Analysis, York, England, April, 1990.
68. R. W. Schmid, C. Wolf, and R. Kupferschmidt, 13th Symposium on Column Liquid Chromatography, Stockholm, Sweden, June, 1989.
69. B. M. Patel, H. A. Moye, and R. Weinberger, *J. Agric. Food Chem.*, **38**, 126 (1990).
70. B. M. Patel, H. A. Moye, and R. Weinberger, *Talanta*, **38**, 913 (1991).

CHAPTER

4

APPLICATIONS OF ORGANIZED BILE SALT MEDIA FOR LUMINESCENCE ANALYSIS

LINDA B. McGOWN

P. M. Gross Chemical Laboratory
Department of Chemistry
Duke University
Durham, NC 27705

4.1. Introduction
    4.1.1. Fluorescence Analysis
    4.1.2. Organized Media in Luminescence Analysis
4.2. Bile Salts
    4.2.1. Bile Salt Structure and Aggregation
    4.2.2. Bile Salts for Luminescence Analysis
4.3. Specific Studies and Applications
    4.3.1. Metal Cation Enhancement of Luminescence
    4.3.2. Energy Transfer Between Probes in Bile Salt Media
    4.3.3. Effects of Bile Salts on Total Luminescence Spectra of Complex Samples
    4.3.4. Bile Salt Microenvironments for Solubilized Fluorescent Probes
4.4. Future Directions: Chiral Selectivity, Mixed Micelles, and Lyotropes
Acknowledgment
References

4.1. INTRODUCTION

4.1.1. Fluorescence Analysis

Fluorescence spectroscopy is a sensitive and selective tool for chemical analysis that can be used to study complex samples such as those of biological and environmental origin. Dynamic intermolecular interactions including

---

Molecular Luminescence Spectroscopy, Part 3, Edited by Stephen G. Schulman. Chemical Analysis Series, Vol. 77.
ISBN 0-471-51580-9 © 1993 John Wiley & Sons, Inc.

aggregation, macromolecular association, collisional quenching, and excited state complexation can all affect such fluorescence characteristics, as peak maxima, intensity, lifetime, anisotropy, and vibronic band ratios. Fluorescence characteristics can vary even for a single compound if individual molecules of that compound are experiencing more than one microenvironment in the sample. Heterogeneity of microenvironments in the sample may arise from self-aggregation of the compound itself or from the presence of multiple binding sites in macromolecular structures or assemblies (e.g., proteins, membranes, nucleic acids, micelles, particulates, or microcrystalline aggregates).

Microenvironmental sensitivity makes fluorescence a powerful tool for studying macromolecular structures, molecular aggregation, and complex samples. Moreover, because of its susceptibility to matrix effects, binding microenvironments, and dynamic intermolecular interactions, fluorescence is excellently suited to sample fingerprinting and characterization. These same features become a serious handicap, however, in the determination of analytes in a complex sample. Accurate calibration is often not possible if the fluorescence characteristics of the analyte are highly sample dependent. One solution to this problem is to perform a physical seperation of the sample components through extraction and/or chromatography.

Recently, a different approach to this problem has been pursued in which organized media are used to provide a uniform microenvironment for the analyte *within* the sample, thereby accomplishing an in situ extraction of the analyte. The isolation of each individual molecule of a given analyte compound in an identical microenvironment in the sample will minimize matrix effects and increase the accuracy of calibration without necessitating a physical separation or extraction of the bulk sample. Moreover, the organized medium can be modified to create an optimal binding microenvironment to maximize the sensitivity and/or selectivity of the determination.

4.1.2. Organized Media in Luminescence Analysis

Micellar media, primarily those formed by synthetic detergent molecules, have been used in fluorescence analysis to solubilize hydrophobic molecules in aqueous solution and to enhance or otherwise modify the luminescence properties of molecules. Specific applications include solubilization of organic molecules in aqueous solution (1,2), enhancement of absorption and luminescence (3-9) and micellar stabilization of room temperature phosphorescence in solution (10-18). However, conventional micelles formed by detergents often bind more than one molecule per micelle in a relatively fluid interior, which may actually *promote* energy transfer and other nonradiative processes that interfere with fluorescence determinations. Cyclodextrins have also been explored as solubilization reagents, but their use is limited by strict size limitations for the guest molecule and relatively low binding constants for all but a few guest compounds. Thus, cyclodextrins are useful as highly selective reagents but are limited to certain sizes and types of analytes.

Recent studies have explored a different class of amphiphilic compounds, the bile salts, as an alternative to synthetic detergents or cyclodextrins for luminescence analysis. Studies have shown that bile salt aggregates are effective at individually solubilizing fluorescent molecules, thereby minimizing their interactions with each other and with sample matrix constituents. The result is an increase in the accuracy and dynamic range of fluorescence analysis of complex samples.

4.2. BILE SALTS

4.2.1. Bile Salt Structure and Aggregation

Bile salts are biological detergents that are synthesized from cholesterol in the liver. They are typically composed of a steroidal backbone with one or more α-oriented hydroxy groups, congugated to an anionic side chain, or "tail." Structures of some common bile salts are shown in Fig. 4.1. Synthetic derivatives, also shown in Fig. 4.1, include zwitterionic and nonionic compounds.

The three-dimensional structure of a bile salt is influenced by several factors (19). The α-orientation of the hydroxy groups places them on the concave side of the steroid skeleton, with the methyl groups positioned on the opposite, convex side. The polar or charged group on the aliphatic tail will therefore interact with the hydroxy groups on the concave surface. The resulting bile salt conformation (Fig. 4.2) provides a hydrophobic (methyl-containing) surface on one side and a hydrophilic (hydroxy-containing) surface on the other. This voluminous structure is markedly different from that of a conventional detergent monomer, which has a hydrophilic head group and a long hydrophobic tail.

Because the molecular structure of bile salts is very different from that of conventional detergents, the bile salts exhibit unique behavior with respect to self-association and molecular solubilization (20–23). In conventional detergents, solubilization sites include the micellar surface, the palisade layer, and the hydrophobic inner core (24). Analogous binding sites are not present in the smaller, more rigid bile salt micelles. Instead, solubilization of hydrophobic compounds is accomplished through favorable interactions with the hydrophobic surfaces of the bile salt micelles (20,25,26). The resulting solubilization microenvironments in the bile salt micelles are often highly apolar. Trihydroxy bile salts also have significantly lower aggregation

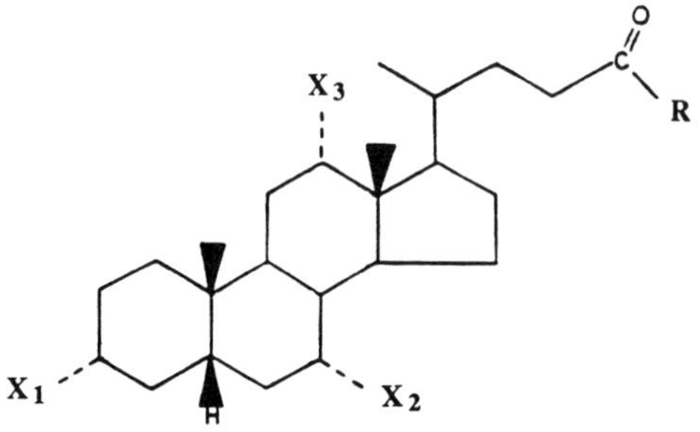

	X_1	X_2	X_3	R
Anionic compounds:				
Lithocholate	OH	H	H	
Deoxycholate	OH	H	OH	O^-
Glycodeoxycholate	OH	H	OH	$NHCH_2CO_2^-$
Taurodeoxycholate	OH	H	OH	$NHCH_2CH_2SO_3^-$
Chenodeoxycholate	OH	OH	H	O^-
Cholate	OH	OH	OH	O^-
Glycocholate	OH	OH	OH	$NHCH_2CO_2^-$
Taurocholate	OH	OH	OH	$NHCH_2CH_2SO_3^-$
Zwitterionic compounds:				
CHAPS	OH	OH	OH	$NH(CH_2)_3(N^+(CH_3)_2)(CH_2)_3SO_3^-$
CHAPSO	OH	OH	OH	$NH(CH_2)_3(N^+(CH_3)_2)CH_2CHOHCH_2SO_3^-$
Nonionic compounds:				
BigCHAP	OH	OH	OH	$N(CH_2CH_2CH_2NHCO(CHOH)_3CH(OH)_2)_2$
DeoxybigCHAP	OH	H	OH	$N(CH_2CH_2CH_2NHCO(CHOH)_3CH(OH)_2)_2$

Fig. 4.1. Structures of some common bile compounds.

numbers than detergent micelles; estimates range from 2 to 32, compared to 30–200 for typical detergents. It is also very interesting that bile salt aggregates show unusually high microviscosity [$\geqslant 100$ cP, versus 15–30 cP for conventional detergent micelles (27)].

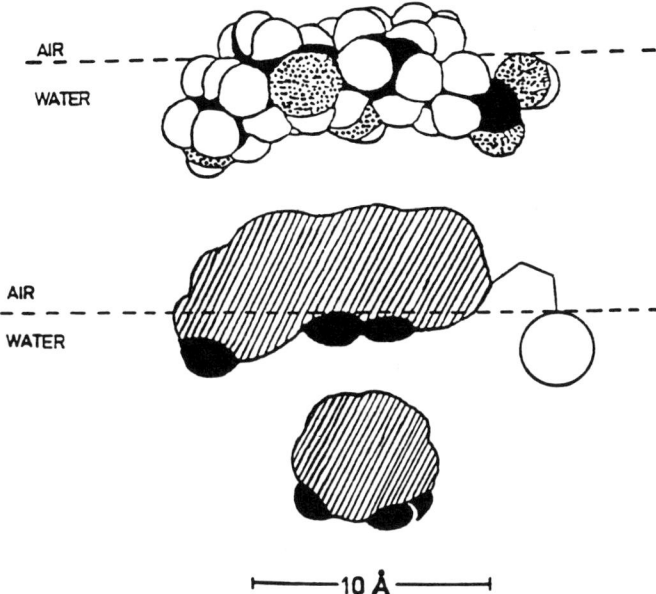

Fig. 4.2. Three-dimensional model of cholic acid: Stuart–Briegleb space filling model (top); longitudinal (middle) and transverse (bottom) shorthand representations. [From O'Connor and Wallace (19a), with permission.]

4.2.2. Bile Salts for Luminescence Analysis

From the viewpoint of luminescence analysis, bile salts offer the following advantages relative to conventional detergents:

- Bile salts offer significantly greater surface area and more functional groups for intermolecular interactions, providing a variety of sites for selective interactions with analyte molecules.
- Bile salts form smaller aggregates than do conventional detergents and therefore cause less light scattering in spectroscopic measurements.
- The models for conventional detergents place the solubilized guest molecules in a hydrophobic core, mingling with the long hydrophobic detergent tails. In contrast, models of bile salt aggregates suggests that individual guest molecules are tightly inserted between the bulky hydrophobic bile salt surfaces (22), providing better isolation of the solubilized molecules from each other and from the bulk solution.
- Bile salt phases (especially those of trihydroxy salts) are highly reproducible, more stable, and less likely to precipitate in complex sample

matrices and in the presence of metal cations, compared to conventional micellar phases.
- Unlike conventional detergents, bile salts are not soapy and do not form suds or bubbles; bile salt solutions are clear and free of bubbles even at concentrations as high as 10^{-2} M.
- The asymmetric binding microenvironments in bile salt aggregates offer unique possibilities for chiral selectivity in the detection of guest molecules.

To date, most studies of bile salts for luminescence analysis have focused on the trihydroxy salt sodium taurocholate (NaTC). Trihydroxy bile salts have the advantages of greater solubility and smaller aggregate size relative to monohydroxy and dihydroxy salts. NaTC reportedly exhibits relatively constant aggregation properties over a wide range of pH and counterion concentration (28,29); micellar NaTC solutions should therefore have high reproducibility and relatively constant characteristics over a wide range of experimental conditions.

4.3. SPECIFIC STUDIES AND APPLICATIONS

4.3.1. Metal Cation Enhancement of Luminescence

Metal cations are known to enhance the aggregation of bile salts and lower their critical micelle concentration (cmc) values. In NaTC, metal cations have been shown to enhance the fluorescence intensity of solubilized probes (30,31). Trivalent cations have a much greater effect than have divalent or monovalent cations, and heavier trivalent cations are more effective than lighter trivalent cations. The expected heavy atom quenching of probe luminescence by the heavier cations is observed only for more hydrophilic probes that are not solubilized in protected bile salt environments. In fact, there is a definite trend toward increased fluorescence enhancement as the aqueous solubility of the probe decreases (30).

The metal cation enhancement of probe fluorescence in NaTC has been attributed, at least in part, to the effects of the cations on the organization of the NaTC media. The effects include stabilization of NaTC aggregation as well as association of the cations with the anionic taurine group in the NaTC tail, thereby reducing access of the solubilized probes to external quenchers and increasing the microviscosity of the probe microenvironment. In a study of the effects of metal cations on the fluorescence of benzo[k]fluoranthene (BkF) (31), a highly insoluble probe showing a high degree of enhancement, it was shown that the enhancement by trivalent cations was at least twice as large

Table 4.1. Metal Cation Enhancement of BkF Fluorescence in NaTC Solutions[a]

[NaTC] (mM)	Na$^+$	Mg^{2+}	Al^{3+}	Tb^{3+}
1	1.1	1.1	1.1	1.2
4	1.4	1.3	2.9	2.9
6	3.1	3.7	12.5	15.0
8	7.0	6.1	13.6	17.6
10	6.9	5.2	10.6	14.8
12	4.4	4.2	4.6	6.0
21	1.5	1.6	1.8	1.8
32	1.0	1.0	1.1	1.1
40	1.0	1.0	1.1	1.1

Source: From Meyerhoffer and McGown (31).
[a] Metal cation concentrations of 60 mM Na$^+$, 20 mM Mg^{2+}, 10 mM Al^{3+}, and 10 mM Tb^{3+}; all salts used were nitrates.

as the enhancement by divalent cations at concentrations near the cmc of NaTC in water [ca. 8–12 mM (32)]. As shown in Table 4.1, the effect becomes negligible at NaTC concentrations much lower and much higher than the cmc. In the same study, measurements of fluorescence anisotropy indicated that the microviscosity of the probe binding environment in NaTC is greater in the presence of lanthanide cations than in the presence of the lighter Al^{3+} or divalent cations, or in the absence of metal cations. This suggests that cation size and electronic structure may be important factors, in addition to cationic charge, in determining the nature of the binding microenvironments in NaTC aggregates.

4.3.2. Energy Transfer Between Probes in Bile Salt Media

The effects of NaTC on energy transfer between fluorescent molecules have been studied for several donor–acceptor pairs and compared to the effects observed for the same probes in sodium dodecyl sulfate (NaDS) micelles (33). Results for phenanthrene donor and several different acceptor molecules suggest that the solubilities of the probes are an important factor in energy transfer interactions in micellar media. For example, very little energy tranfer occurs from phenanthrene to perylene in either NaDS or NaTC. Phenanthrene, which is more than a thousandfold more soluble in water than perylene, is likely to be located in a less hydrophobic environment than the highly insoluble perylene and may not be close enough to facilitate energy transfer. The NaDS strongly promotes energy transfer to both 9,10-dimethylanthracene and 9-methylanthracene, which have solubilities closer to that of phenanthrene (energy transfer is highest for 9-methylanthracene, which is closest in solubility). In contrast to NaDS, little energy transfer to either 9,10-dimethylanthracene or 9-methylanthracene is exhibited in NaTC. Results for 9-phenylanthracene and 9,10-diphenylanthracene, for which solubilities have not been reported, are similar to those for 9-methylanthracene and perylene, respectively. Thus, relative solubilities of the donor and acceptor and their proximity to each other both appear to play an important role in energy transfer processes in the organized media.

The energy transfer experiments support the use of NaTC to solubilize fluorescent molecules in aqueous solution without promoting energy transfer interactions that tend to occur in other detergent micelles.

4.3.3. Effects of Bile Salts on Total Luminescence Spectra of Complex Samples

The ability of bile salts to minimize matrix effects and intermolecular interactions in complex samples has been demonstrated in studies of the total luminescence spectra, or excitation-emission matrices (EEMs), of complex samples such as crude oils (34), petrolatums (35), and coal liquids (36). For example, one study compared the total luminescence spectra of three different crude oils in cyclohexane to the corresponding spectra of the oils in aqueous NaTC (34). The spectral similarity between the three oils is much greater for the oils in NaTC than for the oils in cyclohexane. Also, the spectral intensities in the short-wavelength region are much more intense in NaTC than in cyclohexane, indicating less energy transfer in NaTC. The experiments were also tried in NaDS micellar media, but the formation of precipitates precluded the acquisition of meaningful measurements.

The implications of this study are that (i) the spectral differences between the oils are due not only to differences in their fluorophore composition but also to dynamic intermolecular interactions and matrix effects, and (ii) NaTC reduces the extent of the dynamic interactions and effects. Therefore, these studies support the premise that NaTC could be used to perform an in situ extraction of fluorescent molecules into uniform microenvironments in aqueous solution. The greater stability, e.g., smaller likelihood of precipitation, of the NaTC solutions relative to NaDS is another advantage for chemical analysis, especially when complex sample matrices are involved.

In the study of coal liquids (36), it was found that NaTC and NaDS both greatly increased the luminescence signal of the sample relative to simple solvents such as ethanol (Fig. 4.3A), presumably through increased solubilization of aromatic compounds. Moreover, NaTC showed a much greater ability than ethanol or NaDS to solubilize a high concentration of pyrene that

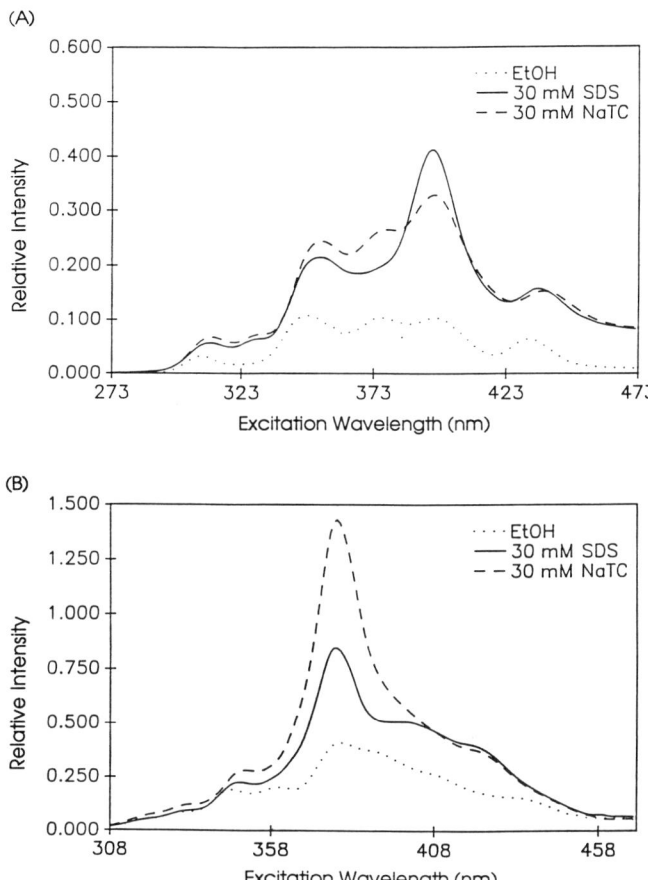

Fig. 4.3. Synchronous excitation fluorescence spectra of a standard coal liquid reference material (SRC-II), diluted to 0.050 μg/mL in ethanol, 30 mM NaDS, and 30 mM NaTC: (A) scanning difference between the emission and excitation monochromators (Δλ) held constant at 3 nm; (B) 15 mM pyrene added to the solutions, Δλ held constant at 38 nm.

was added to the coal liquid. Absorption spectra indicate that the increased luminescence in the organized media is not due to increased absorbance alone. Increases in the quantum yields of the fluorescent compounds, probably due to increased rigidity and decreased quenching and energy transfer upon binding to the micellar microenvironments, play an important role in the enhancement. Energy transfer and other intermolecular processes in the solutions of the coal liquids in the different media are currently under investigation.

4.3.4. Bile Salt Microenvironments for Solubilized Fluorescent Probes

Microenvironmental diversity is of particular interest in the evaluation of bile salts as reagents for chemical analysis because of the potential for enhanced selectivity in luminescence analysis as well as in other areas of analytical chemistry such as chemical separations. Recently, vibronic band ratio measurements were used to study the polarity of binding sites of several fluorescent probes in organized bile salt media (37). Several different probes, including pyrene, phenanthrene, triphenylene, benzo[*e*]pyrene (BeP) and benzo[*ghi*]-perylene (BgP), were measured in three different media, including the NaTC and its dihydroxy analogue, sodium taurodeoxycholate (NaTDC), and—for contrast—the conventional detergent NaDS. The results of the study, described below, indicate that the binding environments in NaTDC are the most apolar whereas the NaTC micellar solutions provide the greatest microenvironmental diversity among probes.

The vibronic band ratios were calculated from emission spectra as the intensity ratio of the vibronically forbidden 0–0 transition of band I to a vibronically allowed band. The intensity of the forbidden band is weak in nonpolar solvents and increases in polar solvents owing to specific solute-solvent interactions, such as dipole–dipole coupling, which serve to reduce the symmetry of the molecular probe (38,39). The intensity of the allowed band, on the other hand, is relatively insensitive to polarity. Thus, the band ratio will increase with increasing polarity of the probe's microenvironment.

Vibronic band ratios are shown in Table 4.2 for three of the probes in simple solvents and in aqueous micellar solutions of NaTC, NaTDC, and NaDS. Values for the probes in NaTC with added metal ion (5 mM Mg^{2+}, Al^{3+}, or Tb^{3+}) are also shown. The polarity "rulers" provided by the simple solvents are not linear, and the positioning of the inserted data for the micellar solutions is only a relative indication of polarity with respect to each other and the simple solvents. Furthermore, in a vibronic band ratio study of a microheterogeneous system, the binding site "polarity" is inferred from macroscopic values of intensity ratios observed in simple solvents. It is important to recognize, however, that it is not polarity of the solvent per se that affects transition probabilities (and therefore intensity ratios) but rather the specific interactions that occur between the molecular probe and its microenvironment. Therefore, although it is interesting and useful to compare band ratios for probes in binding environments to those ratios obtained in simple solvents, caution must be used in interpreting the results.

Differences among the solubilization microenvironments in NaTC, NaTDC, and NaDS are attributed to fundamental structural differences between the micellar aggregates of the three surfactants. The extra hydroxy group of trihydroxy NaTC compared to dihydroxy NaTDC is the source of the

Table 4.2. Vibronic Band Intensity Ratios of Fluorescent Probes in Organized Media

	Intensity Ratio		
Solvent	Pyrene (I/III)	Benzo[e]pyrene (I/IV)	Benzo[ghi]perylene (I/III)
Water	1.88	—[a]	—[a]
Acetonitrile	1.84	1.41	1.64
		NaTC 1.19	NaTC 1.57
			Mg^{2+}-NaTC 1.48
Methanol	1.40	1.11	1.40
		NaDS 1.10	Al^{3+}-NaTC 1.39
		Mg^{2+}-NaTC 1.09	Tb^{3+}-NaTC 1.38
Ethanol	1.25	1.06	1.34
	NaDS 1.21	Al^{3+}-NaTC 1.02	NaDS 1.27
		Tb^{3+}-NaTC 1.00	
2-Propanol	1.12	0.90	1.26
tert-Butanol	1.03	0.76	1.17
	NaTC 1.03		
	Mg^{2+}-NaTC 0.97		
	Al^{3+}-NaTC 0.93		
	Tb^{3+}-NaTC 0.92		
1-Decanol	0.89	0.72	1.00
	NaTDC 0.77	NaTDC 0.60	NaTDC 0.76
Cyclohexane	0.59	0.47	0.49
n-Heptane	0.58	0.36	0.48

Source: From Hertz and McGown (36).
[a] Ratios could not be determined owing to limited solubility of the probes in water.

differences between the two bile salts, increasing the solubility of the NaTC monomer and influencing the aggregate structure (19). The NaDS monomers, in contrast to the bile salts, self-associate to form the conventional, spherical micelles of large aggregation number (approximately 62) that are associated with detergent amphiphiles (40).

In addition to the observed differences in the solubilization microenvironments due to structural differences between the micellar hosts, the microenvironments experienced by the probe molecules are also influenced by the physical properties of the probes themselves. For example, in NaTC smaller probes experience relatively apolar microenvironments whereas larger probes experience relatively polar microenvironments. While both NaTC and NaTDC are capable of providing apolar binding sites to hydrophobic molecules by solubilizing the molecules between the hydrophobic surfaces of the bile salts, larger molecules solubilized in NaTC may protrude into the aqueous solution or disrupt the structure of the relatively small NaTC micelle. Such effects would account for the high intensity ratios observed for BeP and BgP in NaTC. In contrast to NaTC and NaTDC, solubilization of probes in NaDS often occurs in the micelle palisade layer (24), allowing interaction with ionic groups of the NaDS micelle and resulting in high intensity ratios (41).

Addition of the metal cations to the NaTC media decreased the vibronic band intensity ratios in the order: (least decrease) $Mg^{2+} < Al^{3+} < Tb^{3+}$ (most decrease). The increased hydrophobic character of the binding environments in the presence of these metals is consistent with the observations of the enhanced intensity and increased anisotropy that result from the addition of metal cations to NaTC media. Because NaTC provides a diverse range of microenvironments for the solubilized probe molecules, it is not unreasonable to expect the metal ions to affect the probe microenvironments to different extents. It is difficult, however, to compare relative changes in hydrophobicities of binding environments between probes from measurements of intensity ratios alone since the intensity ratio scales of the different probes are not linearly correlated to one another.

The vibronic band ratio measurements show that the microenvironments of solubilized probes in bile salts are very different from those in conventional detergents. The bile salts are capable of providing rigid, apolar binding environments to solubilized probes, whereas the binding environments in NaDS are more fluid and polar. Moreover, the microenvironments of the solubilized probes show a much greater probe-to-probe diversity in NaTC than in either NaTDC or NaDS and are much more influenced by the structure and size of the probe itself. This is consistent with the smaller size and aggregation number of the NaTC micelles, where the smaller probes are more easily isolated in well-protected apolar binding environments than are the less readily accommodated larger probes. Metal salts enhance the aggregation behavior of NaTC and could provide a means for selectively influencing the microenvironments of hydrophobic molecules in the media.

4.4. FUTURE DIRECTIONS: CHIRAL SELECTIVITY, MIXED MICELLES, AND LYOTROPES

Circular dichroism (CD) can be induced or modified in a probe molecule through strong association with a chiral, helical, or otherwise asymmetric binding microenvironment. The ability of some bile salt phases to accomplish enantiomeric separations in chromatography suggests that these phases may offer such environments. In fact, a helical model for bile salt aggregation

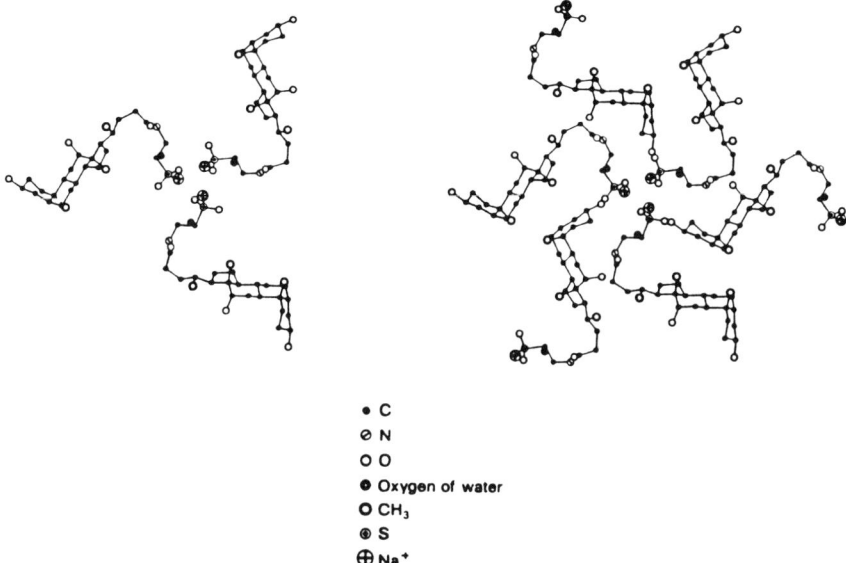

Fig. 4.4. Cross-sectional view of helical aggregates proposed for sodium taurodeoxycholate: three-membered (left) and six-membered (right) helices. [From D'Alagni et al. (42), with permission.]

(Fig. 4.4) has been proposed (42,43). Induced or modified CD, measured by absorption CD spectroscopy and fluorescence-detected CD spectroscopy (FDCD), could be used to study the binding interactions between bile salts and probe molecules in processes such as chiral chromatography. Furthermore, such bile salt phases might well be an important means for incorporating chiral selectivity into luminescence analysis.

An interesting aspect of bile salt aggregation is the formation of mixed micellar phases and lyotropic phases. Mixed micelles have been reported to form between trihydroxy bile salts and detergents, including nonionic (44), anionic (45), and cationic (46) detergents. Figure 4.5 shows a model that has been proposed for mixed micelles formed by cholate and cetyltrimethylammonium ions (46). In mixed micellization, as in complex samples, solutions containing mixtures of detergents and bile salts are less likely to form a precipitate with trihydroxy bile salts than with dihydroxy bile salts.

Lyotropic liquid crystalline phases (47–49) are formed by the action of a solvent on an amphiphilic compound alone or in a mixture, in a scheme that generally includes three different levels of liquid crystalline organization. Soluble amphiphilic compounds can be divided into two groups: those that form lyotropic liquid crystals from the pure compound and those that do not.

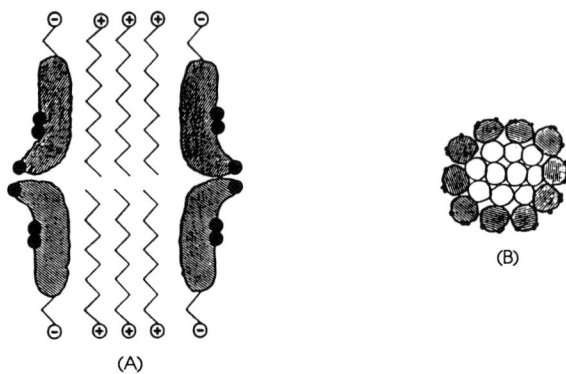

Fig. 4.5. Longitudinal view (A) and cross-sectional view (B) of proposed model for alkyltrimethylammonium cholate mixed micelle. [From Barry and Gray (46), with permission.]

Common detergents of the "polar head–hydrophobic tail" variety belong to the former group, whereas bile salts belong to the latter. Therefore, liquid crystallinity is not expected to occur in solutions of individual bile salts. However, the formation of lyotropic phases in mixtures of bile salts with other lipids is a well-described phenomenon. For example, the following systems exhibit all three of the common lyotropic phases as well as mixed micellar phases: sodium cholate–lecithin–water (50), monoolein–bile salt–water (51), and lecithin–cholesterol–bile salt–water (48).

A few studies have explored analytical applications of lyotropic liquid crystalline phases, including some very recent work on bile salt–alcohol phases in water for HPLC (52). However, lyotropic liquid crystals in chemical analysis have not been widely explored. As with bile salt micellar phases and mixed micellar phases, most of the investigations of lyotropic bile salt–detergent phases have focused on biologically significant systems. In the growing field of organized media as reagents for chemical analysis, bile salts offer unique and exciting possibilities that have only recently begun to be explored.

ACKNOWLEDGMENT

This work was supported by the U.S. Department of Energy (Grant No. DE-FG05-88ER13931).

REFERENCES

1. P. H. Elworthy, A. T. Florence, and C. B. Macfarlane, *Solubilization by Surface Active Agents and Its Application to Chemistry and the Biological Sciences*, Chapman & Hall, London, 1968.

REFERENCES

2. K. L. Mittal, Ed., *Micellization, Solubilization and Microemulsions*, Plenum Press, New York, 1977.
3. W. L. Hinze, in K. L. Mittal, Ed., *Solution Chemistry of Surfactants*, Vol. 1, Plenum Press, New York, 1979, p. 79.
4. H. N. Singh and W. L. Hinze, *Anal. Lett.*, **15**, 221 (1982).
5. H. N. Singh and W. L. Hinze, *Analyst*, **107**, 1073 (1982).
6. T. Taketatsu and A. Sato, *Anal. Chim. Acta*, **108**, 429 (1979).
7. T. Taketatsu, *Talanta*, **29**, 397 (1982).
8. J.-H. Yang, G.-Y. Zhu, and B. Wu, *Anal. Chem. Acta*, **198**, 287 (1987).
9. C. D. Tran, *Anal. Chem.*, **60**, 182 (1988).
10. K. Kalayanasundaram, F. Greiser, and J. K. Thomas, *Chem. Phys. Lett.*, **51**, 501 (1977).
11. R. A. Femia and L. J. C. Love, *Anal. Chem.*, **56**, 327 (1984).
12. R. A. Femia and L. J. C. Love, *Spectrochim. Acta*, **42A**, 1239 (1986).
13. G. R. Ramos, I. M. Khasawneh, M. C. Gracia-Alvarez-Coque, and J. D. Winefordner, *Talanta*, **35**, 41 (1988).
14. L. J. C. Love, M. Skrilec, and J. G. Habarta, *Anal. Chem.*, **52**, 754 (1980).
15. L. J. C. Love, M. Skrilec, and J. G. Habarta, *Anal. Chem.*, **53**, 437 (1981).
16. M. Skrilec and L. J. C. Love, *Anal. Chem.*, **52**, 1559 (1980).
17. M. Skrilec and L. J. C. Love, *J. Phys. Chem.*, **85**, 2047 (1981).
18. P. H. Baudst and J. C. Andre, *Anal. Lett.*, **15**, 471 (1982).
19. D. M. Small, in P. P. Nair and D. Kritchevsky, Eds., *The Bile Acids*, Vol. 1, Plenum Press, New York, 1971, p. 249.
19a. C. V. O'Connor and R. G. Wallace, *Adv. Colloid Interface Sci.*, **22**, 1 (1985).
20. R. Zana and G. Guveli, *J. Phys. Chem.*, **89**, 1687 (1985).
21. L. Fisher and D. Oakenfull, *Aust. J. Chem.*, **32**, 31 (1979).
22. M. Chen, M. Grätzel, and J. K. Thomas, *J. Am. Chem. Soc.*, **97**, 2052 (1975).
23. G. Sugihara, K. Yamakawa, Y. Murata, and M. Tanaka, *J. Phys. Chem.*, **86**, 2784 (1982).
24. M. J. Rosen, *Surfactants and Interfacial Phenomena*, 2nd ed., Wiley, New York, 1988.
25. G. Conte, R. Di Blasi, E. Giglio, A. Parretta, and N. V. Pavel, *J. Phys. Chem.*, **88**, 5720 (1984).
26. E. Kolehmainen, *J. Colloid Interface Sci.*, **105**, 273 (1985).
27. M. Chen, M. Gratzel, and J. K. Thomas, *Chem. Phys. Lett.*, **24**, 65 (1974).
28. D. M. Small, *Adv. Chem. Ser.*, **84**, 31 (1968).
29. M. C. Carey and D. M. Small, *J. Colloid Interface Sci.*, **31**, 382 (1969).
30. K. Nithipatikom and L. B. McGown, *Anal. Chem.*, **60**, 1043 (1988).
31. S. M. Meyerhoffer and L. B. McGown, *J. Am. Chem. Soc.*, **113**, 2146 (1991).
32. S. M. Meyerhoffer and L. B. McGown, *Langmuir*, **6**, 187 (1990).

33. K. Nithipatikom and L. B. McGown, *Anal. Chem.*, **61**, 1405 (1989).
34. L. B. McGown and D. S. Kreiss, *SPIE Proc.*, **909**, 360 (1988).
35. P. M. R. Hertz and L. B. McGown, *Appl. Spectrosc.*, **45**, 73 (1991).
36. P. M. R. Hertz and L. B. McGown, unpublished results.
37. S. M. Meyerhoffer and L. B. McGown, *Anal. Chem.*, **63**, 2082 (1991).
38. A. Nakajima, *J. Lumin.*, **8**, 266 (1974).
39. K. Kalyanasundaram and J. K. Thomas, *J. Am. Chem. Soc.*, **99**, 2039 (1977).
40. J. H. Fendler and E. J. Fendler, *Catalysis in Micellar and Macromolecular Systems*, Academic Press, New York, 1975.
41. K. Kalyanasundaram, *Photochemistry in Microheterogeneous Systems*, Academic Press, Orlando, FL, 1987.
42. M. D'Alagni, E. Giglio, and S. Petriconi, *Colloid Polym. Sci.*, **265**, 517 (1987).
43. A. R. Campanelli, S. Candeloro De Sanctis, E. Chiessi, M. D'Alagni, E. Giglio, and L. Scaramuzza, *J. Phys. Chem.*, **93**, 1536 (1989).
44. H. Asano, A. Murohashi, and M. Ueno, *J. Am. Oil Chem. Soc.*, **67**, 1002 (1990).
45. M. M. Velazquez, I. Garcia-Mateos, F. Lorente, M. Valero, and L. J. Rodriguez, *J. Mol. Liq.*, **45**, 95 (1990).
46. B. W. Barry and G. M. T. Gray, *J. Colloid Interface Sci.*, **52**, 314 (1975).
47. A. Helenius and K. Simons, *Biochim. Biophys. Acta*, **415**, 29 (1975).
48. G. H. Brown and P. P. Crooker, *Chem. Eng. News*, Jan. 31, p. 24 (1983).
49. G. J. T. Tiddy and M. F. Walsh, in E. Wyn-Jones and J. Gormally, Eds., *Aggregation Processes in Solution*, Elsevier, New York, 1983, pp. 151–185.
50. D. M. Small, M. Bourges, and D. G. Dervichian, *Nature (London)*, **211**, 816 (1966).
51. M. Svard, P. Schurtenberger, K. Fontell, B. Jonsson, and B. Lindman, *J. Phys. Chem.*, **92**, 2261 (1988).
52. W. L. Hinze, private communication.

CHAPTER

5

SPECTRAL HOLE-BURNING

KEITH HOLLIDAY AND URS P. WILD

*Physical Chemistry Laboratory
Swiss Federal Institute of Technology
CH-8092 Zürich, Switzerland*

5.1. Introduction
5.2. Experimental Techniques
    5.2.1. Persistent Spectral Hole-Burning
        5.2.1.1. Absorption and Excitation Spectroscopy
        5.2.1.2. Holography
        5.2.1.3. Other Techniques
    5.2.2. Transient Spectral Hole-Burning
5.3. Spectroscopic Applications
    5.3.1. Impurity and Defect Sites
        5.3.1.1. Uranium-Doped Strontium Tungstate
        5.3.1.2. Curium-Doped Cerium Fluoride
        5.3.1.3. The 607-nm Color Center in Sodium Fluoride
        5.3.1.4. Holmium-Doped Calcium Fluoride
        5.3.1.5. Chlorin-Doped n-Octane Shpol'skii Matrix
        5.3.1.6. Nile Red–Doped Poly(vinyl butyral)
        5.3.1.7. Cresyl Violet–Doped Poly(vinyl butyral)
    5.3.2. Electronic Interactions
        5.3.2.1. Nitrogen-Vacancy Color Center in Diamond
        5.3.2.2. Europium-Doped Yttrium Aluminum Perovskite
        5.3.2.3. Erbium-Doped Yttrium Lithium Fluoride
    5.3.3. Tunneling Processes
        5.3.3.1. Thioindigo-Doped Benzoic Acid
        5.3.3.2. Dimethyl-s-tetrazine–Doped Tetramethylbenzene
        5.3.3.3. Praseodymium- and Deuterium-Doped Strontium Fluoride
    5.3.4. Optical Dephasing
        5.3.4.1. Porphyrin-Doped n-Octane
        5.3.4.2. Magnesium Porphyrin–Doped n-Octane
        5.3.4.3. Samarium-Doped Barium Fluorochlorobromide

Molecular Luminescence Spectroscopy, Part 3, Edited by Stephen G. Schulman. Chemical Analysis Series, Vol. 77.
ISBN 0-471-51580-9 © 1993 John Wiley & Sons, Inc.

            5.3.4.4. *Porphyrin-Doped Glycerol:Ethanol Glass*
            5.3.4.5. *Porphyrin-Doped Poly(ethylene)*
            5.3.4.6. *Octaethylporphyrin-Doped Poly(styrene)*
            5.3.4.7. *Praseodymium-Doped Silicate Glass*
    5.3.5. Spectral Diffusion in Glasses
            5.3.5.1. *Quinizarin-Doped Ethanol:Methanol Glass*
            5.3.5.2. *Cresyl Violet–Doped Ethanol Glass*
    5.3.6. Vibronics
            5.3.6.1. *Cresyl Violet–Doped Poly(vinyl alcohol)*
            5.3.6.2. *Porphyrin-Doped n-Decane*
            5.3.6.3. *1,2-Difluoroethane-Doped Solid Krypton*
            5.3.6.4. *Selenium Hydride Centers in Germanium–Arsenic–Selenium Glass*
    5.3.7. Special Materials
            5.3.7.1. *Biological Compounds*
            5.3.7.2. *Thin Films*
            5.3.7.3. *Photon-Gated Spectral Hole-Burning Materials*
5.4. **Technological Applications**
    5.4.1. Optical Data Storage
    5.4.2. Holography
            5.4.2.1. *Image Storage*
            5.4.2.2. *The Molecular Computer*
Acknowledgments
References

5.1. INTRODUCTION

Before the invention of the laser, fluorescence spectroscopy was largely limited to the resolution available from monochromators. In the case of the study of impurity molecules, ions, or color centers in solids, optical spectra were interpreted as being due to a series of electronic transitions coupled to local mode vibrations and phonon bands. Selective excitation and emission studies, analyzed through the application of group theory, revealed a great deal of information about the structure of materials, but the low resolution of the measurements frequently necessitated ambiguous conclusions. Such measurements were essentially of bulk properties in that each spectral feature was composed of contributions from all impurities in the solid.

The electronic states of impurities in solids are influenced by the local electric, magnetic, and strain fields generated by surrounding atoms, ions, or molecules. In a real solid there is a variation in environment for each impurity that causes a distribution of energies for each electronic level. Each spectral line corresponding to an electronic transition is therefore composed of the sum of contributions of all impurities, each centered at a slightly different wavelength. The spectral line is said to be inhomogeneously broadened.

The contribution of each impurity to an inhomogeneously broadened spectral line is homogeneously broadened as a direct result of the uncertainty principle. The excited state lifetime, T_1, and dephasing time of the optical transition, T_2^*, determine the homogeneous line width, Γ_{hom}, of the transition:

$$\Gamma_{\text{hom}} = 1/\pi T_2' = 1/2\pi T_1 + 1/\pi T_2^* \tag{5.1}$$

where T_2' is the total optical dephasing time of the system; T_2^* is composed of contributions due to interactions within the solid that cause fluctuations in the transition frequency. Most dephasing processes are strongly temperature dependent and, consequently, so are homogeneous line widths. Spectral lines due to electronic transitions are usually dominated by homogeneous broadening at room temperature. That is, the dephasing of the optical transition is so fast that the homogeneous line width of the transition is greater than the spread of the spectral line due to site variation. As the solid is cooled, dephasing processes slow and the spectral line becomes dominated by inhomogeneous broadening. This is schematically illustrated in Fig. 5.1, in which a series of Lorentzian, homogeneously broadened lines, representing the transitions of each impurity molecule or ion, are normally distributed, giving rise to an inhomogeneously broadened Gaussian spectral line.

When the resolution of measurement is sufficiently high, it is possible to interact with a subset of impurities, each of which contributes to the same portion of a spectral line. For instance, energy selection may be carried out through the use of a highly monochromatic laser that excites only a small proportion of impurities whose transitions are resonant with the exciting radiation.

Detection of fluorescence line narrowing, in ruby (1), was the first experimental demonstration that the inhomogeneous line width of electronic transitions in solids could be penetrated. Fluorescence line narrowing uses energy selection to examine a subset of impurities but is still restricted by the resolution of the fluorescence detection equipment.

Spectral hole-burning overcomes this limitation. A highly monochromatic laser is used to selectively excite impurities within the solid. Through a variety of mechanisms, all of which rely on the removal of a proportion of absorbing impurities from resonance, the subsequent absorption at the same wavelength may decrease. Measurement of the fluorescence intensity, at any wavelength, will drop correspondingly, independent of detection resolution, and the spectral hole lineshape may be recorded by scanning the excitation wavelength while measuring the fluorescence intensity or while simply measuring the

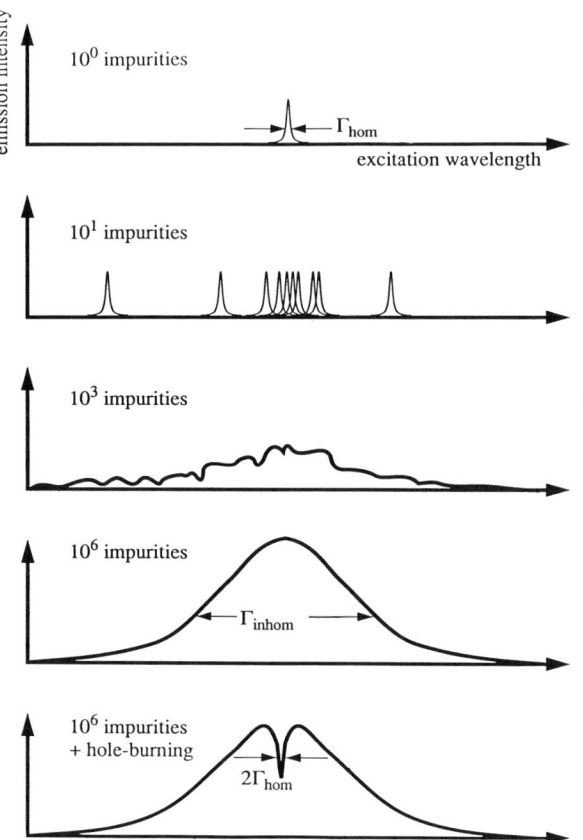

Fig. 5.1. A transition of a single impurity in a host matrix has a homogeneously broadened Lorentzian lineshape. Site variation causes the transition energy to vary for different impurities. The distribution of a large number of impurities is Gaussian. When a laser with a line width less than the homogeneous line width of the transition interacts with a subset of impurities such that their transition energy changes, a Lorentzian "spectral hole" is superimposed on the Gaussian distribution.

strength of the transmitted beam. The spectral resolution is thus determined by the laser resolution alone.

The possibility of detecting weak perturbation effects is immediately obvious, and results from excitation energy selection techniques have removed ambiguities from conclusions derived using traditional methods. Furthermore, these methods have the capacity to reveal information about homogeneous processes. When the line width of the exciting radiation is smaller than the homogeneous line width and the excited state lifetime is known, information about dephasing processes in the solid may be directly obtained.

In an ideal case, a spectral hole created by the photoreaction of a single molecule due to the absorption of a single photon of light, read using a light source of negligible power and line width, has a width of twice the homogeneous line width (2). In reality, single molecule holes cannot easily be detected (3), causing deviations due to saturation after finite burning times and broadening due to statistical considerations associated with the power of the exciting source (4).

It is usual to divide the mechanisms by which spectral hole-burning can take place into three categories based on the persistence of the absorption change.

In principle, it is possible to observe saturation hole-burning in any inhomogeneously broadened transition. By irradiating the material with high enough power densities it is possible to maintain a substantial proportion of impurities in their excited state. A probe beam may examine the absorption of the material simultaneously or after the removal of the saturating beam but during the excited state lifetime, as the hole refills. The systems studied and techniques used are often very different from those for transient and persistent spectral hole-burning and are not discussed in detail here. This chapter is largely restricted to phenomena experimentally observable on a time scale slower than microseconds.

Transient spectral hole-burning refers to the reduction of absorption of a material due to the process of relaxation from the excited state back to the original ground state being slower than the excited state lifetime. For instance, molecules may decay to the ground state via a bottleneck such as a long-lived triplet level. Alternatively, the impurity may have a spin-split ground state causing relaxation to progress, in a proportion of cases, via an alternative ground state. In both cases the persistence of the hole is determined by the lifetimes of the intermediate states, which are typically between milliseconds and seconds. This technique was also demonstrated for the first time in ruby (5).

At about the same time, persistent spectral hole-burning was first observed in perylene and 9-aminoacridine in ethanol glass (6) and in free base phthalocyanine in an *n*-octane Shpol'skii matrix (7). Persistent spectral hole-burning refers to an optical process in which energy-selected molecules are excited but fail to relax to their original ground state. This is due to a pseudopermanent change in the electronic or physical structure of the impurity or host matrix. In some cases the hole may slowly refill, but at low temperatures such holes are usually persistent over a period of hours. It is then a simple matter to make accurate measurements of hole widths to reveal information about dephasing processes, and to monitor the effect of external perturbations, such as electric or magnetic fields or uniaxial stress, to deduce information about the electronic and physical structure of the material.

Mechanisms for persistent spectral hole-burning are traditionally divided into two categories, photochemical and photophysical (sometimes referred to as nonphotochemical). Photochemical mechanisms involve an electronic or structural change within the individual impurity ion or molecule induced by the absorption of one or more photons. Photophysical mechanisms involve a change in the local structure of the matrix caused by the excitation of an impurity. Such a local change may be dominated by the matrix itself being perturbed or by the impurity moving within the matrix.

After such a photophysical or photochemical process the absorbing wavelength of the impurity changes, causing a spectral hole at the original absorption wavelength. The new absorption wavelength is characteristic of the hole-burning mechanism (8). A photochemical change to the impurity typically radically alters the absorption wavelength and the new absorption, or antihole, may be many nanometers removed. A photophysical process merely alters the environment of an impurity and usually displaces the absorption elsewhere within the existing inhomogeneously broadened band, modifying the lineshape but leaving the total absorption of the band unchanged.

The efficiency of spectral hole-burning is also connected to the hole-burning mechanism, photochemical mechanisms being generally more efficient. Hole-burning efficiency is defined as being the proportion of excitations that cause the transition frequency of the chromophore to change (9) and varies from greater than 10^{-2} to less than 10^{-6}.

The power of spectral hole-burning is in the enormous increase in spectral resolution, which leads to information about the physical structure and dynamic processes within the solid, in particular to the extremely accurate data that may be obtained about homogeneous processes. The range of materials for which spectral hole-burning experiments have been successfully performed has become large (of the order of hundreds of different impurities in hundreds of different hosts) since the effect was first observed, yielding information of wide-ranging relevance. Note, however, that there are systems for which neither transient nor persistent hole-burning may be observed. Nevertheless, it is often possible to investigate host matrices by adding a small quantity of an impurity molecule known to exhibit spectral hole-burning in other materials.

A review of some hole-burning materials is given in Section 5.3 of this chapter. Details of studies are given, chosen to demonstrate the nature of the information deducible from hole-burning measurements but also to indicate some of the physical processes that lead to spectral hole-burning and to illustrate experimental techniques. Section 5.2 reviews these hole-burning techniques. Finally, in Section 5.4, a review of possible technological applications is given. Previous reviews of spectral hole-burning are also recommended to the reader (10–13).

5.2. EXPERIMENTAL TECHNIQUES

In this section, the experimental techniques that have proven useful for the creation and detection of spectral holes are discussed. Both persistent and transient hole-burning are dealt with, placing special emphasis on the newer techniques that reduce background, thus allowing weak signals to be detected.

5.2.1. Persistent Spectral Hole-Burning

The persistence of a spectral hole affords experimental luxury. The burn and read periods may be separated by a sufficiently large time to allow, for example, the application of external fields or the completion of temperature cycles. Persistent spectral hole detection has been developed from simple methods that involve monitoring fluorescence or absorption in the spectral region of the hole to holographic, polarizaton, and modulation techniques that provide very low background and therefore allow weaker holes to be detected. When accurate information regarding, for instance, optical dephasing (see Section 5.3.4) or spectral diffusion (see Section 5.3.5) is required, it is necessary to minimize laser powers to prevent power broadening and local heating of the sample. In such cases the low background techniques are ideal.

5.2.1.1. Absorption and Excitation Spectroscopy

It is often the case that the hole width is not of paramount importance to the experimentalist, for instance, when the hole is laser line width limited or being used as a probe of the effect of an external field. Here it is necessary to employ only simple techniques, as relatively high powers may be used to burn and read deep holes.

The frequency-dependent transmitted intensity, I_t of a beam of intensity, I_0, through a material with an absorption coefficient, $\alpha(v)$, and thickness, d is given by

$$I_t(v) = I_0 \exp(-d\alpha(v)) \tag{5.2}$$

In the region of a narrow spectral hole $\alpha(v)$ may be taken to be a constant, α, modified by the Lorentzian hole profile, $L(v)$, given by

$$L(v) = D(\Gamma/2)^2 / \{(v - v_b)^2 + (\Gamma/2)^2\} \tag{5.3}$$

where D, Γ, and v_b are the hole depth, width, and center frequency, respectively, so that the transmitted intensity becomes

$$I_t(v) = I_0 \exp(-d\alpha) \exp\left(dD(\Gamma/2)^2/\{(v-v_b)^2 + (\Gamma/2)^2\}\right) \qquad (5.4)$$

and the signal appears as a Lorentzian peak of width, Γ, superimposed on a background corresponding to the initial transmission conditions.

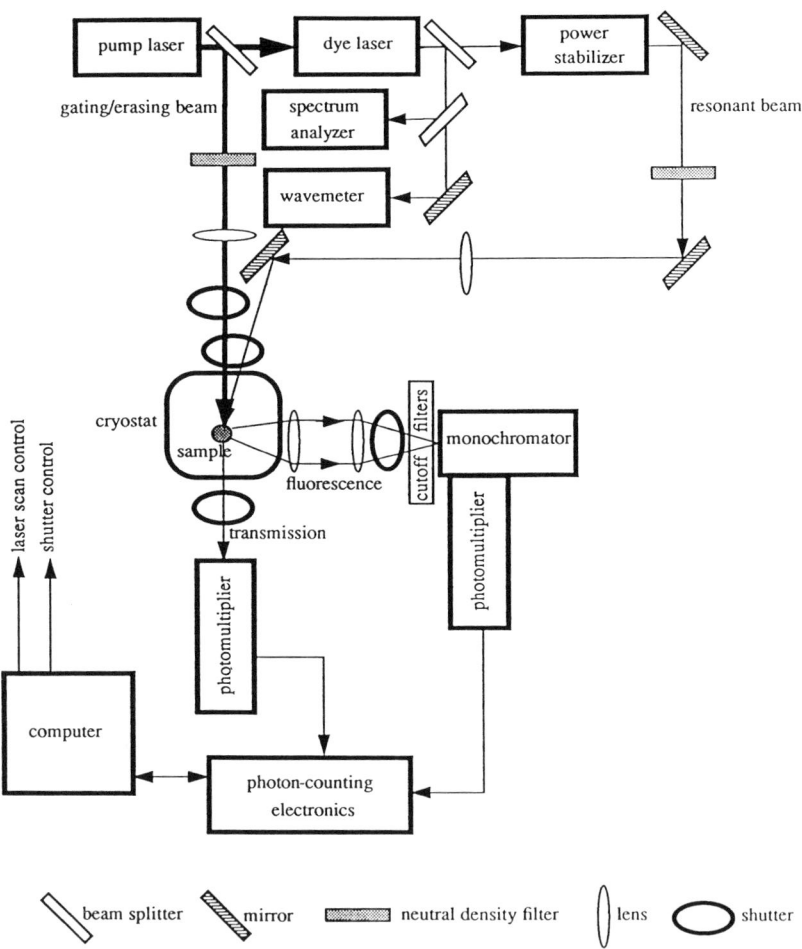

Fig. 5.2. A typical experimental arrangement for spectral hole-burning. Usually, only one of the transmission or excitation photomultipliers is used, depending on the properties of the sample being studied. The gating/erasing beam is also only of use with certain materials.

The fluorescence excitation intensity, $I_e(v)$, corresponds closely to the transmission intensity as the fluorescence signal is proportional to the light absorbed by the same, $I - I_t$, or

$$I_e(v) \propto I_0 [1 - \exp(-d\alpha)\exp(dD(\Gamma/2)^2/\{(v - v_b)^2 + (\Gamma/2)^2\})] \quad (5.5)$$

The signal appears as a Lorentzian dip of width, Γ, in a background corresponding to the initial fluorescence intensity.

While the recording of transmission and excitation spectra is straightforward with a suitable high-resolution dye laser and detection equipment, a more typical experimental arrangement is shown in Fig. 5.2. In this scheme a second beam, which may be used for gating or erasing (see Section 5.3.7.3), is provided by splitting off a small proportion of the pump laser. The dye laser output is power stabilized and monitored for mode stability and wavelength. Neutral density filters provide the option to attenuate the laser power by large factors, as is necessary after the burn stage, to prevent further photoreactions that would degrade the spectral hole from occurring during reading. Typical burn conditions require laser power of the order of milliwatts per square centimeter for some seconds or minutes, whereas the read beam should be attenuated by a factor of at least 10. The transmission signal may be detected directly, whereas the excitation signal requires nonresonant fluorescence to be filtered from parasitic laser scatter by cutoff filters, a monochromator, or both. The photomultiplier output pulses may then be interfaced to a computer via a collection of preamplifiers, discriminators, and counters. It is also convenient to operate the sequence of shutter manipulations and external field applications through computer control.

5.2.1.2. Holography

When the burn beam is split and recombined on a suitable material, the sinusoidal interference fringes create absorption and refractive index gratings due to spatially selective hole-burning. Subsequent irradiation with one of the beams causes some of the light to be diffracted collinear with the other. Using an experimental arrangement similar to that shown in Fig. 5.3, the profile of the spectral hole may be recorded as either a transmission or diffraction signal. It can be shown (14) that, for a purely sinusoidal grating, the diffraction or holographic intensity, I_h, is given by

$$I_h = I_0 \exp(-d\alpha) N^2 \sigma^2 d^2 (\Gamma/2)^2 / 16\{(v - v_b)^2 + (\Gamma/2)^2\} \quad (5.6)$$

where N is the number of molecules burnt, the absorption cross section of the molecules being σ. From this expression it can be seen that holography has

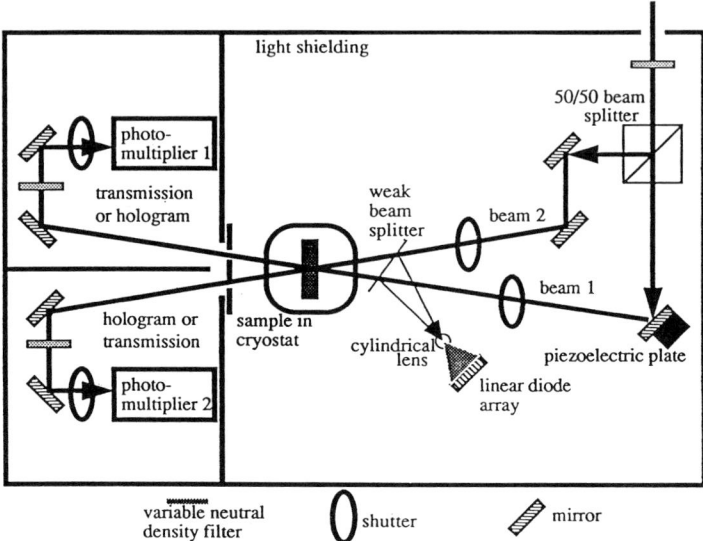

Fig. 5.3. A typical experimental arrangement for burning and detecting spectral holes using the holographic technique. Both beams are used for burning but only one for reading. The piezoelectric plate is used to control the spatial phase of the grating, which the linear diode array monitors.

the advantage of being a zero background technique, in that there is no diffracted signal outside the range of the spectral hole, but that the impurity must have a large value of σ to achieve large diffraction efficiencies. Furthermore, the material must have excellent optical quality to prevent parasitic laser scatter from degrading the signal. This has been achieved for dye molecules in polymer films (Fig. 5.4).

The parameters governing the creation of single plane wave holograms produced as spectral holes have been investigated experimentally and explained theoretically (14). As burning time increases, the hologram efficiency increases linearly while the spatial distribution of removed impurities is largely sinusoidal, corresponding to the intensity of the interference fringes of the burn beams. However, as the absorbing concentration of impurities decreases, the grating becomes increasingly more anharmonic. After a well-defined period the holographic efficiency at the exciting frequency reaches a maximum and begins to decrease. It has been shown (14) that when the holographic efficiency reaches a maximum the frequency broadening of the diffracted signal is twice the equivalent parameter for the transmission signal. By burning a hole to the maximum holographic efficiency at the laser wavelength and measuring the

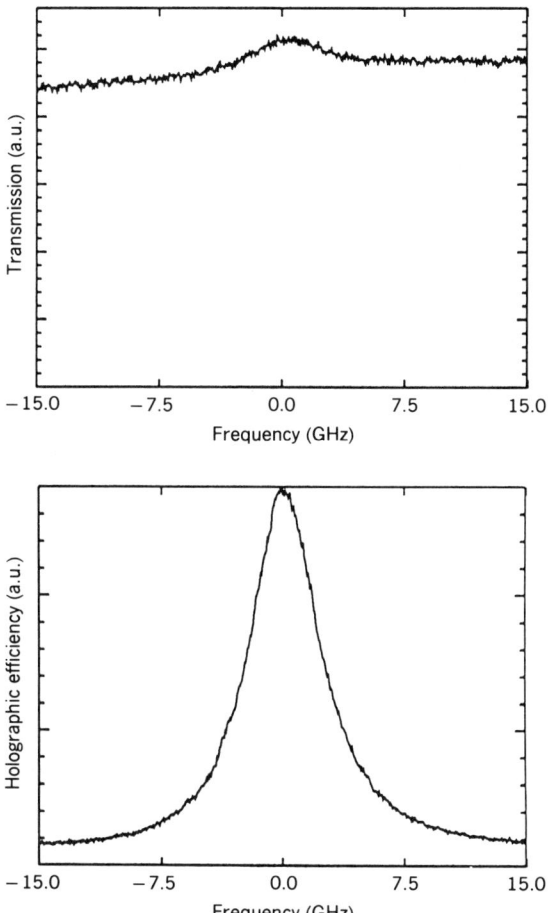

Fig. 5.4. Comparison of transmission (left) and holographic (right) signals for spectral holes burned in Nile red–doped poly(vinyl butyral) at 1.6 K (16).

relative widths of the transmission and holographic signals, the homogeneous line width of free-base chlorin in a poly(vinyl butyral) film was determined to be 170 MHz. Such a technique determines homogeneous line widths from deep holes with good signal-to-noise ratios. This precludes the measurement of very weak holes as is necessary when extrapolating to zero fluence to eliminate the effect of power broadening using the absorption or excitation methods.

A study of the diffraction properties of spectrally adjacent holograms (15) has revealed the phase dependence of the interference caused by the overlap of hole wings. Plane wave holograms burned as spectral holes may be considered to be composed of absorption and refractive index gratings each of which causes a diffracted beam $\pi/2$ out of phase with the other. When hologram wings spectrally overlap, the diffraction efficiency is determined by the vector sum of the absorption and refractive index components. If the adjacent holograms are burnt with the same relative phase, that is, at different frequencies but with interference grating maxima in the same spatial positions, the vector sum is of absorption parts on the "real" axis and refractive index parts on the "imaginary" axis. When the phase of the holograms is unequal, that is, with interference grating maxima in different spatial positions, absorption and refractive index parts interfere on both axes and thus the diffraction efficiency of the holograms is changed. The result is phase-dependent diffraction efficiency for spectrally adjacent holograms.

Application of an electric field to some materials causes a linear Stark shift of the resonant optical transition energy which may split the spectral hole. Split components of spectrally adjacent holes may be overlapped by applying a particular electric field resulting in a superposition of the associated gratings at the center frequency. Nile red (see Section 5.3.1.6) is an example of an appropriate impurity for electric field–induced interference investigations, as the small angle between the transition moment and the dipole moment difference between ground and excited states and the large magnitude of the dipole moment difference give rise to an appreciable hole splitting when the electric field vector of the light field is parallel to an applied electric field (16,17).

A rigorous derivation of the frequency and electric field dependence of the diffraction efficiency, $\eta(v; E)$, of a pair of spectrally adjacent holograms has been obtained (18). It is given by

$$\eta(v, E) = A_1(v, E)^2 + B_1(v, E)^2 + A_2(v, E)^2 + B_2(v, E)^2 \\ + 2[A_1(v, E) \cdot A_2(v, E) + B_1(v, E) \cdot B_2(v, E)] \cos(\phi_2 - \phi_1) \\ + 2[B_1(v, E) \cdot A_2(v, E) - A_1(v, E) \cdot B_2(v, E)] \sin(\phi_2 - \phi_1) \quad (5.7)$$

This expression contains terms, A_1 and A_2 describing contributions due to modulation of the absorption coefficient caused by holograms 1 and 2, respectively, and terms, B_1 and B_2, describing contributions due to modulation of the propagation constant (proportional to the refractive index (14)). The nature of the interference between the gratings when the Stark components are overlapped is determined by the spatial phase difference, $\phi_2 - \phi_1$, between the holograms, selected during burning. Good agreement between experiment and theory was obtained for pairs of holograms burned at different frequencies and zero electric field and for pairs of holograms burned at the same frequency but at different electric field strengths. Figure 5.5 illustrates the latter case for a pair of holograms burned with a phase difference of $\pi/2$. Small differences

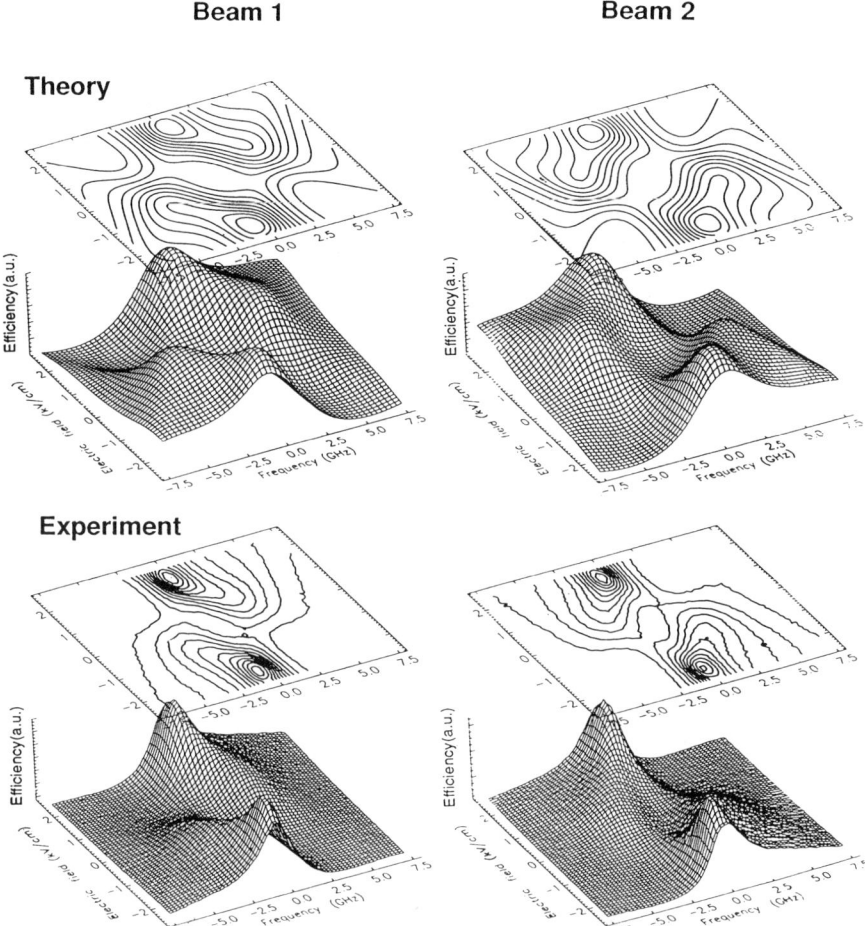

Fig. 5.5. Comparison of experimental and theoretical data for the interference of a pair of holograms recorded as spectral holes (19). Both holes were burned at the same frequency (0) but at different electric field strengths ($\pm 2\,\text{kV/cm}$). The same pair of holograms are reconstructed using each beam (see Fig. 5.3). The data are plotted as an 11 contour plot and a surface plot with arbitrary units of holographic efficiency.

between experimental data and numerical simulations are explained by slight phase fluctuations during burning and by matrix-induced contributions to the dipole moments not accounted for in the computer model.

The dependence of holographic efficiency on the choice of reconstruction beam (19) is also illustrated by Fig. 5.5. To understand the origin of the dependence of holographic efficiency on that choice, consider Eq. (5.7). The beams are incident upon the gratings at angles to the normal of θ and $-\theta$, respectively. The effective spatial phase difference between the holographic gratings is therefore reversed when the reconstruction beam is changed. For one beam the spatial phase difference is $\phi_2 - \phi_1$, whereas for the other it is $\phi_1 - \phi_2$. The absorption coefficient and the propagation constant modulations are physical properties of the gratings and are therefore not influenced by the choice of reconstruction beam. When the gratings are asymmetrically displaced from each other, however, they will appear differently to radiation incident from opposite sides of the normal. When the gratings are displaced symmetrically, for $\phi_2 - \phi_1 = 0$ or π, they will appear identical to each beam and the diffraction efficiency as a function of electric field and frequency is equivalent regardless of which is chosen for reconstruction.

A typical transmission signal is shown in Fig. 5.6. Interference effects arise from the properties of the holographic gratings and are therefore manifested

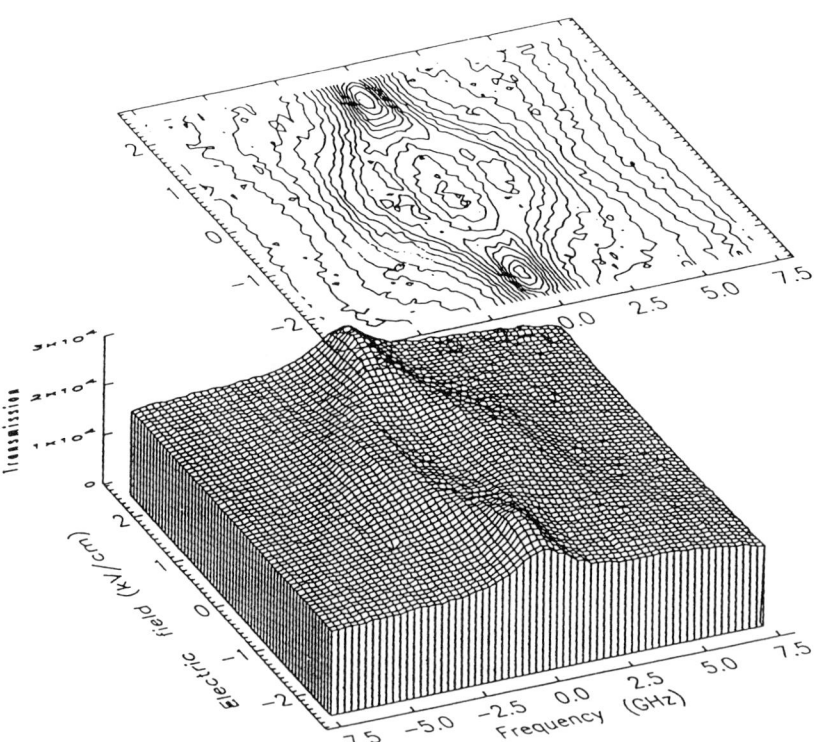

Fig. 5.6. Experimental plot of transmission strength measured concurrently with the holographic efficiency plot shown in Fig. 5.5 for a spatial phase change of $\pi/2$ and with beam 1. Note symmetry of signal, as seen for all combinations of phase and beam in transmission.

only in the holographic signal. The transmission signal therefore has the same symmetry properties regardless of reconstruction beam or grating properties. The large background signal associated with the recording of spectral holes in transmission is clearly seen.

A full understanding of the reconstruction properties of holographic gratings and the interference effects between them is essential to the development of multiple image storing holographic systems (20) and for the realization of a working "molecular computer" (21) (see Section 5.4.2). An analogue of this technique that involves examining the response of a persistent spectral hologram to a pulsed reconstruction source has also been developed (22).

5.2.1.3. Other Techniques

A second low-background technique that may be applied to detect weak spectral holes in strongly absorbing materials with good optical quality involves the application of polarization spectroscopy. When a hole is burned with polarized radiation, as is typically the case when the light source is a laser, frequency-dependent birefringence is induced in the sample. Crossed polarizers allow this birefringence to be detected as the read beam frequency is scanned. Early application of this technique (23) has been supplemented by a thorough theoretical and experimental investigation (24). Furthermore, the technique has been extended by the application of phase modulation (25) to polarization detection (26).

Frequency modulation polarization spectroscopy (26) is one technique by which holes may be probed rapidly after burning to allow hole broadening to be observed on a fast time scale. Other methods to achieve this have also been developed. To investigate spectral diffusion on a time scale of microseconds a system of counterpropagating burn, read, and erase beams, controlled and frequency scanned by the switching of a series of acousto-optic modulators, has been used (27). An earlier way to examine time evolution of persistent spectral holes relied on a rotating mirror that directed probe light onto the sample (28). The light reaching the sample was Doppler shifted by an amount proportional to the speed of rotation of the mirror, allowing a spectrum to be recorded as the angular speed was varied.

Several other hole detection techniques that rely on modulation may be employed. Spatial modulation (29) consists of a system in which the beam is transmitted alternately through burned and unburned parts of the sample using a mirror mounted on a piezoelectric plate. The signal is detected using a lock-in amplifier tuned to the switching frequency of the piezoelectric plate. The resonant frequency of the choromophores may also be modulated and the resulting signal detected using lock-in amplifiers. This may be achieved using, for instance, acousto-optic modulation (30), Stark modulation (31), or Zeeman modulation (32).

5.2.2. Transient Spectral Hole-Burning

Methods employed to burn and detect persistent spectral holes may usually be amended to detect transient spectral holes. When parameters such as hole width are of interest, the period between reading and writing and the laser frequency sweep time must both be completed within the hole lifetime, typically on the order of milliseconds. In these cases the methods described for persistent spectral hole-burning investigations of high-speed spectral diffusion (see Section 5.2.1) may be directly utilized. On the other hand, the automatic refilling of the hole allows further experimental possibilities whereby the hole depth may be monitored as a continuous function of the strength of an externally applied field.

Transient holes are usually examined using a pump-probe procedure in which the hole is burned, followed by rapid scanning of the laser frequency while absorption or emission is monitored. This may be achieved using an intracavity electro-optic modulator. Here the hole may be burned and the frequency scanned by holding the applied voltage to the modulator steady and then quickly varying it. Although the signal available from a rapid scan may be small, the transitory nature of the hole means that the experiment may be repeated many times in a short period and the spectra accumulated. Alternatively, acousto-optic modulators may be used. Here, the zero-order unmodulated beam is used as the pump beam and the first-order modulated beam as the probe. When the two beams are overlapped on the same part of the sample and the modulation frequency varied, the hole spectrum may be recorded.

The depth of a transient hole is a measure of the proportion of chromophores not in their ground state. Consequently, hole depth varies as the balance between optical pumping and relaxation processes is changed. This allows simple experiments to be conducted in which an experimental parameter is changed and the hole depth monitored. Temperature and power variation may reveal information about the hole-burning process itself, while variation of an external field may perturb the energy levels of the chromophore, causing interactions within the solid to occur at particular frequencies. These resonances may be reflected in the transient hole depth.

5.3. SCIENTIFIC APPLICATIONS

This section is organized in such a way as to demonstrate the many areas of physics and chemistry that may be investigated using spectral hole-burning. Rather than attempting to list a large proportion of the systems and phenomena studied we have chosen a number of examples. These examples illuminate the wide range of materials, techniques, and hole-burning mechanisms. Wherever possible, more detailed reviews of each area of interest are referenced.

5.3.1. Impurity and Defect Sites

To obtain information about impurity and defect sites and dependent properties such as electronic wavefunction symmetry and photophysical hole-burning mechanisms, it is not always necessary to measure the myriad of spectral hole parameters as is the case for studies of, for instance, optical dephasing (see Section 5.3.4). The presence of a spectral hole allows high-resolution measurements of the effect of externally applied fields or forces on the electronic eigenenergies, which in turn reveal information on, for instance, impurity sites in crystals, electronic wavefunction symmetry, and solvation effects. In such deterministic studies the most important implication of the spectral hole width is in defining the resolution of the measurements. Furthermore, holes burned with intentionally large depths and widths, limited by the large bandwidth of an exciting laser, may be used to enhance measurement of the photoproduct absorption, or antihole, which gives information on the hole-burning mechanism, also relevant to impurity and defect site studies.

5.3.1.1. Uranium-Doped Strontium Tungstate

Material: actinide ion–doped inorganic crystal
Techniques: persistent spectral hole-burning; quadratic Zeeman effect; line-shape analysis
Properties studied: impurity site; wavefunction symmetry; hole-burning mechanism

It was found that persistent spectral holes could be burned in the zero phonon lines of the excitation spectrum of uranium ion impurities in a strontium tungstate crystal (33). Straightforward excitation techniques (see Section 5.2.1.1) were used to obtain hole-burning spectra.

Figure 5.7 illustrates the line shapes obtained for the two lowest energy zero phonon lines, referred to as β and γ, before and after burning a single broad spectral hole in the center of the β line. This simple experiment immediately reveals a great deal of information and demonstrates the power of the hole-burning technique. The overall absorption of the zero phonon lines are unchanged, suggesting a photophysical mechanism. The hole-burning spectrum is complicated by the presence of two nonresonant holes. The deep hole in the center of the γ line indicates that the effect of local environment on the β and γ energy levels is equivalent. The effect of crystal field variation on all

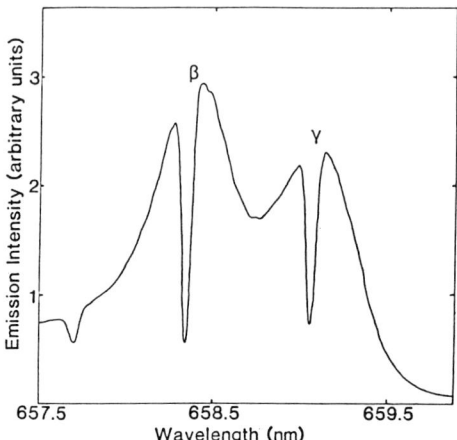

Fig. 5.7. The lineshapes of the β and γ zero phonon transitions after spectral hole-burning at 658.3 nm in SrWO$_4$:U^{6+} at 10 K (33). Symmetrically displaced nonresonant holes are also observed (see the text).

uranium ions is to shift the two lowest excited state energy levels equally, to within the accuracy available from the excitation bandwith used here. Those ions that absorb at the wavelength corresponding to the center of the β spectral line also absorb at the center of the γ spectral line, and so a spectral hole in the center of the β line necessitates a similar hole in the center of the γ line. The small hole in the high-energy wing of the β line corresponds to a similar effect. Though the resonant hole appears to be burned in the β line, inhomogeneous broadening causes the high-energy tail of the γ line to overlap with the β line. A hole burned in the high-energy tail of the γ line thus causes a corresponding hole in the high-energy tail of the β line. This indicates the degree of inhomogeneous broadening of the spectral lines, a measure of crystal disorder.

To further investigate the nature of the photophysical effect, higher resolution measurements were required. The minimum widths of holes burnt in the β and γ lines by a single mode ring dye laser at 5 K were 10 and 2 MHz (laser line width limited), respectively, a very sensitive probe to measure the effects of external perturbations. The holes were found to be displaced quadratically with external magnetic field strength (Fig. 5.8) and to split for most orientations of the field. Findings from traditional fluorescence studies of this material (34,35) had been consistent with a direct substitution of the U^{6+} ion for the W^{6+} ion in a site of S_4 symmetry (Fig. 5.9). However, the hole-burning technique improves the spectroscopic resolution by 5 orders of magnitude and the quadratic Zeeman effect data showed that the uranium ion is actually displaced along the c-axis of the crystal to a site of C_2 symmetry.

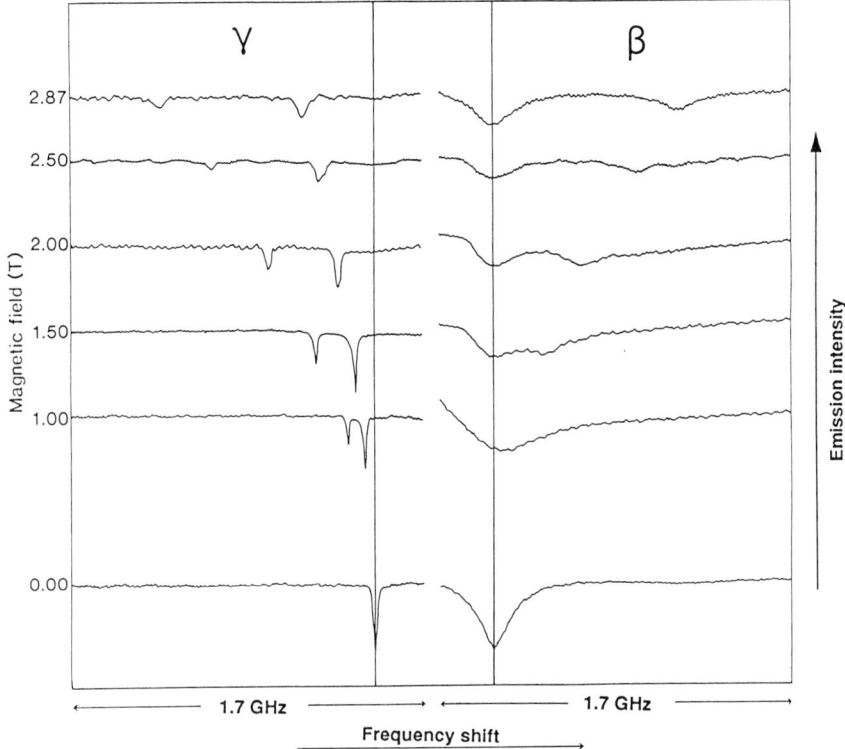

Fig. 5.8. Quadratic Zeeman shifts of spectral holes in the β and γ zero phonon lines in $SrWO_4:U^{6+}$ (33). The magnetic field is applied in the direction of maximum splitting, perpendicular to the crystal c-axis at 30° to $\langle 100 \rangle$.

Ions displaced in one direction along the c-axis have secondary symmetry axes oriented perpendicularly with respect to ions displaced in the opposite direction (Fig. 5.9). The splitting of the hole is due to a difference in the size of the quadratic shift for ions displaced in different directions along the c-axis. That is, there is a difference in g-values for Zeeman interactions between electronic levels dependent on whether the magnetic field is oriented along the local x-axis or local y-axis.

It was concluded that the photophysical hole-burning mechanism is due to a movement of the uranium ion along the c-axis to its alternative C_2 site during the process of excitation and relaxation, probably through coupling to local model vibrations. This was substantiated by the observation that the mean absorption is shifted to lower energy after hole-burning. The uranium ions tend to settle in the C_2 site at lowest ground state potential energy upon

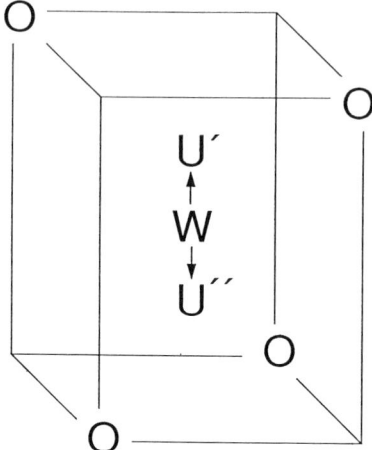

Fig. 5.9. The two C_2 uranium-substitutional sites, U' and U", displaced along the c-axis from the S_4 site of the tungsten ion in the perfect scheelite crystal (33). Hole-burning is thought to occur owing to a change of site during the excitation and relaxation processes.

cooling, and thus a photoinduced change of site is most likely to cause a decrease in the transition energy, assuming that the excited state has a lower potential barrier than the ground state.

5.3.1.2. *Curium-Doped Cerium Fluoride*

Material: actinide ion–doped inorganic crystal
Techniques: reversible persistent spectral hole-burning; selective excitation and emission
Properties studied: impurity site

A similar effect was subsequently observed in another actinide-doped crystal, $CeF_4:Cm^{4+}$ (36). Unusually, the light-induced site distortion creates a new series of thermally unstable excitation and emission lines, removed in energy by only a few tens of wavenumbers. Holes could also be bleached in the newly formed excitation lines, repopulating the original spectra. Though high-resolution investigations have not been performed to determine the nature of the distortion, it may be deduced that the change in crystal field is due to a movement of the Cm^{4+} ion to an intrinsically less stable position within the lattice rather than to a crystallographically equivalent position. Such a conclusion is supported by the reverse hole-burning process having an efficiency an order of magnitude higher than for the forward process.

5.3.1.3. The 607-nm Color Center in Sodium Fluoride

Material: color center in an inorganic crystal
Techniques: persistent spectral hole-burning; Stark effect
Properties studied: center symmetry; permanent electric dipole moments

Application of uniaxial stress to samples of γ-irradiated NaF crystals was found to split the color center zero phonon line at 607 nm (37). The form of the splitting was consistent with that which would occur for the N_1 center previously predicted (38). Hole-burning was reported in this center (39), and the splitting of the holes in an externally applied electric field was studied later (40). The crystallographic orientation of the field and single-mode laser polarization were applied in seven combinations. Analysis of this data proved the center to have C_s symmetry, showing that the zero phonon line is not associated with the N_1 center which has C_{2h} symmetry. The tremendous improvement in the accuracy of measurement is demonstrated by a comparison of Stark effect results obtained through conventional absorption techniques and through spectral hole-burning (Fig. 5.10). The hole-burning data also

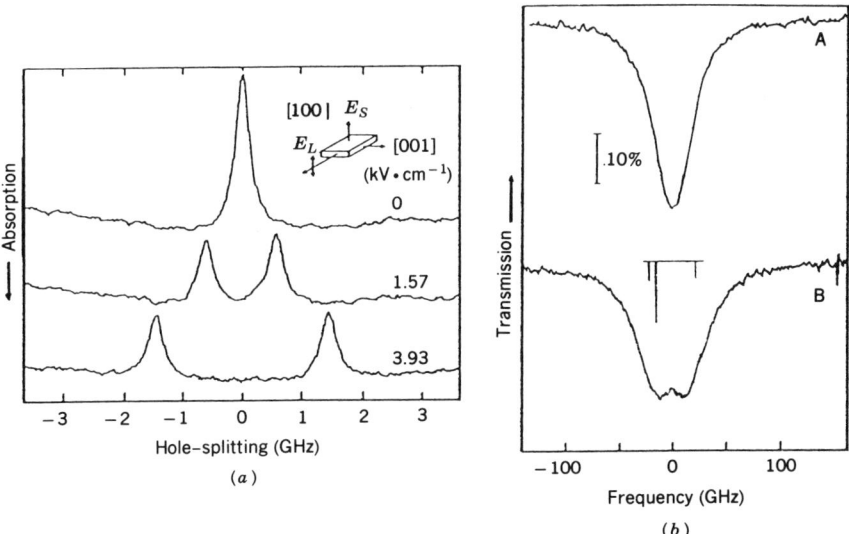

Fig. 5.10. (a) The Stark splitting of a spectral hole burned in the zero phonon line of the 607-nm color center in the configuration indicated (40). (b) Curve A, the total absorption of the same zero phonon line; curve B, the Stark splitting, experimental and theoretical, of the total line using unpolarized light and an electric field strength of 41.1 kV/cm parallel to $\langle 100 \rangle$.

accurately revealed the direction of the transition dipole and the direction and magnitude of the difference between ground and excited state permanent dipoles from an analysis of the splitting magnitudes and split hole component relative intensities.

5.3.1.4. Holmium-Doped Calcium Fluoride

Material: lanthanide ion–doped inorganic crystal
Techniques: transient spectral hole-burning; optically detected magnetic resonance
Properties studied: wavefunction symmetry

Low-resolution Zeeman studies showed that holmium ions substitute into the calcium fluoride lattice in a position of C_{3v} symmetry (41). The symmetry of the $^5F_5$ excited state wavefunction could not be determined, however, due to the overlap of adjacent absorption lines in the structure of the $^5I_8 \rightarrow {}^5F_5$ excitation spectrum. Transient hole-burning was subsequently shown to take place in this transition owing to population redistribution in both the hyperfine and superhyperfine ground state levels during resonant optical excitation (42). Simultaneous microwave excitation increases the relaxation rate between ground state hyperfine levels when resonant with the ground state hyperfine splitting, thus decreasing the transient hole depth and increasing the emission intensity. The ground state hyperfine splittings were directly measured in this way (43) and found to be different to the splittings in the $^5I_8 \rightarrow {}^5F_5$ excitation spectrum, showing that the excited state and ground state hyperfine splittings are unequal and giving conclusive evidence that the excited state wavefunction has E symmetry.

5.3.1.5. Chlorin-Doped n-Octane Shpol'skii Matrix

Material: organic molecule–doped organic crystal
Techniques: persistent spectral hole-burning; Stark effect
Properties studied: ground and excited state electric dipole moments

Highly efficient persistent spectral hole-burning as the result of photo-reversible tautomerism was first observed in porphyrin in n-octane (44). When a particular double bond in one of the pyrrole rings is hydrogenated, 2,3-dihydroporphyrin (chlorin) is formed. This molecule is of particular interest as the chlorophyll systems are derived from it (see Section 5.3.7.1). Figure 5.11 illustrates the structure of chlorin and its photoproduct, the absorption bands of which are removed by $\sim 1600\,\mathrm{cm}^{-1}$ to higher energy. Holes may be burned in the $S_1 \leftarrow S_0$ absorption bands of both the stable tautomer and the

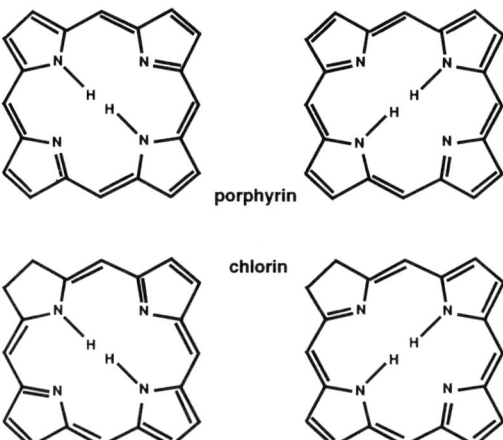

Fig. 5.11. Porphyrin and its photoproduct (above) and chlorin (2,3-dihydroporphyrin) and its photoproduct (below). Porphyrin and its photoproduct are indistinguishable except for the effect of local variation in the lattice structure, which causes transition frequencies to vary slightly. Chlorin has increased asymmetry due to the hydrogenation of a double bond, which causes the absorption band of its photoproduct to be shifted by $\sim 1600 \, \text{cm}^{-1}$.

photoproduct, the reverse reaction having a substantially increased hole-burning efficiency (45).

Chlorin may substitute into Shpol'skii matrices in a number of sites, and the absorption spectra of these have been investigated for n-hexane hosts (46) and n-octane hosts (47). Subsequently, the Stark effect was investigated for holes burned in all centers in both n-hexane and n-octane hosts (48). Using a novel Stark cell that allowed rotation of the sample to any orientation relative to the electric field, the investigators calculated the directions and magnitudes of the change of dipole moment, $\overline{\Delta\mu}$, by measuring the size of the linear splitting of the holes. Stereographic projections were produced that indicated the orientations of the dipole moment with respect to the crystallographic axes (a, b, and c). The stereonet for n-octane is shown in Fig. 5.12.

From a comparison of data obtained using electron spin resonance experiments it was deduced that the plane through the $\overline{\Delta\mu}$ coordinates is equivalent to the chlorin molecular plane. Furthermore, the two groups of chlorin $\overline{\Delta\mu}$ orientations, corresponding to centers labeled 1,2b,2c and 2a,4, were found to be oriented along perpendicular N–N axes of the chlorin molecules, at 45° to the direction of the alkane chains, denoted as r. It was concluded that the ground and excited state dipole moments are parallel, as would be expected for a molecule with C_{2v} symmetry. The $\overline{\Delta\mu}$ orientations for the less stable tautomer, denoted by primes in Fig. 5.12, are considerably

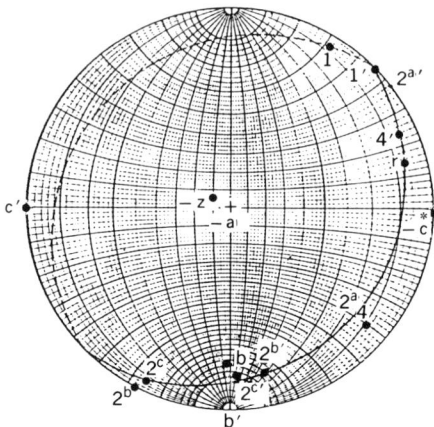

Fig. 5.12. Stereonet indicating the directions of $\overline{\Delta\mu}$ in chlorin and its photoproduct in *n*-octane (48). See text for nomenclature.

shifted, and this was interpreted as being due to a change of dipole moment direction in the excited state, S_1, since it is highly unlikely that the molecule undergoes a substantial reorientation in the lattice during photoisomerization.

5.3.1.6. Nile Red–Doped Poly(vinyl butyral)

Material: organic molecule doped polymer
Techniques: persistent spectral hole-burning; Stark effect
Properties studied: effects of solvation on molecular parameters

Spectral hole-burning in polymers doped with organic dyes (Fig. 5.13) has been intensively studied. The mechanism for hole-burning is usually photophysical and is discussed below (see Section 5.3.6.1). The case of production of films of high optical quality makes these materials ideal for study with the holographic detection technique (see Section 5.2.1.2).

The wavelength dependence of the effect of an external electric field on holes burnt in the absorption spectrum of Nile red (NR) in poly(vinyl butyral) (PVB) films has been investigated using holography (16). Figure 5.14 shows three-dimensional surface plots of holographically detected holes burned in the absorption spectrum of NR in PVB. The holes were burned in the center of the frequency tuning range with no externally applied electric field. A cut parallel to the frequency axis shows the hole profile at the corresponding electric field strength. The external electric field was applied parallel to the polarization of the laser light (left), and so the splitting of the hole indicates a

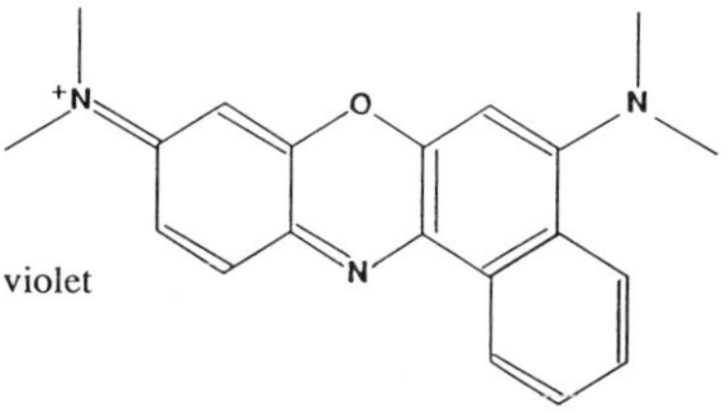

Fig. 5.13. The organic dyes cresyl violet and Nile red. Hole-burning in materials containing these compounds is thought to occur owing to a rearrangement of the local environment during the process of excitation and relaxation (see Section 5.3.6.1).

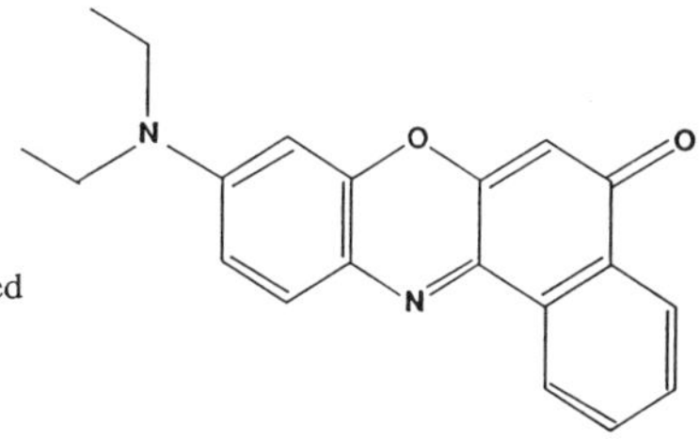

small angle, θ, between the effective dipole moment difference and the transition moment (49). When the external field was applied perpendicular to the laser polarization (right), the hole broadened but did not split.

From these two sets of Stark effect data, the effective magnitude of the molecular dipole moment difference between the ground and the excited state, $\overline{\Delta\mu_e}$, could also be determined. This contains contributions from the isolated molecular dipole moment difference, $\overline{\Delta\mu_M}$, and a matrix induced part. In previous investigations (17), the matrix-induced contributions were described by the mean value of an isotropic distribution.

In the case of NR in PVB, however, fits of the sets of data to the theoretical functions (49) for each orientation could be made using a single parameter to represent $\overline{\Delta\mu_e}$. The value of $\overline{\Delta\mu_e}$ was found to vary linearly with wavelength, from 4.1 at 610 nm to 4.7 at 580 nm, while θ was constant ($\theta = 29 \pm 2°$).

The dependence of $\overline{\Delta\mu_e}$ upon wavelength is an indication of the effect of the matrix on the molecular dipole moment difference. NR is a dye that shows large solvent shifts dependent on the polarity of the solvent, the transition energy decreasing with increasing solvent polarity, suggesting that NR has a large value of $\overline{\Delta\mu_M}$. Semiempirical CNDO (complete neglect of differential overlap) calculations showed that NR also has a large ground state dipole moment of 6 D. This dipole moment causes local orientation of the impurity

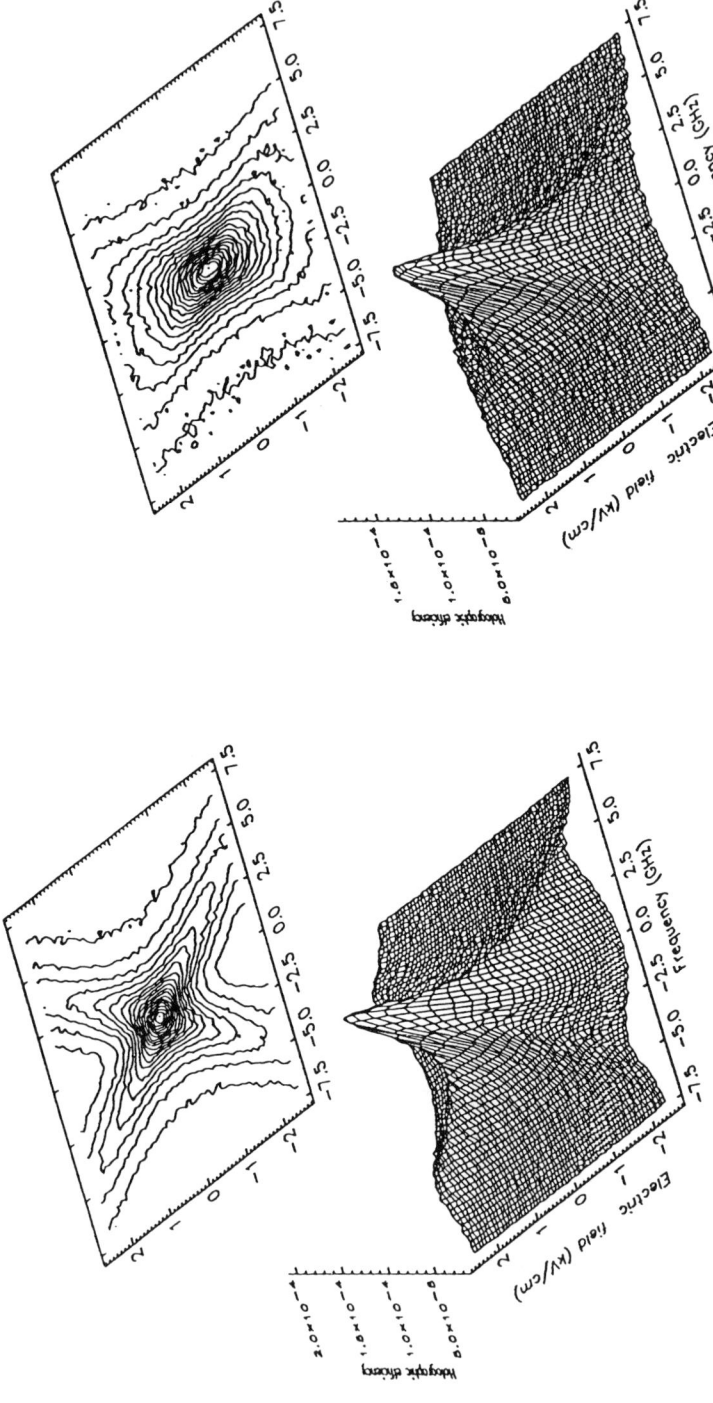

Fig. 5.14. Comparison of the behavior of spectral holes in an externally applied electric field burned using the parallel (left) and perpendicular (right) geometries (16) for Nile red in PVB.

environment, and thus local fields are strongly correlated with the orientation of the guest molecules. The dipole moment decreases with the strength of the interaction, indicating that the matrix-induced part of $\overline{\Delta\mu_e}$ points in the opposite direction to that of the purely molecular part.

The same procedure was performed for cresyl violet in PVB. The effective dipole moment difference was again found to decrease with increasing wavelength, from 3.1 D at 593 nm to 1.1 D at 632 nm. In this case, a distribution of matrix-induced dipole moments had to be introduced to fit the data. The large isotropy of the induced dipole moment observed for cresyl violet was explained by the smaller calculated ground state dipole moment of 1 D, which cannot cause such strong local orientation of the impurity environment.

The magnitude of the induced dipole moment, $\Delta\mu_i$ $(= |\overline{\Delta\mu_M} - \overline{\Delta\mu_e}|)$, depends on the local electric field and on the polarizability difference between the excited and the ground state of the molecule. The transition energy of the molecule also depends on the local field. This is why $\Delta\mu_i$ varies with transition energy. According to a simple electrostatic model (16), $\Delta\mu_i$ is given by

$$\Delta\mu_i = -\Delta\mu_M + (\Delta\mu_M^2 - 2\cdot\Delta\omega\cdot\Delta\alpha\cdot h\cdot c)^{1/2} \quad (5.8)$$

where $\Delta\omega$ is the frequency shift of the transition, and the isotropic polarizability difference is given by $\Delta\alpha$, having a negative value in this case. Hence, for equal values of $\Delta\omega$ and $\Delta\alpha$, the induced dipole moment difference is smaller for molecules with larger $\overline{\Delta\mu_M}$. The smaller change with wavelength of $\overline{\Delta\mu_e}$ in NR was attributed to this.

5.3.1.7. Cresyl Violet–Doped Poly(vinyl butyral)

Material: organic ion–doped polymer
Techniques: persistent spectral hole-burning; constructive interference between spectrally adjacent holograms; Stark effect
Properties studied: detection of specifically oriented impurities in amorphous hosts

The application of an external electric field to a spectral hole causes a splitting or broadening of the profile depending on the relative orientation of the external electric field, the polarization of the burn laser, particular impurity molecular parameters, and the bulk properties of the host (see Section 5.3.1.6). When holes are burned at adjacent frequencies, the application of an electric field causes the profiles to overlap. When the holes are burned using the holographic technique, the superposition causes constructive or destructive interference depending on the spatial phase difference between the gratings (see Section 5.2.1.2).

In the case of cresyl violet in PVB, when the external electric field is applied parallel to the laser polarization the hole profile broadens (49). When two holes are burned at the same frequency but at different electric field values, a region of constructive interference for reconstruction electric fields between the original burn values is expected. Experimentally, it was found that a dip in the hologram efficiency, with a width of the order of the original hole widths, is superimposed on the constructive interference region (50). This was interpreted as being due to the nonlinearity of the hologram burn process. The dip corresponds to those molecules, oriented in a specific direction with respect to the electric field, for which no Stark shift occurs. Other orientations could be selected by burning pairs of holographic gratings at different combinations of frequency and electric field. This technique isolates specifically oriented impurities in amorphous hosts and should be widely applicable. The application of a second perturbation would allow this subset of molecules to be studied in greater detail.

5.3.2. Electronic Interactions

When hole-burning takes place as a result of optical pumping between levels of a split ground state, a great deal of information can be deduced about the nature of the splitting and the dynamics of energy transfer within the system. Inevitably such systems return to equilibrium and the hole decays with a lifetime symptomatic of the restoring processes. This lifetime may vary enormously at liquid helium temperatures, from times less than the excited state lifetime, in which case the ground state processes may not be observed, to periods of many hours, for instance, in rare earth ion–doped crystals (51). The recovery time of the material's absorption may be influenced by the application of external perturbations that will distort the ground state electronic configuration. Measurement of the variation in hole lifetime with respect to the magnitude and direction of the external perturbation may reveal information on the ground state electronic configuration and the size of ground state interactions.

Alternatively, the hole spectrum may be recorded. Optical dephasing information may be deduced from the hole width, while antiholes reveal information on the nature of the optical pumping and the ground state electronic configuration. Excited state interactions may also be observed when they occur on the time scale of burning manifested as nonresonant hole structure.

5.3.2.1. Nitrogen-Vacancy Color Center in Diamond

Material: color center in an inorganic crystal
Techniques: transient spectral hole-bleaching; Zeeman effect
Properties studied: ground state spin–spin cross-relaxation; level anticrossing

Transient spectral hole-burning in the 638-nm zero phonon line of the N-V (nitrogen-vacancy) center in diamond arises from optical pumping between levels of the spin triplet ground state (52,53). The lifetime and structure of the hole at 4 K (Fig. 5.15) was measured using two highly monochromatic lasers—one to burn the hole and one to probe the profile immediately afterward (54,55). The hole-burning spectrum gives a deep central hole accompanied by significant antihole structure on either side, including a sharp feature at 2.88 GHz due to increased population of the alternative ground state level, the triplet being split into a doublet and a singlet by the crystal field. Such a splitting had earlier been reported from electron spin resonance studies (56), though this study attributed the splitting to an excited state triplet rather than the ground state.

During continuous illumination the depth of the hole depends on the balance between the optical pumping and the spin relaxation within the triplet levels. In the presence of an external magnetic field the optical pumping is between all three triplet levels. It was found that sudden changes in the optical hole depth, measured simply as the transmission of the diamond sample, occurred when magnetic fields of specific magnitudes and orientations are applied (57). Spin-lattice relaxation attempts to reestablish the thermal population between the optically pumped spin levels. However, it is common for spin–spin cross-relaxation involving energy-conserving spin flip-flops of neighboring centers to be a faster process (58). For the energy selected subgroup of centers interacting with the laser, the flip-flop processes tend to bring the optically excited defects into equilibrium with the vast majority of nonexcited triplet centers that will already be in equilibrium with the lattice. Efficient cross-relaxation requires resonant conditions for the cross-relaxing spins. This is always the case for centers oriented equivalently with respect to the magnetic field, but the centers are oriented trigonally within the diamond lattice; thus an arbitrary field direction will cause different splittings in centers oriented in each of the four directions. However, there are situations, at particular field strengths and orientations, for which there is an accidental coincidence between ground state splittings of the probed center and those of inequivalent centers or entirely different types of centers contained within the diamond lattice. The coincidence brings the two separate ensembles of centers into resonance, thus providing an extra decay channel. Since the relaxation

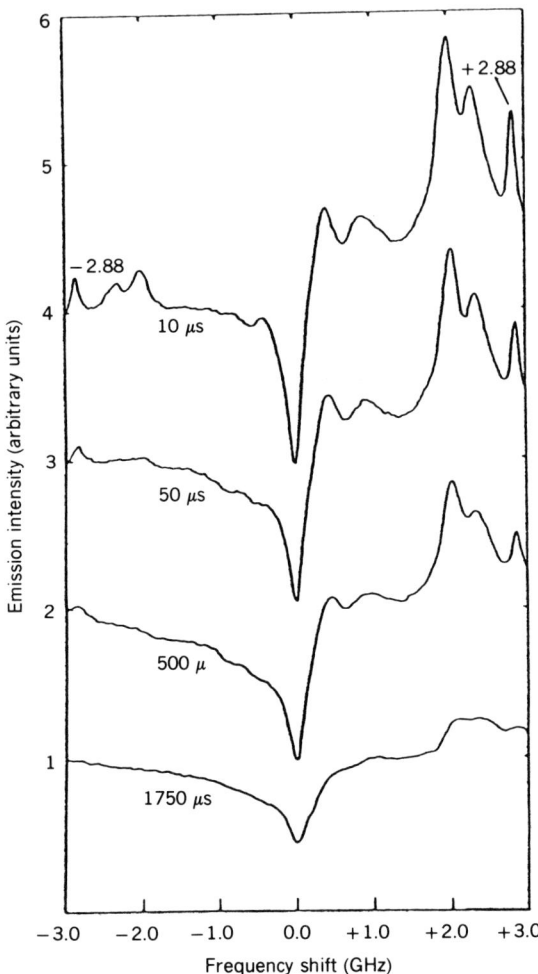

Fig. 5.15. Time dependence of the hole-burning spectrum for the N-V center in diamond (54, 55). The hole recovery rate is indicative of the rate of relaxation between ground state triplet levels split by 2.88 GHz.

Fig. 5.16. (a) Transmission at 638 nm of a diamond sample containing N-V and P1 centers as a function of magnetic field strength applied along the $\langle 111 \rangle$ direction at 10 K. (b) Zeeman splitting for (i) the P1 center, (ii) the N-V center with field along the principal axis, and (iii) the N-V center with the field at an angle of 70.5° to the principal axis. The vertical dashed lines indicate the origin of the hole-burning features, and the solid arrowed lines indicate the transitions involved in the cross-relaxation (57).

SCIENTIFIC APPLICATIONS

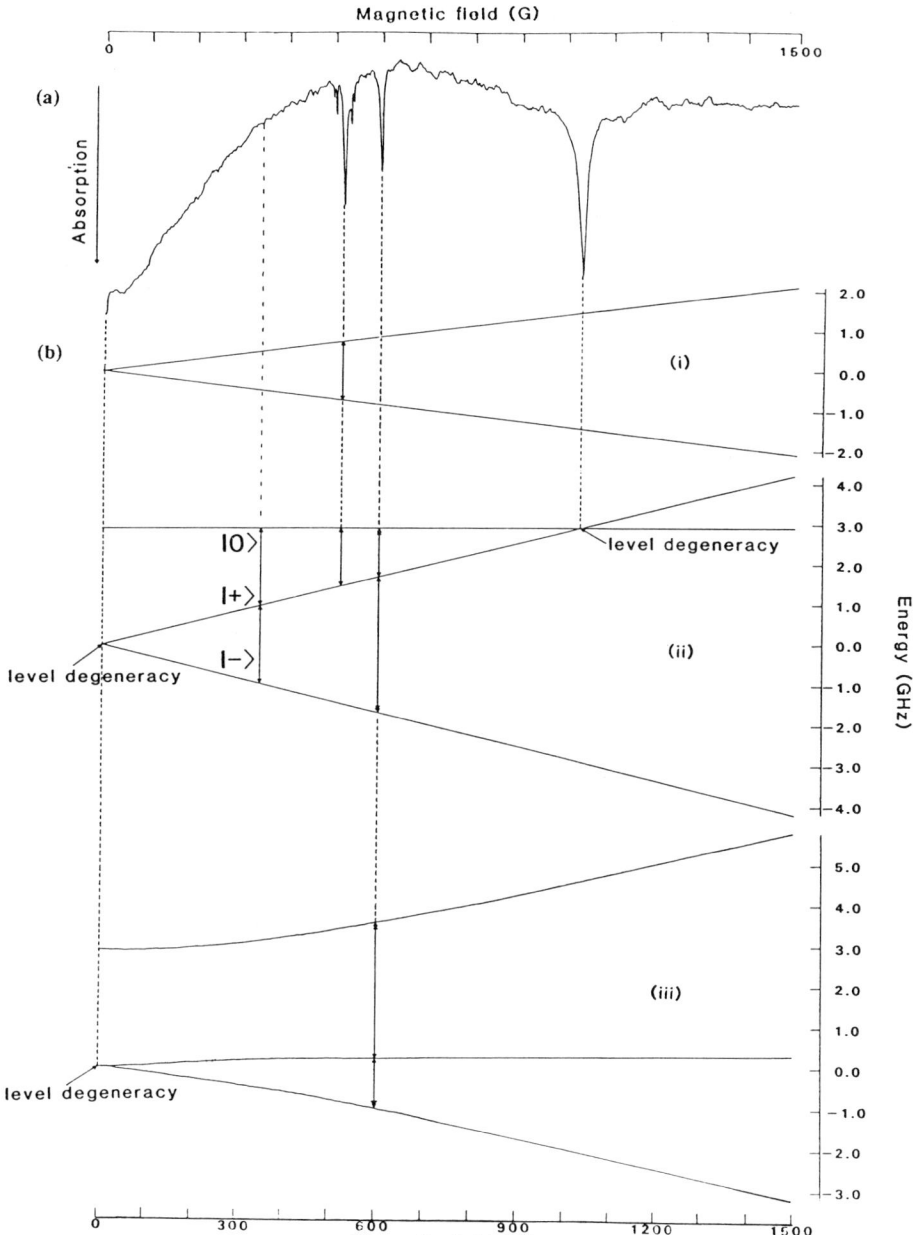

of the combined system is always faster than that of the separate ensemble of centers, the result of any matching of energy separations is a reduction in the optical hole depth, that is, hole bleaching, manifested as a decrease in transmission of the sample.

Figure 5.16 shows the transmission of the sample at the center of the zero phonon line at 638 nm as a function of the magnetic field applied along a $\langle 111 \rangle$ direction. The three principal features (and one "missing" feature) occur as a result of different phenomena. When the field is applied in this direction, the principal axes of the N-V centers are all aligned either along the magnetic field or at 70.5° to it. Both sets split differently in a magnetic field. At 600 G the splittings become degenerate, causing an extra spin–spin cross-relaxation pathway, a reduction in hole depth, and consequently an increase in absorption. The feature at 511 G is attributed to spin–spin cross-relaxation with an entirely different color center in the diamond sample. The P1 center has an $S = 1/2$ electron spin ground state giving EPR signals corresponding to a trigonal center with a predominant hyperfine interaction with an integral nitrogen $(I = 1)$ nuclear spin (59). The hyperfine structure surrounding the feature at 511 G corresponds to that which would be expected for interaction between the N-V and P1 centers. At a field strength of 340 G the three triplet levels are equally spaced for the centers aligned along the $\langle 111 \rangle$ direction. In this case spin quantization is good and so, despite the additional energy splitting degeneracy, spin–spin cross-relaxation involving simultaneous flips between $|+\rangle \leftrightarrow |-\rangle$ states and $|0\rangle \leftrightarrow |+\rangle$ states are forbidden, as they do not conserve angular momentum. As the field is gradually varied from the $\langle 111 \rangle$ direction, the corresponding signal appears and becomes stronger owing to increased mixing of the eigenstates. The final feature, at 1020 G, occurs when the $|0\rangle$ and $|+\rangle$ states approach the same energy for centers aligned with the magnetic field. Strain perturbations at the N-V center will mix the spin states, and the states do not actually cross. It is thought that transient hole bleaching is caused by optical pumping to preferential ground states being reduced as a result of eigenstate mixing. This is substantiated by the broadening of the "level anticrossing" signal as the angle between the magnetic field and N-V center principal axis is increased.

5.3.2.2. *Europium-Doped Yttrium Aluminum Perovskite*

Material: lanthanide ion–doped inorganic crystal
Techniques: persistent spectral hole-burning; optically detected magnetic resonance
Properties studied: nuclear magnetic moment reduction due to hyperfine-induced magnetic shielding; quadrupole interactions

Spectral hole-burning due to optical pumping between split ground states in rare earth ion–doped crystals is a common phenomenon (51). Owing to the slow relaxation rate between nuclear spins and the lattice in $YAlO_3:Eu^{3+}$, spectral holes may be burned with a lifetime of greater than 1 h (60). Europium has two almost equally abundant isotopes each of which has a nuclear spin of 5/2 but with different intrinsic nuclear magnetic and quadrupole moments. The hole spectrum for the $^7F_0 \to {}^5D_0$ transition is shown in Fig. 5.17. The six pairs of side holes (up the page) are due to decreased absorption to the excited state quadrupole levels and therefore occur at frequencies representing the two excited state quadrupole splittings and their sum, for each isotope. The splittings are larger for ^{153}Eu than for ^{151}Eu by a factor corresponding to the different sizes of the intrinsic nuclear quadrupole moments, that is, 2.54. Subsequent application of an external magnetic field splits these side holes (Fig. 5.17, inset) by an amount consistent with the intrinsic nuclear magnetic moments.

Earlier it had been predicted that the ground state effective nuclear moment would be partially canceled by the electronic contribution to the moment owing to second-order hyperfine coupling with the $^7F_1$ levels (61). This effect could not be studied using conventional nuclear quadrupole resonance because of the completeness of the cancellation. Simultaneous radio-frequency

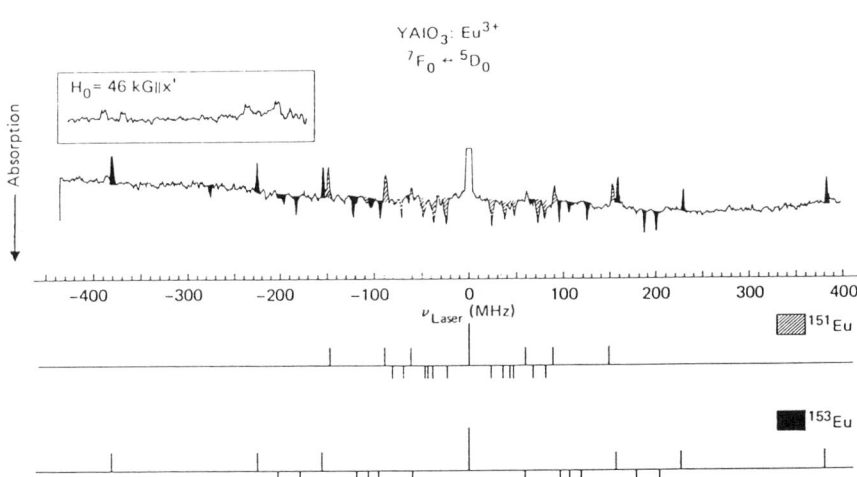

Fig. 5.17. Hole-burning spectrum of $YAlO_3:Eu^{3+}$ at 1.5 K. The features are assigned to population redistribution among ^{153}Eu (solid) and ^{151}Eu (hatched) ground state nuclear quadrupole levels (60). The positions of the holes and antiholes predicted from the quadrupole splittings are shown below. Inset: the effect of a magnetic field on two of the ^{153}Eu features is shown.

excitation increases the relaxation rate between ground state quadrupole levels when resonant with the ground state quadrupole splitting, thus decreasing the transient hole depth and increasing the emission intensity. The ground state quadrupole splittings were directly measured in this way. Subsequent application of an external magnetic field split these optically detected nuclear quadrupole resonance signals by an amount consistent with a reduction of the nuclear magnetic moments to 21% of their intrinsic values, as predicted. Thus excited state quadrupole splittings were measured from side hole positions and ground state quadrupole splittings were measured using a hole-refilling method. The antiholes observed in Fig. 5.17 correspond to the redistributed absorption of the material, and the splittings represent differences between ground and excited state quadrupole levels, fitting calculations well and completing the picture (60).

5.3.2.3. *Erbium-Doped Yttrium Lithium Fluoride*

Material: lanthanide ion–doped inorganic crystal
Techniques: saturation spectral hole-burning; Zeeman effect
Properties studied: superhyperfine coupling in the excited state

Hole-burning due to optical pumping between ground state levels might be expected in $YLiF_4:Er^{3+}$, but studies have shown that dephasing and hole refilling, owing to interactions between impurity ions, is too rapid for this effect to be observed (62). Saturation hole-burning has been observed in the

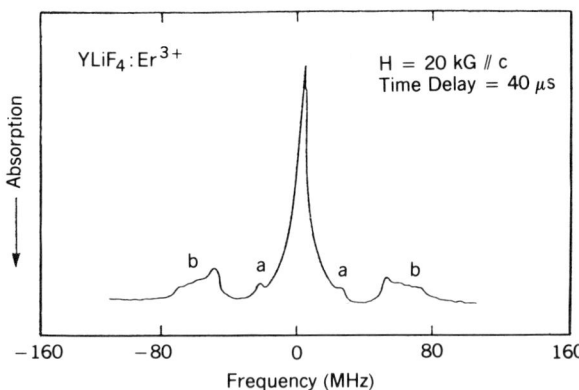

Fig. 5.18. The saturation hole spectrum at 1.6 K obtained by rapidly varying an external magnetic field and keeping the laser frequency fixed. The side holes marked *a* and *b* are assigned to spin-flips on ^7Li and ^{19}F host ions (64).

presence of a large external magnetic field that reduces the rate of electron spin flips on Er^{3+} ion pairs. The time-dependent broadening of the holes due to coupling between the impurities and host nuclear magnetic moments was measured in this way (63). Furthermore, nuclear spin flip side holes have been observed (64). Figure 5.18 shows the hole spectrum obtained by rapidly varying the external magnetic field and keeping the laser frequency fixed, thus scanning the optical resonance frequency. The two pairs of side holes were attributed to excited state superhyperfine interactions involving nuclear spin flips of nearby ^7Li and ^{19}F host ions, the latter signal being broadened owing to contributions from inequivalent locations.

5.3.3. Tunneling Processes

The observation of tunneling phenomena relies on the use of a spectral hole as an energy-selective probe to follow the behavior of a subset of molecules. The hole spectrum may evolve with time or, for instance, upon thermal annealing. The tunneling processes may be observed as the resonant hole refills and satellite holes appear, shifted by an amount characteristic of the material and nature of the tunneling. Translational tunneling along hydrogen bonds, important in many biological and chemical systems, and methyl group rotational tunneling have both been observed in this way. Tunneling may also be investigated through the hole-burning process itself for materials in which lattice hydrogen ions tunnel to alternative sites during the excitation process, thus creating the spectral hole.

5.3.3.1. Thioindigo-Doped Benzoic Acid

Material: organic molecule–doped organic crystal
Techniques: transient spectral hole-burning; Zeeman effect; lineshape analysis; uniaxial and hydrostatic pressure
Properties studied: intermolecular tunneling along hydrogen bonds

Photochemical hole-burning due to proton tautomerism has been discussed above (see Section 5.3.1.5). Photophysical hole-burning may also occur through a rearrangement of protons in the host lattice rather than in the chromophore. Such a change in environment occurs through a photoinduced process, not one necessarily involving tunneling, but one which subsequently reverts to the original configuration in the dark. Crystalline benzoic acid is a particularly important host for observing such phenomena. The crystal is composed of planar dimers, and at high temperatures the exchange of protons between dimers (Fig. 5.19) has been investigated by nuclear magnetic resonance (NMR) and inelastic neutron scattering. At low temperatures this exchange

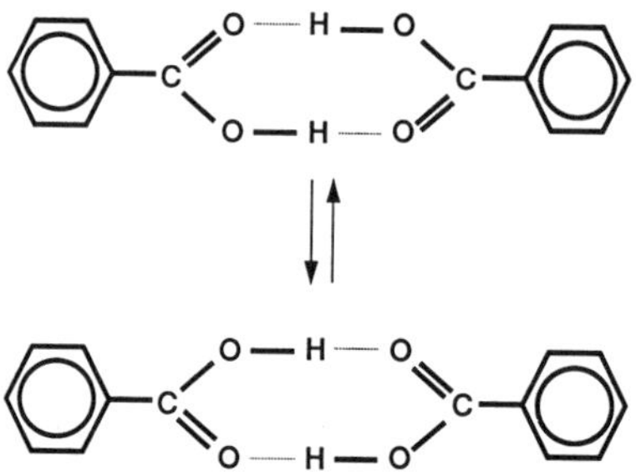

Fig. 5.19. The two tautomers of a benzoic acid dimer. The transition energy of a coupled chromophore may change upon excitation and relaxation if this process causes the dimer to change form.

may not be thermally activated and occurs only through proton tunneling. Hole-burning in several benzoic acid systems has been observed. It was originally postulated that the mechanism was due to dimer hydrogen transfers induced by optical absorption of the impurity (65), but subsequent studies on pentacene-doped samples showed the photoreaction to be more violent, involving temporary hydrogen abstraction from the host lattice by the chromophore (66).

For intermolecular proton tunneling along hydrogen bonds to be observable, it is necessary to find an impurity for which localization due to environment-induced asymmetry is weak. One such molecule is thioindigo. In this case the asymmetry of the crystal potential is accidentally reduced such that the protons in neighboring dimers are sufficiently delocalized in the ground state. This delocalization was first indicated through absorption spectra temperature dependence measurements (67) and later through fluorescence line narrowing investigations (68), but the full structure could only be resolved using spectral hole-burning (69). Spectral hole-burning is a particularly good technique to use for such a study owing to the high spectral resolution available and the fact that, in this case, the protons are localized in the excited state, thus preventing tunneling (68) and allowing ground state tunneling splittings to be directly and unambiguously observable.

Transient holes with a lifetime of about 1 min may be burned in the inhomogeneously broadened absorption bands. The tunneling rate in the ground state was measured through a comparison with the hole widths of deuterated samples, it being assumed that the difference may be attributed to the tunneling ground state lifetime, present only for protonated samples (70). The tunneling rate was found to be rapid (10^6–10^9 s^{-1}, depending strongly on temperature between 1–4 K), and consequently immediate measurement of the hole spectrum after burning revealed a strong resonant hole accompanied by symmetrically displaced but asymmetrically shaped satellite holes (Fig. 5.20) (69). The satellite holes represent those molecules having dimer neighbors for which tunneling has taken place, the frequency shifts of which give a direct measurement of the ground state energy splittings. From these splittings and previously measured parameters (67) the tunneling matrix element for intermolecular hydrogen bonds in the condensed phase could be determined

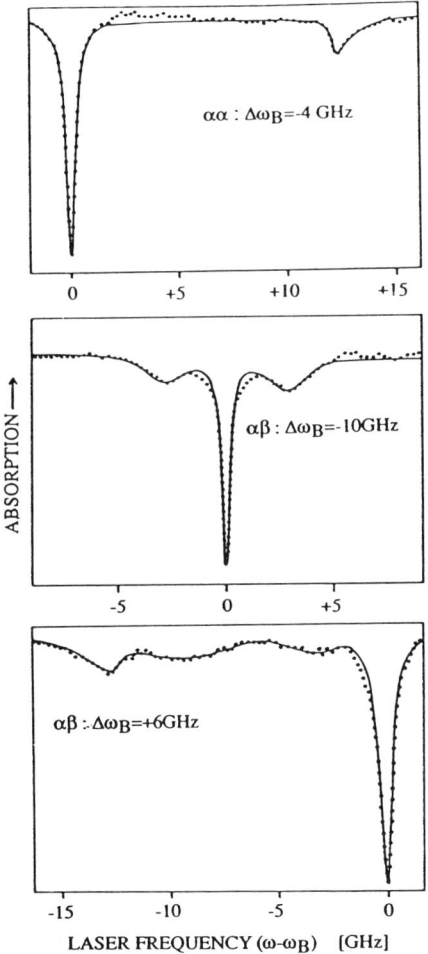

Fig. 5.20. Transient hole-burning spectra of thioindigo in benzoic acid at 1.35 K (69). The excited state localized proton levels ($\alpha\alpha$ and $\alpha\beta$) and the difference, $\Delta\omega_B$, between the burn frequency, ω_B, and the absorption line center are indicated. The solid line is a fit to the experimental data shown as dots.

for the first time. The value of $\langle J \rangle = 8.4$ GHz was close to that calculated for an isolated benzoic acid dimer (71), it being concluded that effects due to the introduction of the dimer into the condensed phase cancel coincidentally. The unusual shape of the side holes was also accounted for. By introducing a Gaussian statistical variable to the tunneling and asymmetry parameters, excellent fits could be obtained (Fig. 5.20). The inhomogeneity of the parameters was attributed to crystal imperfections, and indeed the strength of the satellite holes was found to be highly dependent on crystal quality, in line with previous theoretical predictions (72). A study of the effect of uniaxial and hydrostatic pressure on the proton tunneling potential (73) further substantiated this model, showing a linear dependence that is attributed to a linear increase in the asymmetry while the tunneling matrix element remains almost constant.

5.3.3.2. Dimethyl-s-tetrazine–Doped Tetramethylbenzene

Material: organic molecule–doped organic crystal
Techniques: persistent spectral hole-burning; temperature cycling
Properties studied: methyl group rotational tunneling

Dimethyl-s-tetrazine in crystalline tetramethylbenzene was one of the first materials for which hole-burning was observed (74). The mechanism was ascribed to a photodissociation of the guest molecule. Other hole-burning processes may also occur, but at temperatures below 2 K persistent hole-burning is thought to occur only through consecutive two-photon absorption, leading to photodissociation (75).

Spectral holes created at low temperatures have been used as a probe of the evolution of the system between tunneling levels (76). The tunneling system in this case is a pair of methyl groups coupled to the chromophore, though it is not known whether the methyl groups belong to the host or guest. Nevertheless, the tunneling system may be categorized from geometric considerations. In a solid the potential influencing a methyl group consists of a symmetric triple well with C_3 symmetry. A methyl group may tunnel through the barrier between wells, a rotation of 120°, as a consequence changing its spatial symmetry and therefore energy. The spatial symmetry of the tunneling levels is associated with nuclear spin $3/2(A)$ or $1/2(E)$. As two methyl groups are coupled to the chromophore, four ground states and four excited states exist, defined by combinations of A and E spatial symmetries where the AE and EA states are degenerate to a first approximation. The terms that mix spatial and spin coordinates of the methyl protons are very small in such a system, and spin conversion between A and E states is therefore very slow. The evolution of the hole spectrum is thus a direct measure of the rate of ground state rotational tunneling. However, the splittings record the

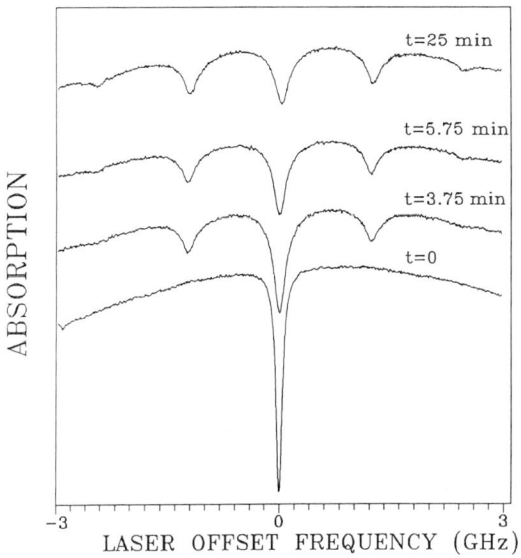

Fig. 5.21. Evolution of a spectral hole burned and read at 1.75 K. The sample is cycled to 6 K for the periods indicated. Thermal equilibrium of the methyl group pair tunneling levels is reached by 25 min (77).

difference between ground and excited state tunneling splittings, because the energy changes may only be measured as an excitation spectrum.

Holes burned at 1.35 K show no spectral evolution when maintained at this temperature, but upon thermal cycling to 6 K for 25 min full thermal equilibrium is reached and a splitting pattern with five components is revealed (Fig. 5.21) (77). Five holes appear as transitions between all three tunneling levels are resonant with the burn laser owing to accidental degeneracy caused by inhomogeneous broadening. Subsequent tunneling is therefore to higher and lower energies for the same spectral hole. The relative intensities of the holes fit theory to within experimental accuracy. Though this is the first demonstration of a measurement of rotational tunneling in solids it is thought many other systems may show similar effects.

5.3.3.3. *Praseodymium, and Deuterium-Doped Strontium Fluoride*

Material: lanthanide ion- and deuterium-doped inorganic crystal
Techniques: persistent spectral hole-burning
Properties studied: hydrogen intersite tunneling; ground state pseudoquadrupole interactions

A number of hydrogenated trivalent rare earth ion centers in calcium and strontium fluoride have been shown to exhibit bleaching upon resonant optical excitation of electronic transitions (78). From the polarized bleaching behavior, the centers could be divided into two groups. In the first, bleaching of those centers that absorb either horizontally or vertically polarized light, corresponding to perpendicularly oriented centers, results in an increased absorption for the opposite polarization at the same wavelength. In the second, the absorption of perpendicularly absorbing centers may be bleached independently but new absorption lines appear close by. Subsequent irradiation at the wavelengths of the new absorption features bleaches the lines, and the original absorption spectra return. Both behaviors are symptomatic of photophysical hole-burning, and the mechanism was ascribed to the reorientation of hydrogen ions within the crystal lattice. One such scheme is indicated in Fig. 5.22 (79). Here the C_{4v} centers (80) are reoriented through 90°, and (based on a consideration of selection rules) educt and product absorb light oriented perpendicularly to each other though at the same wavelength.

A high-resolution study of this same behavior in $SrF_2:Pr^{3+}:D^-$ (81) attributed the hydrogen motion to a tunneling process, though no calculations of barrier heights or dynamics was made. Holes can be burned in the zero phonon lines of two centers, each of which has C_s symmetry. Many studies of hole-burning in rare earth ion–doped crystals rely on a temporary redistribution of the ground state population between spin levels upon optical excitation and subsequent relaxation. The hole spectra reveal a series of holes and antiholes that decay with the spin-lattice relaxation time of the material (see Section 5.3.2.2). In this case, however, the hole spectra consist of several

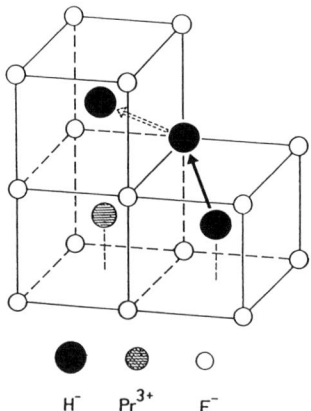

Fig. 5.22. Possible mechanism for the reorientation of the $SrF_2:Pr^{3+}:^2H^-$ center site configuration to give an equivalent center oriented at 90°, causing hole-burning (79).

persistent features, showing the mechanism to be due to a permanent change in the lattice structure as was predicted from low-resolution measurements. The splitting of the holes was analyzed, and their origin was shown to be dominated by a large pseudoquadrupole [second-order hyperline (51)] ground state splitting.

5.3.4. Optical Dephasing

The expression for the homogeneous line width of a transition (see Section 5.1) is repeated here:

$$\Gamma_{\text{hom}} = 1/\pi T'_2 = 1/2\pi T_1 + 1/\pi T_2^* \tag{5.1}$$

The radiative lifetime of a fluorescent or phosphorescent electronic state can usually be easily measured, and so, if the homogeneous line width of the transition from the ground state can be determined from hole-burning measurements, the optical dephasing time may be deduced. An investigation of the temperature dependence of T_2^* indicates the nature of the dephasing.

It follows that studies of optical dephasing depend on accurate determination of spectral hole widths and consequently, though the experiments are conceptually simple, great care must be taken to ensure that conclusions drawn from data are valid in the reference frame of the measuring parameters. Hole widths may be influenced by local heating of the sample by the exciting radiation, a particularly serious problem when one is working at liquid helium temperatures where materials have very low thermal conductivities. The influence of the laser with the quantum mechanical system itself also distorts the hole widths through power (or fluence) broadening. If the nature of the hole-burning system is known, it is possible to model the effect of the variation of burning power on the hole width by a density matrix or kinetic approach (4). In many cases, however, the expression for the hole width, Γ_{hole}, may be simplified to

$$\Gamma_{\text{hole}} = \Gamma_{\text{laser}} + \Gamma_{\text{hom}} + \Gamma_{\text{hom}}(1 + kP)^{1/2} \tag{5.9}$$

where Γ_{laser} and P are the bandwidth and power density of the laser, and k is a constant representing various physical parameters of the system. By varying the burning power, within the restraints that local heating of the sample does not take place and the holes do not become saturated, the homogeneous linewidth may be deduced.

There exist many theories dealing with the variation of homogeneous line width with temperature, and comprehensive reviews of this field have been published (82,83). Here, only the basic principles and conclusions will be considered. Different models are used to deal with crystalline and amorphous materials. As dephasing is usually dominated by phonon coupling, those theories concerned with ordered systems, having well-defined vibrational properties, tend to give more consistent approximations of experimental data than those concerned with disordered systems.

Spectral hole-burning investigations of optical dephasing in crystals have largely been confined to mixed molecular crystals. For simple ion impurity–doped crystals, in particular those doped with rare earth ions, homogeneous broadening is often dominated by weak magnetic interactions such that hole widths are dominated by laser jitter, Γ_{laser}, and other techniques become more appropriate (51). The theory most commonly applied to ordered systems for which dephasing is dominated by phonon coupling is the exchange mechanism (84,85). In this model, a single low-frequency librational mode is coupled to the optical transition. For low temperatures, the homogeneous line width is predicted to broaden exponentially with temperature. The shift in the optical transition energy, further revealed by such hole-burning studies, also has an exponential behavior according to exchange mechanism theory. Extensions to this model, required to explain data from systems for which assumptions made in the original version are no longer valid, have subsequently been made (86,87), and both predict a biexponential temperature dependence of the homogeneous linewidth on temperature. A parallel theory, which considers resonant scattering of a vibrational mode (88,89), predicts a single exponential function for homogeneous line broadening at low temperatures but a quadratic dependence above the Debye temperature.

Theories for homogeneous broadening of impurity transitions in amorphous solids are complicated by the addition to the model of the phenomenological concept of host two-level systems (90,91). A distribution of these two-level systems, asymmetric double well potentials in configuration space, is used to represent the nonperiodic nature of the host potential. Various models consider the coupling between the guest electronic system and the host two-level systems to be engaged in different ways. The lattice phonons are considered to interact with this combination and, again, various forms for this coupling have been considered. A review of these models is beyond the scope of this chapter, but brief (12,92), and comprehensive (82,83,93,94) descriptions have been published previously. Many of these models are able to explain the commonly observed $T^{1.3}$ dependence of the homogeneous line width at low temperatures, but each makes assumptions that have yet to be experimentally verified and the interpretation of such data remains controversial. Furthermore, materials have been found that lie somewhere between the ordered and disordered domains, as determined by their optical dephasing behavior, and some of these will be discussed below.

Other experimental techniques are also of importance in the study of optical dephasing, but these will not be described here. Comparisons between data obtained from hole-burning and data obtained from photon echo investigations and other coherent transient techniques as well as from fluorescence line narrowing are dealt with in the aforementioned reviews.

5.3.4.1. Porphyrin-Doped n-Octane

Material: organic molecule–doped organic crystal
Techniques: persistent spectral hole-burning, sample temperature variation
Properties studied: optical dephasing

Photochemical hole-burning through the rearrangement of the inner hydrogens in porphyrins has been discussed above (Section 5.3.1.5). Porphyrin (or

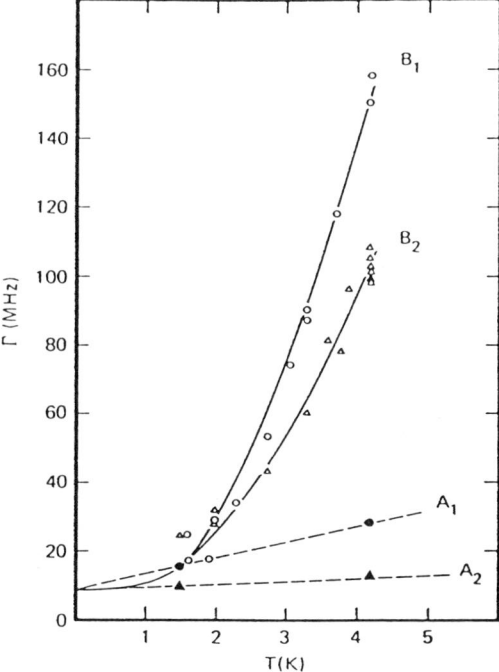

Fig. 5.23. The temperature dependence of the homogeneous line width of the $S_1 \leftarrow S_0$ transitions in porphyrin-doped n-octane as determined from spectral hole-burning (95). Solid lines are best fits to the exchange mechanism theory.

"free-base porphin") (Fig. 5.11), added as an impurity, has been used as a probe of optical dephasing in many materials. One of the earliest studies was for n-octane crystals (88,95). Porphyrin may substitute into the n-octane lattice in one of two crystallographically inequivalent sites, A and B, each of which may be oriented in two perpendicular directions, 1 and 2, with respect to the pair of inner hydrogens. Optical dephasing was investigated using all four of the resulting inhomogeneously broadened $S_1 \leftarrow S_0$ zero phonon transitions. The effect of temperature upon hole broadening and hole frequency shift was measured and the dephasing found to be different between sites A and B but similar for orientations 1 and 2 (Fig. 5.23) (95). This was interpreted as being due to the difference in the tightness of fit of the guest in the host matrix as the A site replaces two octane molecules and the B site three (95). Little dependence on orientation within the site would therefore be expected. The coupled phonon causing the dephasing was assumed to be a pseudolocal phonon, a sharp peak in the phonon density of states caused by the distortion of the host lattice by the guest.

From a comparison with the predictions of the exchange mechanism theory (85), the exponential behavior of both the hole broadening and frequency shift in the B sites showed the pseudolocal phonon energy to be approximately 15 cm^{-1} and the lifetime to be of the order of 3 ps. The data for the A site was less precise owing to the weakness of the hole broadening and frequency shift. Nevertheless, this allowed the A site pseudolocal phonon energy to be estimated as being around double that for the B site, as would be expected for the relative tightness of the fit within the host lattice. Despite the large differences in dephasing behavior, the hole widths for all four sites extrapolate to the same lifetime limited hole width at 0 K (Fig. 5.23).

5.3.4.2. Magnesium Porphyrin–Doped n-Octane

Material: metalloorganic molecule–doped organic crystal
Techniques: transient spectral hole–burning; sample temperature variation
Properties studied: optical dephasing

Optical dephasing may also be studied using transient hole-burning techniques, thus extending the number of systems that may be studied but preventing hole frequency shift measurements from being made. A number of magnesium porphyrin–pyridine centers in n-octane crystals were studied using spectral hole-burning (96). Here only the site arbitrarily labeled 5_x will be discussed. The mechanism for hole-burning in this center is through population storage in a triplet level that decays after excitation of the first excited singlet ceases. Consequently two lasers were employed simultaneously to record hole profiles, the first to deplete the ground state and the second, with a power of

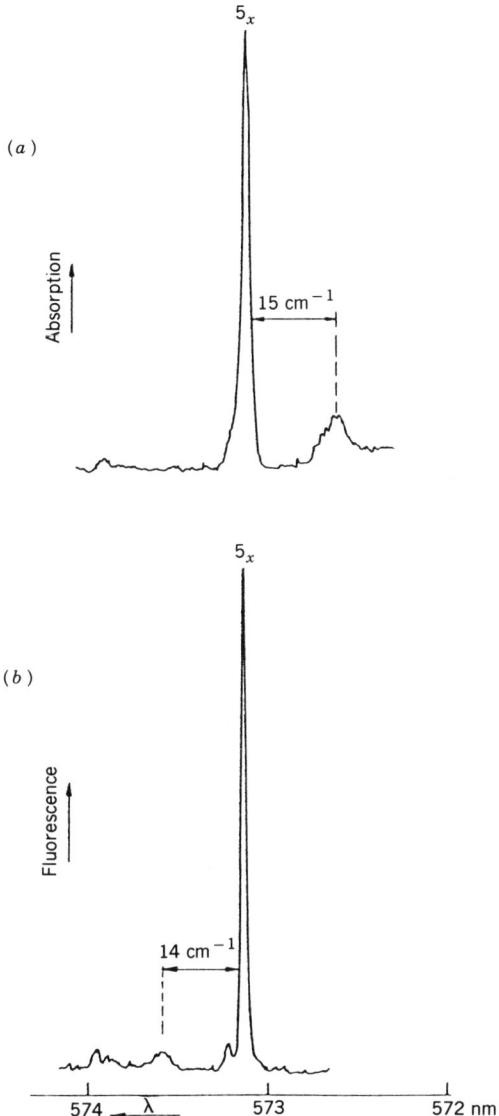

Fig. 5.24. Excitation and fluorescence spectra of the $S_1 \leftarrow S_0$ zero phonon line region for magnesium porphyrin–doped n-octane (96). The hole width in this transition was found to broaden exponentially with temperature, with an activation energy calculated from a fit to the exchange mechanism theory of $14 \pm 1 \, cm^{-1}$.

about half that of the pump beam, to be scanned over a suitable frequency range. The hole width was found to broaden exponentially with temperature, with an activation energy calculated from a fit to the exchange mechanism theory of 14 ± 1 cm^{-1}. This model was further vindicated by the observation of local modes 15 cm^{-1} to higher energy in the excitation spectrum of the center and 14 cm^{-1} to lower energy in the emission spectrum (Fig. 5.24).

5.3.4.3. Samarium-Doped Barium Fluorochlorobromide

Material: lanthanide ion–doped inorganic microcrystals
Techniques: persistent spectral hole-burning; sample temperature variation
Properties studied: optical dephasing

Narrow inhomogeneous line widths and thermal instability of spectral holes have made investigations of optical dephasing in crystals at high temperatures difficult using hole-burning techniques. It was shown, however, that the inhomogeneous line widths of the $4f^6$ $^7F_0$–^5D_J transitions in BaFCl:Sm^{2+}, known to exhibit thermally stable hole-burning through oxidation to Sn^{3+} (97), are broadened through the addition of bromine to the melt without

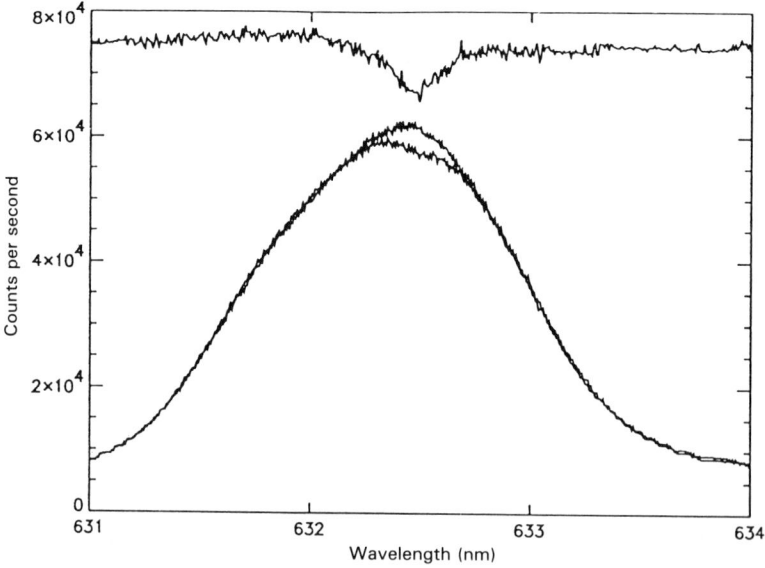

Fig. 5.25. The $^7F_0$–D_1 line in Sr$_{0.5}$Mg$_{0.5}$FCl$_{0.5}$Br$_{0.5}$:Sm^{2+} before and after room temperature hole-burning at 632.5 nm (laser scatter $\sim$ 8000 cps) (101). The difference signal, above, is magnified by a factor of 2.

appreciably splitting or shifting the transition energies (98). This has been attributed to substitutional ligand disorder (99).

Holes with a width of 1.1 cm^{-1} were burnt in the $^7F_0-^5D_2$ transition in BaFCl$_{0.5}$Br$_{0.5}$:Sm^{2+} at 77 K (98), prompting a study of the temperature dependence of the hole width (100). The highest temperature for which spectral hole burning had been observed was raised to 133 K before materials with even greater substitutional disorder, such as Sr$_{0.5}$Mg$_{0.5}$FCl$_{0.5}$Br$_{0.5}$:Sm^{2+}, were investigated and spectral holes were burned at room temperature (101,102) (Fig. 5.25).

The hole width temperature dependence for the $^7F_0-^5D_1$ Sm^{2+} transitions of Sr$_{0.5}$Mg$_{0.5}$FCl$_{0.5}$Br$_{0.5}$ and Sr$_{0.65}$Ba$_{0.35}$FCl$_{0.5}$Br$_{0.5}$ (101) show little variation from that of the same transition in BaFCl$_{0.5}$Br$_{0.5}$ (Fig. 5.26). Powers an order of magnitude greater were required to burn holes in Sr$_{0.65}$Ba$_{0.35}$FCl$_{0.5}$Br$_{0.5}$:Sm^{2+}, but the hole width was little affected, suggesting power broadening to be insignificant and the holes to be good probes of dephasing processes. As the hole widths for all compounds are similar, it was concluded that the magnitude and source of optical dephasing in this class of materials are equivalent.

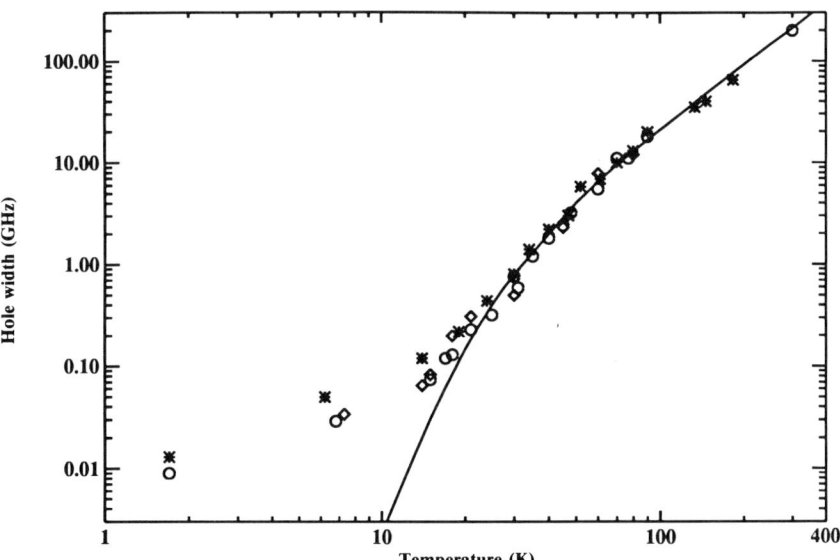

Fig. 5.26. The temperature dependence of hole widths for the $^7F_0-^5D_1$ transitions in BaFCl$_{0.5}$·Br$_{0.5}$:Sm^{2+} (∗) (100,108), Sr$_{0.5}$Mg$_{0.5}$FCl$_{0.5}$Br$_{0.5}$:Sm^{2+} (○) (101), and Sr$_{0.65}$Ba$_{0.35}$FCl$_{0.5}$Br$_{0.5}$:Sm^{2+} (◇)(101). The line is a fit to a Raman phonon scattering model (103) discussed in the text.

Dephasing due to phonon coupling is described by the following equation (103):

$$\Gamma_{\text{hom}}(T) - \Gamma_{\text{hom}}(0) = \bar{\alpha}\left(\frac{T}{T_D}\right)^7 \int_0^{T_D/T} \frac{x^6 \cdot e^x}{(e^x - 1)^2} dx$$

$$+ \sum_{j<i} \beta_{ij}\left(\frac{e^{\Delta E_{ij}/k_B T}}{e^{\Delta E_{ij}/k_B T} - 1}\right) + \sum_{j>i} \beta_{ij}\left(\frac{1}{e^{\Delta_{ji}/k_B T} - 1}\right) \qquad (5.10)$$

The first term describes contributions due to Raman phonon scattering, and the second two to direct processes such as phonon emission and absorption. The hole width between 20 and 90 K was shown to be independent of electronic transition in $BaFCl_{0.5}Br_{0.5}:Sm^{2+}$ (100), though the nonradiative relaxation rates vary considerably (104). It was concluded that the dephasing is not influenced by direct phonon processes and β_{ij} was set to zero. The homogeneous line width, Γ_{hom}, at temperature T is therefore dependent on the Debye temperature, T_D, and the electron–phonon coupling constant, $\bar{\alpha}$. In fitting the hole-burning data shown in Fig. 5.26, $\Gamma_{\text{hom}}(0)$ is neglected, as it is determined by the lifetime of the electronic state (105), corresponding to a value of less than 1 kHz. The value of T_D was estimated from $BaCl_2$ heat capacities between 5 and 10 K (106) to be 130 ± 10 K, leaving the electron–phonon coupling constant as the only adjustable parameter. The best fit to the data was obtained using a value of 100 GHz for $\bar{\alpha}$. The success of this model was further demonstrated by the fact that it also provides a good fit to the fluorescence line width (107), after the transition in $BaFCl:Sm^{2+}$ has become dominated by homogeneous broadening, above 100 K. The discrepancy is still small at the highest values for which fluorescence line widths have been recorded, that is, at 550 K.

At temperatures below about 15 K the dephasing is dominated by a different process such as, perhaps, magnetic coupling between the halide ions and the Sm^{2+} impurities. The temperature dependence of the hole width in this regime is approximately linear for the $^7F_0-^5D_1$ transition (Fig. 5.26), but this has been shown to vary with transition (100). A similar temperature dependence has been observed for Pr^{3+} in silicate glass between 2 and 20 K (see Section 5.3.4.7), and $T^{1.3}$ dependencies are common for organic impurities in amorphous materials (12). The dual crystal/glass nature of $BaFCl_{0.5}Br_{0.5}:Sm^{2+}$ has been demonstrated previously (108). The temperature-induced hole shift was found to obey a T^4 law, as expected for a crystal, but the coupling constant was more typical of a rare earth–doped glass. Such behavior may be attributed to local disorder in the vicinity of the impurity ions within the bulk of the crystalline matrix. Conversely, a spectral hole-burning study of Eu^{3+}

in organic glasses and polymers provided evidence of a crystal-like environment in the vicinity of the impurity ion (109).

5.3.4.4. *Porphyrin-Doped Glycerol:Ethanol Glass*

Material: organic molecule-doped organic glass
Techniques: persistent spectral hole-burning; sample temperature variation
Properties studied: optical dephasing

For organic impurity molecules in amorphous hosts, the temperature broadening of the homogeneous line width of optical transitions, measured by spectral hole-burning, has almost uniformly been found to obey a $T^{1.3}$ relationship (12). This has been explained by several theories involving different forms of coupling between the guest electronic system, the two-level systems of the host, and thermally activated phonons (see above). Again, porphyrin (Fig. 5.11) has proved to be an ideal chromophore for such investigations in organic glasses (110–113).

When the concentration of free-base porphin in glycerol:ethanol glass is increased, a new absorption band appears about 300 cm^{-1} to the red of the $S_1 \leftarrow S_0$ transition. As the absorption strength increases with the square of the guest concentration, the new band was attributed to the $S_1 \leftarrow S_0$ transition of porphyrin dimers (110). The temperature dependence of holes burned and read with low powers was measured for both absorption bands between 1 and 10 K. Despite broadening of holes burned in the dimer transition by more than a factor of 2, both sets of hole widths followed a $T^{1.3}$ relationship (Fig. 5.27).

5.3.4.5. *Porphyrin-Doped Poly(ethylene)*

Material: organic molecule–doped polymer
Techniques: persistent spectral hole-burning; sample temperature variation
Properties studied: optical dephasing

Homogeneous line widths of electronic transitions in organic impurities in polymer films have also been found to obey a $T^{1.3}$ relationship at low temperatures (12). The relative width of the holes varies from host to host and from impurity to impurity, but the $T^{1.3}$ relationship remains constant. Porphyrin (Fig. 5.11) has again proved to be an ideal chromophore for such investigations. Of the polymers studied, the narrowest hole widths are observed in poly(ethylene) hosts (112). It was concluded that hole widths are small for poly(ethylene) owing to the lack of side groups on the polymer chains, which allows tighter packing around the impurity. This effect is analogous to

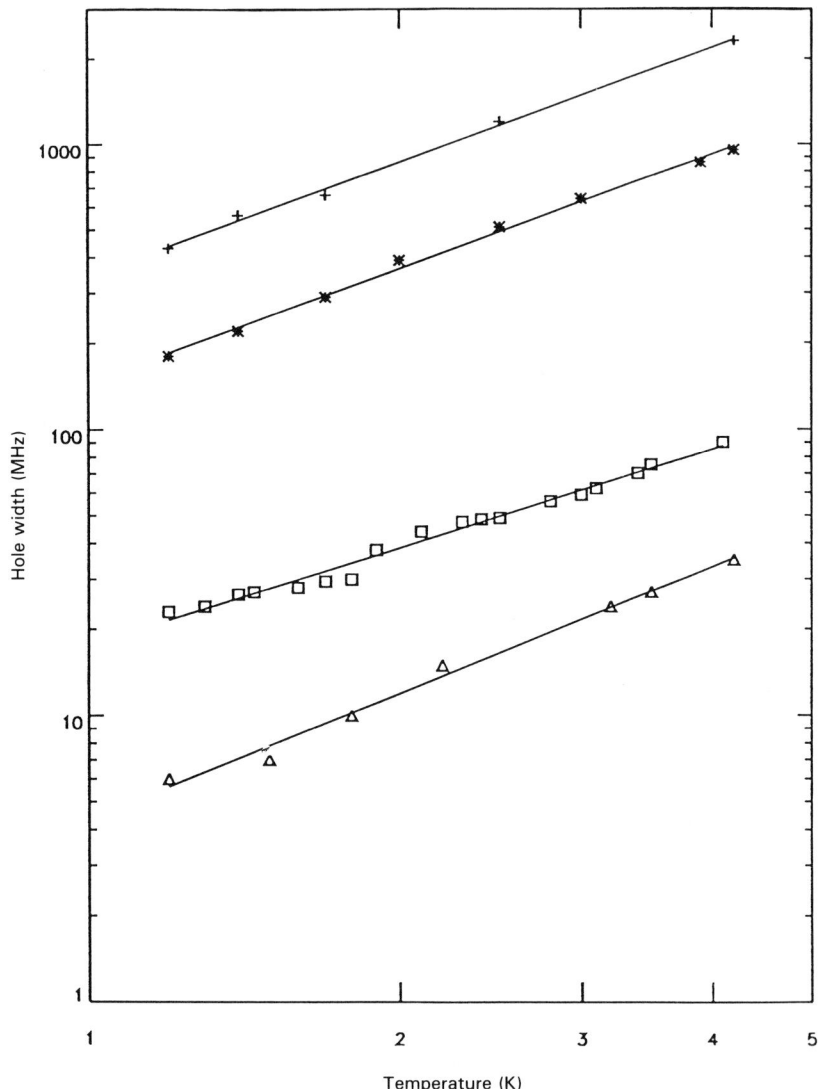

Fig. 5.27. The temperature dependence of the $S_1 \leftarrow S_0$ homogeneous line width determined from hole-burning experiments on porphyrin in various amorphous hosts. The upper two lines are for monomers and dimers (top) in glycerol glass (110). The lower pair are for poly(ethylene) hosts, with and without (bottom) traces of the solvent *p*-xylene (112). Fits ignore error bars and follow a $T^{1.3 \pm 0.2}$ dependence.

the variation in optical dephasing behavior observed for the same impurity in n-octane crystals (see Section 5.3.4.1). The presence of only a small amount of an additional solvent, p-xylene in this case, causes the hole width to broaden considerably (Fig. 5.27), but the $T^{1.3}$ dependence remains.

A later study (114) succeeded in obtaining a good fit to one of the models proposed for optical dephasing in amorphous materials (115). This model predicts a linear contribution to homogeneous line broadening due to interaction with the two-level systems and a more complex term that accounts for the effect of a Gaussian distribution of low-frequency localized modes. The best fit to the data using this model suggested a local mode peak energy that increases with temperature but remains between 2 and 3 cm^{-1} in the range 0.3–17 K. Although correlation between experiment and theory was high, the problem remains that no experimental evidence exists for the predicted local mode vibrations at 2–3 cm^{-1}. Indeed, the phonon sideband for holes burned in porphyrin-doped poly(ethylene) are displaced from the hole by about 18 cm^{-1}. A similar study of porphyrin in semicrystalline polymer hosts has also been conducted. In poly(ethylene) hosts, a $T^{1.3}$ dependence was found and attributed to local disorder. However, a variation in optical dephasing behavior, dependent on the degree of crystallinity of the host, was observed when the impurity was changed to dimethyl-s-tetrazine (116).

5.3.4.6. Octaethylporphyrin-Doped Poly(styrene)

Material: organic molecule-doped polymer
Techniques: persistent spectral hole-burning; sample temperature variation
Properties studied: optical dephasing

That the $T^{1.3}$ homogeneous line width broadening dependence is not universal for organic impurities in amorphous hosts was demonstrated by a photochemical hole-burning experiment conducted on the $S_1 \leftarrow S_0$ absorption band of octaethylporphyrin in poly(styrene) in the range 0.05–1.50 K (117). While the temperature dependence of the hole width above about 0.3 K is close to that observed for most amorphous materials at $T^{1.2}$, an entirely different behavior was observed below this. Two crossovers were observed. The positions of the crossovers and temperature dependencies between them were found to be wavelength dependent. It was concluded that no optical dephasing theory could accurately reproduce the experimental data, but comparisons were made. Dipole–quadrupole interaction between the impurity and the two-level systems, consistent with the fact that octaethylporphyrin has a center of inversion, predicts a single crossover in temperature dependence (118). For poly(styrene), however, the crossover from $T^{1.75}$, due to interactions between two level systems and phonons, to $T^{1.33}$, representing interactions between two-level systems and fractons, should occur at about 8 K. A reasonable fit to the data was obtained by combining the contributions due to dipole–quadrupole interaction between the impurity and the two-level systems, interaction between two-level systems and very-low-frequency pseudolocal vibrations (119), and spectral diffusion (see Section 5.3.5). This fit relies on the existence of unobserved very-low-frequency vibrations and a temperature dependence on spectral diffusion, $T^{1.6}$, not predicted by theory (120,121). Clearly, more work is required both theoretically and experimentally before dephasing in amorphous systems is fully understood.

5.3.4.7. *Praseodymium-Doped Silicate Glass*

Material: lanthanide ion–doped inorganic glass
Techniques: persistent spectral hole-burning; sample temperature variation
Properties studied: optical dephasing

Studies of optical dephasing in inorganic glasses doped with small ions have largely been confined to fluorescence line narrowing techniques. Temperature dependencies for broadening of the homogeneous transition close to T^2 have been measured. For instance, a $T^{1.8}$ dependence was found for Eu^{3+} in silicate glass between 8 and 90 K (122). A further point was added, in agreement with this relationship, at 1.6 K through a transient hole-burning study. In this case, the hole-burning mechanism was found to be the familiar population redistribution among ground state nuclear quadrupole levels.

In the same host, Pr^{3+} shows hole-burning through a photophysical mechanism, attributed to a rearrangement of the local environment (123). The temperature dependence of the hole width could thus be investigated between about 2 and 20 K, revealing a linear relationship. That experimental conditions were not responsible for the anomalous behavior was proven using accumulated photon echo techniques, which provided results consistent with the hole-burning data for homogeneous line widths. Such materials would, according to theory, be expected to behave like organic materials but do not, providing more evidence of the need for improved models that can simulate experimental results more generally.

5.3.5. Spectral Diffusion in Glasses

Persistent spectral holes provide a marker on a subset of impurity molecules within an inhomogeneously broadened band and may be used to accurately examine the response of the system to subsequent changes, either external, in the case of electric or magnetic field studies (see Sections 5.3.1 and 5.3.2), or internal, in the case of spontaneous rearrangements of environment. The latter effect has been observed in glasses and is one of several processes referred to as spectral diffusion. Studies of spectral diffusion rely on the accurate measurement of hole parameters over extremes in the period between burning and reading from microseconds to weeks. Such demanding experimental conditions require great care, and the comments made above with regard to optical dephasing experiments (see Section 5.3.4) also apply here. Spectral diffusion is perhaps the most controversial phenomenon that has been investigated by spectral hole-burning owing to the conflict between interpretation of the available data as physically realistic effects (124) or as artifacts due to the experimental procedure (12,125). The details of the debate are beyond the scope of this chapter, in which only the basic principles and a few studies will be discussed. It should be noted that strong evidence has been found for systems that show spectral diffusion, such as the crystalline system pentacene-doped p-terphenyl studied using a single molecule fluorescence detection technique (126), and systems that do not, such as porphyrine-doped poly(ethylene) (125).

In amorphous materials, spectral diffusion is considered to be a manifestation of tunneling between the potential wells of two-level systems (90,91) (see Section 5.3.4). As the chromophore is coupled to these two-level systems, the tunneling processes may be monitored by the evolution of the spectral hole. The development of the spectral hole profile is dependent on the distribution of tunneling rates in the two-level systems caused by the variations in double well asymmetry and barrier potential. The distribution of rates is material dependent. Furthermore, theory predicts time-dependent hole broadening and area reduction to depend on the time scale of the experiment (27,124,127). By definition, a glass, even at very low temperatures, does not occupy an absolute minimum of its free energy but merely a convenient isolated well. Tunneling processes may therefore take place, and consequently the evolution of the spectral hole shape is due to constant variations in transition frequency of spectrally adjacent unburned molecules as well as through back-reactions of burned centers.

5.3.5.1. *Quinizarin-Doped Ethanol:Methanol Glass*

Material: organic molecule–doped organic glass
Techniques: persistent spectral hole-burning; time-dependent lineshape analysis
Properties studied: spectral diffusion

Photochemical holes may be burned in the $S_1 \leftarrow S_0$ absorption band of quinizarin (1,4-dihydroxyanthraquinone) when incorporated in alcoholic glasses (128). The lack of hole-burning observed for the same impurity in

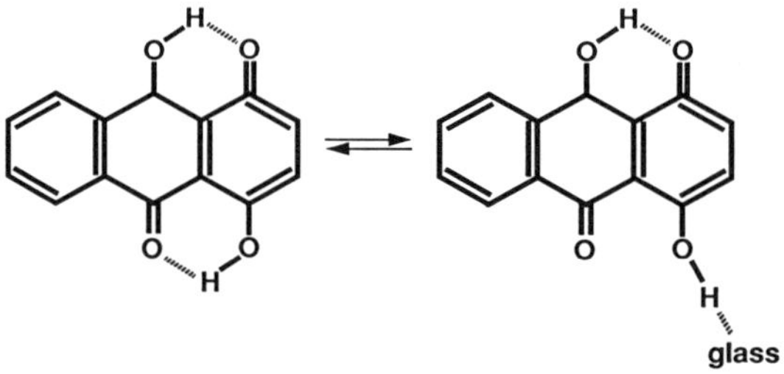

Fig. 5.28. Photochemical hole-burning occurs in quinizarin-doped alcohol glasses owing to conversion between intramolecular and intermolecular hydrogen bonds.

nonpolar hosts suggests that the mechanism is due to the photoinduced conversion of an intramolecular to an intermolecular hydrogen bond (Fig. 5.28). The time evolution of such spectral holes was studied over periods up to a week (129,130). The hole areas were found to decrease on a logarithmic time scale, the rate depending on impurity deuteration but not on small changes in temperature (Fig. 5.29) (129). The influence of impurity deuteration was interpreted as evidence that the principal two-level systems involved in such slow spectral diffusion processes are those that are photochemically induced as opposed to those in the neighbourhood of the impurity comprising solely of matrix molecules. The temperature independence is due to the two-level system barrier potentials being significantly greater than the thermal energy available at 4.2 K. That is, the back-reaction is a tunneling process. Further studies of the same system (130) showed that the hole width increases, also on a logarithmic time scale, while remaining Lorentzian in shape. As the time scale of the hole width increase and hole refilling were seen to be of the same order, it was concluded that this spectral diffusion process also involved mainly photochemically induced two-level systems. The relaxation processes associated with back-reactions change the local geometry of the relevant hydrogens, giving rise to strain fields and thus to strain-mediated spectral diffusion. Calculations based on the application of strain-mediated spectral diffusion to two-level systems (121,131) provided a semiquantitative explanation of the data.

5.3.5.2. *Cresyl Violet–Doped Ethanol Glass*

Material: organic ion–doped organic glass
Techniques: transient and persistent spectral hole-burning; time-dependent lineshape analysis
Properties studied: spectral diffusion

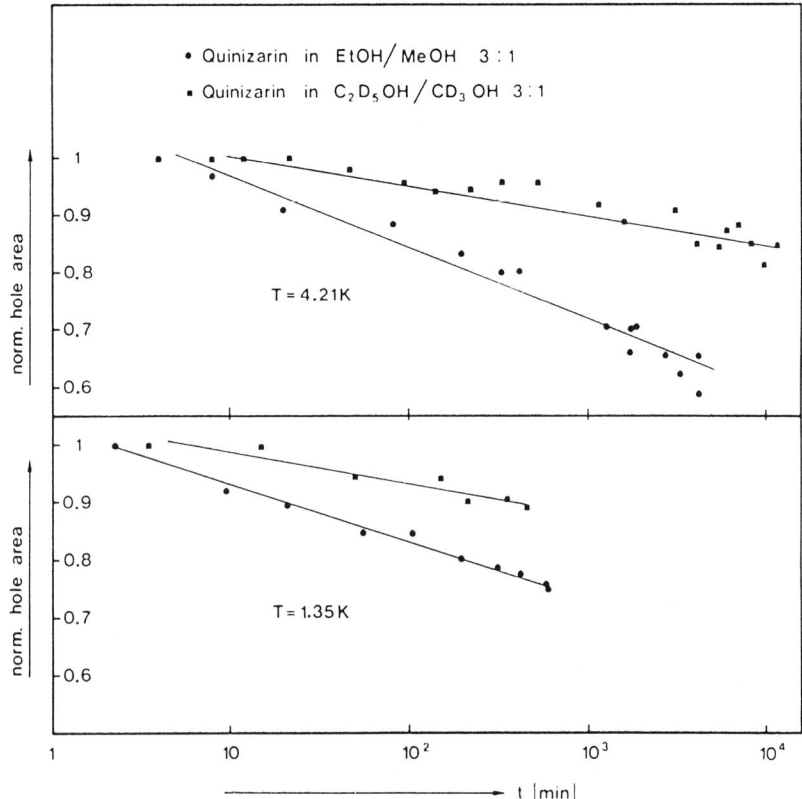

Fig. 5.29. Decay of the normalized hole area for protonated and deuterated samples as indicated (129).

The time development of the width of holes burned in the $S_1 \leftarrow S_0$ absorption band of cresyl violet–doped ethanol glass has been investigated over more than 8 orders of magnitude, from 10 μs to 4000 s (27,132,133). For the long time regime, persistent holes were burned (see Section 5.3.6.1) and measured using standard methods. A special technique, utilizing acousto-optic modulator switches to control counter propagating burn and read beams and to scan the read beam frequency (see Section 5.2.1.1), was used to enable transient holes to be burned and read in the short time regime. These transient holes were attributed to triplet bottleneck hole-burning on the basis of good fits to triple exponentials, representing three decay times as is often observed for triplet decay at low temperatures (134).

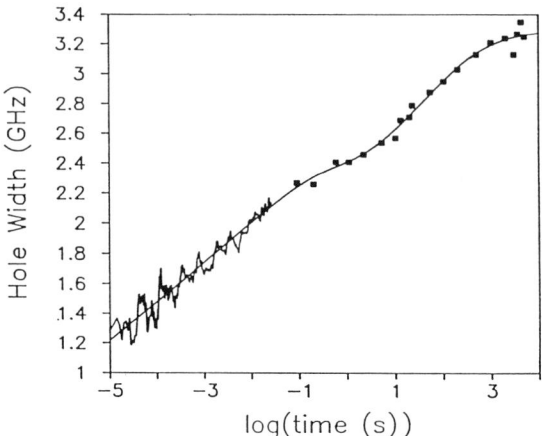

Fig. 5.30. Increase of hole width with time for holes burned in the $S_1 \leftarrow S_0$ absorption band in cresyl violet–doped ethanol glass (27). The two regions correspond to data obtained using different experimental arrangements (see the text).

The data for hole width broadening is shown in Fig. 5.30. The fit to the data relies on theory to describe spectral diffusion in terms of fluctuations in chromophore transition frequency caused by tunneling in nearby coupled two-level systems (127). These fluctuations have a very large range of rates owing to the similarly large spread in the two-level system parameters that determine tunneling probabilities. The four-point dipole correlation function (135) relates the two-level system parameters to actual line broadening, and a fluctuation rate distribution could be calculated for this system from the hole width time evolution data (27). The best fit rate distribution (fit shown in Fig. 5.30) comprises two contributions: log normal for slow rates (10^{-6}–$10\,\text{s}^{-1}$), and rate reciprocal for faster processes. The distribution in fluctuation rates was thus shown to extend over at least 16 orders of magnitude. As for quinizarin-doped ethanol:methanol glass (see Section 5.3.5.1), spectral diffusion in this material was found to be temperature independent at low temperatures because the two-level system barrier potentials are significantly greater than the thermal energy available.

5.3.6. Vibronics

When a subset of molecules is removed from resonance with a high-resolution light source, owing to a photoinduced process, the entire contribution of these molecules to the absorption spectrum of the host material is removed or changed. A resonant spectral hole is produced and, necessarily, nonresonant features corresponding to other absorption features of the same molecules.

An example has been given above, for the case of nonresonant hole-burning in electronic transitions (see Section 5.3.1.1). Information on vibronic structure may also be obtained using similar low-resolution hole-burning techniques. In the case of photophysical hole-burning the disturbed molecules are likely to absorb within the original absorption band but be distributed across it, whereas for photochemical hole-burning the new absorption is unlikely to coinside with the old absorption features (see Section 5.1). In both cases this may facilitate observation of the vibronic structure of the individual molecule and its phonon sideband where they had previously been obscured by bulk properties.

Impurity spectra of amorphous solids consist of broad absorption bands composed of inhomogeneously broadened superpositions of electronic transitions and associated phononic and vibronic structure. When a deep hole is burned into the center of such a band, symmetric phononic and vibronic structure may be observed to both lower and higher energy (136). A broad feature to higher energy, the phonon sideband hole, is due to the removal of molecules from resonance at the burn wavelength via direct excitation to the electronic level. The pseudo–phonon sideband hole is due to the removal of molecules from resonance at the burn wavelength via excitation to the phonon sideband, causing a reduction in purely electronic absorption to lower energy. Similarly, excited state vibronic structure appears as sharp features to both higher and lower energy.

High-resolution information on individually selected excited state vibrations is also attainable using spectral hole-burning. Measurement of hole widths burned into excited state local mode vibrations extrapolated to zero temperature and burn power allow the vibrational state lifetime to be determined.

Similarly, under certain circumstances it is possible to burn holes directly into vibronic transitions (137). This phenomenon is entirely analogous to hole-burning in electronic transitions. A highly monochromatic infrared light source promotes a molecule to its lowest vibrational excited state. In a proportion of cases, a photophysical or photochemical reaction occurs and subsequent relaxation may be to an alternative ground state, separated from the original by an energy barrier. A spectral hole results in the case that the vibrational transition is inhomogeneously broadened. Such behavior has been observed for matrix-isolated molecules in van der Waals and ionic solids (137), in glasses (138), and in amorphous semiconductors (139).

5.3.6.1. Cresyl Violet–Doped Poly(vinyl alcohol)

Material: organic ion–doped polymer
Techniques: persistent spectral hole-burning
Properties studied: intramolecular impurity vibrations

Ionic dyes may simply be embedded in thin polymer films to produce materials of excellent optical quality, many of which exhibit highly efficient hole-burning in their broad, featureless absorption bands. The $S_1 \leftarrow S_0$ absorption band of cresyl violet perchlorate in poly(vinyl alcohol) is typical. Figure 5.31 shows spectra of this absorption band before and after hole-burning (140), showing the pseudo–phonon side band hole centered at $32\,\text{cm}^{-1}$ and accurately revealing the S_1 vibrational frequencies of the cresyl violet ion. The sharp features derive from intramolecular vibrations of the dye ions, as has been shown by the invariance of the mode separations upon variation of the host polymer. In the example shown, the pseudo–phonon hole and vibronic features to lower energy are much stronger than the structure to higher energy. This implies that phononic and vibronic contributions dominate the absorption at the burn wavelength.

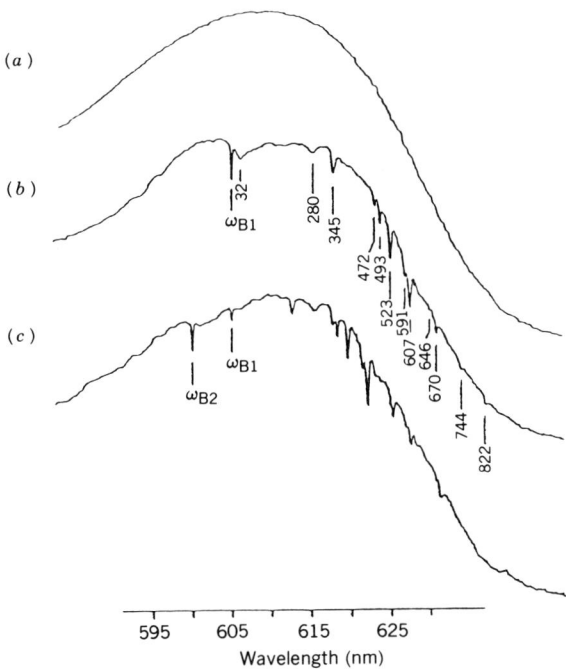

Fig. 5.31. Hole-burning and refilling in cresyl violet–doped poly(vinyl alcohol) (140): (a) before burning; (b) after burning at 605 nm (positions of satellite holes corresponding to S_1 vibrational structure are marked in wavenumbers); (c) after burning a second hole at 600 nm, causing partial refilling of the first.

The lower spectrum shows the effect of burning a second hole at a different wavelength. A similar pattern of side holes results, but the first hole spectrum is partially refilled, suggesting that the hole-burning mechanism is photophysical. Subsequent studies have concluded that the hole-burning mechanism is actually a combination of photophysical and photochemical effects (113,141) involving a charge redistribution within the dye molecule that causes a local redistribution of the host matrix.

5.3.6.2. Porphyrin-Doped n-Decane

Material: organic molecule–doped organic crystal
Techniques: persistent spectral hole-burning
Properties studied: local mode lifetime; optical dephasing

Hole widths in the $S_1 \leftarrow S_0$ 0–0 absorption band of porphyrin B sites in n-decane show an exponential dependence on temperature (142) (see Section 5.3.1.5). From a comparison of the data with the predictions of the exchange mechanism model (85) (see Section 5.3.4), the dephasing was attributed to interaction with a local mode vibration of energy $7\,\text{cm}^{-1}$ and lifetime 140 ps. Such a local mode vibration was indeed observed in the fluorescence spectrum at $7\,\text{cm}^{-1}$ and in the absorption spectrum at $5.6\,\text{cm}^{-1}$. Holes were burnt into the latter at 1.2 K with a minimum width of 3 GHz, corresponding to a lifetime of 120 ps, assuming dephasing processes to be negligible. Thus, the exchange mechanism model fit the data well, and the technique of excited state vibrational lifetime determination through spectral hole-burning was also shown to be accurate.

5.3.6.3. 1,2-Difluoroethane-Doped Krypton

Material: organic molecule–doped van der Waals solid
Techniques: persistent infrared spectral hole-burning; line shape analysis; polarization analysis
Properties studied: molecular rotation; vibrational excited state lifetime; dephasing

A thorough investigation of persistent infrared vibrational hole-burning in 1,2-difluoroethane isolated in van der Waals solids has been made (143,144). Here, by way of example, the results for a solid krypton host matrix are considered, being particularly interesting as both photochemical and photophysical effects are observed for the same transition, the v_{17} vibration (a C—F stretching mode) of the trans conformer. Spectral holes were burnt in the material, deposited on BaF_2 windows at an impurity concentration of 10^{-3}, in both low- and medium-burning-power regimes, using a tunable diode laser and a CO_2 laser, respectively.

Holes burned in the low-power regime ($10\,\mu W/cm^2$) were found to leave the integrated absorption unchanged (143), suggesting a photophysical mechanism. The hole-burning process was shown to be a result of the absorption of a single photon by a simple calculation indicating the probability of two-photon absorption at the power levels used to be negligible on the time scale of the lifetime of the vibrational state. This lifetime was deduced from the hole width at temperatures below about 5 K, where dephasing processes appear to play no role. The hole-burning efficiency was found to be high, of the order of 10^{-2}, and the mechanism was attributed to a rotation of the impurity within the host lattice. Holes burned with specific polarizations were found to retain their absorption polarizations over extended periods, thus showing that the impurity molecules may not rotate freely within the matrix, i.e., that the back-reaction may not occur freely. Calculations were made of the rotational potential barrier that confirmed the credibility of the hole-burning mechanism model. Variation of hole width with temperature was also considered. Unlike the usual weak temperature dependence of lowest electronic excited state lifetimes, T_1, the vibronic analogue would be expected to show a strong variation over the temperature range studied, 2–20 K. However, as the highly efficient hole-burning mechanism is attributed to rotational motion of the impurity, it follows that much of the vibrational energy is dissipated through such channels. Calculations for single and multiphonon relaxation result in a broadening of the hole width of only a factor of 2 in the temperature range studied and, as the observed broadening was by a factor of 10, it was concluded that dephasing processes make a significant contribution above 5 K. The data could be well reproduced using models of dephasing caused by anharmonic interactions with thermally excited phonons, using the Debye model, or with localized modes.

Holes burned in the medium-power regime ($100\,mW/cm^2$) were found to have different properties (144). Again the observed hole-burning effects were shown to be single-photon processes but were characterized by a reduction in the integrated absorption of the v_{17} vibration of the trans conformer. A simultaneous increase in the absorption of gauche features was observed, giving clear evidence that the mechanism in this case is a conformer conversion, that is, a rotation of the 1,2-difluoroethane molecule about its C—C bond, a photochemical process. Calculations of the intramolecular rotation were made, and these validated the proposed mechanism. It should be noted that the medium-power experiments were conducted using lower resolution than the low-power experiments and, therefore, despite the much smaller efficiency of the photochemical mechanism (10^{-6}), photophysical hole-burning effects were obscured. Conversely, photochemical effects were not observed using low-power irradiation owing to the low efficiency of the process.

5.3.6.4. Selenium Hydride Centers in Germanium–Arsenic–Selenium Glass

Material: hydride centers in amorphous semiconductor
Techniques: persistent and transient infrared spectral hole-burning
Properties studied: effect of network topology on matrix relaxation

SeH centers may be incorporated into amorphous Ge–As–Se by heating the glass to well above the glass transition in a hydrogen atmosphere. Upon cooling, the Se—H stretching mode may be observed in absorption at about $2200\,\text{cm}^{-1}$. Holes may be burned and read in this spectral feature at 1.5 K by using a tunable lead salt diode laser. In order to study matrix relaxation dynamics, the composition of the glass was varied and the decay rate of the spectral hole measured (145). The nonexponential recovery rate was found to be dependent only on the average coordination number of the glass, $\langle r \rangle$, and not on the specific chemical composition. Glasses having different compositions but the same value of $\langle r \rangle$ had almost identical hole decay rates, but as $\langle r \rangle$ was varied from 2.0 to 2.8 the decay rate increased by more than 3 orders of magnitude. The network topology alone appears to determine spectral hole relaxation in this case, though a full theory to explain this behavior has yet to be developed.

5.3.7. Special Materials

Rather than being used to investigate a particular phenomenon through selection of appropriate materials, spectral hole-burning techniques may be employed to examine the properties of special types of materials. Here, three such groups of topical interest are mentioned: biomolecules, thin films, and substances that exhibit photon-gated hole-burning. Photochemical reactions such as photosynthesis are being intensively studied by biologists, chemists, and physicists alike, and hole-burning has proved to be a useful method for gaining accurate information about the molecules involved in these processes. Thin film technology is becoming important in molecular electronics, engineering, and instrumentation, though the structure and dynamics of materials such as Langmuir–Blodgett films is still not well understood. Spectral hole-burning is helping to provide such information. Materials that exhibit photon-gated hole-burning are being actively sought for implementation in high-density optical memories (see Section 5.4.1) and are also of more general scientific interest.

5.3.7.1. Biological Compounds

Spectral hole-burning investigations of porphyrins have been discussed in earlier sections. Porphyrin molecules are closely related to chlorophylls, which play an important role in photosynthesis. Although hole-burning experiments must usually be carried out at temperatures at which life is impossible, information can still be gained on the electronic and vibronic structure of the molecules and on the processes in which they may take part. The first evidence of hole-burning in chlorophyll-like molecules came from a low-resolution study of protochlorophyll in organic hosts (146). Higher resolution studies of chlorophyll *a* (Fig. 5.32) have subsequently been made, using ether–butanol as a host (147). Both transient and unstable photophysical hole-burning, with a refill time of about 10 s, were observed. The former was attributed to bottleneck hole-burning through intersystem crossing to the triplet state, which has a lifetime of several milliseconds. Photochemical hole-burning has also been observed in chlorophyll-like molecules. For the cases of magnesium porphyrin–ethanol (148) and magnesium porphyrin–pyridine (96) complexes, the mechanism was attributed to axial ligand reorientation within the host lattice.

Many of the hole-burning studies on biological compounds have focused on physical phenomena such as spectral diffusion and optical dephasing at

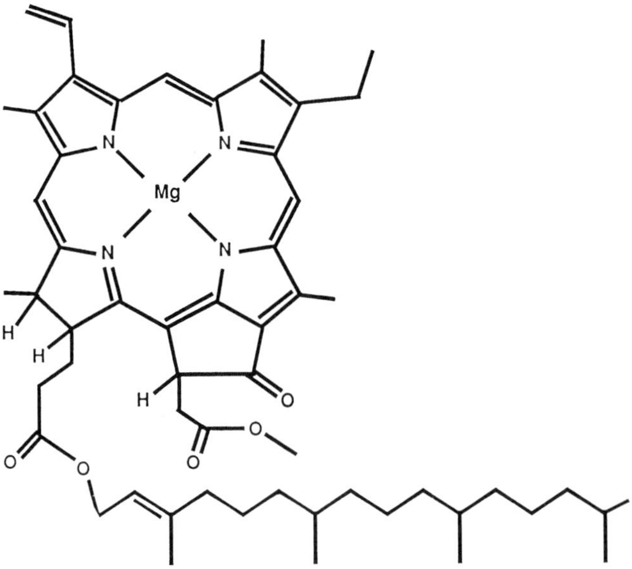

Fig. 5.32. Structure of chlorophyll *a*. Note similarity to porphyrin structure (Fig. 5.11).

low temperature (149), rather than the life processes occurring within the biomolecule. An exception to this is the attempt to determine the mechanism for primary charge separation that is known to occur after excitation but before electron transfer has taken place during the first stage of photosynthesis in both plants and bacteria. This event takes place in a special pair of strongly coupled chlorophylls within the reaction center protein. Holes with a width of the order of 500 cm^{-1} were burned in the transition to the S_1 state in the bacteriochlorophyll pairs in the purple bacteria *Rhodopseudomonas sphaeroides* and *Rhodopseudomonas viridis* (150,151). Various models to explain these very large hole widths have been suggested (150–152), but work continues to determine an unequivocal solution to the question. The hole width has been determined in both the isolated antenna complex and the intact membrane of *Rhodobacter sphaeroides* to be independent of wavelength and temperature, between 1 and 30 K (153). It was concluded that this is a manifestation of energy transfer between the bacteriochlorophylls on a time scale of about 2 ps. Similar work has been conducted on plant reaction centers from spinach chloroplast for which holes with two component widths, one of which is much narrower than those in bacteria, have been found (154). It was subsequently suggested (155) that the narrow component may be due to hole-burning in an unrelated chlorophyll, though further work appears to have removed this concern (156). Furthermore, this study (156) produced comprehensive data on the excited state vibrational structure of the native antenna complex of photosystem I using techniques similar to those used to investigate cresyl violet in poly(vinyl alcohol) films (see Section 5.3.6.1). The conservation of the total absorption suggests a photophysical mechanism, and this was ascribed to protein-pigment configurational tunneling (154,156).

Experimentation on complex biological systems not normally associated with photochemistry have also been investigated using spectral hole-burning. For instance, photophysical hole-burning has been observed in zinc-substituted myoglobin (157) and in the long wavelength absorption band of daunomycin intercalated into a DNA oligonucleotide (158). It is expected that the range of biomolecules for which hole-burning occurs is much larger than that so far studied and that a great deal of information about the structure and dynamics of proteins will become accessible in the near future. More thorough reviews of this field have been published (159).

5.3.7.2. *Thin Films*

By incorporating a chromophore in an optically inactive film, the dynamics and structure of the material may be investigated. Optical signals are necessarily small, owing to the dimensions of the host, but, when the impurity is chosen to be a molecule with a high quantum efficiency and which is known to exhibit spectral hole-burning in other hosts, high-resolution data may still be obtained.

One of the first such studies was of chlorin and cresyl violet perchlorate adsorbed onto porous silica surfaces (160). The hole widths were found to be much broader than for the same chromophores embedded in the bulk of amorphous silica, reflecting the different dephasing processes of the dye when confined to a surface. Subsequent studies of similar systems (161) and of quinizarin adsorbed onto γ-alumina powder (162) have confirmed these findings. Both studies attributed the increased dephasing rate to low-frequency adsorbate motions coupled to the electronic transitions of the chromophores. Such motions were observed in an earlier fluorescence line narrowing investigation (163).

Langmuir–Blodgett films were first produced several decades ago (164), and their optical properties have been extensively studied since (165). Recently, resorufin (166) and a free-base derivative of porphyrin (167) have been incorporated into Langmuir–Blodgett film structures and hole-burning investigations conducted. Monolayer and multilayer samples may be prepared, and the information obtained provides insight into the nature of pseudo-two-dimensional structures. Wavelength dependencies on both hole width and hole width temperature dependence were observed and explained in terms of energy transfer between chromophores. Further studies, for instance, of impurity concentration dependence (168) and of Stark effect to determine chromophore orientation (169), were undertaken to clarify these early findings.

5.3.7.3. *Photon-Gated Spectral Hole-Burning Materials*

A number of materials exist for which the efficiency of spectral hole-burning has been shown to be enhanced by the addition of a second excitation source, operating at a different wavelength. There is no requirement for the second source to be highly monochromatic. The scheme for photon-gated hole-burning is shown in Fig. 5.33. A high-resolution laser selectively excites a subset of molecules within the inhomogeneously broadened absorption band of the material, as for single photon hole-burning. This causes little or no photochemistry, but excited state absorption of a photon from the second source, sometimes after relaxation to a different metastable level, causes a persistent change in the molecule. Hence, the process is known as photon-gated spectral hole-burning, as the second light source acts as a "gate" to allow selective photochemistry to progress. The "gating ratio" is defined as being the proportional increase in hole-burning efficiency caused by the addition of the second beam. Typically, the gating photon has a higher energy than the selective photon, and the hole-burning mechanism involves the movement of an electron to a trap elsewhere in the lattice or between donor and acceptor

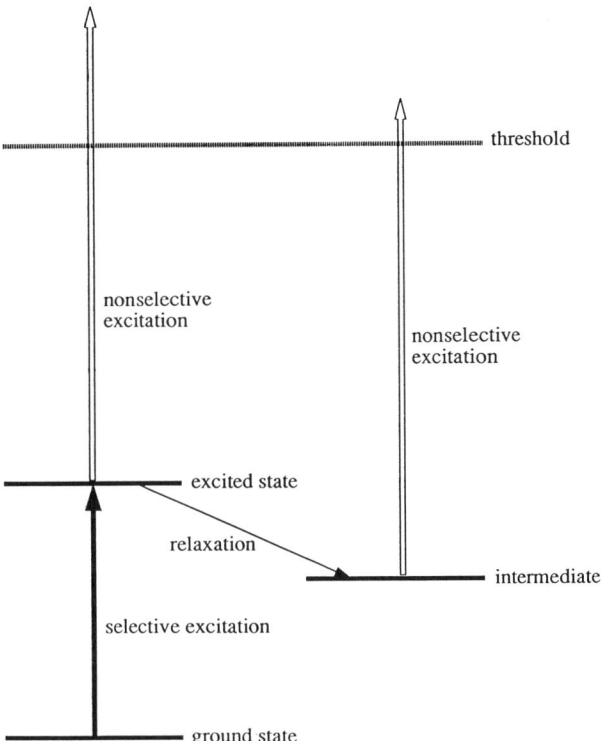

Fig. 5.33. Photon-gated hole-burning. A high-resolution laser selectively excites a subset of molecules. A second laser, operating at a different wavelength, excites this same subset from the excited state or, after relaxation, from an intermediate state to beyond a threshold for photochemistry. Hence, hole-burning occurs only in the presence of the second laser but may subsequently be read using only the high-resolution beam.

molecules. Reviews summarizing the properties of photon-gated spectral hole-burning materials stress both inorganic materials (170) and organic materials (171). Such materials have important potential applications in data storage (see Section 5.4.1).

The first report of photon-gated spectral hole-burning was for $BaFCl:Sm^{2+}$ (97). A subsequent investigation added bromine to the melt to produce $BaFCl_xBr_{1-x}:Sm^{2+}$. Gated hole-burning has been observed in all three $4f^6$ $^7F_0-^5D_J$ transitions (100) (see Section 5.3.4.3). The complexity of the competing processes occurring in such systems is illustrated by Fig. 5.34. It is frequently the case that hole erasure, transient hole-burning, and single-color hole-burning make parameterization of such systems difficult. In the case of the

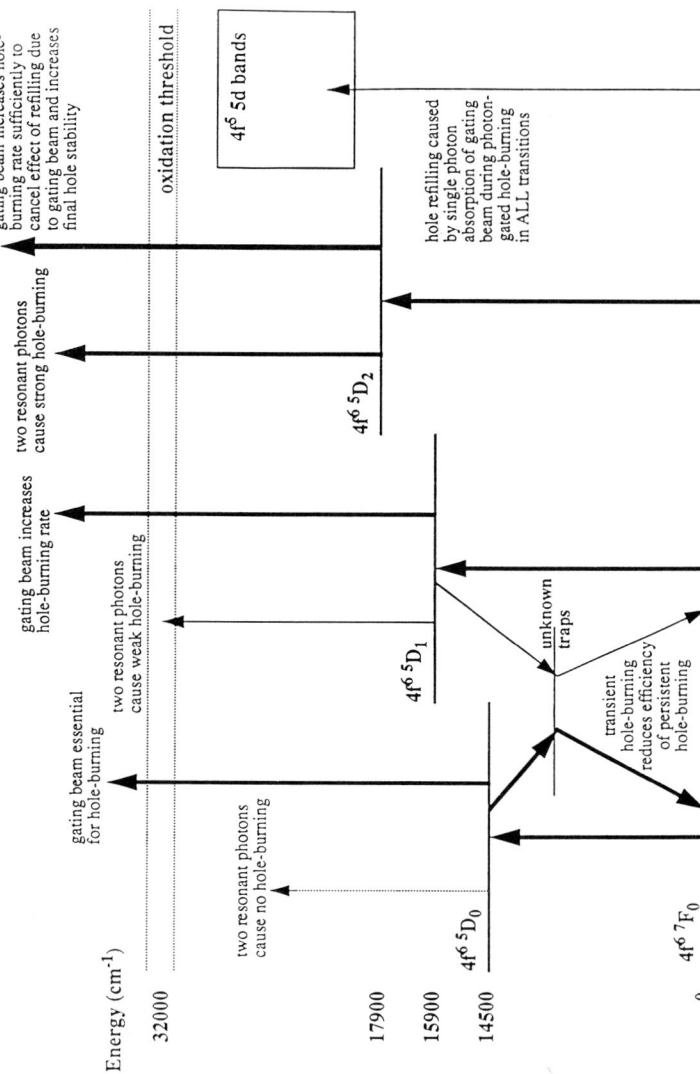

Fig. 5.34. Summary of single-color hole-burning, photon-gated hole-burning, and hole-erasing processes that occur in $BaFCl_{0.5}Be_{0.5}:Sm^{2+}$ (100). Line thicknesses qualitatively represent relative strengths.

$^7F_0-^5D_2$ transition in $BaFCl_{0.5}Br_{0.5}:Sm^{2+}$, holes appear to be uninfluenced by the addition of gating light. However, the holes burned with gating light show greater resistance to refilling than those burned with resonant light only, suggesting that two different mechanisms take place, gating light allowing more energetically remote traps to be reached by the released electron. The system $CaSO_4:Sm^{2+}$ (172) and $LiGa_5O_8:CO^{2+}$ (173) are other examples of materials that show this multiple hole-burning behavior.

Organic glasses also demonstrate photon-gated spectral hole-burning that is complicated by simultaneous single-photon processes. One such system is the chromophore, *meso*-tetra-*p*-tetrabenzoporphyrin, co-doped with halomethane electron acceptors in poly(methyl methacrylate). Here, single-color persistent photophysical hole-burning occurs upon excitation of the $S_1 \leftarrow S_0$ transition at about 630 nm. Such excitation also populates the triplet state through intersystem crossing with high efficiency. Excitation from the triplet state with a gating beam at about 500 nm causes electron transfer from chromophores to nearby halomethane acceptors, resulting in persistent photochemical hole-burning (174). A subsequent study (175) examined hole-burning parameters for this system, using chloromethane (chloroform) as the electron acceptor, between 1.5 and 90 K, revealing an approximate T^2 hole width dependence and a gating ratio which increases with temperature. The latter is due to a decrease in single-photon hole-burning efficiency at higher temperatures, unique to this material, while the photon-gated hole-burning efficiency remains approximately constant.

5.4. TECHNOLOGICAL APPLICATIONS

Spectral hole-burning materials have several potential technological applications (176). The most important ideas, relevant to optical computing, are reviewed here. As before, ultrafast phenomena are not discussed, but it should be noted that analogues in the time domain exist for optical data storage (177,178), holography (2,179), and molecular computing (180).

5.4.1. Optical Data Storage

Spectral hole-burning was proposed as a technique to optically store digital information some time ago (181,182). The concept involves recording a bit of information as a hole, nominally a "1," or the absence of a hole, a "0." Thus, digital information can be recorded throughout the inhomogeneously broadened absorption bands of hole-burning materials and recovered by measuring transmission at the relevant wavelength. A device would operate in a similar way to a compact disc, utilizing two spatial dimensions, but having storage increased by a factor equal to the ratio of inhomogeneous to homogeneous line widths, which may be as high as 10^6, yielding a data storage density of 10^{14} bits/cm^2. In the case of spectral holes that split upon application of an electric field, a further storage dimension is added (183,184), though it should be noted that any individual molecule may now be required to make contributions to more than one hole at different values of electric field. The storage density is determined by fundamental statistical limits implied by the number of photoreactive molecules rather than by the number of possible storage dimensions.

The parameters required for a material to perform in a device efficiently have been analyzed (185), but the search for a material possessing the required properties continues. A major difficulty is the inevitable degradation of data upon information retrieval, i.e., hole-burning that takes place during the readout process. Photon-gated hole-burning (see Section 5.3.7.3) circumvents this problem. Here a photon of different energy enhances the hole-burning rate. Only light at the initial frequency is required to probe for the existence of a hole, and therefore the information may be accessed repeatedly with little degradation. Fundamental statistical considerations are still a limiting factor when one is optimizing a photon-gated hole-burning storage device (186), however.

High-density data storage relies on the ratio of inhomogeneous to homogeneous line widths for the optical transition being large, as occurs at low temperatures. The homogeneous line widths of optical transitions rise rapidly with temperature (see Section 5.3.4), however, and this limits the systems for which discrete holes may be burned at temperatures above 100 K. Indeed holes burned in most systems are erased upon cycling to these temperatures. Materials have now been found in which spectral holes are stable at much higher temperatures. Holes burned in the first photon-gated material to be discovered, BaFCl:Sm^{2+}, were found to be cyclable from 2 K to room temperature with little degradation (97). The addition of bromine to the melt, to produce BaFCl$_{0.5}$Br$_{0.5}$:Sm^{2+}, broadens the inhomogeneous widths to greater than 50 cm^{-1}, owing to increased disorder (98). This enabled holes to be burned at temperatures well in excess of 100 K (100,108), and room temperature holes 10 times narrower than inhomogeneous line widths have subsequently been burned in electronic transitions in Sr$_{0.5}$Mg$_{0.5}$FCl$_{0.5}$Br$_{0.5}$:Sm^{2+} (Fig. 5.25) (101).

An entirely new concept for spectral hole-burning at room temperature has recently been developed that relies on selective bleaching of dyed microspheres (187). In this case, an ensemble of particles with sizes on the order of micrometers are irradiated with high-power light from a dye laser. Photolysis occurs for all dye molecules but is enhanced for those incorporated in particles for which a morphologically dependent resonance (188) with the exciting radiation occurs. Hence, though the optical transition may be homogeneously broadened at room temperature, a spectral hole is created through particle size selection.

5.4.2. Holography

The techniques required to produce plane wave holograms using spectral hole-burning have been outlined above (see Section 5.2.1.2). Holograms recorded in this way have proved to be important investigative tools for pure research (see Sections 5.3.1.6 and 5.3.1.7). Owing to the imaging properties of holograms, it is possible to produce more complex grating patterns that contain spatial information, or pictures. In the same way that binary data storage increases the information that can be stored in a single spot, many holograms may be recorded in a single piece of hole-burning material, each one stored by a different subset of molecules. These images may be manipulated by utilizing interference effects that cause interaction between spectrally adjacent holes (see Section 5.2.1.2). Upon the application of a particular external electric field the images overlap and are combined in a way determined by the relative spatial phase of the original holograms, making parallel processing applications a possibility.

5.4.2.1. Image Storage

To record an image holographically, the experimental arrangement shown in Fig. 5.3 requires only the addition of an object in one of the burn beams. The object may be a solid structure from which light is reflected to record a three-dimensional image, or a partially transparent material through which the beam passes to record a two-dimensional image. The latter technique has been demonstrated using thin films of chlorin-doped poly(vinyl butyral) (189) (see Section 5.3.1.5). Monochrome slides were placed in the object beam before burning the spectral hole. The images were then recovered by irradiating the sample with only the reference beam and recording the diffracted signal collinear with the object beam using an image intensifier and video camera.

The application of an external electric field to such a sample causes the optical transition energy to change for most molecules. Consequently, more than one hole can be burned at the same frequency by choosing different electric field strengths such that the laser interacts with a different subset of molecules each time. In order to minimize cross talk and maximize storage density it is necessary to control the spatial phase of adjacent gratings. Spatial phase refers to the position of the grating maxima on the hole-burning material. Figure 5.35 demonstrates the effect of spatial phase on the background signal. Burning with the same phase causes constructive interference between overlapping

(a) phase difference (0,0)

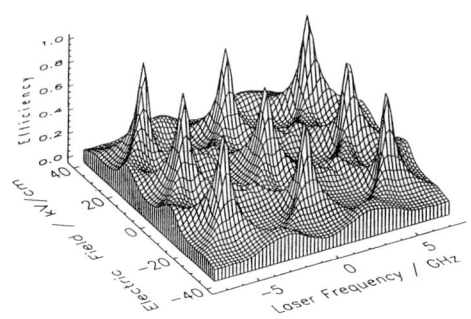

(b) phase difference (π,π)

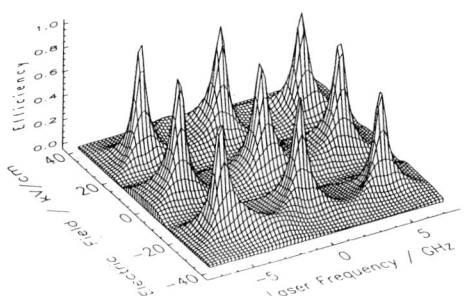

Fig. 5.35. Multiple storage of holograms in the frequency and electric field dimensions. The diffraction efficiency of nine holograms are illustrated. Selecting a spatial phase difference of $\pi/2$ between holograms burned in both the frequency and electric field dimensions minimizes cross talk (189).

wings of the Lorentzian holes. When a spatial phase difference of half a period is selected between each hole in both the frequency and electric field domains, destructive interference results in a large reduction in background. The spatial phase may be controlled using a piezoelectric plate and monitored by splitting off part of the burn beams before the sample and observing the expanded interference pattern using a linear diode array (see Fig. 5.3). This technique has been used to store 100 images within a single wavenumber (190).

In a more recent development to improve speed and flexibility, the slides used as objects have been replaced by a computer-controlled liquid crystal

TECHNOLOGICAL APPLICATIONS   219

screen through which the beam passes. By means of this technique 2000 individual frames from a video have been stored in single-polymer films (191), still utilizing only a small part of the available inhomogeneous bandwidth.

5.4.2.2. The Molecular Computer

Whereas image storage utilizes the Stark effect to increase spectral density while maintaining low cross talk, applications to data manipulation rely on the Stark effect to intentionally cause interference between spectrally adjacent holograms (see Section 5.2.1.2). The nature of the interference, constructive or destructive, may be selected upon burning by choosing the spatial phase difference between the holograms. Figure 5.36 schematically illustrates the principle. Two holes are burned at the same frequency but at different externally applied electric field values. When the electric field is varied, both holes split as indicated. Note, however, that this splitting occurs in amorphous systems only when the dye molecule impurity has a change of angle between ground and excited state dipole moments close to $0°$ or $90°$ and the orientation of laser polarization and electric field direction is appropriate

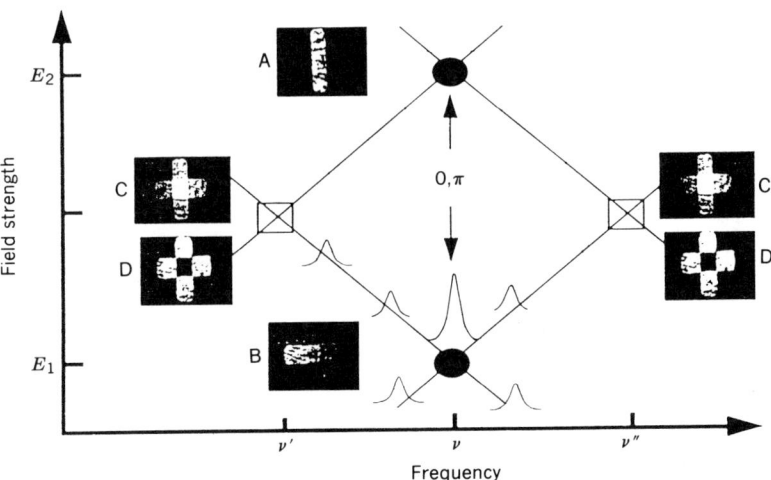

Fig. 5.36. The principle of operation of the molecular computer (192). Two images are stored using spectral hole-burning at the same frequency, v, but at different electric field strengths, E_1 and E_2, positions represented by black ovals. The stored images are also shown. When the holograms are reconstructed at the intermediate electric field strength and at frequencies v' or v'', represented by open rectangles, the superposition of holograms is dependent on the spatial phase difference selected between holograms during burning. For zero spatial phase difference, constructive interference results (upper reconstructed images). For π spatial phase difference, destructive interference results (lower reconstructed images).

(see Section 5.3.1.6). The hole-burning positions are indicated by dark spots, and the areas of overlap by rectangles. When the holograms are reconstructed using the values of frequency and electric field at the overlap position, a combined image is observed by the video camera.

Figure 5.36 further illustrates an example of the dependence of the reconstruction strength on spatial phase difference between holograms (192). Pictures A and B are the reconstructed images at the burn positions, (v, E_1) and (v, E_2). Pictures C and D are the reconstructed images at the overlap positions, (v_1, E) or (v_2, E), for the cases of zero phase difference and a half-period phase difference, respectively. When zero phase difference is selected (picture C), constructive interference results where the images overlap. As the amplitudes, rather than intensities, add in the interferometric holographic operation, the intensity of the overlapping area becomes four times that of those areas for which only a single image is reconstructed. When a half-period phase difference is selected (picture D), destructive interference results in zero reconstruction intensity for the overlapping area. The resulting intensities are summarized in the accompanying table, where the original intensities are regarded as being 0 or 1.

	Constructive Interference		Destructive Interference	
	Image 1: 0	0	0	1
Image 2:				
0	0	1	0	1
1	1	4	1	0

If the detection system is now set to detect light intensities of greater than or less than 0.5 to produce an output of 1 or 0, that is, a digital discriminator level of 0.5, the table becomes the truth tables for the logical operations OR and XOR (exclusive or).

	Constructive Interference OR		Destructive Interference XOR	
	Image 1: 0	1	0	1
Image 2:				
0	0	1	0	1
1	1	1	1	0

Alternatively, if a discriminator level of 2.5 is set, the constructive interference table becomes the truth table for the AND logical operation.

	Constructive Interference AND	
	Image 1: 0	1
Image 2: 0	0	0
1	0	1

These logical operations are performed in an entirely parallel manner, that is, each area of image is manipulated synchronously with and independently of every other area. The images may be extended to large arrays of pixels, representing digital 1's and 0's, to produce a large-scale parallel processor or molecular computer. It is hoped to extend this concept to the simulation of an error-correcting neural network by introducing feedback between the video camera and the liquid crystal object screen.

ACKNOWLEDGMENTS

We wish to thank all authors and publishers for permission to reproduce figures as referenced in figure captions, in particular those authors who provided us with original drawings: Jean-Claude Vial, Glynn Jones, Dietrich Haarer, Michael Fayer, Neil Manson, Eric Vauthey, and Alois Renn.

REFERENCES

1. A. Szabo, *Phys. Rev. Lett.*, **25**, 924 (1970).
2. K. K. Rebane and L. A. Rebane, in W. E. Moerner, Ed., *Persistent Spectral Hole-Burning: Science and Applications*, Springer, Berlin, 1988, Sect. 2.8.
3. M. Orritt and J. Bernard, *Phys. Rev. Lett.*, **65**, 2716 (1991).
4. H. de Vries and D. A. Wiersma, *J. Chem. Phys.*, **72**, 1851 (1980).
5. A. Szabo, *Phys. Rev. B*, **11**, 4512 (1975).
6. B. M. Kharlamov, R. I. Personov, and L. A. Bykovskaya, *Opt. Commun.*, **12**, 191 (1974).
7. A. A. Gorokhovskii, R. K. Khaarli, and L. A. Rebane, *JETP Lett.* (*Engl. Transl.*), **20**, 216 (1974).
8. J. M. Hayes and G. J. Small, *Chem. Phys.*, **27**, 151 (1978).
9. W. E. Moerner, M. Gehrtz, and A. L. Huston, *J. Phys. Chem.*, **88**, 6460 (1984).

10. W. E. Moerner, Ed., *Persistent Spectral Hole-Burning: Science and Applications*, Springer, Berlin, 1988.
11. O. Sild and K. Haller, Eds., *Zero-Phonon Lines*, Springer, Berlin, 1988.
12. S. Völker, in J. Fünfschilling, Ed., *Relaxation Processes in Molecular Excited States*, Kluwer, Dordrecht, The Netherlands, 1989, Chap. 3.
13. A. Renn and U. P. Wild, in H. Dürr and H. Bouas-Laurent, Eds., *Photochromism: Molecules and Systems*, Elsevier, Amsterdam, 1990, Chap. 29.
14. A. J. Meixner, A. Renn, and U. P. Wild, *J. Chem. Phys.*, **91** 6728 (1989).
15. A. Renn, A. J. Meixner, and U. P. Wild, *J. Chem. Phys.*, **92**, 2748 (1990).
16. E. Vauthey, K. Holliday, C. Wei, A. Renn, and U. P. Wild, to be published.
17. A. J. Meixner, A. Renn, S. E. Bucher, and U. P. Wild, *J. Phys. Chem.*, **90**, 6777 (1986).
18. A. Renn, A. J. Meixner, and U. P. Wild, *J. Chem. Phys.*, **93**, 2299 (1990).
19. K. Holliday, C. Wei, A. J. Meixner, and U. P. Wild, *J. Lumin.*. **48**, 329 (1991).
20. A. Renn, C. De Caro, and U. P. Wild, *Jpn. J. Appl. Phys.*, **28**, 257 (1989).
21. U. P. Wild, C. De Caro, S. Bernet, M. Traber, and A. Renn, *J. Lumin.*, **48**, 335 (1991).
22. P. Saari, R. Kaarli, and A. Rebane, *J. Opt. Soc. Am.*, B, **3**, 527 (1986).
23. J. H. Lee, J. J. Song, M. A. F. Scarparo, and M. D. Levenson, *Opt. Lett.*, **5**, 196 (1980).
24. B. Dick, *Chem. Phys.*, **136**, 413 (1989).
25. M. Romagnoli, W. E. Moerner, F. M. Schellenberg, M. D. Levenson, and G. C. Bjorklund, *J. Opt. Soc. Am. B*, **1**, 341 (1984).
26. M. Romagnoli, M. D. Levenson, and G. C. Bjorklund, *J. Opt. Soc. Am. B*, **1**, 571 (1984).
27. K. A. Littau and M. D. Fayer, *Chem. Phys. Lett.*, **176**, 551 (1991).
28. K. K. Rebane and V. V. Palm, *Opt. Spectrosc.*, **57**, 229 (1984).
29. S. Saikan, Y. Kanematsu, R. Shiraishi, T. Nakabayashi, and T. Kushida, *J. Lumin.*, **38**, 15 (1987).
30. A. L. Houston and W. E. Moerner, *J. Opt. Soc. Am. B*, **1**, 349 (1984).
31. O. N. Korotaev, A. I. Yurchenko, and V. P. Karpov, *Opt. Spectrosc.*, **61**, 474 (1986).
32. R. Wannemacher, R. M. Macfarlane, Y. P. Wang, D. Sox, D. Boye, and R. S. Meltzer, *J. Lumin.*, **48**, 309 (1991).
33. K. Holliday and N. B. Manson, *J. Phys.: Condens. Matter*, **1**, 1339 (1989).
34. A. M. Morozov, L. G. Morozova, and P. P. Feofilov, *Opt. Spectrosc.*, **32**, 50 (1972).
35. R. U. E. 't Lam and G. Blasse, *J. Chem. Phys.*, **72**, 1803 (1980).
36. G. K. Liu and J. V. Beitz, *Chem. Phys. Lett.*, **171**, 335 (1990).
37. G. Baumann, *Z. Phys.*, **203**, 464 (1967).
38. H. Pick, *Z. Phys.*, **159**, 69 (1960).

39. M. D. Levenson, R. M. Macfarlane, and R. M. Shelby, *Phys. Rev. B*, **22**, 4915 (1980).
40. R. T. Harley and R. M. Macfarlane, *J. Phys. C: Solid State Phys.*, **16**, 1507 (1983).
41. M. B. Seelbinder and J. C. Wright, *Phys. Rev. B*, **20**, 4308 (1979).
42. Z. Hasan, R. Danby, and N. B. Manson, *J. Lumin.*, **40**, 397 (1988).
43. Z. Hasan, H. Ghafoori Fard, and N. B. Manson, *J. Lumin.*, **45**, 304 (1990).
44. A. I. M. Dicker, M. Noort, S. Völker, and J. H. van der Waals, *Chem. Phys. Lett.*, **73**, 1 (1980).
45. S. Völker and R. M. Macfarlane, *Mol. Cryst. Liq. Cryst.*, **50**, 213 (1979).
46. A. I. M. Dicker, M. Noort, H. P. H. Thijssen, S. Völker, and J. H. van der Waals, *Chem. Phys. Lett.*, **78**, 212 (1981).
47. R. M. Hochstrasser and D. A. Wiersma, *J. Chem. Phys.*, **73**, 156 (1980).
48. A. I. M. Dicker, L. W. Johnson, M. Noort, and J. H. van der Waals, *Chem. Phys. Lett.*, **94**, 14 (1983).
49. A. Renn, S. E. Bucher, A. J. Meixner, E. C. Meister, and U. P. Wild, *J. Lumin.*, **39**, 181 (1988).
50. K. Holliday, E. Vauthey, M. Croci, A. Renn and U. P. Wild, *J. Opt. Soc. Am. B*, **9**, 982 (1992).
51. R. M. Macfarlane and R. M. Shelby, in A. A. Kaplyanskii and R. M. Macfarlane, eds., *Spectroscopy of Solids Containing Rare Earth Ions*, Elsevier, Amsterdam, 1987, Chap. 3.
52. N. R. S. Reddy, N. B. Manson, and E. R. Krausz, *J. Lumin.*, **38**, 46 (1987).
53. K. Holliday, X.-F. He, P. T. H. Fisk, and N. B. Manson, *Opt. Lett.*, **15**, 983 (1990).
54. N. R. S. Reddy, Ph.D. thesis, Australian National University, Canberra, 1988.
55. W. A. Runciman, N. B. Manson, and K. Holliday, *Proc. Electrochem. Soc.*, **88-24**, 90 (1988).
56. J. H. N. Loubser and J. A. Van Wyk, *Diamond Res.*, **11**, 4 (1977).
57. K. Holliday, N. B. Manson, M. Glasbeek, and E. van Oort, *J. Phys.: Condens. Matter*, **1**, 7093 (1989).
58. A. Abragam and B. Bleaney, *Electron Paramagnetic Resonance of Transition Ions*, Oxford University Press (Clarendon), London and New York, 1970.
59. R. J. Cook and D. H. Whiffen, *Proc. R. Soc*, **A295**, 99 (1966).
60. R. M. Shelby and R. M. Macfarlane, *Phys. Rev. Lett.*, **47**, 1172 (1981).
61. R. J. Elliott, *Proc. Phys. Soc. London*, **B70**, 119 (1957).
62. R. Wannemacher, D. Boye, Y. P. Wang, R. Pradhan, W. Grill, J. E. Rives, and R. S. Meltzer, *Phys. Rev. B*, **40**, 4237 (1989).
63. R. Wannemacher, R. S. Meltzer, and R. M. Macfarlane, *J. Lumin.*, **45**, 307 (1990).
64. R. Wannemacher, R. M. Macfarlane, Y. P. Wang, D. Sox, D. Boye, and R. S. Meltzer, *J. Lumin.*, **48**, 309 (1991).
65. R. W. Olson, H. W. H. Lee, F. G. Patterson, M. D. Fayer, R. M. Shelby, D. P. Burum, and R. M. Macfarlane, *J. Chem. Phys.*, **77**, 2283 (1982).

66. R. Casalegno and H. P. Trommsdorff, *Photochem. Photobiol.* **2**, 1167 (1983).
67. J. M. Clemens, R. M. Hochstrasser, and H. P. Trommsdorff, *J. Chem. Phys.*, **80**, 1744 (1987).
68. R. M. Hochstrasser and H. P. Trommsdorff, *Chem. Phys.*, **115**, 1 (1987).
69. A. Oppenländer, C. Rambaud, H. P. Trommsdorff, and J. C. Vial, *Phys. Rev. Lett.*, **63**, 1432 (1989).
70. C. Rambaud, A. Oppenländer, M. Pierre, H. P. Trommsdorff, and J. C. Vial, *Chem. Phys.*, **136**, 335 (1989).
71. R. Meyer and R. R. Ernst, *J. Chem. Phys.*, **86**, 784 (1987).
72. J. L. Skinner and H. P. Trommsdorff, *J. Chem. Phys.*, **89**, 897 (1988).
73. C. Rambaud, A. Oppenländer, H. P. Trommsdorff, and J. C. Vial, *J. Lumin.*, **45**, 310 (1990).
74. H. De Vries and D. A. Wiersma, *Phys. Rev. Lett.*, **36**, 91 (1976).
75. D. M. Burland, F. Carmona, and J. Pacansky, *Chem. Phys. Lett.*, **56**, 221 (1978).
76. C. von Borczyskowski, A. Oppenländer, H. P. Trommsdorff, and J. C. Vial, *Phys. Rev. Lett.*, **65**, 3277 (1990).
77. C. von Borczyskowski, A. Oppenländer, H. P. Trommsdorff, and J. C. Vial, *J. Lumin.*, **48**, 179 (1991).
78. N. J. Cockroft, T. P. J. Han, R. J. Reeves, G. D. Jones, and R. W. G. Syme, *Opt. Lett.*, **12**, 36 (1987).
79. N. J. Cockroft, Ph.D. thesis, University of Canterbury, New Zealand, 1988.
80. A. Edgar, G. D. Jones, and M. R. Presland, *J. Phys. C: Solid State Phys.*, **12**, 1569 (1979).
81. R. M. Macfarlane, R. J. Reeves, and G. D. Jones, *Opt. Lett.*, **12**, 660 (1967).
82. R. Silbey, in J. Fünfschilling, Ed., *Relaxation Processes in Molecular Excited States*, Kluwer, Dordrecht, The Netherlands, 1989, Chap. 4.
83. J. M. Hayes, R. Jankowiak, and G. J. Small, in W. E. Moerner, Ed., *Persistent Spectral Hole-Burning: Science and Applications*, Springer, Berlin, 1988, Chap. 5.
84. C. B. Harris, *J. Chem. Phys.*, **67**, 5607 (1977).
85. R. M. Shelby, C. B. Harris, and P. A. Cornelius, *J. Chem. Phys.*, **70**, 34 (1979).
86. P. de Bree and D. A. Wiersma, *J. Chem. Phys.*, **70**, 790 (1979).
87. D. Hsu and J. L. Skinner, *J. Chem. Phys.*, **83**, 2107 (1985).
88. S. Völker, R. M. Macfarlane, A. Z. Genack, H. P. Trommsdorff, and J. H. van der Waals, *J. Chem. Phys.*, **67**, 1759 (1977).
89. A. A. Gorokhovskii and L. A. Rebane, *Opt. Commun.*, **20**, 144 (1977).
90. P. W. Anderson, B. I. Halperin, and C. M. Varma, *Philos. Mag.* [8], **25**, 1 (1972).
91. W. A. Phillips, *J. Low Temp. Phys.*, **7**, 351 (1972).
92. D. Haarer, in W. E. Morner, Ed., *Persistent Spectral Hole-Burning: Science and Applications*, Springer, Berlin, 1988, Chap. 3.
93. I. Zschokke-Gränacher, Ed., *Optical Spectroscopy of Glasses*, Kluwer, Dordrecht, The Netherlands, 1986.

94. M. J. Weber, Ed., "Special Issue on Optical Linewidths in Glasses," *J. Lumin.*, **36**, 179 (1987).
95. S. Völker, R. M. Macfarlane, and J. H. van der Waals, *Chem. Phys. Lett.*, **53**, 8 (1978).
96. A. I. M. Dicker, L. W. Johnson, S. Völker, and J. H. van der Waals, *Chem. Phys. Lett.*, **100**, 8 (1983).
97. A. Winnacker, R. M. Shelby, and R. M. Macfarlane, *Opt. Lett.*, **10**, 350 (1985).
98. C. Wei, S. Huang, and J. Yu, *J. Lumin.*, **43**, 161 (1989).
99. L. Zhang, J. Yu, and S. Huang, *J. Lumin.*, **45**, 301 (1990).
100. C. Wei, K. Holliday, A. J. Meixner, M. Croci, and U. P. Wild, *J. Lumin.*, **50**, 89 (1991).
101. K. Holliday, C. Wei, M. Croci, and U. P. Wild, *J. Lumin.*, **53**, 227 (1992).
102. R. Jaaniso and H. Bill, *Europhys. Lett.*, **16**, 569 (1991).
103. D. E. McCumber and M. D. Sturge, *J. Appl. Phys.*, **34**, 1682 (1963).
104. G. C. Gâcon, J. C. Souillat, J. Seriot, F. Gaume-Mahn, and B. Di Bartolo, *J. Lumin.*, **18/19**, 244 (1979).
105. G. C. Gâcon, State doctoral thesis, University of Lyons, 1978.
106. R. M. Goodman and E. F. Westrum, Jr., *J. Chem. Eng. Data*, **11**, 294 (1966).
107. B. Birang, A. S. M. Mahbub'ul Alam, and B. Di Bartolo, *J. Chem. Phys.*, **50**, 2750 (1969).
108. A. Oppenländer, F. Madeore, J. C. Vial, and J.-P. Chaminade, *J. Lumin.*, **50**, 1 (1991).
109. R. van den Berg and S. Völker, *Chem. Phys. Lett.*, **150**, 491 (1988).
110. H. P. H. Thijssen, A. I. M. Dicker, and S. Völker, *Chem. Phys. Lett.*, **92**, 7 (1982).
111. H. P. H. Thijssen, R. van den Berg, and S. Völker, *Chem. Phys. Lett.*, **97**, 285 (1983).
112. H. P. H. Thijssen and S. Völker, *Chem. Phys. Lett.*, **120**, 496 (1985).
113. H. P. H. Thijssen, R. van den Berg, and S. Völker, *Chem. Phys. Lett.*, **120**, 503 (1985).
114. R. van den Berg, A. Visser, and S. Völker, *Chem. Phys. Lett.*, **144**, 105 (1988).
115. B. Jackson and R. Silbey, *Chem. Phys. Lett.*, **99**, 331 (1983).
116. H. P. H. Thijssen and S. Völker, *J. Chem. Phys.*, **85**, 785 (1986).
117. A. Gorokhovskii, V. Korrovits, V. Palm, and M. Trummal, *Chem. Phys. Lett.*, **125**, 355 (1986).
118. S. K. Lyo and R. Orbach, *Phys. Rev. B*, **29**, 2300 (1984).
119. S. K. Lyo, *Phys. Rev. Lett.*, **48**, 688 (1982).
120. V. Hizhnyakov and I. Tehver, *Phys. Status Solid B*, **95**, 65 (1979).
121. T. L. Reinecke, *Solid State Commun.*, **32**, 1103 (1979).
122. P. J. Selzer, D. L. Huber, D. S. Hamilton, W. M. Yen, and M. J. Weber, *Phys. Rev. Lett.*, **36**, 813 (1976).
123. R. M. Macfarlane and R. M. Shelby, *Opt. Commun.*, **45**, 46 (1983).

124. J. Friedrich and D. Haarer, in I. Zschokke-Gränacher, Ed., *Optical Spectroscopy of Glasses*, Kluwer, Dordrecht, The Netherlands, 1986, Chap. 4.
125. P. J. van der Zaag, J. P. Galaup, and S. Völker, *Chem. Phys. Lett.*, **166**, 263 (1990).
126. W. P. Ambrose and W. E. Moerner, *Nature (London)*, **349**, 225 (1991).
127. Y. S. Bai and M. D. Fayer, *Phys. Rev. B*, **39**, 11066 (1989).
128. F. Graf, H. K. Hong, A. Nazzal, and D. Haarer, *Chem. Phys. Lett.*, **59**, 217 (1978).
129. W. Breinl, J. Friedrich, and D. Haarer, *Chem. Phys. Lett.*, **106**, 487 (1984).
130. W. Breinl, J. Friedrich, and D. Haarer, *J. Chem. Phys.*, **81**, 3915 (1984).
131. J. L. Black and B. I. Halperin, *Phys. Rev. B*, **16**, 2879 (1977).
132. K. A. Littau, Y. S. Bai, and M. D. Fayer, *Chem Phys. Lett.*, **159**, 1 (1989).
133. K. A. Littau, Y. S. Bai, and M. D. Fayer, *J. Chem. Phys.* **92**, 4145 (1990).
134. R. A. Avarmaa, *Chem. Phys. Lett.*, **46**, 279 (1977).
135. S. Mukamel and R. F. Loring, *J. Opt. Soc. Am. B*, **3**, 595 (1986).
136. J. Friedrich, J. D. Swalen, and D. Haarer, *J. Chem. Phys.*, **73**, 705 (1980).
137. A. J. Sievers and W. E. Moerner, in W. E. Moerner, Ed., *Persistent Spectral Hole-Burning: Science and Applications*, Springer, Berlin, 1988, Chap. 6.
138. S. P. Love and A. J. Sievers, *J. Lumin.*, **45**, 58 (1990).
139. S. P. Love and A. J. Sievers, in H. Fritzsche, Ed., *Advances in Disordered Semiconductors*, Vol. III, World Scientific, Singapore, 1990.
140. B. L. Fearey, T. P. Carter, and G. J. Small, *J. Phys. Chem.*, **87**, 3590 (1983).
141. R. van den Berg and S. Völker, *Chem. Phys.*, **128**, 257 (1988).
142. A. I. M. Dicker, J. Dobkowski, and S. Völker, *Chem. Phys. Lett.*, **84**, 415 (1981).
143. M. Dubs, L. Ermanni, and H. H. Günthard, *J. Mol. Spectrosc.*, **91**, 458 (1982).
144. P. Felder and H. H. Günthard, *Chem. Phys.*, **85**, 1 (1984).
145. S. P. Love, A. J. Sievers, B. L. Halfpap and S. M. Lindsay, *Phys. Rev. Lett.*, **65**, 1792 (1990).
146. R. A. Avarmaa, K. Mauring, and A. Suisalu, *Chem. Phys. Lett.*, **77**, 88 (1981).
147. K. K. Rebane and R. A. Avarmaa, *J. Photochem.*, **17**, 311 (1981).
148. R. J. Platenkamp, *Mol. Phys.*, **45**, 113 (1982).
149. T. P. Carter and G. J. Small, *Chem. Phys. Lett.*, **120**, 178 (1985).
150. S. R. Meech, A. J. Hoff, and D. A. Wiersma, *Chem. Phys. Lett.*, **121**, 287 (1985).
151. S. G. Boxer, D. J. Lockhart, and T. E. Middendorf, *Chem. Phys. Lett.*, **200**, 237 (1986).
152. J. M. Hayes and G. J. Small, *J. Chem. Phys.*, **90**, 4928 (1986).
153. H. van der Laan, T. Schmidt, and S. Völker, *J. Lumin.*, **48**, 199 (1991).
154. J. K. Gillie, B. L. Fearey, J. M. Hayes, G. J. Small, and J. H. Golbeck, *Chem. Phys. Lett.*, **134**, 316 (1987).
155. K. J. Vink, S. de Boer, J. J. Plijter, A. J. Hoff, and D. A. Wiersma, *Chem. Phys. Lett.*, **142**, 433 (1987).

156. J. K. Gillie, G. J. Small, and J. H. Golbeck, *J. Phys. Chem.*, **93**, 1620 (1989).
157. A. Kurita, Y. Kanematsu, and T. Kushida, *J. Lumin.*, **45**, 317 (1990).
158. G. Flöser and D. Haarer, *Chem. Phys. Lett.*, **147**, 290 (1988).
159. J. Friedrich, *Mol. Cryst. Liq. Cryst.*, **183**, 91 (1990).
160. R. Locher, A. Renn, and U. P. Wild, *Chem Phys. Lett.*, **138**, 405 (1987).
161. T. Basché, B. Sauter, and C. Bräuchle, *Ber. Bunsenges. Phys. Chem.*, **93**, 1055 (1989).
162. T. Basché and C. Bräuchle, *J. Mol. Struct.*, **218**, 387 (1990).
163. T. Basché and C. Bräuchle, *J. Chem. Phys.*, **91**, 1944 (1989).
164. K. B. Blodgett and I. Langmuir, *Phys. Rev.*, **51**, 964 (1937).
165. J. D. Swaten, *Thin Solid Films*, **160**, 197 (1988).
166. M. Orrit, J. Bernard, and D. Möbius, *Chem. Phys. Lett.*, **156**, 233 (1989).
167. J. Bernard, M. Orrit, R. I. Personov, and A. D. Samoilenko, *Chem. Phys. Lett.*, **164**, 377 (1989).
168. Y. V. Romanovskii, R. I. Personov, A. D. Samoilenko, K. Holliday, and U. P. Wild, *Chem. Phys. Lett.*, in press.
169. M. Orrit, J. Bernard, A. Mouhsen, H. Talon, D. Möbius, and R. I. Personov, *Chem. Phys. Lett.*, **179**, 232 (1991).
170. R. M. Macfarlane, *J. Lumin.*, **38**, 20 (1987).
171. W. E. Moerner, *Jpn. J. Appl. Phys.*, Part 1, **28-3**, 221 (1989).
172. R. J. Danby, K. Holliday, and N. B. Manson, *J. Lumin.*, **42**, 83 (1988).
173. R. M. Macfarlane and J. C. Vial, *Phys. Rev. B*, **34**, 1 (1986).
174. T. P. Carter, C. Bräuchle, V. Y. Lee, M. Manavi, and W. E. Moerner, *J. Phys. Chem.*, **91**, 3998 (1987).
175. W. P. Ambrose and W. E. Moerner, *Chem. Phys.*, **144**, 71 (1990).
176. W. E. Moerner, W. Lenth, and G. C. Bjorklund, in W. E. Moerner, Ed., *Persistent Spectral Hole-Burning: Science and Applications*, Springer, Berlin, 1988, Chap. 7.
177. T. W. Mossberg, *Opt. Lett.*, **7**, 77 (1982).
178. M. Mitsunaga and N. Uesugi, *Opt. Lett.*, **15**, 195 (1990).
179. A. Rebane and J. Feinberg, *Nature (London)*, **351**, 378 (1991).
180. A. Rebane and O. Ollikainen, *Opt. Commun.*, **83**, 246 (1991).
181. A. Szabo, U.S. Pat. 3,896,420 (1975).
182. G. Castro, D. Haarer, R. M. Macfarlane, and H. P. Trommsdorff, U.S. Pat. 4,101,976 (1978).
183. U. P. Wild, S. E. Bucher, and F. A. Burkhalter, *Appl. Opt.*, **24**, 1526 (1985).
184. U. Bogner, K. Beck, and M. Maier, *Appl. Phys. Lett.*, **46**, 534 (1985).
185. W. E. Moerner and M. D. Levenson, *J. Opt. Soc. Am. B*, **2**, 915 (1985).
186. W. E. Moerner, in *Polymers for Microelectronics: Science and Technology*, Kodansha, Tokyo, 1990.
187. S. Arnold, C. T. Liu, W. B. Whitten, and J. M. Ramsey, *Opt. Lett.*, **16**, 420 (1991).

188. P. W. Barber and R. K. Chang, Eds. *Optical Effects Associated with Small Particles*, Wiley, New York, 1986.
189. A. Renn, C. De Caro, and U. P. Wild, *Jpn. J. Appl. Phys., Part 1*, **28-3**, 257 (1990).
190. C. De Caro, A. Renn, and U. P. Wild, *Appl. Opt.*, **30**, 2890 (1991).
191. B. Kohler, S. Bernet, A. Renn, and U. P. Wild, to be published.
192. U. P. Wild, A. Renn, C. De Caro, and S. Bernet, *Appl. Opt.*, **29**, 4329 (1990).

Note added in proof: The reader's attention is brought to a special issue of the *Journal of the Optical Society of America B* [**9**, 711 (1992)] on "Persistent Spectral Hole Burning," and two technical digests on similar topics: *Persistent Spectral Hole-Burning: Science and Applications*, Optical Society of America, Washington, DC, 1991; and *Spectral Hole-Burning and Luminescence Line Narrowing*, Optical Society of America, Washington, DC, 1992.

CHAPTER
6

NEAR-INFRARED LUMINESCENCE SPECTROSCOPY

SHUZO AKIYAMA

School of Pharmaceutical Sciences
Nagasaki University
Nagasaki 852, Japan

6.1. Introduction
6.2. Semiconductor Laser Luminometry in the NIR Region
6.3. Ultramicro Flow Cell for NIR Semiconductor Laser Fluorometry
6.4. Pulsed NIR Semiconductor Laser Fluorometry
6.5. High-Performance Liquid Chromatography with NIR Semiconductor Laser Fluorometric Detection
    6.5.1. Detectors Based on NIR Semiconductor Laser Fluorometry
    6.5.2. Application of HPLC with NIR Semiconductor Laser Fluorometric Detection
6.6. NIR Semiconductor Laser Fluorometry for Enzymatic Assays
6.7. Dyes with Absorption in the NIR Region
References

6.1. INTRODUCTION

Near-infrared (NIR) luminescence (fluorescence) spectroscopy should be a quite useful new tool in numerous scientific areas, including physical and analytical chemistry, biology, and medicine. Nevertheless, NIR spectroscopy has been little used in the above fields. It is almost exclusively utilized by only a few research groups. The reason is that the expense of the laser as a strong light source and the complexity of its operation restrict practical applications. Laser luminometry has great potential for sensitive and selective determinations of ultratrace species concentrations (1).

Recently reported papers on NIR luminescence spectroscopy deal with inorganic compounds such as vanadium (V^{3+})-doped aluminum oxide,

---

Molecular Luminescence Spectroscopy, Part 3, Edited by Stephen G. Schulman. Chemical Analysis Series, Vol. 77.
ISBN 0-471-51580-9 © 1993 John Wiley & Sons, Inc.

V^{3+}-doped yttrium metaphosphate (2), V^{3+}-doped cesium sodium yttrium chloride bromides (3), and beryllium and carbon doped in silicon (4). Brittain has described materials suitable for IR-to-visible up-conversion in Part 2 of this monograph series (5).

Very few analytical applications of the NIR semiconductor laser have been reported. The first report was its use as a light source in photoacoustic spectrometry for the determination of phosphorus, based on the molybdenum blue method (6). Laser fluorometry for analysis of organic compounds has been demonstrated, using a continuous wave semiconductor laser, by Imasaka et al. (7).

This chapter will be devoted exclusively to semiconductor NIR luminometry for organic compounds, because the analytical advantages of the semiconductor laser are extremely valuable in biochemistry.

6.2. SEMICONDUCTOR LASER LUMINOMETRY IN THE NIR REGION

Laser luminometry is a highly sensitive method of molecular analysis for organic compounds in biochemistry. Semiconductor lasers are very small, relatively inexpensive, and powerful (> 1 W). For practical spectrophotometry such a laser is a suitable light source (6,8,9).

Unfortunately, the oscillating wavelengths of semiconductor lasers are restricted to the NIR (750–1500 nm), a serious disadvantage in luminometry. In order to overcome this disadvantage and to get high sensitivity, the development of fluorescent dyes in the NIR region is urgently required. For this region there are a large number of polymethine dyes (> 2000), some of which are used as laser materials and have large molar absorptivities ($\varepsilon \times 10^5$) and high fluorescent quantum yields. In the NIR region the blank fluorescence from a solvent can be largely reduced, which correlates with the increase of the detection limit in fluorometry. It is quite necessary that a biomedical assay, based on a fluorescence labeling technique, be carried out in the long-wavelength region to exclude blank fluorescence from the various fluorescent impurities in biosamples.

In 1984 Imasaka et al. reported a study that dealt with the inherent sensitivity of the method (10). They used a 5-mW AlGaAs laser of 786 nm and a photomultiplier with a GaAs photocathode and compared the sensitivity of detection using a lock-in amplifier of the analog signal and a photon counter, with the results obtained by a conventional spectrofluorometer. They demonstrated trace analysis of surfactants based on ion–pair extraction with a polymethine dye. The polymethine test dye was 3,3'-diethyl-2,2'-(4,5,4',5'-dibenzo)thiatricarbocyanine iodide (DDTC). As DDTC has a positive charge, it forms an ion pair with a large molecule with a negative charge and is

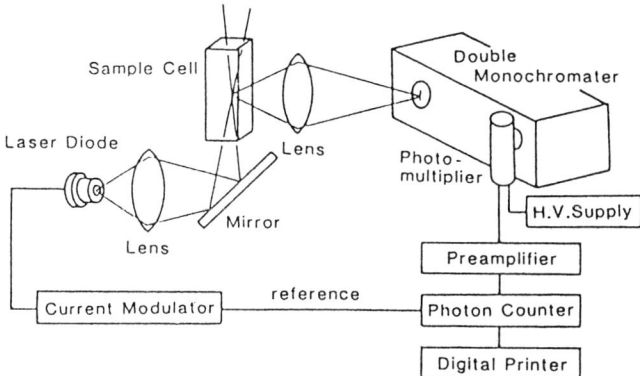

Fig. 6.1. Block diagram of semiconductor laser fluorometer. H.V. = high-voltage (power supply). [From Imasaka et al. (10), with permission of the American Chemical Society.]

extracted into organic solvents. Therefore, a sample with a negative charge can be sensitively detected by measuring fluorescence from the polymethine. The study is briefly introduced below. A block diagram of the experimental apparatus is shown in Fig. 6.1.

The light source is a double hetero structure AlGaAs laser [5 mW, 786 nm, fwhm (full width at half-maximum) = 1 nm]. A lens with short focal length (7 mm) is placed 8 mm from the semiconductor laser. Fluorescence is collected by a lens [f(focal length) = 20 mm] and imaged onto the entrance slit of a double monochromator. A photomultiplier with a GaAs photocathode (186–930 nm, 1400 V) is used. The output analog signal was measured by a lock-in amplifier. The data were recorded by a photon counter. A He–Ne laser (1 mW) was used for comparison with a semiconductor laser. A commercial fluorescence spectrometer was also used.

The experimental procedure is as follows: For the determination of the surfactants, the samples of sodium lauryl sulfate (or sodium dodecyl sulfate, SDS) and sodium dodecylbenzenesulfonate (SDBS) were prepared by adding the aqueous solution of the surfactant and the saturated solution of DDTC (absorbance = 0.8) to a test tube with a ground-glass stopper. After the samples were mixed, the solvent, benzene (7 mL) was added to the aqueous solution (13 mL). The mixture was shaken (5 min), and the ion pair of the surfactant and DDTC was extracted into the organic phase. The organic phase separated was taken into a quartz cell, and the fluorescence intensity was measured by a laser or conventional fluorometer.

DDTC, having a large conjugated structure, has a strong absorption band [$\varepsilon = 150000$ (MeOH)] in the NIR region. Figure 6.2 shows the excitation and emission spectra of DDTC in methanol. The detection limit of this dye in

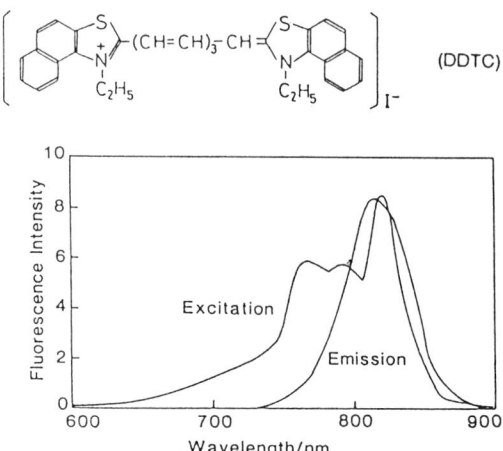

Fig. 6.2. Excitation and emission spectra of DDTC in methanol: excitation wavelength, 700 nm; emission wavelength, 860 nm [From Imasaka et al. (10), with permission of the American Chemical Society.]

MeOH was 3×10^{-10} M. Semiconductor laser fluorometry with the lock-in amplifier yielded enhanced sensitivity to 5×10^{-11} M, and by using photon counting the detection limit was lowered to 5×10^{-12} M. Imasaka et al. pointed out emphatically that blank fluorescence was almost negligible in the measurement and that minimum detectability was determined by the dark current noise of the photomultiplier. This result must arise from the fact that very few substances are fluorescent in the NIR region. This can result in an apparent analytical advantage of NIR fluorometry. For the trace analysis of surfactant, the ion pair of SDS–DDTC was extracted into benzene and the fluorescence intensity was measured. Benzene gave the largest signal intensity among various organic solvents. The calibration curve for SDS was a straight line from 0 to 7×10^{-7} M and observed blank fluorescence corresponded to 3×10^{-7} M. The blank seems to arise from impurity surfactant that remained on the surface of the glassware. The minimum detectability of SDS was 1×10^{-7} M. A linear analytical curve for SDBS was obtained from 0 to 6×10^{-7} M. The background signal corresponded to 2×10^{-7} M, and the detection limit of DBS was 1×10^{-7} M. The present detection limit is similar to or slightly better than that obtained using the conventional method and related methods based on spectrophotometry (11) using methylene blue as the chromophoric counter ion.

6.3. ULTRAMICRO FLOW CELL FOR NIR SEMICONDUCTOR LASER FLUOROMETRY

Lasers produce good beam coherence and a very high photon flux. Laser beams can be accurately positioned and focused, which makes them ideal radiation sources in combination with very-low-volume flow cells (12). Various types of flow cell have been designed to reject background emission from the cell windows (13–15). The use of an optical fiber for fluorescence collection was proposed by Sepaniak and Yeung (16) and Todoriki and Hirakawa (17).

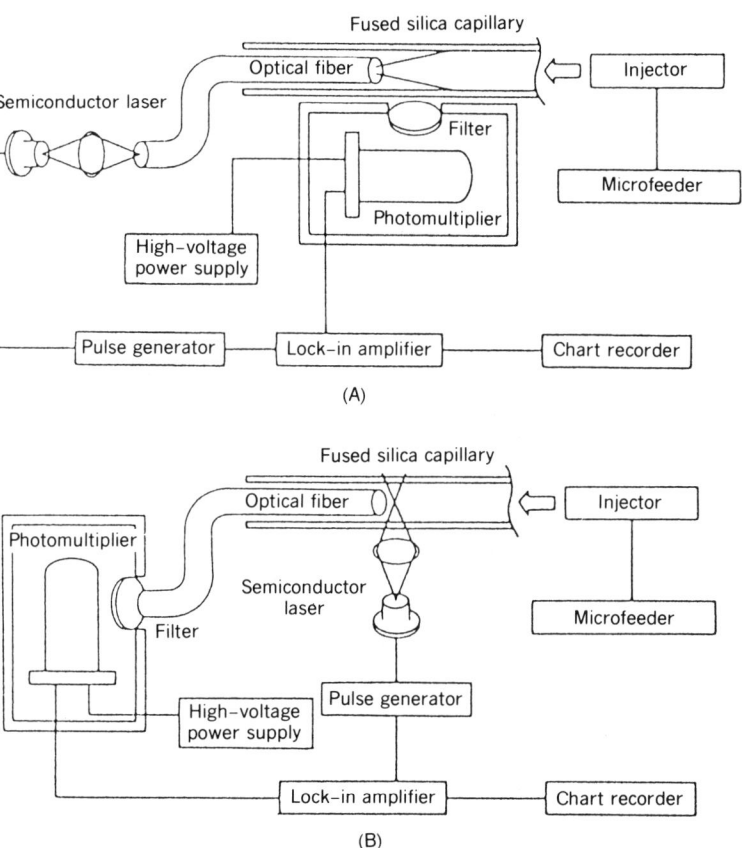

Fig. 6.3. Schematic diagram of experimental apparatus: (A) sample excited through the optical fiber; (B) fluorescence detected through the optical fiber. [From Kawabata et al. (18), with permission of Pergamon Press (Oxford).]

In high-performance liquid chromatography (HPLC), a small detector is essential for getting good separation resolution. Recently, Kawabata et al. (18) developed a nanoliter flow cell using a capillary combined with an optical fiber for laser fluorescence detection and constructed an ultramicro flow cell for fluorometry. They used a semiconductor laser as an excitation source and an optical fiber as a waveguide for light introduction or fluorescence collection. The apparatus is shown in Fig. 6.3. The solvent was pumped by a microfeeder at a flow rate of $2\,\mu\text{L/min}$. The sample was injected into the stream by a microloop injector, the capacity of which had been modified to 60 nL. The excitation source of the semiconductor laser ($\lambda = 780$ nm, 3 mW) was modulated to square waves at 100 Hz by a pulse generator. Fluorescence from the sample was detected by a photomultiplier. The output signal was fed through a lock-in amplifier to a recorder. A flow cell was made from a fused silica capillary (bore, 200 μm), into which a quartz optical fiber (core diameter 50 μm, cladding diameter 125 μm) was inserted. Kawabata and colleagues considered two optical arrangements: (1) the semiconductor laser was focused onto the side surface of the optical fiber by a ball lens (diameter, 3 mm) and the sample was irradiated from the end of the fiber, fluorescence being detected through the capillary wall (Fig. 6.3A); (2) the semiconductor laser was focused into the fused silica capillary perpendicularly and fluorescence was measured through the optical fiber (Fig. 6.3B). As an NIR fluorophore, 3,3'-diethyl-2,2'-(4,5,4',5'-dibenzo)thiatricarbocyanine iodide (Dye No. NK427; by Nippon Kanko-Shikiso Kenkyusho, Japan) was used in 2-butanol. The signal intensities and the detection limits for two optical arrangements are listed in Table 6.1.

Kawabata and co-workers suggest that the difference in sensitivity (a factor of 28) for the two optical arrangements may be atrributed to the difference in cell volume (a factor of 20). The detection limits for NK427 were 12 and 90 fg for methods 1 and 2, respectively. The absolute amount of the detection volume was 140–370 ag. The fluorometry developed is expected to be widely used in many practical applications using various fluorescers.

Table 6.1. Comparison of the Optical-Fiber Fluorescence Detectors

Item	Excitation Through the Optical Fiber (Method 1)	Detection Through the Optical Fiber (Method 2)
Cell volume, nL	60	3
Background signal, μV	8.5	Negligible
Noise level, μV	0.6	0.2
Detection limit for NK427m, fg	12	90

6.4. PULSED NIR SEMICONDUCTOR LASER FLUOROMETRY

A semiconductor laser can be operated in a pulsed mode (19), but it has not yet been used for the purpose of spectroscopy. The laser fluorometer in a pulsed mode was also developed by Imasaka et al. (20) to explore fluorescence lifetimes in the NIR.

They presented the fluorescence lifetime measurements of a polymethine dye in various solvents determined with the novel pulsed semiconductor in a time-correlated photon-counting system and reported a relationship between the fluorescence lifetime and solvent polarity. Moreover, the authors discussed its potentiality for use in evaluation of microenvironmental hydrophobicity. A block diagram of their laser fluorometer is shown in Fig. 6.4. The light source is a pulsed semiconductor laser that has a pulse width of 136 ps and a peak output power of 1.3 W. The laser was usually operated at a repetition rate of 10 kHz. Laser emission was directed to a conventional quartz sample cuvette with an optical fiber and a lens. The fluorescence emission was collected with a lens and then passed through an interference filter and a color filter. The fluorescence signal was finally focused by a lens onto the photocathode of a red-sensitive photomultiplier, which was typically operated at 1500 V. The outlet signal was fed sequentially to a constant fraction discriminator, a time-to-amplitude converter, and a multichannel analyzer. The synchronous

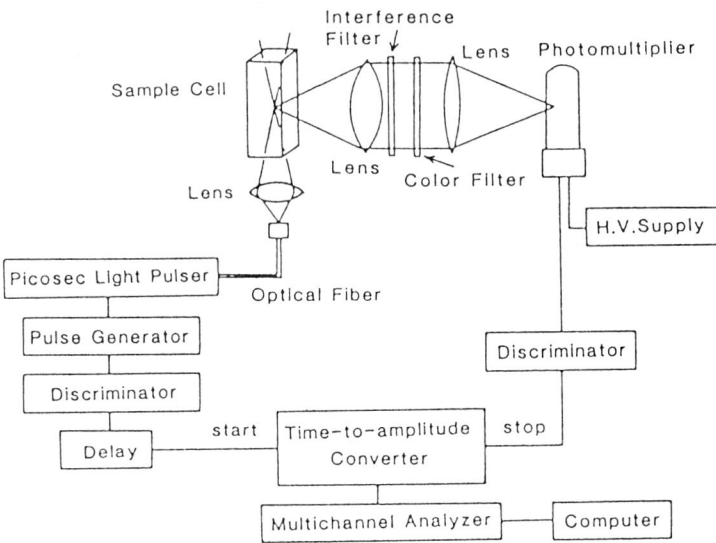

Fig. 6.4. Block diagram of experimental apparatus for measurement of fluorescence lifetime. [From Imasaka et al. (20), with permission of the American Chemical Society.]

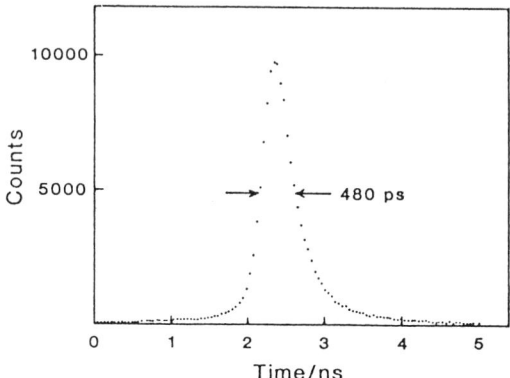

Fig. 6.5. Excitation time profile of laser pulse recorded by NIR semiconductor laser fluorometric system. [From Imasaka et al. (20), with permission of the American Chemical Society.]

output signal from the semiconductor laser was delayed and inverted by using a pulse generator and fed to a constant fraction discriminator.

Figure 6.5 shows the time profile of the laser pulse as measured by the present laser fluorometer. The fwhm is 480 ps. Imasaka et al. believe that, for an improvement of the time resolution, a laser fluorometric system consisting of a mode-locked semiconductor laser and avalanche photodiode incorporated into a time-correlation photon-counting system would be quite attractive for the measurement of picosecond fluorescence decay. They measured the fluorescence decay curve for NK427 and reported that the fluorescence lifetimes varied with the dielectric constant of the solvent in which the dye was dissolved, such that a straight-line relationship was found, with a correlation coefficient of 0.987.

6.5. HIGH-PERFORMANCE LIQUID CHROMATOGRAPHY WITH NIR SEMICONDUCTOR LASER FLUOROMETRIC DETECTION

6.5.1. Detection Based on NIR Semiconductor Laser Fluorometry

Laser fluorometry has recently been used for sensitive detection in HPLC (12,13,21–23).

In 1986, an HPLC system with a detector based on NIR semiconductor laser fluorometry was reported and its use in ultratrace determination of polymethine dyes was demonstrated (24). A block diagram of the typical apparatus is shown in Fig. 6.6. The light source of the semiconductor laser is modulated to square waves at 100 Hz by a pulse generator. Fluorescence

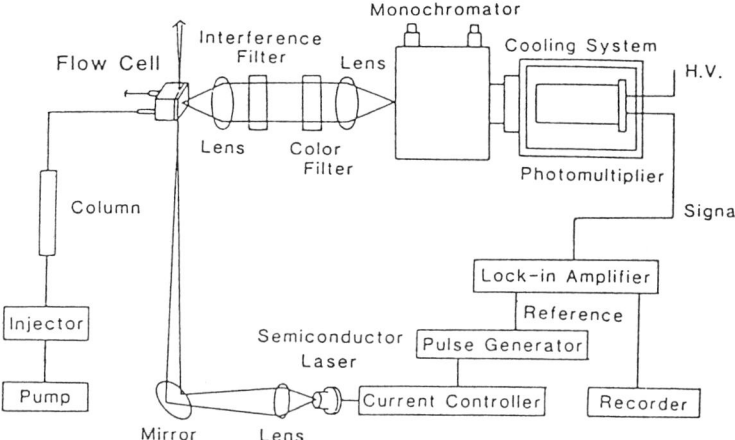

Fig. 6.6. Block diagram of the apparatus. [From Sauda et al. (24), with permission of Elsevier Publ. B.V. (Amsterdam).]

from the sample is collected by a lens and passed through an interference filter and a color filter. Fluorescence is focused by a lens to a monochromator equipped with a photomultiplier with a red-sensitive photocathode. A photomultiplier cooling system is used to reduce the temperature to $-20\,°C$. The output signal is fed to a lock-in amplifier combined with a chart recorder. The sample (110 µL) is injected into the reversed-phase HPLC column [octadecylsilylsilica gel (ODS)] from a loop injector, the flow rate is 1 mL/min. The specially designed flow cell is made entirely of quartz (path length, 1 cm; total volume, 18 µL). The polymethine dyes used [anhydro-3,3,3'3'-tetramethyl-1,1'-bis(4-sulfomethyl)-4,5,4',5'-dibenzoindotricarbocyanine hydroxide sodium salt (NK2611); 3,3,3',3'-tetramethyl-1,1'-dimethyl-4,5,4',5'-dibenzoindotricarbocyanine perchlorate (NK2014); and NK427] fluoresce and have absorption maxima at around 780 nm, which agrees with the emission wavelength of the semiconductor laser. The emission maxima of these dyes in MeOH are located at around 810 nm. Figure 6.7 shows a chromatogram obtained for the mixture of the dyes used with the semiconductor laser fluorometric detector.

The detection limits for NK2611 in MeOH obtained with various spectroscopy detectors are compared in Table 6.2, from which the value of 0.3 pg obtained using the semiconductor laser and cooled photomultiplier is almost 3 orders of magnitude better than obtained by using conventional methods. The use of such a laser is very attractive for ultratrace detection of suitable analytes.

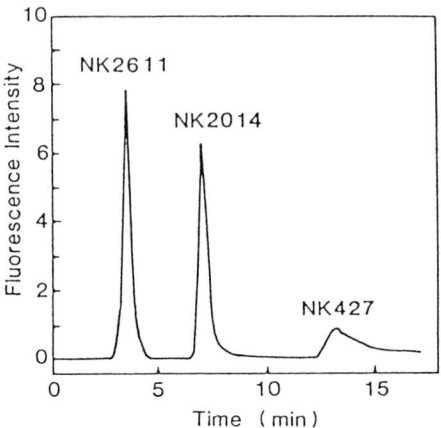

Fig. 6.7. Chromatogram for an equimolar mixture of the three polymethine dyes with semiconductor laser fluorometry for detection. [From Sauda et al. (24), with permission of Elsevier Publ. B.V. (Amsterdam).]

Table 6.2. Detection Limits for NK2611

Method	Detection Limit (pg)
Semiconductor laser fluorometry[a]	0.3
Semiconductor laser fluorometry[b]	1.9
Conventional fluorometry	190
Conventional spectrophotometry[c]	810

[a] Photomultiplier cooled to $-20°C$.
[b] Photomultiplier operated at room temperature.
[c] At 740 nm, the longest wavelength available with respect to spectrometer used.

6.5.2. Application of HPLC with NIR Semiconductor Laser Fluorometric Detection

As previously stated, among the various spectroscopy methods, semiconductor laser fluorometry is most sensitive and allows sample detection at 10^{-12} M levels when a 5-mW laser is used (10). Routine use of the laser fluorometric HPLC detector is expected in commercial instruments in the near future.

It is generally agreed that one of the most important analytical tasks is the trace analysis of proteins. The efforts in this area (25) have regrettably been unsuccessful by the ion pair extraction technique (10). The reason is probably that the positively charged dye will not combine with positively charged proteins. However, fortunately Imasaka et al. found that indocyanine green (ICG) is exceptional in that it carries a negative charge and forms a complex with proteins (25). It is desirable to use a red fluorescence tag for labeling protein since the fluorescence intensities of native proteins are weak in this wavelength region and no impurity fluorescence can be observed in NIR fluorometry. Figure 6.8 shows the excitation and emission spectra for free ICG in aqueous solution as well as for ICG combined with bovine serum albumin (BSA). The ICG molecule was found to have a large molar absorptivity (180,000 at 780 nm), comparable to that of porphin. As a result of examining the fluorescence behavior of the IGA–BSA complex, Imasaka et al. (25) showed that ICG is useful as an labeling reagent and can be utilized in the chromatographic determination of protein.

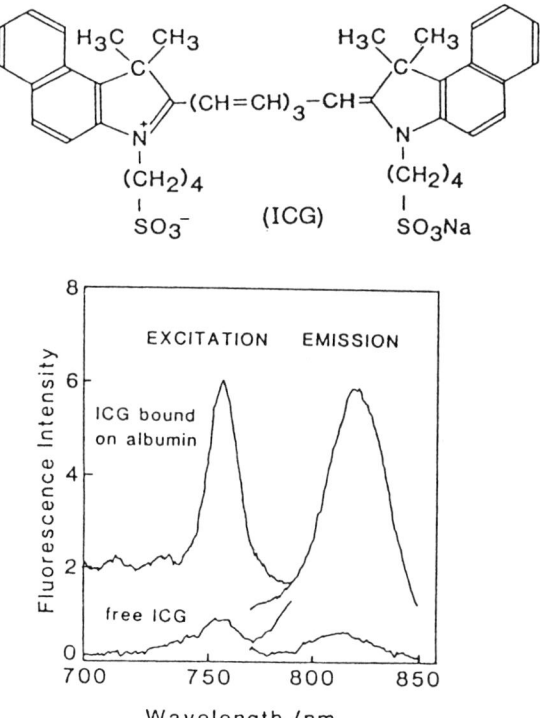

Fig. 6.8. Excitation and emission spectra for ICG: excitation wavelength, 765 nm; emission wavelength, 820 nm. The concentrations of ICG and BSA are adjusted to 3.2×10^{-5} M. [From Sauda et al. (25), with permission of the American Chemical Society.]

These investigators first constructed an HPLC system with a semiconductor laser fluorometric detector and used the polymethine dye ICG as a labeling reagent for protein. Thereafter, they carried out the trace analysis of protein in human serum after sample separation by HPLC and compared the performance of the semiconductor laser detector with that of conventional spectrophotometric detectors. The experimental apparatus with an HPLC column shown in Fig. 6.6 can be used here. ICG (25 mg in a bottle) was dissolved in distilled water (10 mL) and diluted stepwise to the specific concentration. Freeze-dried human serum was dissolved in 5 mL of water and treated by ultrasonic agitation for 3 min. The pH of the eluant containing 0.1 M sodium sulfate was adjusted to 6.8 with a buffer solution of 0.2 M sodium dihydrogen phosphate and 0.2 M potassium dihydrogen phosphate. Protein in human serum (concentration for albumin, 1.2×10^{-2} M) was labeled with 2×10^{-5} M aqueous ICG and then injected into the HPLC system. The chromatogram obtained, which illustrates sharp and intense signal peaks originating from ICG-labeled protein, is shown in Fig. 6.9. From the retention times, the three peaks are concluded to be α_1-lipoprotein, a group of γ-globulin, and albumin, respectively. As summarized in Table 6.3, the detection limits for BSA labeled with ICG were investigated by the constructed HPLC system with various

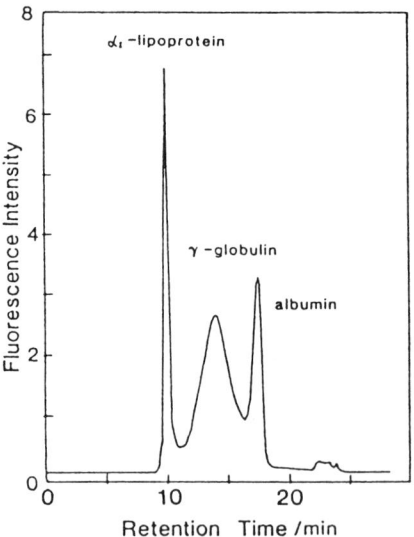

Fig. 6.9. Chromatogram for human serum labeled with ICG. Semiconductor laser fluorometry is used for sample detection. The wavelength of the monochromator was adjusted to 840 nm. [From Sauda et al. (25), with permission of the American Chemical Society.]

Table 6.3. Detection Limit for BSA Labeled with ICG

Method	Detection Limit (pmol)
Semiconductor laser fluorometer[a]	1.3
Conventional fluorometer	150
Conventional spectrometer	230[b]
	270[c]

[a] The photomultiplier is cooled down to $-20\,°C$.
[b] Absorption of ICG bound to protein is measured at 740 nm.
[c] Absorption of native protein is measured at 280 nm.

spectrometric detectors. These results suggest that the semiconductor laser fluorometric detector should be useful for ultratrace analysis of protein.

The present system can certainly detect a very small amount of protein in a pure aqueous solution, but it may be difficult to detect specified trace protein in biological fluid, since ICG is nonspecifically bound to protein. Therefore, it is necessary in the next step that greater selectivity for protein be added to the reagent.

6.6. NIR SEMICONDUCTOR LASER FLUOROMETRY FOR ENZYMATIC ASSAYS

In clinical chemistry, many biological samples are analyzed by enzymatic methods owing to their great selectivity. If an enzyme reaction could be monitored by semiconductor laser fluorometry, it would be a powerful weapon for trace analysis of bilogical fluids. However, no dye fluorescing in the NIR and usable as a substrate for enzymatic assays has yet been reported.

Imasaka et al. (26) have recently reported an enzymatic assay based on NIR semiconductor laser fluorometry, in which the fluorescence of ICG is quenched by hydrogen peroxide, and the enzymatic reaction, producing hydrogen peroxide, is monitored by measuring this quenching. This method was applied to the assays of xanthine and xanthine oxidase.

The procedure is as follows: For the study of fluorescence quenching, the concentration of hydrogen peroxide was 2×10^{-2} to 2×10^{-7} M, and that of ICG was 1.1×10^{-7} M. For the calibration graph for xanthine, 0.8 mL of 1.1×10^{-5} M ICG, 0.2 mL of 2×10^{-3} M sodium molybdate, 0.2 mL of 2×10^{-3} M iron(II) sulfate, 0.05–5 mL of 10^{-3} M sample, and 0.45 unit of xanthine oxidase were added, in that order, to a 100-mL volumetric flask. The fluorescence intensity was measured 1–5 min after the reagents were mixed,

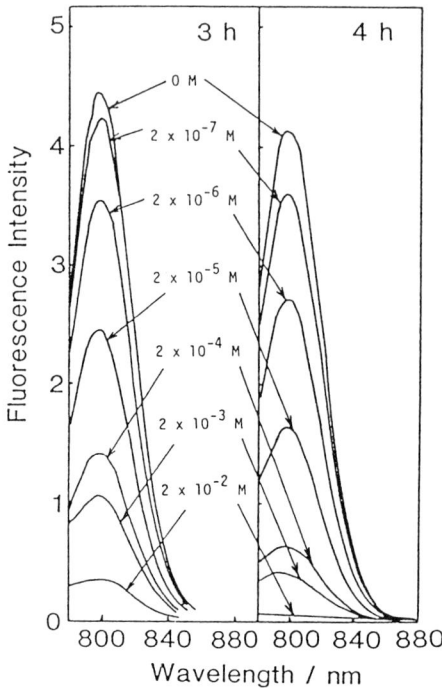

Fig. 6.10. Fluorescence spectra for ICG measured after reaction with hydrogen peroxide at the concentrations indicated on the curves. Spectra were measured after 3 h (left) and 4 h (right). [From Imasaka et al. (26), with permission of Elsevier Publ. B.V. (Amsterdam).]

during which time the fluorescence intensity was almost constant. The apparatus described in Fig. 6.1 was used for the experiment. As shown in Fig. 6.10, the fluorescence intensity decreased with increasing concentrations of hydrogen peroxide to be determined.

Xanthine is well known to be converted to uric acid by xanthine oxidase:

$$\text{xanthine} + O_2 + H_2O \rightarrow \text{uric acid} + H_2O_2$$

Consequently, the xanthine can be determined by measuring the decrease in the fluorescence intensity of ICG. Imasaka et al. (26) considered that ICG forms a nonfluorescent product, P, with hydrogen peroxide. For the reaction

$$H_2O_2 + ICG \rightarrow P$$

the stability constant (K) can be written as

$$K = [P]/[H_2O_2][ICG] = Y/(S_0 - Y)(Y_0 - Y)$$

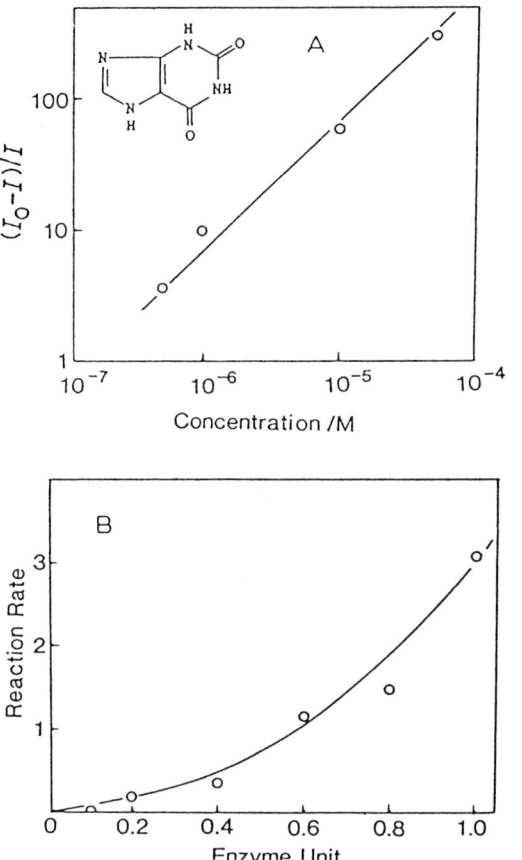

Fig. 6.11. Calibration graphs: (A) for xanthine; (B) for xanthine oxidase, with 10^{-7} M ICG and 10^{-4} M xanthine. [From Imasaka et al. (26), with permission of Elsevier Publ. B.V. (Amsterdam).]

where S_0 and Y_0 are the initial concentrations of hydrogen peroxide and the dye, respectively; Y represents the concentration of the product after the reaction. If $S_0 \gg Y$, then $Y/(Y_0 - Y) = KS_0$, and because S_0 is proportional to the initial concentration of xanthene, X_0, this equation can be rewritten as $(I_0 - I)/I = KX_0$. The left side is plotted against the concentration of xanthine (X_0), as shown in Fig. 6.11A. The calibration graph is straight from 5×10^{-5} to 5×10^{-7} M. Their proposed method is the first demonstration of an enzymatic assay in the NIR and should be useful not only for the determination of xanthine but for monitoring of many enzymatic reactions producing hydrogen peroxide. The present approach is also useful for determination

6.7. DYES WITH ABSORPTION IN THE NIR REGION

New compounds absorbing light in the NIR region, i.e., NIR dyes, have recently been described as laser dyes or as materials for storing information with the help of diode lasers. The application of organic NIR dyes to analytical chemistry is at present very limited. It was noted earlier that Imasaka et al. have skillfully used NIR dyes as fluorescence dyes in HPLC laser fluorometry and in laser fluorometry for enzyme studies and enzymatic assays.

The development of compounds absorbing at extremely long wavelengths can be carried out on the basis of either empirical or theoretical knowledge (27). Fabian and Zahradnik (28) have quite recently proposed that it is also possible to begin from a completely different conceptual starting point: in this approach they consider diradicaloid compounds are vary their molecular structure in order to obtain compounds that absorb at long wavelengths. Consequently, they declared that one should commence by noting that the change in the reference point is intended less as a new type of classification of the deeply colored compounds than as a method for simulating chemical intuition in the search for deeply colored systems.

It has been observed that the absorption wavelength cannot be increased beyond a certain limit simply by increasing the size of the π-electron system (28). A film of pure *trans*-poly(acetylene) (**1**) absorbs at $\lambda \simeq 650$ nm (29). In annulenes [**2** (30) and **3**] and annulenoannulenes (**4**), the long-wavelength absorption band reaches into the NIR region as shown in Table 6.4. (31–44).

Cyclic hydrocarbons with a fixed geometry, which makes for extensive electron delocalization, may exhibit absorption bands in the NIR region. As examples, **5–7** (45–47) are shown.

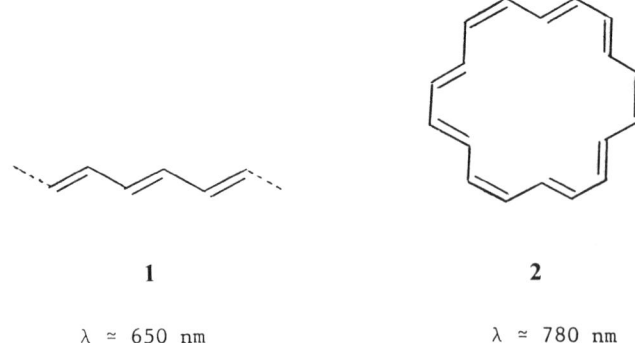

1

$\lambda \simeq 650$ nm

2

$\lambda \simeq 780$ nm

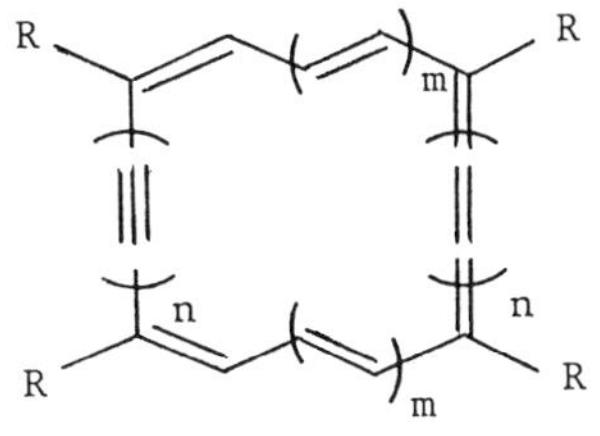

3

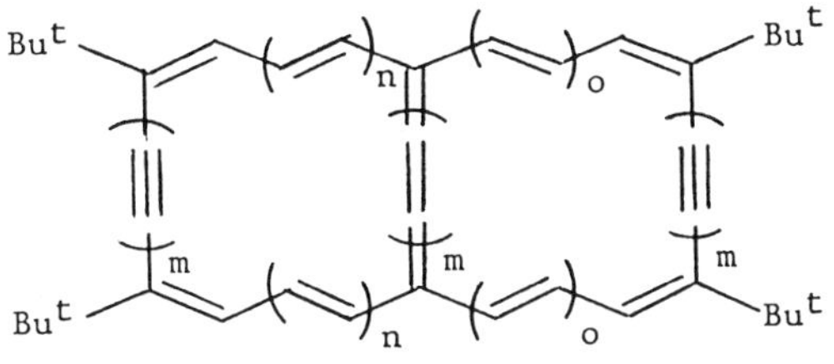

4

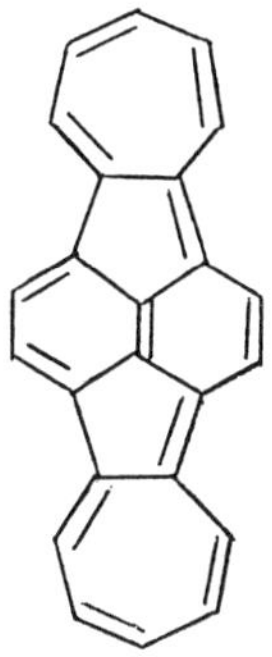

5

$\lambda \simeq 900 - 1500$ nm

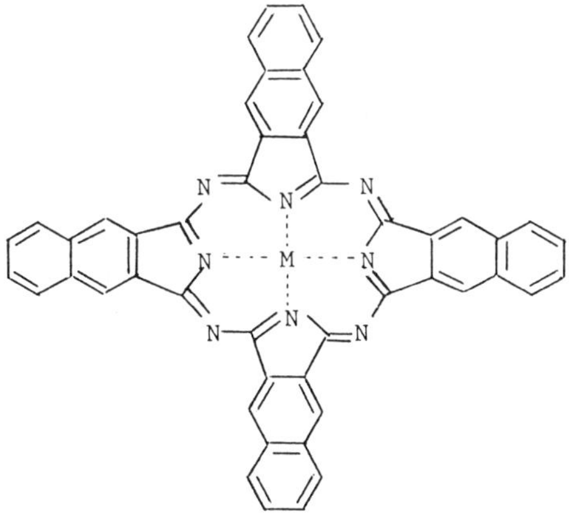

6 (M = AlOH)

$\lambda = 1204$ nm
in $CH_3COCl/AlBr_3$

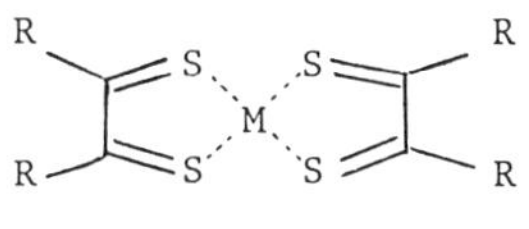

7

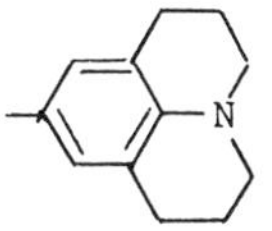

M = Ni

$\lambda = 1298$ nm
in dichloromethane

Table 6.4. The Absorption Maxima of Dehydroannulene (3) and Dehydroannulenoannulens (4) in the NIR Region in THF

	m	n	o	R	λ	Ref.
3	1	1	—	t-Bu	590	31
	1	1	—	Ph	658	31
	1	2	—	Me	720	32
	1	2	—	t-Bu	707	33
	1	2	—	Ph	772	34
	2	1	—	t-Bu	751	35
	2	1	—	Ph	785	35
	2	2	—	t-Bu	859	36
	2	2	—	Ph	942	37
	3	1	—	t-Bu	895	38
	4	1	—	t-Bu	975	39
4	1	1	1	—	742	40
	1	1	2	—	845	41
	1	2	2	—	932	42
	1	1	3	—	945	43
	2	1	1	—	855	44

The polymethines **8** have extremely deeply colored representatives such as merocyanines **8a**, cyanines **8b**, or oxonol dyes **8c** (48,49). Squarylium dyes, e.g., **9** (hydroxysquarylium) (50) are of interest with **7**. They all absorb at the long wavelengths up to 1200 nm.

NIR dyes of other types are the xanthenylium, thioxanthenylium, selenoxanthenylium (e.g., **10a**) (51), and fluorenylium dye ethynologues (**10b**) (52), azo dyes (e.g., **11**) (53), naphtho- and anthraquinones (e.g., **12**) (54–58), indophenols and analogues (e.g., **13**) (59), and indophenol–metal complexes (e.g., **14**) (60–62).

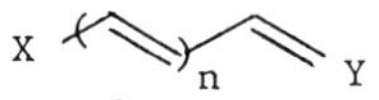

8

8a: X = NR_2, Y = O
8b: X = Y = $NR_2^{1/2+}$
8c: X = Y = $O^{1/2-}$

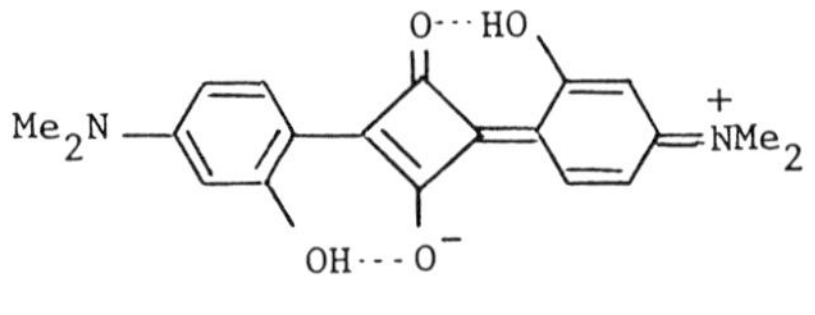

9

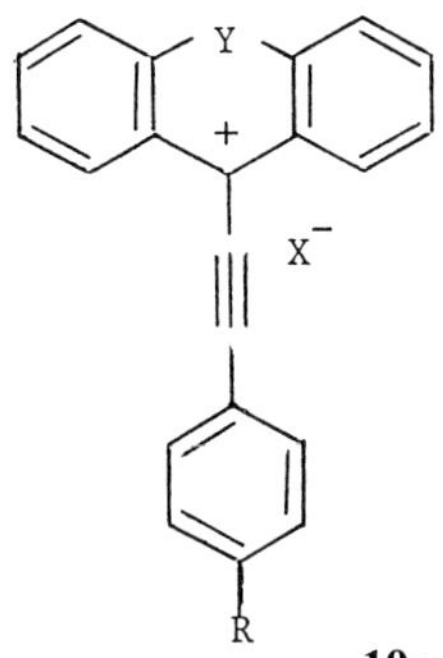

DYES WITH ABSORPTION IN THE NIR REGION

10a

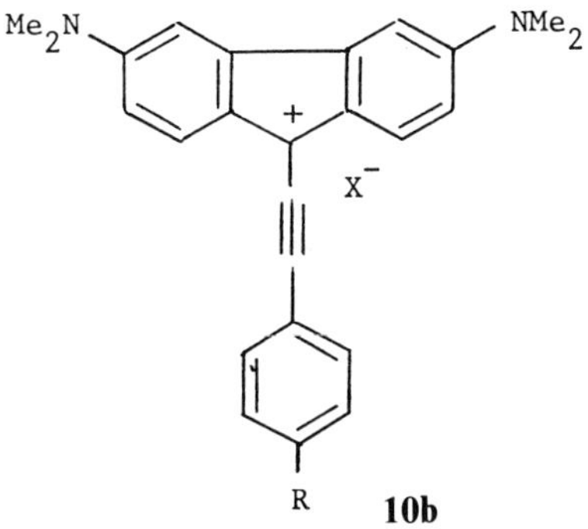

Y = O, S, Se

X = ClO_4, BF_4

R = NMe_2, OMe, H, NO_2

10b

X = ClO_4, BF_4

R = NMe_2, OMe, Me, H, Br, NO_2

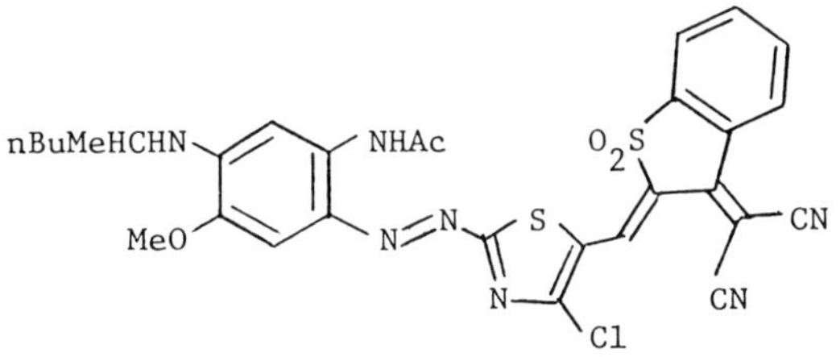

11

778 nm in dichloromethane

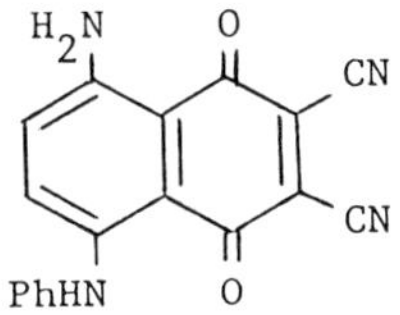

12

760 nm in cyclohexane

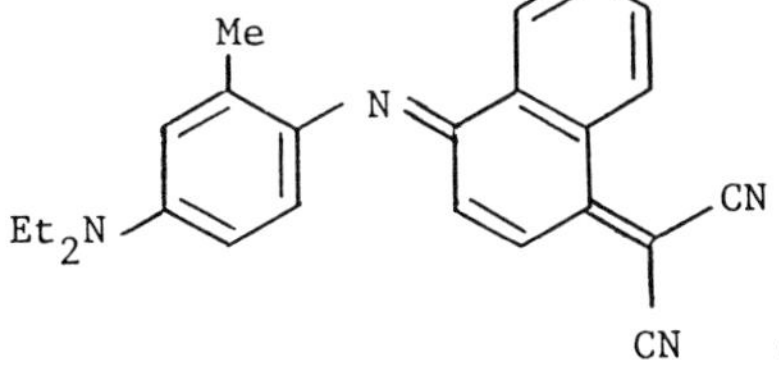

13

761 nm in chloroform

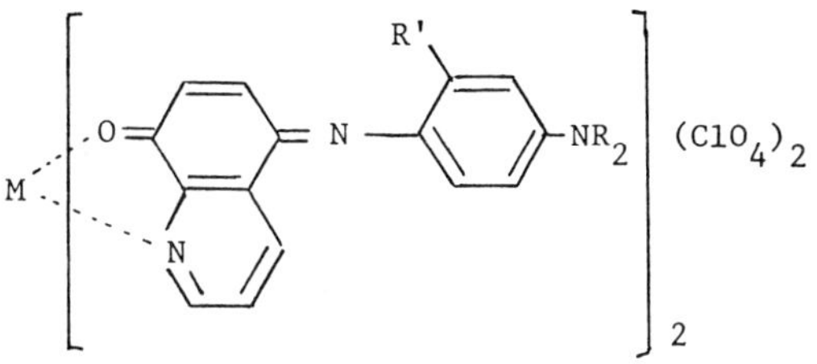

14

a: M = Cu, R' = Me, R = Et; 776 nm
b: M = Ni, R' = Me, R = Et; 778 nm
c: M = Cu, R' = H, R = Me; 772 nm
d: M = Ni, R' = H, R = Me; 745 nm in EtOH

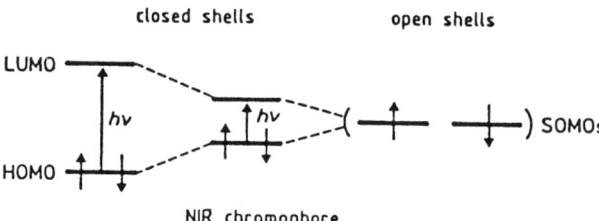

Fig. 6.12. Highest occupied molecular orbital (HOMO) energy separation and spectral excitation energy of singlet configurations; SOMO = singly occupied molecular orbital. [From Fabian and Zahradrik (28), with permission of VCH Verlagsgesellschaft mbH (Weinheim, Germany).]

For the design of such new dyes, it should always be considered what options are chemically available to decrease the energy difference between the ground state and the first excited state. Otherwise, from a more conventional starting point, it will be concluded that the synthesis is required of extensive π-electron-containing molecular structures having only a very small energy difference between the lowest energy electronic states. Fabian and Zahradnik (28) suggested that such diradicaloid molecules occupy a special position among the various types of organic compounds and that it is possible by means of suitable structural modification to stabilize such molecules in a singlet form that absorbs light at very long wavelengths. This can be illustrated by the color-determining electron transition in polymethines shown in Fig. 6.12.

Many kinds of laser and NIR dyes are now commercially available (63). The development of NIR dyes bearing functional group(s) for derivatization of a target compound is being actively pursued for the application of diode laser luminescence and absorption spectroscopy.

Of course, the term "NIR dye" is really a subclass of the term "functional dye," first used in 1981 in Japan and recently a subject of renewed interest and activity in the field of dye chemistry. The first international symposium on the chemistry of functional dyes was held in Osaka, Japan, in 1989, and the subsequent book, *Chemistry of Functional Dyes*, presented an overview of the latest developments in the field (64). Quite recently, more than 190 charts of absorption spectra of dyes for diode lasers have been published in Japan (65). Further work on the discovery, characterization, and application of new functional dyes will surely continue unabated in the future.

REFERENCES

1. D. S. Kliger, *Ultrasensitive Laser Spectroscopy*, Academic Press, New York, 1983.
2. C. Reber and H. U. Guedel, *Chem. Phys. Lett.*, **154**, 425 (1989).

3. C. Reber and H. U. Guedel, *J. Lumin.*, **42**, 1 (1988).
4. D. Labrie, I. J. Booth, S. P. Watkins, and M. L. W. Thewalt, *Solid State Commun.*, **63**, 115 (1988).
5. H. G. Brittain, in S. G. Schulman, Ed., *Molecular Luminescence Spectroscopy: Methods and Applications*, Part 2, Wiley, New York, 1988, pp. 440–459.
6. Y. Kawabata, T. Kamikubo, T. Imasaka, and N. Ishibashi, *Anal. Chem.*, **55**, 1419 (1983).
7. T. Imasaka, A. Yoshitake, and N. Ishibashi, *Anal. Chem.*, **56**, 1077 (1984).
8. T. Imasaka, T. Kamikubo, Y. Kawabata, and N. Ishibashi, *Anal. Chim. Acta*, **153**, 153 (1983).
9. K. Nakanishi, T. Imasaka, and N. Ishibashi, *Anal. Chem.*, **57**, 1219 (1985).
10. T. Imasaka, A. Yoshitake, and N. Ishibashi, *Anal. Chem.*, **56**, 1077 (1984).
11. K. Kimura, Ed., *Handbook of Analytical Chemistry*, Maruzen, Tokyo, 1971.
12. A. Hulshoff and H. Lingeman, in S. G. Schulman, Ed., *Molecular Luminescence Spectroscopy: Methods and Applications*, Part 1, Wiley, New York, 1985, p. 660.
13. G. J. Diebold and R. N. Zare, *Science*, **196**, 1439 (1977)
14. L. W. Hershberger, J. B. Callis, and G. D. Christian, *Anal. Chem.*, **51**, 1444 (1979).
15. N. J. Dovici, J. C. Martin, J. H. Jett, M. Trkula, and R. A. Keller, *Anal. Chem.*, **56**, 348 (1984).
16. M. J. Sepaniak and E. S. Yeung, *J. Chromatogr.*, **190**, 377 (1980).
17. H. Todoriki and A. Y. Hirakawa, *Chem. Pharm. Bull.*, **32**, 193 (1984).
18. Y. Kawabata, T. Imasaka, and N. Ishibashi, *Talanta*, **33**, 281 (1986).
19. Y. Tsuchiya, A. Takeshita, and M. Hosoda, *Rev. Sci. Instrum.*, 52, 579 (1981).
20. T. Imasaka, A. Yoshitake, K. Hirata, Y. Kawabata, and N. Ishibashi, *Anal. Chem.*, **57**, 947 (1985).
21. R. N. Zare, *Science*, **226**, 298 (1984).
22. E. S. Yeung and M. J. Sepaniak, *Anal. Chem.*, **52**, 1465A (1980).
23. R. B. Green, *Anal. Chem.*, **55**, 20A (1983).
24. K. Sauda, T. Imasaka, and N. Ishibashi, *Anal. Chim. Acta*, **187**, 353 (1986).
25. K. Sauda, T. Imasaka, and N. Ishibashi, *Anal. Chem.*, **58**, 2649 (1986).
26. T. Imasaka, T. Okazaki, and N. Ishibashi, *Anal. Chim. Acta*, **208**, 325 (1988).
27. G. Griffiths, *Chem. Ber.*, **22**, 997 (1986).
28. J. Fabian and R. Zahradnik, *Angew. Chem., Int. Ed. Engl.*, **28**, 677 (1989).
29. C. R. Fincher, Jr., M. Ozaki, M. Tanaka, D. Peebles, L. Lauchlan, A. J. Heeger, and A. G. MacDiarmid, *Phys. Rev. B*, **20**, 1589 (1979).
30. H.-R. Blattmann, E. Heilbronner, and G. Wagniere, *J. Am. Chem. Soc.*, **90**, 4786 (1968).
31. K. Fukui, T. Nomoto, S. Nakatsuji, S. Akiyama, and M. Nakagawa, *Bull. Chem. Soc. Jpn.*, **50**, 2758 (1977).
32. J. Ojima, K. Katakami, G. Nakaminami, and M. Nakagawa, *Tetrahedron Lett.*, 1115 (1971).

33. T. Katakami, S. Tomita, and M. Nakagawa, *Chem. Lett.*, 225 (1972).
34. K. Fuki, T. Okamoto, and M. Nakagawa, *Tetrahedron Lett.*, 3121 (1971).
35. M. Iyoda, S. Akiyama, and M. Nakagawa, *Bull. Chem. Soc. Jpn.*, **51**, 3559 (1978).
36. M. Iyoda, H. Miyazaki, and M. Nakagawa, *Bull. Chem. Soc Jpn.*, **46**, 2565 (1973).
37. S. Akiyama, T. Nomoto, M. Iyoda, and M. Nakagawa, *Bull. Chem. Soc. Jpn.*, **49**, 2579 (1976).
38. M. Iyoda and M. Nakagawa, *J. Chem. Soc., Chem. Commun.*, 1003 (1972).
39. M. Iyoda and M. Nakagawa, *Tetrahedron Lett.*, 4253 (1972).
40. S. Akiyama, M. Iyoda, and M. Nakagawa, *J. Am. Chem. Soc.*, **98**, 6410 (1976).
41. S. Nakatsuji, S. Akiyama, and M. Nakagawa, *Tetrahedron Lett.*, 3723 (1977).
42. M. Osuka, Y. Yoshikawa, S. Akiyama, and M. Nakagawa, *Tetrahedron Lett.*, 3719 (1978).
43. S. Nakatsuji, S. Akiyama, and M. Nakagawa, *Tetrahedron Lett.*, 1486 (1976).
44. T. Kashitani, S. Akiyama, M. Iyoda, and M. Nakagawa, *J. Am. Chem. Soc.*, **97**, 4424 (1975).
45. S. Itoh, *Pure Appl. Chem.*, **54**, 957 (1982).
46. W. Freyer, *Z. Chem.*, **26**, 263 (1986).
47. W. Freyer, *J. Prakt. Chem.*, **328**, 253 (1986).
48. Cf. H. Zollinger, *Color Chemistry*, VCH Verlagsgesellschaft, Weinheim, Germany, 1987, pp. 45–58.
49. Cf. M. Okawara, T. Kitao, T. Hirashima, and M. Matsuoka, *Handbook of Dyestuffs*, Kodansha Scientific, Tokyo, 1986.
50. K. Y. Law, *Can. J. Chem.*, **64**, 1607, 2267 (1986).
51. S. Akiyama, S. Nakatsuji, K. Nakashima, and M. Watanabe, *J. Chem. Soc., Chem. Commun.*, 710, 1420 (1987); S. Akiyama, S. Nakatsuji, K. Nakashima, M. Watanabe, and H. Nakazumi, *J. Chem. Soc., Perkin Trans.* **1**, 3155 (1988).
52. S. Nakatsuji, H. Nakazumi, H. Fukuma, T. Yahiro, K. Nakashima, M. Iyoda, and S. Akiyama, *J. Chem. Soc., Chem. Commun.*, 489 (1990); *J. Chem. Soc., Perkin Trans.* **1**, 1881 (1991).
53. K. A. Bello and J. Griffiths, *J. Chem. Soc., Chem. Commun.*, 1639 (1986).
54. N. Ito and H. Aiga, *Jpn. Kokai Tokkyo Koho*, **JP 62132963** (1987); *Chem. Abstr.*, **107**, 238195v (1987).
55. K.-Y. Chu and J. Griffiths, *J. Chem. Res., Synop.*, 180 (1978); *J. Chem. Rev., Miniprint*, 2319 (1978).
56. K. Takagi, M. Kawabe, M. Matsuoka, Y. Kubo, and T. Kitao, *Dyes Pigm.*, **6**, 117 (1985).
57. S. H. Kim, M. Matsuoka, Y. Kubo, T. Yoshida, and T. Kitao, *Dyes Pigm.*, **7**, 93 (1986).
58. S. H. Kim, S. Minami, M. Matsuoka, and T. Kitao, *Chem. Express.*, **2**, 73 (1987); *Chem. Abstr.*, **106**, 86211m (1987).
59. Y. Kubo, F. Mori, and K. Yoshida, *Chem. Lett.*, 1761 (1987); Y. Kubo, F. Mori, K.

Komatsu, and K. Yoshida, *J. Chem. Soc. Perkin Trans. 1*, 2439 (1988); Y. Kubo, M. Kuwana, K. Okamoto, and K. Yoshida, *J. Chem. Soc., Chem. Commun.*, 855 (1989).
60. K. A. Bello, L. Cheng, and J. Griffiths, *J. Chem. Soc., Perkin Trans. 2*, 815 (1987).
61. H. Junek, G. Uray, and G. Zusching, *Dyes Pigm.*, **9**, 137 (1988).
62. Y. Kubo, K. Sasaki, and K. Yoshida, *Chem. Lett.*, 1563 (1987).
63. Z. Yoshida and T. Kitao, Eds., *Chemistry of Functional Dyes*, Mita, Tokyo, 1989.
64. U. Brackmann, Ed., *Lambdachrome Laser Dyes*, Lambda Physik, Göttingen, 1986, in which many absorption spectral curves are conveniently inserted.
65. M. Matsuoka, Ed., *Absorption Spectra of Dyes for Diode Lasers*, Bunshin, Tokyo, 1990.

CHAPTER

7

MICROSPECTROFLUOROMETRY ON SUPPORTED PLANAR MEMBRANES

LUKAS K. TAMM

Department of Physiology
University of Virginia, School of Medicine
Charlottesville, Virginia 22908

EDWIN KALB

Department of Biophysical Chemistry
Biocenter, University of Basel
CH-4056 Basel, Switzerland

7.1. Introduction
7.2. Techniques of Microspectrofluorometry
    7.2.1. Total Internal Reflection Fluorescence Microscopy (TIRFM)
        7.2.1.1. Principle
        7.2.1.2. Thin Films
        7.2.1.3. Polarized Evanescent Light Intensities
        7.2.1.4. Quantitation of Fluorescence in TIRFM Measurements
    7.2.2. Fluorescence Recovery After Photobleaching
        7.2.2.1. Spot Photobleaching
        7.2.2.2. Periodic Pattern Photobleaching with a Step Function
        7.2.2.3. Fringe Pattern Photobleaching
    7.2.3. Fluorescence Correlation Spectroscopy
    7.2.4. Polarized Fluorescence
        7.2.4.1. Orientation Distributions
        7.2.4.2. Rotational Diffusion
    7.2.5. Resonance Energy Transfer
    7.2.6. Combinations
7.3. Instrumentation for Microspectrofluorometry
    7.3.1. The System
    7.3.2. Microscope
    7.3.3. Laser (and Conventional) Light Sources

Molecular Luminescence Spectroscopy, Part 3, Edited by Stephen G. Schulman. Chemical Analysis Series, Vol. 77.
ISBN 0-471-51580-9 © 1993 John Wiley & Sons, Inc.

7.3.4. Fluorescence Detection
7.3.5. Electronic Control and Data Handling
7.4. **Preparation of Supported Planar Bilayers for Microspectrofluorometry**
  7.4.1. Substrates
  7.4.2. Langmuir–Blodgett Films
  7.4.3. Vesicle Fusion Techniques
    7.4.3.1. Direct Fusion of Lipid Vesicles on a Hydrophilic Surface
    7.4.3.2. Fusion of Vesicles on a Supported Monolayer (Monolayer Fusion Technique)
7.5. **Binding of Macromolecules to Supported Planar Membranes**
  7.5.1. Monoclonal Antibody–Lipid Hapten Binding
  7.5.2. Antibody–F_c Receptor Binding
  7.5.3. Laminin–Sulfatide Binding
7.6. **Translational and Rotational Diffusion of Lipids and Macromolecules in and on Supported Planar Membranes**
  7.6.1. Lipids
  7.6.2. Antibodies
  7.6.3. Integral Membrane Proteins
7.7. **Polarization Measurements**
7.8. **Resonance Energy Transfer Measurements**
7.9. **Conclusion and Future Developments**
References

7.1. INTRODUCTION

Fluorescence spectroscopy has long been a very powerful and extremely sensitive tool of analytical and physical biochemists who were interested in studying the composition, structure, dynamics, and function of biological membranes. Also, biologists have long exploited the high sensitivity and the specific staining capabilities of fluorescent dyes, and especially of fluorescent immunoconjugates in fluorescence microscopy. It is, however, only during the last two decades or so that advanced techniques of fluorescence spectroscopy and fluorescence microscopy have been wed and applied to membrane research in order to probe the spectroscopic properties of selected fluorophores with the high spatial resolution of the fluorescence microscope.

Probably the first example of microspectrofluorometry in membrane research was the heterokaryon experiment of Frye and Edidin (1), which proved for the first time that proteins in biological membranes are laterally mobile and that they are completely miscible when the membranes of two different cells are fused. The success of this early experiment then triggered the development of more quantitative techniques to measure the lateral diffusion of the constituent molecules of cell and model membranes by fluorescence microphotolysis (2) and fluorescence photobleaching recovery (3–5). Fluorescence photobleaching recovery and various later variants of the original technique have since become indispensable and very popular tools in fundamental membrane research, in cell and developmental biology, as well as in immunology and the neurosciences.

At about the same time as when the first photobleaching experiments were performed, another microspectrofluorometric technique, fluorescence correlation spectroscopy, emerged from the laboratories of E. Elson and W. Webb at Cornell University. Fluorescence correlation spectroscopy was able to provide valuable and unique information about the aggregation behavior of fluorescent molecules in solution (6,7), and already in its earliest stages it was applied to study lateral diffusion in lipid bilayers (8). After these initial experiments, which marked the beginning of microspectrofluorometry in membrane research, many more microspectrofluorometric experiments were designed and lead to a wealth of new knowledge on the behavior and function of all kinds of molecules in biological membranes. In fact, in the past 10–15 years, most fluorescence techniques that have been used in solution have also been tried out on the fluorescence microscope. A large number of these attempts have been successful, and it is now possible to measure many parameters that are well known in conventional fluorescence spectroscopy, on microscopic samples, with the additional advantage of a spatial resolution on a micrometer-to-millimeter scale.

The main emphasis of the present chapter is on a special use of microspectrofluorometry, namely, its application to the study of molecular interactions in or on supported planar membranes. Supported planar membranes are single phospholipid bilayers, with or without protein, which are supported by a flat and hydrophilic substrate such as a quartz microscope slide or an oxidized silicon wafer. These membranes are large in area and can extend over several square centimeters. Compared to other popular model membrane systems, the advantage of supported planar bilayers (SPBs) is that they are unilamellar, stable for extended periods of time (usually several days), and geometrically planar. This latter feature is extremely useful for microspectrofluorometric experiments, because the membranes can easily be oriented on the stage of the fluorescence microscope. They have the further advantage that single cells can be specifically bound to them and that the molecules participating in the contact area between the cell surface and the supported membrane can be investigated by microspectrofluorometry. In fact, supported membranes were first introduced by McConnell and co-workers to study the physical chemistry of the immune recognition in the contact zone between lymphocytes (and other cells of the immune system) and supported planar membranes used as models for the target cells. Two reviews are available that cover these early microspectrofluorometric experiments on supported planar membranes (9,10).

7.2. TECHNIQUES OF MICROSPECTROFLUOROMETRY

7.2.1. Total Internal Reflection Fluorescence Microscopy (TIRFM)

Total internal reflection fluorescence is an optical effect that is particularly well suited to the study of molecular phenomena at solid–liquid interfaces. Fluorophores residing in a membrane or the aqueous buffer near a quartz–buffer interface are selectively excited by a total internally reflected laser beam in the quartz slide. As a technique for selective surface illumination, total internal reflection fluorescence was first introduced by Hirschfeld (11) and has become particularly well known by the work of Axelrod, Burghardt, and Thompson in the early 1980s (12–14).

7.2.1.1. Principle

Total internal reflection excitation is conceptually simple (15). When a light beam that propagates through a medium of high refractive index (e.g., a quartz microscope slide) encounters an interface with a medium of a lower refractive index (e.g., an aqueous solution), it undergoes total internal reflection for incidence angles (measured from the normal to the interface) greater than the

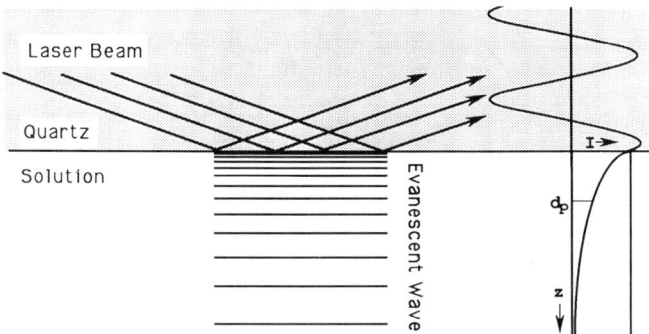

Fig. 7.1. Schematic drawing of the evanescent wave produced by a total internally reflected laser beam at a quartz–buffer interface. The diagram shows the exponential decay of the intensity (I) of the evanescent field and indicates the characteristic penetration depth at the distance from the interface $z = d_p$, where $I(d_p) = I(0)/e$. For the 488-nm laser line and an incidence angle of 72°, d_p is about 100 nm in the quartz–buffer system.

"critical angle." The critical angle (θ_c) for total internal reflection is given by

$$\theta_c = \sin^{-1}(n_3/n_1) \qquad (7.1)$$

where n_1 and n_3 are the refractive indices of the high and low refractive index media, respectively.

For $\theta > \theta_c$, all of the incident light is reflected back at the interface into the solid and creates a standing wave at the interface. This standing wave has a periodic sinusoidal component in the optically denser medium (n_1) and an "evanescent" component in the optically thinner medium (n_3) (Fig. 7.1).

The intensity of the evanescent wave decays exponentially with the distance (z) from the interface.

$$I(z) = I_0 \exp(-z/d_p) \qquad (7.2)$$

with a characteristic penetration depth

$$d_p = \frac{\lambda_0/n_1}{4\pi[\sin^2\theta - (n_3/n_1)]^{1/2}} \qquad (7.3)$$

where λ_0 is the wavelength of the incident light in vacuum. The penetration depth is independent of the polarization of the incident light and decreases with increasing θ. Here d_p is in the order of λ_0 or smaller; I_0, the intensity at the interface ($z = 0$), is a function of θ and the polarization of the incident light, and can be several times higher than the intensity of the incident light (see below).

7.2.1.2. Thin Films

A case of special interest for TIRFM on supported membranes is to consider the effect of a thin film of refractive index n_2 between the solid and liquid bulk media. When the film is much thinner than the penetration depth, i.e., film thickness $d < \lambda_0/(10\pi n_1) \ll d_p$, and the absorption of the thin film is small [attenuation index $\kappa_2 = \alpha\lambda_0/(4\pi n_2) < 0.1$, with α being the absorption coefficient], the electric field can be assumed to be constant over the film thickness and relatively simple and practically useful expressions for the evanescent field strengths can be derived from Fresnel's formulas even in the presence of the film (16,17).

Regardless of the refractive index n_2, total internal reflection will always occur in such a system when $\theta > \theta_c$. The only question is whether TIR occurs at the $n_1:n_2$ or the $n_2:n_3$ interface. There is now a second critical angle, $\theta_{cf} = \sin^{-1}(n_2/n_1)$, determined by the refractive index of the film. For single supported bilayers $n_2 > n_3$ and $\theta_{cf} > \theta_c$. In practice, precisely which interface supports TIR is not critical for the evaluation of the measured parameters in many applications. Also, Eqs. (7.2) and (7.3) are still valid in the presence of an intermediate layer. However, the penetration depth measured from the solid surface is affected by the intermediate layer. The most dramatic effect of a thin and weakly absorbing intermediate layer is a change of the p-polarized evanescent light intensity at the TIR-supporting interface (see below).

7.2.1.3. Polarized Evanescent Light Intensities

Polarized incident light is used in many TIRFM experiments. Therefore, it is important to consider the evanescent electromagnetic field intensities in the different directions at the TIR-supporting interface. We define a coordinate system such that the plane of incidence is the xz-plane, with z-perpendicular to the membrane-supporting surface of the substrate. The y-direction is parallel to the quartz surface and perpendicular to the xz-plane. The incident light is either p-polarized, with the electric field vector parallel to the xz-plane, or s-polarized, with the electric field vector perpendicular ("*senkrecht*") to the xz-plane. The intensities in the different directions of the evanescent field, at $z = 0$ and for unit incident intensities, are given by (15–17).

$$I_x = \frac{4\cos^2\theta\,(\sin^2\theta - n_{31}^2)}{(1 - n_{31}^2)\,[(1 + n_{31}^2)\sin^2\theta - n_{31}^2]} \tag{7.4a}$$

$$I_y = \frac{4\cos^2\theta}{1 - n_{31}^2} \tag{7.4b}$$

$$I_z = \frac{4\cos^2\theta\, n_{32}^4 \sin^2\theta}{(1 - n_{31}^2)\,[(1 + n_{31}^2)\sin^2\theta - n_{31}^2]} \tag{7.4c}$$

where $n_{ij} = n_i/n_j$. The approximations of Eqs. (7.4a–c) are good to a few percent for weakly absorbing thin films, such as supported phospholipid bilayers in the visible region of the electromagnetic spectrum. For other applications, the limits on d and κ given above in Section 7.2.1.2 must be observed. Equations (7.4 a–c) degenerate to the evanescent intensity equations in bulk solution when $n_2 = n_3$ and, therefore, $n_{32}^4 = 1$ in Eq. (7.4c).

The total evanescent intensities for p-polarized and s-polarized incident light are

$$I_\| = I_x + I_z \tag{7.5a}$$

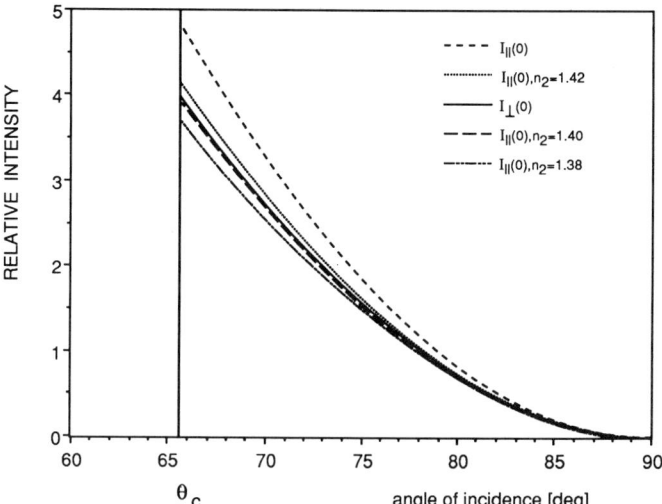

Fig. 7.2. Evanescent intensities $I_\parallel$ (broken lines) and $I_\perp$ (solid line) for p-polarized and s-polarized incident light ($\lambda_0 = 488$ nm) at the quartz–buffer interface ($z = 0$) as a function of the angle of incidence, θ. Plots are shown for cases with and without an intermediate thin film (medium 2) and with refractive indices $n_1 = 1.46$, n_2 variable, and $n_3 = 1.33$; $I_\perp$ is independent of n_2, but $I_\parallel$ is decreased by 10–20% for values of n_2 ranging from 1.42 to 1.38, i.e., reasonable values for thin lipid–protein films. The critical angle is $\theta_c = 65.7°$.

and

$$I_\perp = I_y \qquad (7.5b)$$

respectively. The evanescent intensities $I_\parallel$ and $I_\perp$ are plotted in Fig. 7.2 as a function of the incident angle θ for some combinations on n_1, n_2, and n_3. The most salient features are that these intensities can become severalfold higher than the incident intensities and that the presence of a (non- or weakly absorbing) intermediate layer has quite a distinct effect on $I_\parallel$ but not on $I_\perp$ if n_2 is sufficiently different from n_3. Very large changes in $I_\parallel$ and $I_\perp$ occur when a thin film of metal [with a complex refractive index, $\hat{n}_2 = n_2(1 - i\kappa_2)$] is present on the quartz surface. The reader is referred to the specialized literature on this subject (18,19).

7.2.1.4. *Quantitation of Fluorescence in TIRFM Measurements*

Whereas a relative quantitation of emitted light from bound fluorophores at the surface is very accurate, it is difficult to obtain good absolute values for the actual surface concentrations. There are several approaches to quantitate fluorophore concentration present at the interface. One method that can be used at high fluorophore concentrations in solution and in the absence of fast surface diffusion of adsorbed molecules has been recently introduced by Zimmermann et al. (20). In this method, all the molecules present in the evanescent field are bleached by TIR illumination and the emitted fluorescence intensity is measured after bulk diffusion has exchanged the fluorophores that are bleached in the vicinity of the interface. From the intensity ratios before and after bleaching, together with the intensity function of the evanescent field and the known concentration of fluorophores in solution, one can calculate the absolute amount of bound fluorophores. The amount of bound bovine serum albumin (BSA) on glass has been determined by this method, and the lower concentration limit of this method has been 0.2 mg/mL BSA in solution (20). To increase the accuracy of the method, the distance of the bound layer from the interface and the dependency of the collection efficiency of the microspectrofluorometer on the fluorophore distance from the interface (18) should be known.

A second quite similar approach was used by Pisarchick and Thompson (21). These workers determined the fluorescence intensity of a known concentration of specific fluorescent ligand in the evanescent wave in the absence of specific binding sites and with blocked unspecific sites on the solid surface. Apart from the aforementioned problems of the first approach, an additional drawback of this calibration technique is that even little unspecific binding can introduce quite large artifacts. However, the technique can be used down to much lower solution concentrations compared with the photobleach technique and is not restricted to molecules that are immobilized on the surface.

A third approach for the quantitation of the bound molecules is to calibrate the measured fluorescence intensity with a known amount of bound fluorophores. This method has been applied to measure the amount of fluorescent antibodies that were bound to supported membranes containing specified amounts of lipid hapten (22). For the case of low hapten concentrations in the membrane, the antibody–lipid hapten binding stoichiometry is known and can be used reasonably well for calibration purposes. This method is not very accurate either, but it can be used at low fluorophore concentrations and in the presence of surface diffusion, and it is much less susceptible to nonspecific binding artifacts. One source of inaccuracy results from the intensity decay of the evanescent field. Fluorophores that are farther away from the interface show a smaller fluorescence intensity than those very near to the interface. Therefore, with different proteins, the fluorophores are located at different distances from the interface and different fluorescence intensities will be produced. Also, the fluorescence intensity of a fluorophore bound to a protein depends to some extent on its immediate environment. We believe that, depending on the system, absolute surface concentrations can be estimated within a factor of 2–3. However, the relative accuracy of this method, i.e., comparing the same proteins in the same system, is high. With our present TIRFM instrument, we can quantitate about 100 molecules of fluorescein per square micrometer with an accuracy of better than 10% in a series of different experiments with the same receptor–ligand system.

Finally, a fourth method to roughly estimate absolute amounts of binding has been to measure the depletion of ligand in solution (23). This depletion is usually smaller than a few percent.

7.2.2. Fluorescence Recovery After Photobleaching

Nowadays, the most widely used technique to measure lateral diffusion is fluorescence recovery after photobleaching (FRAP), which has also been termed fluorescence photobleaching recovery or fluorescence microphotolysis. In FRAP, discrete regions of the membrane that contains homogeneously distributed and relatively photolabile fluorophores are irreversibly photobleached. This results in an inhomogeneous distribution of the fluorophores in the membrane without a significant perturbation of the chemical equilibrium. Owing to Brownian (or other) molecular motions, the bleached and unbleached molecules mix, an influx of unbleached fluorophores into the bleached areas occurs, and therefore the fluorescence intensity measured in the bleached area will increase. The kinetics of this fluorescence recovery can be evaluated in terms of transport coefficients (diffusion, exchange reactions, or flow). The most frequent use of FRAP is to determine the lateral diffusion coefficients of lipids or proteins in model membranes or on the surface of single cells. By using polarized light, rotational diffusion coefficients can also be determined by photobleaching techniques (24–27; and see below). The range of diffusion coefficients that can be measured by FRAP is limited by the available bleaching power for fast diffusion and by the mechanical and electronic stability of the photobleach apparatus for slow diffusion. Usually, lateral diffusion coefficients in the range of 10^{-6}–10^{-12} cm^2/s can be measured by FRAP with an accuracy of $\pm 20\%$. The fraction of mobile fluorophores can be calculated from the ratio of the experimental and theoretical total fluorescence recoveries. Irradiation profiles of different geometries have been used in photobleaching experiments on membranes, and they will be briefly discussed in the following subsections.

7.2.2.1. Spot Photobleaching

The theory of single-spot photobleaching is described in detail by Axelrod et al. (3). The irradiated area is a circular spot a few micrometers in diameter with either a Gaussian or uniform intensity profile (Fig. 7.3A). The lateral

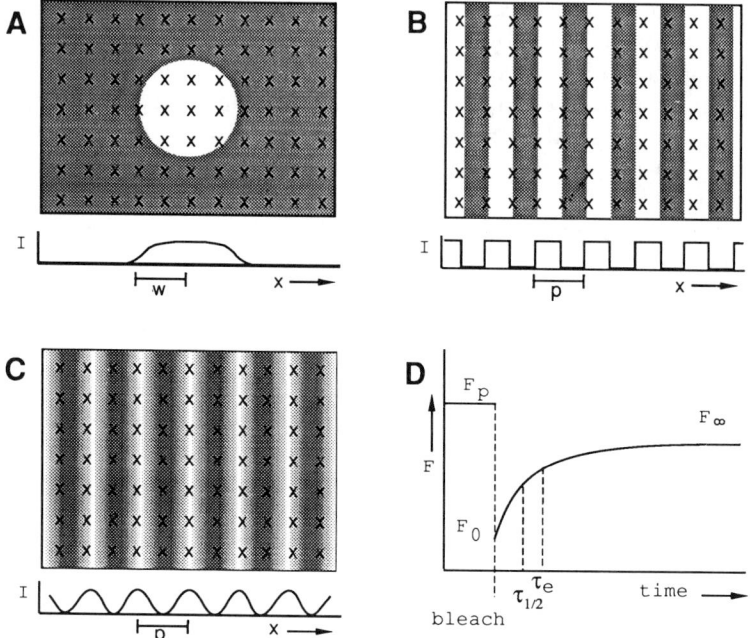

Fig. 7.3. Illumination profiles commonly used in photobleaching: (A) spot photobleaching (Gaussian or circular disk profile); (B) periodic pattern photobleaching with a step function; (C) fringe pattern photobleaching (sinusoidal profile); (D) fluorescence intensities as a function of time in a photobleaching experiment. Here F_p = prebleach intensity; F_0 = intensity at $t = 0$; F_∞ = intensity at $t = \infty$; $\tau_{1/2}$ and τ_e are characteristic time constants after which the fluorescence recovery has gone to 50% and 63.2% completion, respectively. (See text for more details).

diffusion coefficient, D_L is calculated from

$$D_L = \frac{w^2}{4\tau_{1/2}} \gamma_D \qquad (7.6)$$

where w is the radius of the bleached spot where $I = I_{max}/e^2$ for a Gaussian or the geometric radius for a uniform disk profile, respectively; $\tau_{1/2}$ is the time required for 50% recovery; and $\gamma_D = \tau_{1/2}/\tau_D$, the ratio between $\tau_{1/2}$ and the characteristic diffusion time, is a complex function of the bleach profile, which itself depends on the shape of the beam that is used for photobleaching and the depth of photobleaching (3). The mobile fraction, m.f., is defined in percent as

$$\text{m.f.} = \frac{F_\infty - F_0}{F_{pre} - F_0} \cdot 100 \qquad (7.7)$$

where F_{pre}, F_0, and F_∞ are the fluorescence intensities before, immediately after, and a long time after the short bleach pulse, respectively (see Fig. 7.3D). The maximum fluorescence recovery $(F_\infty - F_0)/(F_{pre} - F_0)$ in spot photobleaching is 1.0. The quality of the intensity profile is critical for the accuracy of the resulting value of D_L because the correction factor (γ_D) has to be calculated for each bleaching depth depending on the beam profile. With this technique, relatively high concentrations of fluorophores are needed (e.g., about 10^5 molecules of fluorescein per square micrometer) in order to get enough signal from the small irradiated spot.

7.2.2.2. Periodic Pattern Photobleaching with a Step Function

Periodic pattern photobleaching was introduced by Smith and McConnell (28). An image of alternating bright and dark stripes (e.g., Ronchi ruling) is projected via the microscope objective into the sample plane (Fig. 7.3B). A periodic step concentration profile of fluorophores is established by photobleaching. The corresponding fluorescence intensity profile can be expressed as a Fourier series,

$$F(x, 0) = F_\infty + A \sin(ax) + \frac{A}{3}\sin(3ax) + \frac{A}{5}\sin(5ax) + \cdots \quad (7.8)$$

where the fluorescence intensity after equilibration of the system, F_∞, and the amplitude, $A = F_0 - F_\infty$ are the first two Fourier coefficients, and the spatial frequency of the bleaching pattern a is defined as

$$a = \frac{2\pi}{p} \quad (7.9)$$

with p being the stripe period.

The one-dimensional diffusion equation (Fick's second law) is

$$\frac{\partial c(x,t)}{\partial t} = D_L \frac{\partial^2 c(x,t)}{\partial x^2} \quad (7.10)$$

When the foregoing differential equation is integrated over time with the initial conditions of Eq. (7.8) and with the assumption that the fluorescence intensity is proportional to the concentration of the fluorophore, the time-dependent fluorescence intensity profile becomes

$$F(x,t) = F_\infty + A[\exp(-D_L a^2 t)]\sin(ax)$$
$$+ \frac{A}{3}[\exp(-9D_L a^2 t)]\sin(3ax) + \cdots \quad (7.11)$$

Here, D_L is the lateral diffusion coefficient and can be calculated from

$$D_L = \frac{1}{a^2\tau} \tag{7.12}$$

It is evident from Eq. (7.11) that the third and all further terms decay very rapidly with time and, in practice, can be neglected. By observation with the same pattern that was used for bleaching, Eq. (7.11) to reduces to

$$F(t) = F_\infty + (F_0 - F_\infty)\exp(-D_L a^2 t) \tag{7.13}$$

Therefore, the fluorescence recovery is single exponential to a good approximation and $\tau = \tau_e$ is the exponential time constant at which the amplitude has reached $1/e$ of its final value (Fig. 7.3D). The mobile fraction in pattern photobleaching is

$$\text{m.f.} = \frac{F_\infty - F_0}{F_{\text{pre}} - F_0} \cdot 200 \tag{7.14}$$

and the maximum fluorescence recovery $(F_\infty - F_0)/(F_{\text{pre}} - F_0)$ is 0.5. Three major advantages of pattern photobleaching are (i) that the fluorescence recovery is a simple function of time; (ii) that the determination of D_L is independent of the depth of photobleaching and the exact beam profile; and (iii) that the signal-to-noise ratio is high because of the large area over which the signal is integrated. A disadvantage of this method is that a relatively large irradiation area is needed (diameter $\geq 5p$) to get a negligible difference between the theoretically assumed infinite pattern size and the experimental situation.

7.2.2.3. Fringe Pattern Photobleaching

In 1982, periodic patterns produced by interference of two colliding light beams were introduced for photobleaching studies independently in two laboratories (29,30). In one case (29), interference fringes were produced in a regular fluorescence cuvette to measure the lateral diffusion coefficients of macromolecules in solvents of different viscosities. In the other case (30), the fringe pattern was produced at the interface of a supported membrane and buffer by the interference of two total internally reflected laser beams. The mathematical description of the time evolution of the sinusoidal pattern in fringe pattern photobleaching is similar to the one described above for stripe pattern photobleaching. The light intensity profile is (Fig. 7.3C).

$$I(x) = \bar{I}[1 + \cos(ax)] \tag{7.15}$$

where $\bar{I}$ is the mean light intensity of the observed area and $a = 2\pi/p$. The concentration profile immediately after the bleach pulse and for a first-order bleach reaction is

$$c(x,0) = c_0 e^{-\alpha I(x) \Delta t} \tag{7.16}$$

where c_0 is the concentration of fluorophores before the bleach pulse of duration Δt, and α is the first-order photolytic reaction rate constant. After introduction of the mean bleaching efficiency index, $K = \alpha \bar{I} \Delta t$, the concentration profile of unbleached fluorophores becomes

$$c(x,0) = c_0 e^{-K[1 - \cos(ax)]} \tag{7.17}$$

The diffusion Eq. (7.10) can be integrated with the initial conditions of Eq. (7.17) by first expanding both equations into Fourier series (29). The result is

$$F(t) = B\bar{I}c_0[A_0(K,0) + A_1(K,0)\exp(-D_L a^2 t)] \tag{7.18}$$

where $A_0(K,0)$ and $A_1(K,0)$ are the Fourier coefficients and are functions of the bleaching efficiency K, and B is a constant of proportionality to account for the fluorescence of the fluorophor and the collection efficiency of the microscope. Here $A_0(K,0)$ and $A_1(K,0)$ represent, respectively, the mean concentration of fluorophores and the amplitude of the fluorescence at spatial frequency a just after the bleach pulse. Therefore, Eq. (7.18) in practice reduces to

$$F(t) = F_\infty - (F_\infty - F_0)\exp(-D_L a^2 t) \tag{7.19}$$

which is a single exponential function of time. Then D_L can be calculated in the same way as in stripe pattern photobleaching by

$$D_L = \frac{1}{a^2 \tau_e} \tag{7.20}$$

with τ_e being the experimentally determined exponential time constant of the fluorescence recovery curve (Fig. 7.3D).

In their original work, Davoust et al. (29) used a periodically changing pattern and determined the mobile fraction from the intensity difference between bleached and unbleached stripes at various positions of the patterns. Since changing patterns are not so easy to implement in some applications, the mobile fraction may alternatively also be determined from the calculated bleach profiles at each given depth of photobleaching. The concentration profile of unbleached fluorophores after the bleach pulse [Eq. (7.17)] as observed with the intensity profile of the monitoring light [Eq. (7.15)] can be compared to the same amount of fluorophores, which are homogeneously distributed and illuminated with the same monitoring light. The fractional theoretical fluorescence intensities (normalized to the prebleach intensities) immediately after and a long time after the bleach pulse are

$$f_0 = \int_0^p e^{-K[1-\cos(ax)]}[1-\cos(ax)]\,dx \qquad (7.21)$$

and

$$f_\infty = \int_0^p e^{-K[1-\cos(ax)]}\,dx \qquad (7.22)$$

respectively. The theoretical fluorescence recovery for a 100% fraction of mobile molecules at each bleach depth is given by

$$\Delta f = f_\infty - f_0 \qquad (7.23)$$

where f_∞ and Δf correspond to the previously introduced Fourier coefficients

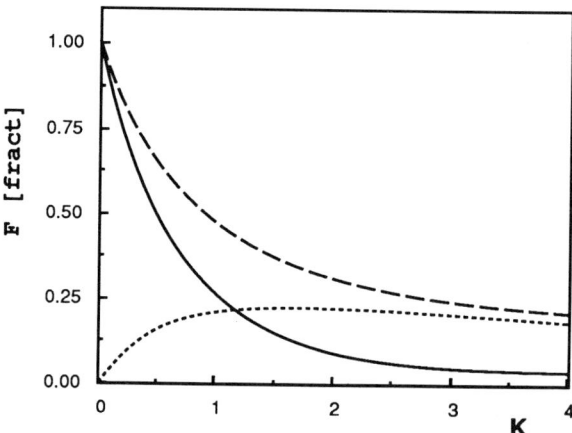

Fig. 7.4. Theoretical fractional fluorescence intensities for different depths of photobleaching (K) in interference pattern photobleaching, with the assumption of a 100% fraction of mobile molecules at each bleach depth: f_0 (solid line); f_∞ (dashed line); and Δf (dotted line). The curves are calculated with Eqs. (7.21)–(7.23). The maximal theoretically achievable fluorescence recovery is about 22% at a value of K of about 1.5.

$A_0(K, 0)$ and $A_1(K, 0)$, respectively. The values of f_0, f_∞, and Δf are plotted in Fig. 7.4 as a function of K for unit prebleach intensities. It is evident from Fig. 7.4 that the maximum Δf which can possibly be achieved is 22% and occurs at a value of K of about 1.5. To avoid photochemical damage to the sample (other than fluorophore bleaching), bleach depths of about 50–60% are often used in practice, which further reduces the theoretical maximal recoveries to about 15%. This explains why fringe pattern photobleaching is always less sensitive (lower signal-to-noise ratio) than stripe pattern photobleaching when used on the same area and fluorophore concentration. The mobile fraction in fringe pattern photobleaching is determined by

$$m.f. = \frac{F_\infty - F_0}{F_{pre}} \cdot \frac{100}{\Delta f} \tag{7.24}$$

7.2.3. Fluorescence Correlation Spectroscopy

In fluorescence correlation spectroscopy (FCS), the fluorescence emission is recorded from a small volume of the sample as a function of time. Owing to diffusion or flow of the fluorescent molecules through the observed volume, the fluorescence emission will fluctuate with Poisson statistics if the average number of fluorescent particles in the measuring volume is sufficiently small. These fluctuations, which can easily be analyzed and quantified by calculating autocorrelation functions, depend very sensitively on the number of independently moving fluorescent units in the sample region, and therefore the magnitudes of the autocorrelation functions can be used to measure aggregation numbers or cluster sizes if the average number of fluorescent molecules in the system is known. If the fluorescent molecules are lipids or proteins in membranes, their lateral diffusion coefficients or flow rates can be determined from the decay rates of the measured autocorrelation functions. If the fluorescent molecules are ligands that bind to membrane receptors, the autocorrelation functions contain additional information on the kinetic association and dissociation rate constants. In addition to conventional FCS, which was first developed by Elson, Magde, and Webb (6, 7, 31), two more advanced techniques, scanning FCS/fluctuation spectroscopy (32–36) and high-order FCS (37, 38), are now available. Temporal and spatial correlations can be carried out in scanning FCS/fluctuation spectroscopy, whereas high-order FCS provides independent additional information about molecular oligomerizations from the distribution of fluorescence fluctuations.

The first-order normalized autocorrelation function is

$$G(\tau) = \frac{\langle \delta F(t+\tau) \cdot \delta F(t) \rangle}{\langle F(t) \rangle^2} \tag{7.25}$$

In this equation, $F(t)$ is the measured fluorescence in the sample volume at time t; $\delta F(t) = F(t) - \langle F(t) \rangle$ is the fluctuation of fluorescence from its average value at time t; and the angular brackets $\langle \cdots \rangle$ signify long time averages. It has been shown (6) that for a monodisperse sample in which the molecules undergo translational diffusion the autocorrelation function can be expressed as a function of the effective average number of fluorescent particles within the sample volume, $\langle N \rangle$, and the correlation time, τ_c,

$$G(\tau) = \frac{h}{\langle N \rangle (1 + \tau/\tau_c)} \tag{7.26}$$

and that the variance of the relative intensity fluctuation is proportional to $\langle N \rangle^{-1}$

$$G(0) = \frac{h}{\langle N \rangle} \tag{7.27}$$

The constant h depends on the beam profile in the active volume of the sample and is 0.5 for a Gaussian and 1.0 for a uniform circular beam profile, respectively. In practice, $G(\tau)^{-1}$ is often plotted versus τ, which (for a single mobile species) yields a straight line with an intercept $\langle N \rangle/h$ and a slope $\langle N \rangle/h\tau_c$. Since

$$\tau_c = \frac{w^2}{4D_T} \tag{7.28}$$

where w is the $1/e^2$ radius of the Gaussian or the geometric radius of the circular beam profile, respectively, the translational diffusion coefficient D_T of the fluorescent particles can be calculated from the measured correlation time. For the case of unidirectional flow with speed v, the correlation function takes an exponential form

$$G(\tau) = G(0) \exp[-(\tau/\tau_c)^2] \tag{7.29}$$

and

$$\tau_c = \frac{w}{v} \tag{7.30}$$

More complex functional forms of $G(\tau)$ have been derived for several cases of fluorescent molecules which are in chemical equilibrium with other species or themselves (6,38).

Fluctuation spectroscopy (32,33) and scanning FCS (34–36) are two similar techniques that are particularly useful in samples with relatively small lateral diffusion coefficients and where photobleaching due to an extended exposure of the same fluorophors to the measuring beam may become a problem. That is, these techniques are perhaps superior to conventional FCS for measurements of fluorescence autocorrelations on cell and some model membranes. In both techniques, fluorescence intensities are autocorrelated in space rather than in time. This is experimentally achieved either by sweeping the observation beam over the sample with a rotating objective (32,33) or by translating the sample on the stage of the microscope (34–36). From the $G(0)$ value and the spatial decay of the correlation function, the average aggregation number and lateral diffusion coefficients or the beam width, respectively, can be determined in a similar way as described above for conventional FCS.

7.2.4. Polarized Fluorescence

As is the case in conventional fluorescence spectroscopy, polarized light can be exploited in microspectrofluorometry to measure various useful parameters of fluorescent molecules in their respective microenvironments. Most notable among these parameters are molecular orientation distributions and rotational diffusion coefficients. However, because of the relatively high numerical aperture (NA) of the microscope objective, the polarizations of the exciting and emitted light are not maintained in their original form but rather are mixed with components from the other principal directions. This problem complicates the analysis microscopic polarized fluorescence data sometimes rather severely (39). The most dramatic effects occur with high NA objectives for fluorophores that have their emission dipoles preferentially oriented along the symmetry axis of the exciting beam, i.e., the optical axis of the microscope, and for rotational diffusion times that are short compared to the fluorescence lifetimes. Some of the most striking differences between conventional and microscopic fluorescence polarization are discussed here very briefly. For a derivation of the conclusions presented here, the original literature should be consulted (24,27,39–41).

A convenient experimental parameter in polarized fluorescence spectroscopy is the fluorescence anisotropy, r, which is defined as

$$r = \frac{F_\| - F_\perp}{F_\| + 2F_\perp} \tag{7.31}$$

Because of photoselection, the maximal fluorescence anisotropy, r_0, in conventional fluorescence spectroscopy is 0.4. This value is found for randomly distributed fluorophores that do not undergo any motion during the lifetime,

τ_F, of the excited state ($\tau_F < 100$ ns, for most fluorophores) and whose absorption and emission transition dipole moments are parallel. For high numerical aperture observation on a microscope, r_0^{NA} is always smaller than r_0:

$$r_0^{NA} < r_0 \tag{7.32}$$

Intuitively this can be understood, because the values of $F_{\parallel}$ and $F_{\perp}$, measured at high numerical aperture, contain not only the conventional contributions from F_z and F_y, respectively (where x is the direction of light propagation, and z is the orientation of the electric field vector of the plane polarized light) but also cross-terms from the two other orthogonal directions.

7.2.4.1. Orientation Distributions

It is often convenient to express orientation distributions in membranes in terms of order parameters. For distributions that are axially symmetric around the membrane normal, it is sufficient to consider a single order parameter:

$$S = \frac{\langle 3\cos^2\theta - 1 \rangle}{2} \tag{7.33}$$

where θ denotes the angle between the principal molecular symmetry axis and the membrane normal, and the angular brackets are a time and ensemble average over the possible fluctuations and distributions of the fluorophore; S can vary between -0.5 (for a rigid alignment parallel to the plane of the membrane) and $+1.0$ (for a rigid alignment perpendicular to the plane of the membrane). Intermediate values of S occur for other orientations and when the molecules undergo rapid orientation fluctuations.

In the following, two extreme cases are considered, namely, carbocyanine dyes (such as diI), which have their absorption and emission transition dipole moments oriented preferentially parallel to the plane of the membrane, and the common membrane probe diphenylhexatriene (DPH), in which the absorption and emission transition dipole moments are coaxial with the rotational symmetry axis in the membrane. For the case of carbocyanine dyes, order parameters may be obtained from steady-state polarized fluorescence intensity distributions on vesicles or single cells (40), or by polarized evanescent illumination on supported membranes (42) (see below). For the case of DPH, S can easily be derived from fluorescence anisotropy measurements (43,44):

$$S = \left(\frac{r_\infty}{r_0}\right)^{1/2} \tag{7.34}$$

where r_∞ is the end value of the fluorescence anisotropy in a time-resolved experiment. In rigid or slowly reorienting systems, i.e., when the rotational correlation time is much larger than the fluorescence lifetime ($\tau_R \gg \tau_F$), r_∞ in Eq. (7.34) can be replaced by the steady-state anisotropy, r, and S can be determined from steady-state experiments. If orientational distributions are to be derived from steady-state (or time-resolved) fluorescence anisotropies that have been measured under conditions of high NA, r_0 in Eq. (7.34) must be replaced by r_0^{NA} in order to avoid an overestimation of S. The effect of high NA observation of DPH in a membrane that is oriented perpendicular to the observation axis of the microscope is expected to be rather dramatic because r_0^{NA} will become very small under these conditions (39). However, for transition dipoles that are less than 45° out of the plane of the membrane and NAs ≤ 1.0, the error in S can be estimated to be smaller than 10% when r_0^{NA} is approximated by r_0, because of the square root dependence of S on r_0 in Eq. (7.34).

7.2.4.2. Rotational Diffusion

A very useful application of time-resolved polarized fluorescence microscopy is the measurement of rotational diffusion of labeled proteins or lipids in membranes. Two different fluorescence techniques are available to measure rotational diffusion, namely, the measurement of fluorescence depletion anisotropy (45,46) and polarized fluorescence photobleaching recovery (24,27).

In fluorescence depletion anisotropy, triplet state probes (such as eosin) are excited by polarized light and the depletion of the fluorescence anisotropy

$$r(t) = \frac{\Delta F_\parallel - \Delta F_\perp}{\Delta F_\parallel + 2\Delta F_\perp} \tag{7.35}$$

$$\Delta F_{\parallel, \perp} = F_0 - F_{\parallel, \perp} \tag{7.36}$$

is measured, where F_0 is the fluorescence in the absence of polarization, and $F_\parallel$ and $F_\perp$ are the fluorescences detected parallel and perpendicular, respectively, to the excitation polarization. The individual fluorescence intensity decays follow the general equation (46)

$$F = F_0\{A_1 - A_2 B[\exp(-k_T t)] - A_3 B[\exp(-k_c t)]\} \tag{7.37}$$

where

$$k_T = \frac{1}{\tau_T} \tag{7.38}$$

is the reciprocal lifetime of the triplet state and

$$k_c = \frac{1}{\tau_R} + \frac{1}{\tau_T} \tag{7.39}$$

depends on the rotational correlation time τ_R and τ_T. The rotational diffusion coefficient, D_R, is related to τ_R is by

$$\frac{1}{\tau_R} = 4D_R \tag{7.40a}$$

and

$$\frac{1}{\tau_R} = 6D_R \tag{7.40b}$$

for two- and three-dimensional diffusion, respectively. The coefficients A_1, A_2, and A_3 depend on the geometry of the sample, the particular polarizer combination, and the observation aperture. Some useful values are tabulated in Yoshida and Barisas (46). The term $B/3$ is the total fractional ground state depletion and is

$$\frac{B}{3} = \frac{2.3031 I \varepsilon \lambda}{hcN} \Phi_T \Delta t \tag{7.41}$$

for a short pulse of duration Δt (I = intensity of depletion pulse; λ = wavelength of light in the vacuum; ε = molar absorption coefficient; h = Plank's constant; c = speed of light; N = Avogadro's number; Φ_T = quantum yield for triplet state formation).

Since oxygen quenches excited triplet states very efficiently, a general limitation of techniques using triplet state probes is that the samples need to be deoxygenated, which is no problem for model systems but can severely affect the physiology of live cells. Polarized photobleaching techniques avoid this problem and the presence of oxygen often accelerates the rate of irreversible photobleaching. Many aspects of the theory of polarized photobleaching recovery are very similar to those outlined above for fluorescence depletion anisotropy. The final expressions for the fluorescence recoveries for the case of two-dimensional diffusion are (27)

$$\Delta F_{\|,\perp}(t) = a \pm b[\exp(-4D_R t)] - c[\exp(-16D_R t)] \tag{7.42}$$

and

$$r_b(t) = \frac{2b[\exp(-4D_R t)]}{3a - b[\exp(-4D_R t)] - 3c[\exp(-16D_R t)]} \tag{7.43}$$

where a, b, and c are complicated functions of the duration and intensity of the bleach pulse and the order parameter of the fluorophore. The order parameter itself is determined by the average orientation and the extent of fast wobble of the fluorophore [see Eq. (7.33)]. Similar expressions have been derived for diffusion in three dimensions (27). Corrections for high aperture observation can usually be neglected when working at NA ≤ 0.75.

7.2.5. Resonance Energy Transfer

When two fluorophores with sufficient overlap between the donor emission and the acceptor absorption spectra are in proximity, (< 100 Å) nonradiative transfer of energy can occur between the excited states of the two molecules (47). This resonance energy transfer (RET) manifests itself in a decreased quantum yield of the fluorescence emission of the donor and an increased fluorescence emission of the acceptor:

$$\frac{Q}{Q_0} = \frac{\tau_F}{\tau_0} = 1 - E \quad (7.44)$$

The transfer efficiency, E, in Eq. (7.44) can range between $0 \leq E \leq 1$ and is measured either from the ratio of the quantum yields Q or from the fluorescence lifetimes τ_F of the donor in the presence and absence of acceptor, respectively. It has been shown that E is a function of the sixth power of the distance r between the donor and acceptor (47):

$$E = \frac{R_0^6}{R_0^6 + r^6} \quad (7.45)$$

$$R_0 = 9.8 \times 10^3 \, (\kappa^2 n^{-4} Q_0 J)^{1/6} \quad (7.46)$$

$$J = \int_0^\infty q(\lambda) \varepsilon(\lambda) \lambda^4 \, d\lambda \quad (7.47)$$

where R_0 is the critical distance (in angstroms); κ^2 is a factor describing the orientation of the emission transition dipole moment of the donor relative to the absorption transition dipole moment of the acceptor (usually assumed to be 2/3); n is the index of refraction of the medium between the donor and the acceptor; J is the spectral overlap integral; $q(\lambda)$ is the spectral quantum yield distribution function of the donor in each interval λ to $\lambda + d\lambda$; and $\varepsilon(\lambda)$ is the corresponding decadic extinction coefficient of the acceptor. Equations (7.45)–(7.47) describe molecular properties of the sample and are independent of the system that is used for fluorescence detection. Since relative ratios are compared in Eq. (7.44), this equation is also valid for the measurement of E under high aperture conditions on a fluorescence microscope. Because of the sharp distance dependence of Eq. (7.45), RET is extremely useful to measure distance changes in the neighborhood of R_0, that is, at around 40 Å for many donor–acceptor pairs. Molecular rearrangements that occur at distances larger than about $2R_0$ are unlikely to be detectable unless there is a marked change in the orientation factor κ^2.

In order to determine the density of fluorophores in membranes, the appropriate rate equations for energy transfer have been solved for two-dimensional fluorophore distributions (48). It is first recalled that

$$\tau_F = \int_0^\infty F(t)/F(0)\,dt \tag{7.48}$$

where the time-dependent fluorescence intensity divided by the initial fluorescence intensity of the donor following a very short excitation light flash is

$$F(t) = F(0)\exp[-(t/\tau_0)]\exp[-\sigma S(t)] \tag{7.49}$$

with

$$S(t) = \int_a^\infty \{1 - \exp[-(t/\tau_0)(R_0/r)^6]\}4\pi r^2\,dr \tag{7.50a}$$

for three-dimensional fluorophore distributions and

$$S(t) = \int_a^\infty \{1 - \exp[-(t/\tau_0)(R_0/r)^6]\}2\pi r\,dr \tag{7.50b}$$

for two-dimensional fluorophore distributions, respectively. Here a is the distance of closest approach of donor and acceptor, and σ is the three- or two-dimensional density of energy acceptor. For the case of a fixed (three-dimensional) distance between the donor and acceptor, the insertion of the integrand of $S(t)$ in Eq. (7.50a) into Eqs. (7.49), (7.48), and finally (7.44) leads to the familiar expression (7.45). No closed-form solution can be found for the random planar distribution of acceptors, but the values of E can be calculated by a numerical solution of Eqs. (7.44), (7.48), (7.49)). and (7.50b). These equations show that the energy transfer efficiency in two dimensions depends only on three variables: R_0, σ, and a. In particular, the transfer efficiency is independent of the surface density of the donor. The surface densities of energy acceptor resulting in 50% transfer have been found at 2.4, 1.0, 0.66, and

0.57 mol% of phospholipid for energy transfer pairs that had characteristic distances, R_0, of 26, 38, 46, and 49 Å, respectively (48). Similar expressions as those presented in Eqs. (7.49) and (7.50) can be derived for other donor–acceptor geometries, such as a donor that is elevated from the membrane surface and a randomly distributed acceptor in the plane of the membrane (49), or the transfer between two fluorophores that are each randomly distributed in two different planes, e.g., the two surfaces of a membrane (48).

RET has also been used in microscopic fluorescence experiments to measure the cell surface distribution of lectin receptors on single cells (50,51) and the T-cell-mediated binding of peptide antigens in planar supported membranes (52,53). Furthermore, RET images of single cells have been used to co-localize two different fluorescent lipid analogues or a fluorescent lipid and a fluorescent lectin in different organelles after their internalization by endocytosis (54).

7.2.6. Combinations

Several of the techniques described above can provide useful information on supported planar membranes when combined with total internal reflection illumination.

TIR/Spot FRAP and TIR/FCS. These combinations have been first described theoretically and experimentally by Thompson, Burghardt, and Axelrod (12,13,55). When these techniques are used with fluorescent molecules that are in chemical equilibrium between the surface and the bulk solution, the theory shows (12) that the TIR photobleaching recovery curves and the TIR autocorrelation functions depend in a complex manner on (a) the adsorption rate constant, (b) the desorption rate constant, (c) the surface diffusion coefficient, (d) the bulk diffusion coefficient, and (e) the size of the illuminated region. However, the experimental conditions can be varied such that the rate constants or the surface diffusion coefficients can be readily obtained.

The following surface binding reaction is considered:

$$A + B \underset{k_{-1}}{\overset{k_1}{\rightleftharpoons}} AB$$

where A is a solute that can bind to free surface sites B to form occupied sites AB; k_1 and k_{-1} are surface adsorption and desorption rate constants, and $\bar{A}$, $\bar{B}$, and $\overline{AB}$ are the equilibrium concentration values of A, B, and AB, respectively. An interesting question is whether this reaction is limited by bulk diffusion or by the reaction rate constants. This can be decided by changing $\bar{A}$ (and therefore, the ratio $\bar{A}/\overline{AB}$) in TIR/spot FRAP experiments. If the recovery curves change their shape, bulk diffusion contributes to the measured signal. If they do not change shape, the reaction is reaction-limited and the rate constants are easier to extract from TIR/spot FRAP or TIR/FCS experiments than in the non-reaction-limited case. Within certain experimental limitations, the recovery and autocorrelation curves can be forced toward the reaction limit by decreasing the size of the observation region and/or increasing $\bar{A}$. For a reaction-limited system, the following equations are useful to determine the rates k_1, k_{-1}, and the lateral diffusion coefficient on the surface, D_L: For TIR/spot FRAP with a linear Gaussian beam profile of a $1/e^2$ width w (for this profile, see Ref. 13),

$$\frac{F_\infty - F(t)}{F_{pre} - F(0)} = \left(1 + \frac{4D_L}{w^2} t\right)^{-1/2} \exp(-k_{-1} t) \quad (7.51)$$

and for TIR/FCS with a circular Gaussian observation profile with a $1/e^2$ radius w,

$$\frac{G(t)}{G(0)} = \left(1 + \frac{4D_L}{w^2} t\right)^{-1} \exp[-(k_1 \bar{A} + k_{-1}) t] \quad (7.52)$$

To obtain both rate constants, the equilibrium binding constant $K(= k_1/k_{-1})$ must be known from independent experiments such as a Langmuir binding isotherm obtained by TIRFM.

TIR/Fringe FRAP. Interference fringes have been produced with two intersected and total internally reflected laser beams at the interface of a quartz-supported membrane and buffer (30,56–58). They were used to set up a periodic pattern for fluorescence recovery after photobleaching in order to determine the lateral diffusion coefficients of the lipids and proteins in these membranes. In the absence of rapid exchange reactions of the fluorophores with the bulk solution, the lateral diffusion coefficients and mobile fractions can be evaluated straight forward as outlined in Sect. 7.2.2.3. The periodicity of the evanescent fringe pattern should be calibrated with a known ruler, e.g., the Ronchi rulings that are used for pattern photobleaching in the epifluorescence mode. This or another type of calibration is required, because the evanescent interfringe distances depend not only on the collision angle but also on the refractive indices of the various media and the polarization angles of the two intersected light beams (59).

TIR Anisotropy. Orientation distributions of fluorophores at interfaces may also be investigated by polarized evanescent waves (60,61). The fluorescence anisotropy depends in a complex fashion on at least two order parameters, which in general cannot be separately determined from the measured polarized TIR fluorescence intensities.

TIR/RET. Energy transfer under TIR illumination has been used to study the conformation of surface-adsorbed BSA (62) and the binding of peptide antigens to histocompatibility antigens (52,53). Since in this technique relative fluorescence intensity changes are observed without changing the evanescent field of the illumination system, no special effects related to the TIR geometry should occur and the normal RET theory (Section 7.2.5) can be applied. However, because of the special polarization features of the evanescent field, conformational changes that result in a change in the orientation factor κ^2 may yield larger changes of the energy transfer efficiency as compared to those measured in solution.

7.3 INSTRUMENTATION FOR MICROSPECTROFLUOROMETRY

7.3.1. The System

In this section, we describe the general design of an instrument for microspectrofluorometry that is capable of using epi- and TIR illumination and with which many of the experimental techniques of Section 7.2 can be performed. There are many ways to design such an instrument, but we will restrict ourselves to describing the instrument that we are presently using. This design is especially adapted to studies of supported planar membranes.

A schematic diagram of our laser fluorescence microscope is shown in Fig. 7.5 (see also Ref. 56). The intensity of a laser beam is regulated with a computer-controlled acousto-optic modulator (AOM). A removable mirror (M*) is used to switch the instrument between TIR illumination and epi-illumination. For TIR illumination, the beam is split with a nonpolarizing beam splitter (BS) and the two beams are focused with two lenses (L) and intersected at the quartz–buffer interface in the image plane of an inverted microscope. In epi-illumination FRAP, a Ronchi ruling (RR) is imaged at the quartz–buffer interface through the dichroic mirror (DM) and the microscope objective (MO). An image plane diaphragm (IPD) in the optical path of the microscope is used to select for areas of interest, and the fluorescence intensities in these areas are measured by a photomultiplier tube (PMT) or a video system. The instrument is mounted on a heavy granite or pneumatic optical table in order to isolate it from external vibrations.

7.3.2. Microscope

In principle, most microscope types can be used for microspectrofluorometry. However, it is more convenient to work with fixed-stage microscopes for many

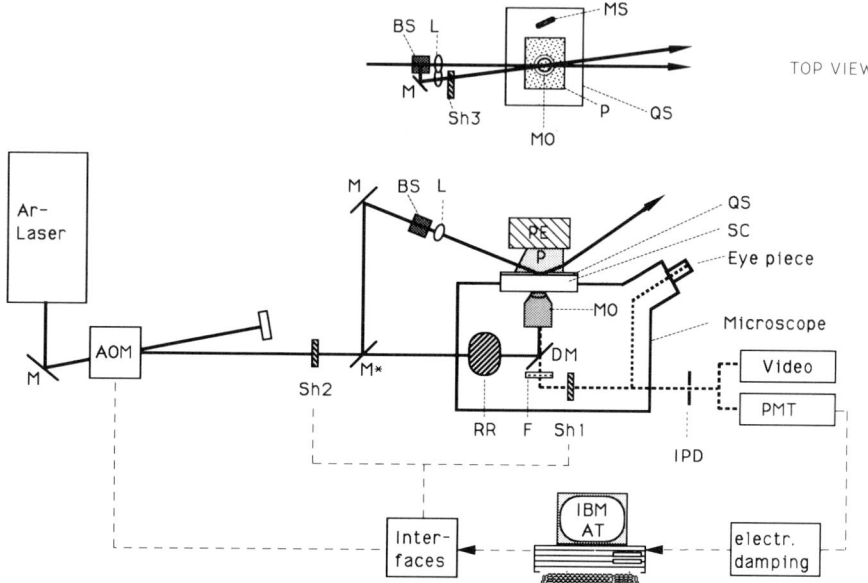

Fig. 7.5. Schematic diagram of the laser fluorescence microscope used in our laboratory. It is designed to perform total internal reflection fluorescence microscopy (TIRFM) and fluorescence recovery after photobleaching (FRAP) experiments. Details of the figure are explained in the text.

TIR applications because the alignment of the beam with respect to the sample remains fixed during focusing. Upright or inverted microscopes may be used. Two important factors for selecting a microscope are its mechanical stability and the provision of enough open working space to implement the required additional optical and electronic parts. Dichroic mirrors and cutoff filters supplied by the microscope manufacturers are generally sufficient for most purposes of microspectrofluorometry. The design of the TIR optics part depends on whether an upright or inverted microscope is used. When an inverted microscope is used, the supported planar membrane is "hanging" below the ceiling (quartz slide) of the TIR cell. This geometry has the advantage that owing to gravitation, unbound cells or protein aggregates, etc., sediment to the bottom of the cell and therefore cannot obstruct the fluorescence measurements.

We use an inverted microscope with a fixed stage in our setup (Carl Zeiss, IM35 or Axiovert 35). A trapezoidal prism (P) with one side normal to the incoming beam is optically coupled with immersion oil to the quartz slide (QS) for TIR illumination. The quartz slide ($40 \times 25 \times 1$ mm) fits tightly into the sample cell (SC) with a glass coverslip bottom that is approximately 1.5 mm below the lower surface of the quartz slide. The walls of the cell are made of Teflon, and the whole cell is clamped with an aluminum frame. Two access holes in the frame and the walls allow the perfusion of buffer and other materials through the cell. The contents of the cell are stirred with a little magnetic stirrer (MS) and the temperature can be controlled with a Peltier element (PE). In TIR illumination, the laser beam is split in a plane (see top view, insert Fig. 7.5) that is inclined at an angle of 72° from the vertical microscope axis. This is more than 6° larger than the critical angle θ_c for the quartz–buffer system and results in a penetration depth of about 75 nm ($\lambda_{ex} = 488$ nm). The size of the area of TIR illumination can be varied by changing the focus of the lenses (L). The advantage of a trapezoidal prism (instead of a cube) is that the beam is not refracted when it enters the prism. This results in less beam distortion by nonideal surfaces. It is our experience that this geometry allows an easy and stable beam adjustment with a good preservation of the beam profile, which is important for the production of "clean" interference fringes. Extreme mechanical stability of the TIR optics part is important for the pointing stability of the beam on the same spot over hours, which is needed in many binding experiments (isotherms and kinetics). A high-precision remounting of the removable mirror (M*) is extremely helpful for adjusting the area of TIR illumination to the center of the optical axis of the microscope.

Many other geometries for TIR illumination are possible and have been reviewed recently (63). For instance, prismless TIR is a setup in which the excitation light propagates along the very edge of the microscope objective aperture and therefore strikes the interface at $\theta > \theta_c$. This technique has the disadvantage that only a small portion of the total optical field is illuminated and that only objectives with a NA > 1.33 can be used. Alternatively, two prisms can be coupled to the substrate such that multiple reflections occur (30), or a hemisphere prism with two parabolic mirrors can be used to produce interference fringes (63).

7.3.3. Laser (and Conventional) Light Sources

In order to measure the fluorescence intensity of a sample, a lamp as a light source would be sufficient. But for studies involving photobleaching only high-intensity lasers have enough power to bleach the sample in a sufficiently short time (the bleach time should be smaller than one-tenth of the exponential recovery time for diffusion). A laser beam can be focused to smaller spots than can a conventional light source, and—most important—a coherent light source is needed to produce interference fringes for TIR photobleaching. Argon ion lasers for fluorophores such as 7-nitrobenz-2-oxa-1,3-diazol (NBD), fluorescein, and rhodamine and krypton ion lasers with more lines between 500 and 600 nm have enough power (more than 1 W for some lines) for photobleaching studies, provide stable intensity outputs, and are easy to operate. Very-high-power lasers are needed for rotational diffusion studies in the millisecond range.

Two different techniques can be used to modulate the intensity of the laser beam: one uses beam splitters, and the other an AOM. The ratio of the bleach and observation beam intensities has to be at least 1000 in order to avoid additional photobleaching during fluorescence observation. When two different light paths are used for bleaching and observation, it is important to correctly align the two beams to the same spot on the sample. Misalignment or small drifts of one optical component can cause severe artifacts in photobleaching experiments. When an AOM is used, the light is diffracted and the intensity of the first-order diffraction beam is modulated. The position of the diffracted beam is independent of its intensity, and therefore the bleach and observation beams are always exactly superimposed. Another advantage of an AOM is that the light intensity can be changed in the submicrosecond time range. This is especially useful for fast rotational diffusion measurements. But along with these advantages, the use of an AOM also has some disadvantages. The intensity of the AOM is controlled by applying voltages between 0 and 1 V, leading to intensity changes of the order of 10^4. Low voltages (of the order of 10–20 mV) are used to produce the observation beam intensities. In this regime, voltage changes of about 0.1–0.2 mV result in observation beam intensity changes of about 5%. Therefore, temperature-stabilized electronic parts and a stable ground level are imparative to avoid large intensity fluctuations.

The spatial modulation of the light in epi-illumination is produced by a set of Ronchi rulings of different periodicities that are imaged in the sample plane through the dichroic mirror (DM) and the microscope objective. Ronchi rulings with 50–250 lines/in. are used to obtain patterns with repeat distances of 12.5–2.5 μm in the sample plane with a 40× microscope objective.

In TIR illumination it is not possible to image an extended Ronchi ruling in the sample plane for pattern photobleaching. Therefore, we used interference fringes, produced by two intersecting laser beams, to create a spatial variation of light for photobleaching. Note that the maximum theoretical percentages of fluorescence recovery are only obtained with an exact 50:50 beam splitting and that they depend critically on the quality of the two beams and their correct alignment in the focal plane of the microscope objective (see Section 7.2.2.3 and Ref. 56).

7.3.4. Fluorescence Detection

To detect fluorescence intensities in a microspectrofluorometer, it is important to remove scattered excitation light with efficient high-wavelength pass filters and to use a sensitive photomultiplier tube or a low-light-level camera. Different types of PMTs can be used for microspectrofluorometry, namely, photoncounting and analog PMTs. Especially when long sampling intervals with low light intensities are needed, photon counting is superior to analog detection.

Although sampling times are often relatively short (typically 1–40 ms), photon-counting PMTs are normally used in microspectrofluorometry. However, in our setup we use an analog PMT with an analog electronic noise reduction system. A PMT with extended red characteristics (EMI 9658 A) has been selected. With this system, we can accurately determine the diffusion of 10^6 NBD or 10^5 fluorescein molecules. For fluorescence intensity measurements, about 10 times fewer molecules are needed. Quantitation of fluorescence intensities on supported planar membranes has the advantage that relatively large areas (we are routinely observing about 4000 μm^2) can be measured, and therefore changes of 1–10 molecules/μm^2 can be detected with our present instrument. A lower limit for the quantitation of fluorophores in epi-illumination is given by the autoluminescence of the filters and the dichroic mirror, which are also excited by the laser beam.

For special purposes, a monochromator may be intercalated between the image plane diaphragm and the detector (52, 64). This configuration allows one to obtain spectral information on the adsorbed molecules.

In order to compare fluorescence intensities measured on different days (even over several months), it is important to carefully control the excitation light intensity. This can be done by measuring the excitation light intensity with a photodiode immediately before it enters the sample. With this procedure, one can correct for differences in the laser intensity, the performance of the AOM, and the alignment of the beam before the sample.

7.3.5. Electronic Control and Data Handling

A computer with dedicated software is an important component of a microspectrofluorometer. For many applications, it is sufficient to use a standard personal computer (PC). Such computers are fast enough to sample data points every 0.2 ms and, at the same time, to control the voltage of the AOM and the two electronic shutters, and to store thousands of data points during data acquisition. The correct steering of the shutters is important to avoid damage of the PMT during the bleach pulses and to avoid unnecessary additional photobleaching with the low-intensity observation beam when data are collected over long time courses (e.g., in TIRFM kinetics). To perform all these functions, we use a program written in ASYST (Asyst Technologies, Rochester, New York). Data analysis is easily performed with subroutines of the same software by using nonlinear least square fits, i.e., based on a

Marquart algorithm. Using all data points for analysis is preferable to procedures that use only 3 point fits, as has been customary in the early days of spot photobleaching (3). Small deviations of the actual data from the function used for fitting can be determined, and different fit functions can be compared with standard statistical methods (e.g., T-tests).

7.4. PREPARATION OF SUPPORTED PLANAR BILAYERS FOR MICROSPECTROFLUOROMETRY

Since the early days of the monolayer technique, Langmuir–Blodgett films have been used to coat various substrates by alternating immersion and withdrawal of the substrate through a monolayer of amphiphilic molecules at the air–water interface (65,66). These layers have often been dried after coating and sometimes in between different coating steps. Depending on the substance, one, a few, or many layers could be transferred onto the solid substrates by these techniques. Multilayer (more than two layers) deposition by the conventional vertical dipping technique has traditionally been difficult with phospholipids. Further, if such layered structures of phospholipids (and proteins) are to be used in biophysical studies as models for biological membranes, it is desirable to engineer fully hydrated and well-ordered lipid structures in which lipids and proteins can freely diffuse in the plane of the membrane. Biophysical studies on fully hydrated supported planar membranes have first been carried out on single phospholipid monolayers supported on silanized glass slides (67,68). These monolayers have been characterized by lateral diffusion measurements and have been used as target membranes for antibody-dependent macrophage attack (67,68).

Supported monolayers are limited in their use as models for biological membranes because (i) half of the membrane is missing, (ii) trans-membrane proteins cannot be incorporated, and (iii) they are coupled directly to the solid substrate without a gap between the membrane and the substrate. In response to the need for a better model system for biological membranes and a system that is suitable for microspectrofluorometry, the supported phospholipid bilayer has been developed. Three techniques are now available to prepare supported phospholipid bilayers: a modified Langmuir–Blodgett technique (69,70); the direct fusion of phospholipid vesicles to plain glass or quartz slides (71); and a combined Langmuir–Blodgett and vesicle fusion technique (56,72). The three techniques are briefly described in Sections 7.4.2. and 7.4.3. below.

7.4.1. Substrates

Supported planar bilayers are now being used for many different purposes. Some techniques require specific substrates, whereas others are less selective.

The only general requirement for supported phospholipid bilayer formation is a clean, hydrophilic solid surface. Useful substrates for microspectrofluorometry are glass, quartz, and oxidized silicon wafers. For special purposes where small penetration depths of the evanescent field are important, optically transparent, high refractive index materials such as sapphire, TiO_2, and Sn-Ti-oxide may be preferred. Oxidized germanium plates are useful for infrared studies (73), optically transparent semiconductor films such as SnO_2, Cd_2SnO_4, or In-Sn-oxide can be used to produce transparent electrodes (74), and mica is an adequate substrate for atomic force microscopy (75). In addition, nonoxidized Si wafers and Al/Al-oxide–coated quartz slides have been used to quench the fluorescence within a few nanometers of the surface/buffer interface and to discriminate between fluorophores that are very near and far, respectively, from the surface of the substrate (18, 76). SPBs have also been deposited on surfaces of SiO_2/TiO_2 with the ultimate aim of manufacturing biosensors (77) or on porous filter membranes (78).

For most purposes of microspectrofluorometry, quartz slides are the best substrates, because different techniques of illumination can be applied (epi-, TIR, and UV) and because their autoluminescence is very small. Adherence to careful pretreatment or cleaning procedures is important to obtain supported phospholipid bilayers of good and reproducible quality. Recent specular neutron reflection measurements on quartz have revealed an amplitude of the surface roughness of quartz of about 1.7 nm (79). This surface roughness seems to have negligible influence on the physical properties of the SPB.

Our routine cleaning procedure for quartz slides consists of 20-min boiling in the detergent Contrad 90 (Linbro 7x can give different results; see Ref. 56), a 30-min treatment in an ultrasonic bath, an extensive rinse with distilled water, treatment with methanol (or isopropylalcohol, p.a. quality), and drying at 120–150 °C. These slides can be stored for several days, and immediately before use they are cleaned in an argon plasma cleaner (Harrick, Ossining, New York). Plasma cleaning removes the last trace impurities (e.g., residual detergents) and may also change the surface charge on the quartz slide. Quartz slides cleaned by this procedure can be reused five to ten times.

7.4.2. Langmuir–Blodgett Films

Supported phospholipid bilayers (SPBs) are prepared with the Langmuir–Blodgett (LB) technique by a consecutive deposition of two monolayers on the substrate (69,70; see also Fig. 7.6). A first lipid monolayer is spread in a Langmuir trough and compressed to an appropriate surface pressure. Usually, a surface pressure of 32 mN/m is used, which corresponds approximately to the monolayer–bilayer equivalence pressure (80–82). A quartz slide is vertically

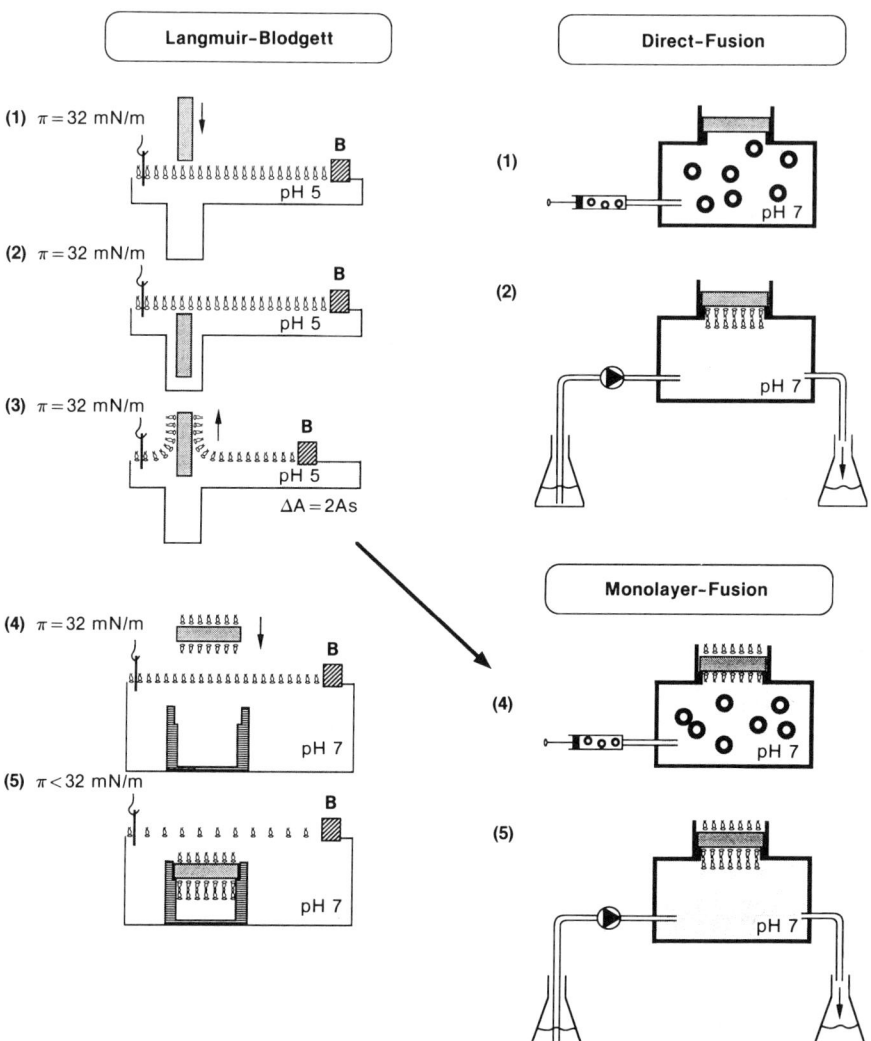

Fig. 7.6. Preparation of supported planar bilayers by the Langmuir–Blodgett, the direct fusion, and the monolayer fusion techniques. Details of each technique are described in the text.

immersed into the buffer (normally: 10 mM Tris–acetic acid, pH 5.0). No lipid is transferred onto the quartz during this step. The first monolayer is then transferred onto the quartz slide by slowly vertically withdrawing the slide from the trough. During this process, the surface pressure of the monolayer is kept constant by an electronic feedback mechanism. The second monolayer is then transferred horizontally by a slow apposition of the supported planar monolayer to the second monolayer, which is usually also compressed to 32 mN/m, but on a neutral pH buffer (e.g., 10 mM Tris/HCl, 0.15 M NaCl, pH 7.4). After the SPB is completed it must never be exposed to air because this results in an irreversible reordering of the lipids, with a complete and irreversible loss of their lateral mobility.

When a pH 5 buffer is used in the first trough, the monolayer is more tightly coupled to the substrate, which consistently results in high-quality SPBs (56). When instead ion-exchanged or glass-distilled water is used, SPBs with defects are produced quite frequently, whereas a pH 7.5 buffer in the first trough results most of the time in a loss of the lipid during the second monolayer deposition. The pH of the second monolayer trough is not critical. Advantages of the LB preparation method are that the surface pressure of the lipid can be adjusted to different desired values and that bilayers with an asymmetric transverse lipid distribution can be produced. Disadvantages are that the preparation is time consuming (1.5–2.0 h) and expensive equipment is needed (one or two monolayer troughs with film lifts, one with an electronic constant-pressure feedback circuit).

SPBs prepared by the LB method are stable for several days. The lateral diffusion coefficient of NBD-labeled lipids remains constant; the fluorescence intensity distribution is homogeneous and decreases only little after 2–3 days. Nevertheless, we normally prefer to perform all our measurements on the day of preparation.

7.4.3. Vesicle Fusion Techniques

7.4.3.1. Direct Fusion of Lipid Vesicles on a Hydrophilic Surface

The direct fusion of unilamellar phospholipid vesicles on glass or quartz slides is a very simple way to prepare supported planar bilayers (71; see also Fig. 7.6). A solution containing unilamellar phospholipid vesicles is brought into contact with a clean hydrophilic substrate. They fuse spontaneously on the hydrophilic surface and form an extended SPB. After a few minutes, the excess vesicles are washed away with buffer. Since it is simple and fast, this is the technique that is probably the currently most frequently used to prepare SPBs. SPBs can be formed with lipid concentrations as low as 50 μM (56). Somewhat higher lipid concentrations are needed if lipid mixtures containing negatively charged lipids are used. The formation of the first lipid bilayer is very fast and depends on the ionic strength of the buffer (56). Further vesicles adsorb (but do not fuse) on the SPB when vesicles composed of zwitterionic lipids are used for fusion. These vesicles are adsorbed reversibly, and they are removed by washing with buffer. The lateral mobilities of the lipids in SPBs prepared by the direct fusion technique are comparable to those in SPBs prepared by the LB technique.

7.4.3.2. Fusion of Vesicles on a Supported Monolayer (Monolayer Fusion Technique)

The monolayer fusion technique is a relatively new way to prepare SPBs. A short account of a rigid asymmetric bilayer of dipalmitoylphosphatidic acid/palmitoyloleoylphosphatidylcholine has been given by Fringeli (72). The technique was then generalized to fluid phospholipid bilayers, and the kinetics and conditions for preparing SPBs by the monolayer fusion technique have been described in detail in a recent study by Kalb et al. (56): A supported monolayer is prepared by the LB technique at 32 mN/m. To complete the bilayer, a suspension of unilamellar vesicles is brought into contact with the supported monolayer (Fig. 7.6). The vesicles fuse spontaneously on and with the supported monolayer. When the fusion process is completed, additional vesicles adsorb which can later be washed away with buffer. Surprisingly, the supported monolayer is stable in the presence of buffer, and a large fraction of the lipid is laterally mobile in the supported monolayer. The addition of the vesicles results in a fast adsorption of the vesicles to the monolayer and, at a certain critical lipid surface concentration (which corresponds to the equivalent of a second monolayer), the vesicles suddenly fuse and form the second leaflet of an extended supported bilayer. The fusion process has been followed and confirmed by FRAP in the TIR mode (56). Additional vesicles adsorb to the SPB only if they are composed of zwitterionic lipids. The adsorbed vesicles are bound reversibly and can be removed by washing procedures. Asymmetric bilayers can be prepared by the monolayer fusion technique, but not by the direct fusion method.

It is interesting and surprising that the adsorption kinetics of vesicles to hydrophilic quartz and to supported monolayers are nearly the same, although the surfaces are extremely hydrophilic in one case and hydrophobic in the other case. When negatively charged vesicles are used, the adsorption kinetics are much slower than with zwitterionic lipids. An increase of the ionic strength of the buffer increases the rate of adsorption of neutral vesicles. At NaCl concentrations smaller than 50 mM, only very slow and incomplete adsorption occurs (56).

7.5. BINDING OF MACROMOLECULES TO SUPPORTED PLANAR MEMBRANES

TIRFM has been used to examine and quantitate the binding of a variety of proteins to binding sites at quartz–buffer interfaces (14,64,83–85).

Suggestions to measure ligand binding to membrane receptors by TIRFM on supported planar membranes have been made in earlier reviews (9,10), but it was only in 1990 that this approach was first used in actual studies of protein binding to membranes (22,23). Here, we briefly review presently available experimental studies that use TIRFM to measure the binding of proteins to specific binding sites (receptors) on supported membranes. We also discuss some of the advantages and limitations of this relatively recent method.

Owing to the spatial extension of the evanescent field, the total fluorescence intensity measured in the microspectrofluorometer is the sum of the fluorescence of molecules that are bound to the membrane and the fluorescence of molecules that are present in the illuminated volume but not bound to the membrane. The ratio of the fluorescence of bound (F_b) to unbound (F_f) ligands is

$$F_b/F_f = c_b[\exp(-z/d_p)]/(c_f d) \qquad (7.53)$$

where d_p is the $1/e$ penetration depth of the evanescent field [Eq. (7.3)]; z is the mean distance of the fluorophores on the bound ligand from the interface; and c_b and c_f are the concentrations of the bound and free ligands, respectively. In this simple first-order approximation, the volume of the thin layer of the supported membrane is not excluded for free fluorescent ligands. Therefore, the ratios F_b/F_f as calculated from Eq. (7.53) are lower limiting estimates for the actual values. Some useful limiting concentrations as calculated by Eq. (7.53) are given in Table 7.1. In these calculations, z was taken to be 20 nm and d_p was 93 nm. It is evident from this table that the concentrations that give rise to equal fluorescence from bound and free ligands are shifted to lower

Table 7.1. Estimate of Useful Concentration Ranges of the TIRFM Method[a]

c_{ligand} (M)	$c_{receptor}$ (molecules·μm^{-2}): Area/Receptor (nm^2):	F_b/F_f		
		20,000 50	5,000 200	500 2,000
10^{-3}		0.27	0.068	0.0068
10^{-4}		2.7	0.68	0.068
10^{-5}		27.2	6.8	0.68
10^{-6}		272.	68.	6.8

[a] The TIRFM signal ratios of bound (F_b) and unbound (F_f) ligands present in the volume illuminated by the evanescent field are tabulated at various receptor densities. A limiting line (dashed) is drawn between ligand concentrations above which the fluorescence intensity arising from the free ligand gets larger than the intensity from the bound ligand. The F_b/F_f ratios are calculated for full saturation of the binding sites.

free ligand concentrations when the receptor surface concentrations are decreased. When the binding site concentration is 270 pmol/dm^{-2} (16,700 molecules/μm^2), which corresponds to tightly packed antibodies on the surface [i.e., 60 nm^2/molecule of immunoglobulin G (IgG)], binding constants as low as 10^4 M^{-1} can be determined (22). However, these theoretical considerations are often overruled in practice because unspecific binding can become the limiting factor.

7.5.1. Monoclonal Antibody–Lipid Hapten Binding

The binding of a monoclonal antibody to lipid haptens in a supported membrane has been used as a test system to critically evaluate TIRFM for measuring binding constants to membrane receptors (22). The binding of monoclonal antibodies to monovalent haptens in solution is a biochemically and biophysically well-defined process. However, although immunochemically very important, the binding of antibodies to membrane-bound target sites is theoretically much more complex and still not fully understood. Nevertheless, the binding reaction of the bivalent monoclonal antibody GK14-1 to the TNP–lipid hapten, which has a trinitrophenol group covalently linked to a phospholipid headgroup via a six methylene segment linker, has been studied in considerable detail and at many different surface concentrations of the lipid hapten in vesicles by a fluorescence resonance energy transfer technique (86). Therefore, this system could serve as an almost ideal reference system for establishing and evaluating the usefulness of TIRFM for quantitative binding studies of membrane receptor–ligand interactions.

Figure 7.7 shows the binding of the monoclonal antibody GK14-1 to SPBs at three different surface concentrations of the TNP–lipid hapten. With increasing antibody concentration, the fluorescence intensities on the surface, as measured with TIRFM, increase up to different saturating values. These saturating values increase when the lipid hapten concentration is increased from 0.2 to 1 mol% and from 1 to 5 mol%. Assuming an occupied area of 60 nm^2 per antibody on the membrane (and a lipid cross-sectional area of 0.7 nm^2), one would expect the entire surface to be covered with antibody at and above 2.3 mol% lipid hapten. In agreement with these considerations, the experimentally determined saturating fluorescence intensities increase by a factor of about 5 when the lipid hapten concentration increases from 0.2 to 1 mol%, but only by a factor of 2–3 from 1 to 5 mol% lipid hapten.

The numerical values of the binding constants obtained with the TIRFM method are in good agreement with those obtained with the resonance energy transfer assay in solution, although light scattering problems limit the application of the latter technique to high lipid hapten concentrations (22). Also, the kinetic data obtained with both techniques are in good agreement.

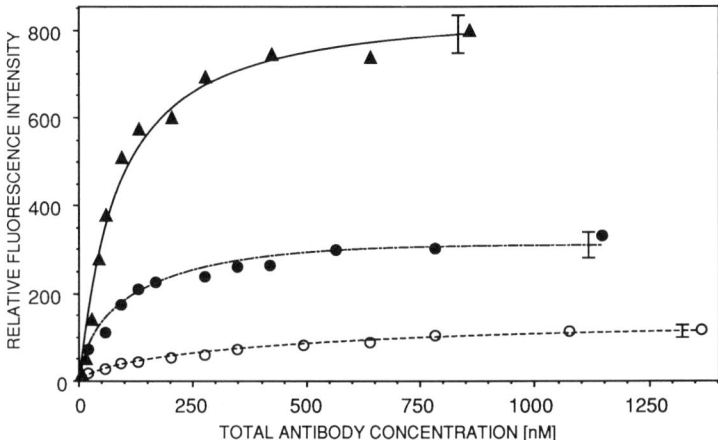

Fig. 7.7. Background-corrected binding isotherms of fluorescein isothiocyanate(FITC)–labeled antibody to 0.2 (○), 1.0 (●), or 5.0 mol% (▲) lipid hapten in supported planar bilayers. The error bars indicate the uncertainty ($\pm 5\%$) in the fluorescence intensity at saturation, estimated from repeated experiments with 1 mol% hapten. The lines represent the best fits of the data to simple Langmuir adsorption isotherms. [From Kalb et al. (22), with permission.]

The overall (apparent) binding constant for the binding of the bivalent antibody GK14-1 to the membrane-bound hapten was $1.5 \times 10^7 \, \text{M}^{-1}$. This value is much lower than the expected value, if the measured binding energies for *two soluble* haptens were additive, that is $(5 \times 10^7)^2 \, \text{M}^{-1} = 2.5 \times 10^{15} \, \text{M}^{-1}$. We do not believe that the relatively low binding constant on membranes is primarily due to monovalent antibody binding to the lipid haptens, because experiments with F_{ab} fragments resulted in still lower binding constants (L. K. Tamm, unpublished results). Rather, a reduced accessibility of the hapten on the membrane surface and unfavorable entropy effects (orientation) could account for the observed discrepancy.

The binding of another monoclonal antibody, ANO2, to supported membranes that contained a nitroxide spin label hapten has also been studied by TIRFM (21). The affinity of ANO2 for the spin label lipid hapten is much lower than that of GK14-1 for the TNP–lipid hapten, and therefore specific binding to supported lipid monolayers was only observable at binding site concentrations greater than 5 mol% (87). The requirement of this antibody for such high binding site concentrations in the membrane, i.e., a site density that is larger than the highest possible antibody packing density, is bound to cause serious problems when the data are evaluated by simple Langmuir adsorption isotherms (cf. Ref. 86). However, when compared on a qualitative, empirical basis, the apparent binding constant for bivalent antibody–ligand binding in solution ($K = 2.7 \times 10^6 \text{ M}^{-1}$) is similar to the one for the binding of the bivalent antibody to membrane-bound target sites ($K = 3 \times 10^6 \text{ M}^{-1}$), as it has also been found for the higher affinity antibody GK14-1. Part of this effect is due to bivalency of the antibody, because the binding of F_{ab} fragments to the lipid hapten is decreased by an order of magnitude or more. Further efforts have been made by the same investigators to measure the relative fractions of mono- and bivalently bound antibodies by competitive binding studies in which membrane-bound bivalent antibodies were displaced with soluble hapten. These data suggest that 50–90% of the antibody is bivalently bound to the membrane, although a clear decision could not be made. The antibody surface concentration at saturation has been estimated to be 600 molecules/μm^2 by taking the ratio of the fluorescence intensities that were obtained in the presence and absence, respectively, of 25 mol% lipid hapten in the supported membranes. This value translates into one antibody per 170 nm^2, which is roughly in agreement with the loose packing of antibodies on monolayer-coated EM grids (88) but much larger than the actual cross-sectional area (about 60 nm^2) of IgGs as observed by electron microscopy (89).

7.5.2. Antibody–F_c Receptor Binding

The first TIRFM binding study in which the membrane-binding sites were integral membrane proteins was reported by Poglitsch and Thompson in 1990 (23). The supported membrane was prepared by directly fusing vesicles (formed from plasma membrane fragments of a macrophage-related cell line) to hydrophilic quartz slides. Among other proteins, these membranes contained an $F_{c\gamma}$ receptor, specific for binding the F_c region of γ-immunoglobulins. The receptor concentration in these membranes was measured to be about 50 molecules/μm^2, which was too low for direct binding studies of the F_c part of the antibodies but adequate for studying the binding of a monoclonal anti-F_c-receptor antibody. However, F_c–$F_{c\gamma}$ receptor interactions could be

---

Fig. 7.8. Aggregation of the laminin/nidogen complex on the surface of lipid bilayers. Different amounts of fluorescein-labeled laminin were bound to supported bilayers that contained 100% sulfatide in the outer leaflet. These membranes were imaged under total internal reflection illumination. A grainy appearance of the fluorescence indicates the aggregation of the protein on the surface. Larger aggregates are visualized as bright spots. Their average size increases with increasing protein concentration. Here A and B are typical fields at 10 and 40 nM laminin, respectively; in C, 40-nM laminin was bound and subsequently treated with a 2-M excess of EDTA over Ca^{2+}. The brightness of the photographs cannot be taken as a measure of the absolute fluorescence intensity, which was much lower in C than in B. The bar corresponds to 10 μm. [From Kalb and Engel (92), with permission.]

BINDING OF MACROMOLECULES TO MEMBRANES 291

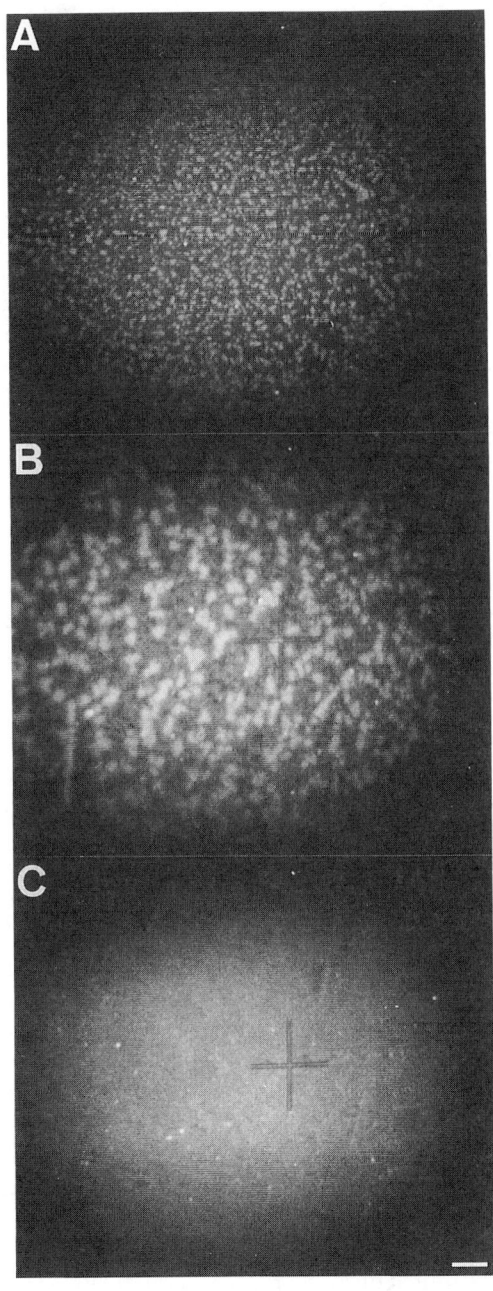

measured with an indirect method in which the inhibition of the binding of the fluorescently labeled monoclonal anti-F_c-receptor antibody by the F_c-domain-containing unlabeled antibodies was followed by TIRFM. A binding constant of about 10^5 M^{-1} was obtained.

Following up on this work, Poglitsch et al. (90) then purified the $F_{c\gamma}$ receptor and reconstituted it into liposomes and finally into SPBs. By this method, they were able to increase the surface concentration of receptors to about 800–1700 molecules/μm^2 and to directly measure the binding of labeled IgGs by TIRFM. Binding constants of 1×10^5 and 9×10^5 M^{-1} were determined, depending on whether polyclonal mouse IgG or the monoclonal ANO2 antibody was used. The binding constant of the monoclonal anti-F_c-receptor antibody was more than 10-fold smaller when measured with purified reconstituted F_c receptors instead of membranes prepared from native membrane fragments. The reasons for this difference are not yet well understood.

7.5.3. Laminin–Sulfatide Binding

Based on conventional solid phase radioimmunoassays, it has been concluded that laminin, a large extracellular matrix glycoprotein, binds specifically to sulfatide lipids (e.g., 3-SO$_4$-galactosylceramide) with a binding constant of the order of 10^9 M^{-1} (91). The undefined lipid deposition in plastic wells that is customary in solid phase assays is rather unsatisfactory for quantitative studies of laminin–sulfatide binding. However, TIRFM on SPBs that contain the sulfatide at specified mole fractions allows characterization of these interactions in greater detail and in a well-defined membrane environment. When fluorescein-labeled laminin was studied in titration experiments with sulfatide-containing SPBs, no saturation could be detected (92). Also, relatively high fluorescence intensities were observed when laminin was bound in the absence of BSA to lipid bilayers that contained no sulfatides. The binding levels were similar for all control lipid systems tested, and even for membranes containing up to 30 mol% sulfatides. In the presence of BSA, no binding was detectable to membranes containing low amounts of sulfatides. The kinetics of the binding and the dissociation of bound laminin after the addition of EDTA (ethylenediaminetetraacetic acid) were comparable to the kinetics of the calcium-dependent self-assembly of laminin in solution (93). When the membrane surface was observed directly by TIR microscopy, the bound laminin displayed an inhomogeneous fluorescence distribution (Figs. 7.8A, B) (92). Distinct areas of high fluorescence intensity could be distinguished from the background fluorescence, and the size of these areas depended on the lipid composition and the laminin surface concentration. Upon the addition of EDTA, these spots almost disappeared and a relatively homogeneous fluorescence was observed (Fig. 7.8C). These observations led to the conclusion that the binding of laminin to lipid membranes is closely linked with the (membrane-independent) laminin self-association and therefore that the binding constant of laminin cannot be correctly determined by either the solid phase or the TIRFM assays. However, this example further demonstrates that TIRFM is a technique that can reveal various aspects of very complex interactions on membrane surfaces, including equilibrium binding, binding kinetics, and surface-induced self-aggregation.

7.6. TRANSLATIONAL AND ROTATIONAL DIFFUSION OF LIPIDS AND MACROMOLECULES IN AND ON SUPPORTED PLANAR MEMBRANES*

7.6.1. Lipids

The first lateral diffusion measurements of lipids in supported planar membranes have been carried out with single supported monolayers of DPPC on alkylated glass slides (67). The DPPC monolayers were doped with 1 mol% NBD-DMPE and transferred to the hydrophobic support at different surface pressures. A sharp transition of the lateral diffusion coefficient from $\sim 10^{-10}$ to $\sim 10^{-8}\,\mathrm{cm^2/s}$ was observed between the solid-condensed ($\pi > 20\,\mathrm{mN/m}$) and the liquid-condensed state of the monolayer. No significant changes in D_L occurred in the liquid-condensed/liquid-expanded transition region and in the liquid-expanded state of the supported monolayer.

Lateral diffusion coefficients were measured as a function of temperature in supported bilayers of DPPC, DMPC, and DOPC (70) that contained 0.5 mol% NBD-eggPE and were prepared by two successive Langmuir–Blodgett dipping steps. Two sharp phase transitions were found in supported DPPC bilayers at 32 and 39 °C, respectively. These transitions are 2–3 °C lower than the calorimetrically observed pretransition and chain melting phase transition

---

*The following abbreviations are used for the phospholipids mentioned in Sections 7.6 and 7.7: DMPC, 1,2-dimyristoyl-3-sn-phophatidylcholine; DNP-DOPE, N-(3,5-dinitrophenyl)-1,2-dioleoyl-3-sn-phosphatidylethanolamine; DOPC, 1,2-dioleoyl-3-sn-phosphatidylcholine; DPPC, 1,2-dipalmitoyl-3-sn-phosphatidylcholine; DSPC 1,2-distearoyl-3-sn-phosphatidylcholine; NBD-(C_6)-PC, 1-acyl-2-[6-(7-nitro-2,1,3-benzoxadiazol-4-yl)amino]caproyl-3-sn-phosphatidylcholine; NBD(C_{12})-PC 1-acyl-2-[12-(7-nitro-2,1,3-benzoxadiazol-4-yl)amino]lauroyl-3-sn-phosphatidylcholine; NBD-DMPE, N-(7-nitro-2,1,3-benzoxadiazol-4-yl)-1,2-dimyristoyl-3-sn-phosphatidylethanolamine; NBD-DPPE, N-(7-nitro-2,1,3-benzoxadiazol-4-yl)-1,2-dipalmitoyl-3-sn-phosphatidylethanolamine; NBD-eggPE, egg N-(7-nitro-2,1,3-benzoxadiazol-4-yl)-phosphatidylethanolamine; POPC, 1-palmitoyl-2-oleoyl-3-sn-phosphatidylcholine; POPG, 1-palmitoyl-2-oleoyl-3-sn-phosphatidylglycerol; TNP-cap-eggPE, egg N-[[(2,4,6-trinitrophenyl)amino]caproyl]-phosphatidylethanolamine.

in large multilayered liposomes (94). The lateral diffusion coefficients were 10^{-8} to 10^{-7} cm^2/s (depending on the temperature and the lipids used) in the liquid-crystalline phase, about 10^{-10} cm^2/s in the intermediate phase, and 10^{-12} to 10^{-10} cm^2/s in the low-temperature phase. Whether the small shift of the transition temperatures is due to the absence of inter-bilayer coupling or to interactions with the solid support is not known. A significant result of these lateral diffusion studies is that the lipids in both leaflets of the bilayer are mobile and exhibit very similar lateral diffusion coefficients. Since single exponential recoveries with mobile fractions of 80–100% were always observed, it was concluded that a *thin water-filled space exists between the bilayer and its hydrophilic support*. The average distance between the substrate and the headgroups of the leaflet of the supported phospholipid bilayer that faces the substrate was recently determined by neutron scattering experiments to be 30 ± 10 Å (79). Lateral lipid diffusion coefficients $(3.7 \pm 0.15) \times 10^{-8}$ cm^2/s with mobile fractions of 80–100% were also observed when supported bilayers of DMPC with 1 mol% NBD-DMPE were prepared by the direct vesicle fusion technique (71). The diffusion coefficient was not significantly changed when purified H-2K^k, an integral membrane protein and class I histocompatibility antigen, was included in the vesicles and reconstituted into the supported bilayers. Similar results were found when a class II histocompatibility antigen (95) or an F_c receptor (90) was incorporated into supported bilayers by the direct fusion technique. In sum, even though small differences in lateral diffusion are found between single supported bilayers and multibilayers, these data support the notion that the physical properties of the lipids in supported bilayers are (almost) the same as in other common model membrane systems. This conclusion is further supported by recent independent measurements of the lipid order parameters in supported bilayers by polarized attenuated total reflection infrared spectroscopy (73).

Supported bilayers with an asymmetric phospholipid distribution between the two leaflets can be prepared with the Langmuir–Blodgett or the monolayer fusion technique. The same lateral diffusion coefficients were measured in DMPC bilayers (3.7×10^{-8} cm^2/s; 30 °C), independent of whether the proximal, distal, or both leaflets of the bilayers were labeled with NBD-eggPE (96). Also, a single diffusion coefficient (3.8×10^{-8} cm^2/s; 22 °C) was measured in both leaflets of POPC bilayers that were formed by vesicle fusion to supported monolayers (56). These results prove that the two leaflets of supported phospholipid bilayers are equivalent as far as lateral diffusion is concerned. It was further shown that a sharp chain melting phase transition occurred in DMPC bilayers at 19 °C (96), i.e., again about 4 °C below the calorimetrically determined transition of the same lipid in multibilayers (see Fig. 7.9). The measured lateral diffusion coefficients at a given temperature in the liquid-crystalline phase depend on the chemical nature of both the host lipid and the

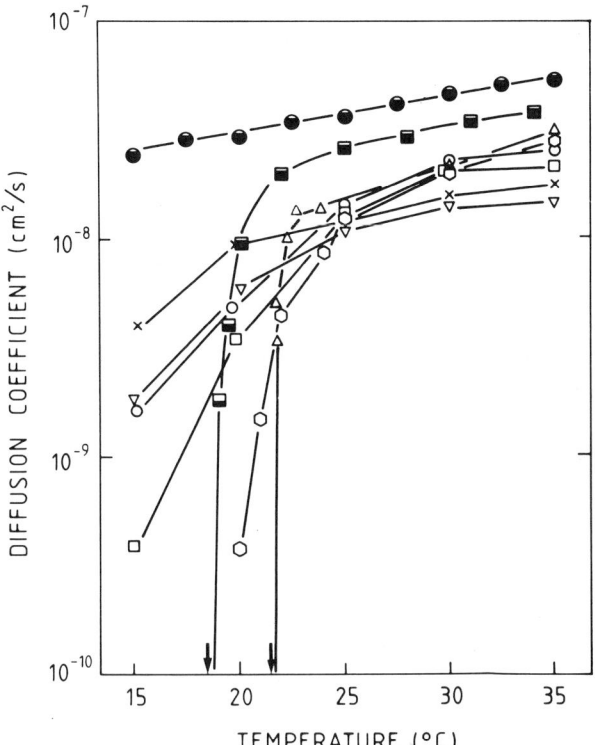

Fig. 7.9. Chain melting phase transitions in supported phospholipid bilayers as detected by the lateral diffusion of fluorescent lipid probes (closed and semiclosed symbols) and of fluorescein-labeled monoclonal anti-TNP antibodies (open symbols). *Key:* (●) POPC plus 2 mol% NBD–eggPE (in both leaflets); (■) DMPC plus 2 mol% NBD–DMPE (in the leaflet facing the large aqueous compartment; the black level in the symbol represents percent mobile fraction of the probe); DMPC plus (△) 2, (◯) 5, (□) 10, (○) 25, (▽) 50, and (×) 100 mol% TNP–cap–eggPE (in the leaflet facing the large aqueous compartment) incubated with 20 μg/mL fluorescent antibody for 5 min. The supported bilayers of DMPC show a sharp phase transition at around 19 °C, and the POPC bilayers are fluid (liquid-crystalline) in the whole displayed temperature range. The temperature-induced phase transition broadens with increasing amounts of lipid hapten, and the transition temperature is shifted by about 3 °C to the higher temperatures in the presence of bound antibodies. [From Tamm (96), with permission.]

fluorescent probe lipid, i.e., they decrease in the order DOPC > POPC ≈ POPC:POPG (4:1) > DMPC, for the host lipids, and NBD-eggPE > NBD-DMPE > NBD(C_6)-PC, for the probe lipids (56,70,96). The activation energies for lateral diffusion of NBD-eggPE in liquid-crystalline bilayers of DMPC, DOPC, and POPC are 43.7, 40.9, and 37.3 kJ/mol (70,96), respectively.

The lateral diffusion coefficient of NBD-eggPE in DMPC at 30 °C is about 40% smaller in single planar bilayers than in stacked multibilayers (96).

Rotational diffusion of dioctadecyl–carbocyanine (diI) was recently measured in supported monolayers of DSPC on alkylated quartz slides by polarized fluorescence recovery after photobleaching (42). At room temperature, these monolayers are in a gel state and rotational diffusion is very slow. In fact, two rotational diffusion coefficients, $3.1 \times 10^{-3}\,\text{s}^{-1}$ and $46 \times 10^{-3}\,\text{s}^{-1}$, were found that accounted for 25% and 38%, respectively, of the total dye population in the membrane. The remaining 37% did not rotate at all in the time frame of the experiment (about 3 min). The higher rotational diffusion coefficient is very similar to the one found in multibilayer samples of the same lipid/dye system under the same temperature conditions (24). The rotational diffusion coefficients were not dramatically changed by inclusion of 20 mol% of the lipid hapten DNP-DOPE (42) or various mole fractions up to 20% cholesterol (24).

7.6.2. Antibodies

Antibodies bound to lipid haptens that were included in the membrane were the first proteins whose lateral and rotational diffusion coefficients were measured on supported membranes. The first experiments were again performed on supported monolayers on alkylated glass slides or silicon wafers (30,68,97). When a polyclonal anti-nitroxide IgG was bound to a spin-labeled lipid hapten in supported monolayers of DMPC or DPPC, lateral diffusion coefficients of $3.5 \times 10^{-9}\,\text{cm}^2/\text{s}$ or $< 10^{-11}\,\text{cm}^2/\text{s}$, respectively, were reported at 37 °C (68). Similar results were found for a monoclonal anti-dinitrophenyl (DNP) IgE on supported monolayers (30). A systematic study on the dependency of the lateral diffusion coefficient on the length of the covalently linked alkyl chains on the solid support and the length of the fatty acyl chains of the lipids in the supported monolayer revealed three diffusion regimes of the monoclonal anti-nitroxide IgG (ANO2) in these systems (97): Relatively high lateral diffusion coefficients ($D_L = 5$–$10 \times 10^{-9}\,\text{cm}^2/\text{s}$; $m.f. = 80$–90%) were found on DMPC and DPPC monolayers supported on decylsilane(C_{10})-derivatized silicon wafers at high and low antibody concentrations; similar lateral diffusion coefficients (3–$7 \times 10^{-9}\,\text{cm}^2/\text{s}$), but lower mobile fractions ($\sim 60\%$) were measured on DMPC and DPPC monolayers on hexadecylsilane (C_{16})- or octadecylsilane(C_{18})-derivatized substrates at high temperatures ($T > 30$–40 °C) and essentially no lateral mobility ($D_L < 10^{-11}\,\text{cm}^2/\text{s}$; $m.f. = 0$–10%) could be detected with antibodies bound to the latter monolayer-substrate combinations but measured at lower temperatures (25 °C $< T <$ a critical temperature between 30 and 40 °C that depended on the particular silane–lipid combination).

Antibody diffusion was also measured on supported bilayers (96,98). In one study, the monoclonal anti-trinitrophenyl IgG, GK14-1 which has a high affinity for the lipid hapten TNP-cap-eggPE in membranes (see Section 7.5.1) was used (96). When bound to DMPC bilayers that contained 2 or 5 mol% lipid hapten, antibody diffusion was fast (2×10^{-8} cm^2/s; 30 °C), i.e., about 50% of the lateral diffusion coefficient of the lipids measured in the same system. This result confirmed the earlier observation that lipid-bound antibodies diffuse nearly as fast as the lipids themselves, an observation made in multilayered lipid systems (99). The twofold reduction of the antibody lateral diffusion coefficient relative to the lipid diffusion coefficient is further consistent with the idea that (a) each antibody crosslinks two lipids which because of the flexibility of the antibody can still undergo relatively independent Brownian motion, (b) the friction of the two lipid haptens in the membrane is additive, and (c) the friction of the antibody with the solvent (water) is negligible compared to the friction of the lipid haptens in the bilayer. When the surface concentration of the antibody was increased to a fractional coverage of the whole surface of about 80% (by either increasing the antibody concentration in solution or the lipid hapten concentration in the membrane), the lateral diffusion coefficient decreased by a factor of 2–3 relative to the value measured at a very small surface concentration of the antibody. Single exponential recoveries and high mobile fractions were always found under these conditions. Apparently, protein–protein interactions are not strong enough in this system to cause large-scale aggregation or to crystallize (cf. Ref. 88) this antibody on supported bilayers. However, when the temperature was lowered below the chain melting phase transition temperature of DMPC, all of the antibody became immobilized on the supported membrane ($D_L < 10^{-10}$ cm^2/s; see Fig. 7.9) and extended areas (several micrometers in diameter) of aggregated antibodies were visualized by epi-fluorescence microscopy. This process was reversible. The phase transition temperature of the bilayer in the presence of the antibody and measured by the lateral diffusion of the antibody occurred at 22 °C, i.e., about 3 °C higher than in the same membrane without the antibody. This small shift of the transition temperature may be caused by weak protein–protein interactions on the membrane surface and is consistent with the concentration-dependent lateral diffusion measurements described above.

The ANO2 antibody has a lower affinity for the spin-label or DNP–lipid haptens in membranes than does GK14-1 for TNP–lipid haptens (see Section 7.5.1). Membranes composed of 68 mol% DMPC and 32 mol% of the spin-label lipid hapten were used for lateral diffusion measurements of ANO2 on supported bilayers (98). Double-exponential recoveries were observed in this system, which indicated the coexistence of a fast and a slow diffusing antibody population. The fast component exhibited a lateral diffusion coefficient of 3.5×10^{-9} cm^2/s at all employed antibody concentrations and the slow component decreased from 1.5×10^{-9} to 2.5×10^{-10} cm^2/s when the antibody concentration was increased from 10 to 100 µg/mL. The lateral diffusion coefficient of the lipid was equal to that of the fast antibody component (3.7×10^{-9} cm^2/s) and was always single-component. Also, the relative proportions of the fast and slow components changed from about 1:0 to 0.3:0.7 as the antibody concentration was increased. The presence of the slow component indicated that the antibodies aggregate into oligomers on the membrane surface in a concentration-dependent fashion in this system. An alternative explanation for this observation would be that the DMPC:spin-label lipid hapten (68:32) mixture gradually phase-separates as more antibody binds to the lipid haptens.

Recently, the rotational diffusion coefficient of ANO2 was measured on supported DSPC monolayers that contained 30 mol% of the lipid hapten DNP-DOPE (42). It was important in these experiments to label the antibody with a fluorescent dye (CY-B) that could be simultaneously attached to the protein with two functional groups and to show that the wobbling of the dye itself did not contribute to the polarized fluorescence recoveries. A rotational diffusion coefficient of $\sim 4 \times 10^{-3}$ s^{-1} was measured (estimated mobile fraction, 27%), which was very close to the smaller of the two rotational diffusion coefficients of the lipid probe diI in the same membrane (see Section 7.6.1). The higher rotational diffusion coefficient, which was found for diI rotation, may also have been present for some of the monolayer-bound ANO2 but would not have been detected in these experiments because bleach pulses of relatively long durations had to be used. Therefore, the rotation of the bivalent antibody on gel phase lipid monolayers is determined most likely and to a large extent by the rotation of the lipid haptens, as is the lateral diffusion of antibodies on phopholipid bilayers in the liquid-crystalline phase. When the monolayer-bound ANO2 was incubated with a rabbit anti-mouse IgG antibody, the mobile fraction for rotational diffusion decreased to about 6%, indicating that the secondary antibody quantitatively cross-linked ANO2 on the surface of these membranes.

7.6.3. Integral Membrane Proteins

Integral membrane proteins were usually incorporated into supported membranes by the direct fusion technique. Although the proteins (receptors) that have been tested in this respect all retained significant specific binding capabilities for their natural ligands, they were unfortunately not laterally mobile. These proteins include purified class I and class II histocompatibility antigens from lymphocytes and F_c receptors from macrophages (71,90,95).

Recently, we have incorporated cytochrome b_5 into supported planar bilayers with the monolayer and the direct fusion techniques (100). About 35%

of all reconstituted protein was mobile but exhibited a reduced lateral diffusion coefficient of 2.6×10^{-10} cm^2/s. The lipid always diffused at normal rates (4×10^{-8} cm^2/s) in these systems. Cytochrome b_5 is a nontypical integral membrane protein because it is inserted into the membrane with a hydrophobic α-helix at its C-terminal end, and the orientation and topology of this helix in the membrane are still controversial (101,102). Therefore, the nature of the interactions that cause (a) the observed reduction of the lateral diffusion coefficient of the mobile fraction of cytochrome b_5 and (b) the immobilization of about 65% of cytochrome b_5 is unknown.

7.7. MEASUREMENTS OF ORIENTATION DISTRIBUTION

Polarization fluorescence microscopy using epi-illumination or TIR illumination has been employed to determine orientation distributions of fluorophores in membranes. In one study the orientation of the chromophore of dioctadecylcarbocyanine (diI) in erythrocyte ghosts was examined (40). It was found that the thermodynamically most plausible model that is consistent with the data was an average orientation of the conjugated bridge chromophore parallel to the surface of the cell membrane.

Orientation distributions of headgroup-labeled NBD-DPPE, chain-labeled NBD(C_{12})-PE, and a chain-labeled lipid–peptide were measured by polarized TIR fluorescence in monolayers of DPPC on alkylated microscope slides as a function of the monolayer transfer pressure (61). The order was high at all pressures for chain-labeled lipids with and without the conjugated peptide and decreased monotonically for the headgroup-labeled peptide when the coating pressure was increased from 10 to 40 mN/m. The same experiment performed on diI in monolayers composed of DSPC:DNP-DOPE (7:3) and supported on alkylated microscope slides revealed an order parameter of -0.3, which again suggested a preferential orientation distribution [weighted as defined by Eq. (7.33)] of the absorption dipole moments near the plane of the membrane (42). This order parameter did not change when antibodies were bound to the surface of the membranes.

7.8. RESONANCE ENERGY TRANSFER MEASUREMENTS

Although resonance energy transfer has been used in several fluorescence microscopic investigations on lipid and protein distributions in single cells (50,51,54,103,104), there are only two reports in which resonance energy transfer microscopy has been used to study molecular interactions in supported membranes (52,53). In these experiments, supported bilayers were prepared which contained the purified and Texas red–labeled class II histocompatibility antigen, I-A^d. A fluorescein-labeled peptide was added and excited with an evanescent wave at 488 nm. No Texas red fluorescence emission was observed under these conditions. Significant resonance energy transfer and Texas red emission occurred only when peptide-specific and I-A^d-restricted T-lymphocytes were specifically bound to the supported planar membrane. This effect could be reversed with an excess of unlabeled peptide, with an antibody against I-A^d, or with irrelevant helper/inducer T-lymphocytes. These experiments demonstrated that a specific ternary complex of histocompatibility antigen, peptide antigen, and T-cell receptor had been formed at the interface between the cell and the supported membrane and that a stable association of the peptide antigen with the histocompatibility antigen only occurred in the presence of the proper T-cell receptor.

7.9. CONCLUSION AND FUTURE DEVELOPMENTS

Microspectrofluorometry on supported planar membranes is still a very young discipline of membrane biophysics and biophysical chemistry. However, it has grown out of its infancy and has developed to a stage where many new exciting and long-awaited experiments on ligand–receptor, protein–lipid and cell surface–planar membrane interactions can be devised and successfully carried out. Furthermore, microspectrofluorometry on supported membranes can be readily used as an analytical tool and for quality control when supported membranes are prepared for other purposes, such as for scanning probe microscopy or the development of biosensors. Total internal reflection fluorometry on supported membranes is superior to more conventional methods of measuring the binding of large protein ligands to membrane receptors in several respects: (i) The membrane receptor and the membrane itself are much better defined and characterized as described here for the TIRFM method than in solid phase assays, such as radioimmunoassays (RIA) or enzyme-linked immunosorbant assays (ELISA). In these latter assays, solubilized receptors are usually bound to plastic or to ligand that themselves are bound to plastic in an ill-defined conformational state and state of aggregation. TIRFM on supported membranes allows the experimenter to control the states of the receptors and ligands much more carefully. (ii) TIRFM on supported membranes further permits access to kinetic studies of ligand–receptor interactions in real time. This is not possible with many other techniques that are currently used to study such interactions. (iii) Finally, compared to luminescence techniques on membranes in solution, TIRFM is relatively free of light-scattering artifacts. This feature extends the range of binding constants that are accessible by this method to values that are about 2 orders of magnitude lower than those that can be measured in solution.

One of the most urgent immediate goals is to improve the currently available methods for the functional reconstitution of integral membrane proteins and receptors into supported bilayers. Although the direct fusion methods can produce supported membranes with receptors that maintain their specific binding functions toward their ligands, it has not yet been possible to reconstitute these proteins into planar membranes in a laterally mobile form. However, the lateral mobility of receptors in cell membranes is believed to be a key element of many signal transduction pathways. Success in the functional reconstitution of integral membrane protein receptors in a laterally mobile form will have at least two immediate benefits for biophysical studies on these proteins. First, it is likely that the fraction of receptors which are in an active conformation, which are properly oriented, and therefore which are able to bind ligand will be significantly increased. This experimental improvement and a precise knowledge of the surface concentration of *active* receptors in the membrane is important in rigorous calculation of the relevant kinetic and equilibrium binding constants. Second, new insights in the signal transduction pathways of the investigated ligand–receptor pairs may be gained from microfluorescence studies on supported membranes. In addition to only observing the binding reaction per se, the fate of the receptors may be followed up one step further: it should be possible to characterize oligomerization reactions and lateral redistributions of the receptors upon stimulation by specific ligands, inhibitors, and other cofactors that might control the biological activity of these receptors.

REFERENCES

1. L. Frye and M. Edidin, *J. Cell Sci.*, **7**, 319–355 (1970).
2. R. Peters, J. Peters, K. Tews, and W. Bähr, *Biochim. Biophys. Acta*, **367**, 282–294 (1974).
3. D. Axelrod, D. E. Koppel, J. Schlessinger, E. Elson, and W. W. Webb, *Biophys. J.*, **16**, 1055–1069 (1976).
4. M. Edidin, Y. Zagyansky, and T. J. Lardner, *Science*, **191**, 466–468 (1976).
5. J. Jacobson, Z. Derzko, E.-S. Wu, Y. Hou, and G. Poste, *J. Supramol. Struct.* **5**, 565–576 (1976).
6. E. L. Elson and D. Magde, *Biopolymers*, **13**, 1–27 (1974).
7. D. Magde, E. L. Elson, and W. W. Webb, *Biopolymers*, **13**, 29–61 (1974).
8. P. F. Fahey, D. E. Koppel, L. S. Barak, D. E. Wolf, E. L. Elson, and W. W. Webb, *Science*, **195**, 305–306 (1977).

9. H. M. McConnell, T. H. Watts, R. M. Weis, and A. A. Brian, *Biochim. Biophys. Acta*, **865**, 95–106 (1986).
10. N. L. Thompson, A. G. Palmer, III, L. L. Wright, and P. E. Scarborough, *Comments Mol. Cell Biophys.*, **5**,109–131 (1988).
11. T. Hirschfeld, *Can. Spectrosc.*, **10**, 128 (1965).
12. N. L. Thompson, T. P. Burghardt, and D. Axelrod, *Biophys. J.*, **33**, 433–454 (1981).
13. T. P. Burghardt and D. Axelrod, *Biophys. J.*, **33**, 455–467 (1981).
14. D. Axelrod, T. P. Burghardt, and N. L. Thompson, *Annu. Rev. Biophys. Bioeng.* **13**, 247–268 (1984).
15. N. J. Harrick, *Internal Reflection Spectroscopy*, Harrick Scientific Corp., New York, 1967.
16. N. J. Harrick, *J. Opt. Soc. Am.*, **55**, 851–857 (1965).
17. N. J. Harrick and F. K. du Pré, *Appl. Opt.*, **5**, 1739–1743 (1966).
18. E. H. Hellen and D. Axelrod, *J. Opt. Soc. Am.*, **4**, 337–350 (1987).
19. D. Axelrod and E. H. Hellen, *Methods Cell Biol.*, **30**, 339–416 (1989).
20. R. M. Zimmermann, C. F. Schmidt, and H. E. Gaub, *J. Colloid Interface Sci.*, **139**, 268–280 (1990).
21. M. L. Pisarchick and N. L. Thompson, *Biophys. J.*, **58**, 1235-1249 (1990).
22. E. Kalb, J. Engel, and L. K. Tamm, *Biochemistry*, **29**, 607–1613 (1990).
23. C. L. Poglitsch and N. L. Thompson, *Biochemistry*, **29**, 248–254 (1990)
24. L. M. Smith, R. M. Weis, and H. M. McConnell, *Biophys. J.*, **36**, 73–91 (1981).
25. W. A. Wegener and R. Rigler, *Biophys. J.*, **46**, 787–794 (1984).
26. D. E. Koppel, *Biochem. Soc. Trans.*, **14**, 842–845 (1986).
27. M. Velez and D. Axelrod, *Biophys. J.*, **53**, 575–591 (1988).
28. B. A. Smith and H. M. McConnell, *Proc. Natl. Acad. Sci. U.S.A.*, **75**, 2759–2763 (1978).
29. J. Davoust, P. F. Devaux, and L. Leger, *EMBO J.*, **1**, 1233–1238 (1982).
30. R. Weis, K. Balakrishnan, B. A. Smith, and H. M. McConnell, *J. Biol. Chem.*, **257**, 6440–6445 (1982).
31. D. Magde, E. L. Elson, and W.W. Webb, *Phys. Rev. Lett.*, **29**, 705–708 (1972).
32. M. Weissman, H. Schindler, and G. Feher, *Proc. Natl. Acad. Sci. U.S.A.*, **73**, 2776–2780 (1976).
33. T. Meyer and H. Schindler, *Biophys. J.*, **54**, 983–993 (1988).
34. N. O. Petersen, *Biophys. J.*, **49**, 809–815 (1986).
35. N. O. Petersen, D. C. Johnson, and M. J. Schlesinger, *Biophys. J.*, **49**, 817–820 (1986).
36. A. G. Palmer III and N. L. Thompson, *Biophys. J.*, **51**, 339–343 (1987).
37. A. G. Palmer III and N. L. Thompson, *Biophys. J.*, **52**, 257–270 (1987).
38. A. G. Palmer III and N. L. Thompson, *Proc. Natl. Acad Sci. U.S.A.*, **86**, 6148–6152 (1989).

39. D. Axelrod, *Methods Cell Biol.*, **30**, 333–352 (1989).
40. D. Axelord, *Biophys. J.* **26**, 557–574 (1979).
41. T. M. Yoshida and T. Asakura, *Optik*, **41**, 281–292 (1974).
42. M. M. Timbs and N. L. Thompson, *Biophys. J.*, **58**, 413–428 (1990).
43. M. P. Heyn, *FEBS Lett.*, **108**, 359–364 (1979).
44. F. Jähnig, *Proc. Natl. Acad. Sci. U.S.A.*, **76**, 6361–6365 (1979).
45. P. Johnson and P. B. Garland, *FEBS Lett.* **132**, 252–265 (1981).
46. T. M. Yoshida and B. G. Barisas, *Biophys. J.*, **50**, 41–53 (1986).
47. T. Förster, *Ann. Phys. (Leipzig)* [6], **2**, 55–75 (1948).
48. B. K. Fung and L. Stryer, *Biochemistry*, **17**, 5241–5248 (1978).
49. N. Shaklai, J. Yguerabide, and H. M. Ranney, *Biochemistry*, **16**, 5585–5592 (1977).
50. S. M. Fernandez and R. D. Berlin, *Nature*, **264**, 411–415 (1976).
51. B. A. Herman and S. M. Fernandez, *Biochemistry*, **21**, 3275–3283 (1987).
52. T. Watts, H. Gaub, and H. M. McConnell, *Nature (London)*, **320**, 179–181 (1986).
53. T. H. Watts and H. M. McConnell, *Proc. Natl. Acad. Sci. U.S.A.*, **83**, 9660–9664 (1986).
54. P. S. Uster and R. E. Pagano, *J. Cell Biol.*, **103**, 1221–1234 (1986).
55. N. L. Thompson and D. Axelrod, *Biophys. J.*, **43**, 103–114 (1983).
56. E. Kalb, S. Frey, and L. K. Tamm, *Biochim. Biophys. Acta*, **1103**, 307–316 (1992).
57. R. D. Tilton, A. P. Gast, and C. R. Robertson, *Biophys. J.*, **58**, 1321–1326 (1990).
58. R. D. Tilton, C. R. Robertson, and A. P. Gast, *J. Colloid Interface Sci.*, **137**, 192–203 (1990).
59. J. R. Abney, B. A. Scalettar, and N. L. Thompson, *Biophys. J.*, **61**, 542–552 (1992).
60. T. P. Burghardt and N. L. Thompson, *Biophys. J.*, **46**, 729–737 (1984).
61. N. L. Thompson, H. M. McConnell, and T. P. Burghardt, *Biophys. J.*, **46**, 739–747 (1984).
62. T. P. Burghardt and D. Axelrod, *Biochemistry*, **22**, 979–985 (1983).
63. D. Axelrod, *Methods Cell Biol.*, **30**, 245–270 (1989).
64. V. Hlady, D. R. Reinecke, and J. D. Andrade, (1985) in J. D. Andrade, Ed., *Surface and Interfacial Properties of Biomedical Polymers*, Vol. 2, Plenum Press, New York, pp. 81–119.
65. K. B. Blodgett and I. Langmuir, *Phys. Rev.* **51**, 964–973 (1937).
66. H. Kuhn, D. Möbius, and H. Bucher (1972) in A. Weissberger and B. Rossiter, Eds., *Physical Methods of Chemistry*, Vol. 1, Part III B, Wiley, New York, pp. 577–702.
67. V. von Tscharner and H. M. McConnell, *Biophys. J.*, **36**, 421–427 (1981).
68. D. G. Hafeman, V. von Tscharner, and H. M. McConnell, *Proc. Natl. Acad. Sci. U.S.A.*, **78**, 4552–4556 (1981).
69. L. K. Tamm, *Klin. Wochenscher.* **62**, 502–503 (1984).

70. L. K. Tamm and H. M. McConnell, *Biophys. J.*, **47**, 105–113 (1985).
71. A. A. Brian and H. M. McConnell, *Proc. Natl. Acad. Sci., U.S.A.*, **81**, 6159–6163 (1984).
72. A. P. Fringeli, in U.P. Schlunegger, Ed., *Biologically Active Molecules*, Springer-Verlag, Berlin, 1989, pp. 241–252.
73. S. Frey and L. K. Tamm, *Biophys. J.* **60**, 922–930 (1991).
74. J. G. E. M. Fraaije, J. M. Kleijn, M. van d. Graaf, and J. C. Dijt, *Biophys. J.*, **57**, 965–975 (1990).
75. J. A. M. Zasadzinski, C. A. Helm, M. L. Longo, A. L. Weisenhorn, S. A. C. Gould, and P. K. Hansma, *Biophys. J.* **59**, 755–760 (1991).
76. M. Nakache, A. B. Schreiber, H. Gaub, and H. M. McConnell, *Nature (London)*, **317**, 75–77 (1985).
77. P. M. Nellen, K. Tiefenthaler, and W. Lukosz, *Sens. Actuators*, **15**, 285–295 (1988).
78. K. Yoshikawa, H. Hayashi, T. Shimooka, H. Terada, and T. Ishii, *Biochem. Biophys. Res. Commun.*, **145**, 1092–1097 (1987).
79. S. J. Johnson, T. M. Bayerl, D. C. McDermott, G. W. Adam, A. R. Rennie, R. K. Thomas, and E. Sackmann, *Biophys. J.*, **59**, 289–294 (1991).
80. R. A. Demel, W. S. M. Geurt van Kessel, R. F. A. Zwaal, B. Roelofson, and L. L. M. van Deenen, *Biochim. Biophys. Acta*, **822**, 97–107 (1975).
81. A. Seelig, *Biochim. Biophys. Acta*, **899**, 196–204 (1987).
82. S. H. Portlock, Y. Lee, J. M. Tomich, and L. K. Tamm, *J. Biol. Chem.*, **267**, 11017–11022 (1992).
83. B. K. Lok, Y. L. Cheng, and C. R. Robertson, *J. Colloid Interface Sci.*, **91**, 87–103 (1983).
84. S. A. Darst, C. R. Robertson, and J. A. Berzofsky, *Biophys. J.*, **53**, 533–539 (1988).
85. V. Hlady, J. Rickel, and J. D. Andrade, *Colloids Surf.*, **34**, 171–183 (1988).
86. L. K. Tamm and I. Bartoldus, *Biochemistry*, **27**, 7453–7458 (1988).
87. M. M. Timbs, C. L. Poglitsch, M. L. Pisarchick, M. T. Sumner, and N. L. Thompson, *Biochim. Biophys. Acta*, **1064**, 219–228 (1991).
88. E. E. Uzgiris and R.D. Kornberg, *Nature (London)*, **301**, 125–129 (1983).
89. P. M. Coleman, J. Deisenhofer, and R. Huber, *J. Mol. Biol.*, **100**, 257–282 (1976).
90. C. L. Poglitsch, M. T. Summer, and N. L. Thompson, *Biochemistry*, **30**, 6662–6671 (1991).
91. D. D. Roberts, C. N. Rao, J. L. Magnani, S. L. Spitalnik, L. A. Liotta, and V. Ginsberg, *Proc. Natl. Acad. Sci. U.S.A.*, **82**, 1306–1310 (1985).
92. E. Kalb and J. Engel, *J. Biol. Chem.*, **267**, 19047–19052 (1991).
93. P. D. Yurchenko, E. C. Tsilibary, A. S. Charonis, and H. Furthmayr, *J. Biol. Chem.*, **260**, 7636–7644 (1985).
94. S. Mabrey and J. M. Sturtevant, *Proc. Natl. Acad. Sci. U.S.A.*, **73**, 3862–3866 (1976).

95. T. H. Watts, A. A. Brian, J. W. Kappler, P. Marrack, and H. M. McConnell, *Proc. Natl. Acad. Sci. U.S.A.*, **81**, 7564–7568 (1984).
96. L. K. Tamm, *Biochemistry*, **27**, 1450–1457 (1988).
97. S. Subramanian, M. Seul, and H. M. McConnell, *Proc. Natl. Acad. Sci. U.S.A.*, **83**, 1169–1173 (1986).
98. L. L. Wright, A. G. Palmer, and N. L. Thompson, *Biophys. J.*, **54**, 463–470 (1988).
99. L. M. Smith, J. W. Parce, B. A. Smith, and H. M. McConnell, *Proc. Natl. Acad. Sci. U.S.A.*, **76**, 4177–4179 (1979).
100. E. Kalb and L. K. Tamm, *Thin Solid Films* **210/211**, 763–765 (1992).
101. Y. Takagaki, R. Radhakrishnan, C. M. Gupta, and H. G. Khorana, *J. Biol. Chem.*, **258**, 9136–9142 (1983).
102. E. Arinc, L. M. Rzepecki, and P. Strittmatter, *J. Biol. Chem.*, **262**, 15563–15567 (1987).
103. A. Kusumi, A. Tsuji, M. Murata, Y. Sako, A. C. Yoshizawa, S. Kagiwada, T. Hayakawa, and S. Ohnishi, *Biochemistry*, **30**, 6517–6527 (1991).
104. U. Kublitscheck, M. Kirchais, R. Schweitzer-Stenner, W. Dreybrodt, T. M. Jovin, and I. Pecht, *Biophys. J.*, **60**. 307–308 (1991).

CHAPTER

8

CLINICAL APPLICATIONS OF LUMINESCENCE SPECTROSCOPY

GEORGE H. SCHENK

Department of Chemistry
Wayne State University
Detroit, Michigan 48202

8.1. Introduction
8.2. Determination of Inorganic Substances
  8.2.1. Methods for Simple Metal Ions
  8.2.2. Methods for Other Inorganic Species
8.3. Determination of Organic Substances
  8.3.1. Methods for Fluorescent Amino Acids and Proteins
  8.3.2. Methods for Other Selected Organic Substances
  8.3.3. Methods for Selected Pharmaceuticals
8.4. Specialized Types of Luminescence
  8.4.1. Fiber-Optic Sensors
  8.4.2. Fluoroimmunoassays
  8.4.3. Total and Low-Temperature Luminescence
  8.4.4. Luminescence in Organized Media
References

8.1. INTRODUCTION

Luminescence analysis is currently used for the determination of a large number of chemical substances, including drugs, found in the body. This chapter will attempt to survey the various classes of these substances. It will not cover the fundamentals of luminescence spectroscopy since this has been covered earlier in this series by Schulman in Chapter 1 of Part 1 (1). It is assumed that the reader can review those fundamentals using that chapter or

Molecular Luminescence Spectroscopy, Part 3, Edited by Stephen G. Schulman. Chemical Analysis Series, Vol. 77.
ISBN 0-471-51580-9 © 1993 John Wiley & Sons, Inc.

other texts. Similarly, this chapter will also not cover the luminescence of the pharmaceuticals surveyed by Baeyens in Chapter 2 of Part 1 (2).

The field of biomedical luminescence is so extensive that no attempt will be made to report the literature in complete form. The reader is referred to reviews, such as those in *Analytical Chemistry*, for this purpose. Especially recommended is the biennial review of molecular fluorescence, phosphorescence, and chemiluminescence spectrometry by McGown and Warner (3), the last of which in 1990 surveyed luminescence techniques in biological systems. Also recommended is the sporadic clinical chemistry review by Stinshoff and colleagues (4), the last of which in 1987 surveyed fluorescence immunoassays as well as other areas. Welcome, too, is the monograph by Taylor et al. (5) on biomedical applications of fluorescence spectroscopy. Finally, Schneckenburger et al. (6) have reviewed fluorescence techniques for biotechnology.

This chapter includes many purely clinical fluorescence methods, but also includes methods that should give researchers ideas for developing new fluorescence methods. It will be divided into three main sections dealing with the following areas: the luminescence of inorganic substances, including systems containing metal ions; the luminescence of systems of wholly organic molecules; and specialized types of luminescence, such as fluoroimmunoassays.

8.2. DETERMINATION OF INORGANIC SUBSTANCES

It is assumed that the reader can review the fundamentals of inorganic luminescence using an earlier chapter in Part 1 of this series by Fernández-Gutiérrez and de la Peña (7), and the fundamentals of bioinorganic spectroscopy in the earlier chapter by Brittain (8). The present chapter will cover only the luminescence of inorganic substances of special importance in body fluids.

Before starting, it will be useful to summarize the inorganic ions occurring in the body that may require trace analysis. In roughly decreasing order of concentration in the whole body, these are

$$NH_4^+, Ca^{2+}, PO_4^{3-}, K^+, Na^+, Cl^-, Mg^{2+},$$
$$Fe^{2+} \text{ and } Fe^{3+}, F^-, Zn^{2+}, Cu^{2+}, Co^{2+}, Mn^{2+}, I^-, \text{ and } BO_2^-$$

Naturally the concentration of these inorganic ions will vary from, say, the blood to other body fluids, but the list does indicate the ions whose analysis may be required, as well as indicate possible interferences.

8.2.1. Methods for Simple Metal Ions

We will focus mainly on fluorescence methods for simple metal ions since these are the most convenient for clinical analysis. Apart from the lanthanides, the only unchelated metal ions that luminesce are UO_2^{2+} and the Tl^+ ions. Thus a metal chelate with an aromatic organic ligand must be formed to determined clinically important metal ions.

Effect of Metal Ions on Ligand Fluorescence. The fluorescence in a metal chelate having an aromatic ligand originates from the aromatic ligand, as modified by the positive metal ion. In most metal ion chelates, the $S_0 \rightarrow S_1$ transition of the aromatic ligand has $\pi-\pi^*$ character. Complexation of the metal ion usually is said to favor the $\pi-\pi^*$ transition over the $n-\pi^*$ transition and thus enhance the fluorescence of the ligand relative to intersystem crossing and internal conversion. However, metal ions also can enhance the rate of intersystem crossing and/or internal conversion. This reduces the quantum efficiency of fluorescence in many cases to near zero. Thus it is important to recognize that not all metal ions will form fluorescent chelates.

Which Metal Chelates Fluoresce? If the chelates of metal ions from all rows of the periodic table are considered, then in general only chelates of metal ions having empty d-subshells or filled d-subshells will fluoresce. It would appear that d^1 through d^9 metal ions do not form fluorescent chelates. However, there are exceptions. In the lightest transition metals (row 4), both Cr(III) and Mn(II) form a few fluorescent chelates. In the heavier transition metals (rows 5 and 6), some diamagnetic d^6 ions form fluorescent chelates that are not of clinical interest. For clinical purposes, then, direct fluorescent chelate analysis is practical mainly for ions in Groups IA, IIA, IIIA, IIIB, and IIB.

Clinically important ions such as Cu(II), Fe(II), and Fe(III) not only should not be amenable to direct fluorometric analysis but also should quench the fluorescence of other chelates. Of course, it is also possible to determine such metal ions by quenching provided that quenching by other ions is not significant. An ion such as Pb^{2+} can also be determined *indirectly* by the fluorescence of zinc(II) protoporphyrin.

Magnesium and Calcium. The metal ions of Group IIA, being d^0 or d^{10} ions, all form fluorescent chelates. Since magnesium is of the greatest physiological importance in this group, we will discuss it first, followed by methods for calcium.

The classic method for magnesium depends on the measurement of its fluorescent 8-hydroxyquinoline chelate (9). The fluorescence intensity of this chelate depends on the pH: it is highest near pH 6.5, and minimal at pH's below 5. An acetate buffer is used to maintain the pH near 6.5. A correction is necessary for the analysis of urine because other metal chelates yield a small amount of fluorescence at this pH. A second measurement at pH 3.5 yields a correction factor for these chelates.

The chelate of magnesium and 8-hydroxyquinoline is not the most intensely fluorescent chelate of magnesium. For example, 8-hydroxyquinoline-5-sulfonic acid is more sensitive, detecting 0.001 ppm as compared to 0.01 ppm for 8-hydroxyquinoline (10). Many other fluorescent chelates are known.

Methods for calcium are based on the fluoresence of fluorescein-type molecules (11). The chelates of calcein are often used for calcium as well as the other Group IIA ions, the quantum efficiencies ranging from 0.79 to 1. The calcium chelate fluoresces in 0.8 M KOH and is free from fluorescence of the unchelated calcein which has no significant fluorescence in this reagent. Measurement of calcium in individual cells by single-cell fluorometry has also been described (12).

Fluorescence-based fiber-optic sensors based on the immobilized salt of 8-hydroxyquinoline-5-sulfonic acid have been reported for magnesium (13) but should be useful for calcium and perhaps the other metal ions in Group IIA.

Zinc and Cadmium. The metal ions of Group IIB, being d^{10} ions, should in theory form fluorescent chelates. However, methods have been reported only for zinc(II) and cadmium(II) ions but not for mercury(II) ion. The latter evidently exerts an efficient heavy atom effect on chelates and so quenches potential fluorescence.

Methods for zinc(II) have been based on such ligands as 8-hydroxyquinoline (14). More recently, West and Pflaum (15) have reported fluorescence of zinc with pyridylbis(quinolylhydrazone), and Fernández et al. (16) have published a method using dibromoquinolinol as a ligand. As mentioned below in the discussion of lead(II), zinc protoporphyin can also be measured fluorometrically.

Although cadmium(II) would not appear to be normally measured in the body, it enters the lungs in tobacco smoke and hence its analysis deserves a brief mention. Methods for cadmium(II) determination have also been based on the use of 8-hydroxyquinoline-5-sulfonic acid ligand (17) and the pyridylbis(quinolylhydrazone) ligand (18).

Lead. Zinc(II) ion reacts rapidly with many unchelated ligands in body fluids to form zinc(II) chelates. Significant levels of lead(II) ion in the blood lead to the formation fluorescent zinc protoporphyrin. This occurs because lead(II) ion blocks the action of an iron-insertion enzyme in the blood. Because zinc(II) is present in the blood and can react much faster than iron(II) with protoporphyrin when the iron-insertion enzyme is blocked, zinc(II) protoporphyrin forms in proportion to the level of the lead(II) ion. Thus measurement of zinc(II) protoporphyrin indirectly yields the concentration of serum lead(II) ion. A drop of blood is transferred to a glass slide and inserted into a "Hematofluorometer"-type fluorometer. The zinc(II) protoporphyrin is excited at 424 nm and fluoresces at 625 nm (18). Standard materials have been developed for use with the Hematofluorometer (19).

Copper. Copper(II) is a d^9 ion and should not form fluorescent complexes. Ritchie and Harris (20) report that tricyanoaminopropene (TRIAP) forms a fluorescent "complex" after 15 min at 40 °C, and they used it for the determination of copper(II), not copper(I), in tissues. It is possible that the copper(II) catalyzes the formation of some type of fluorescent organic product from TRIAP alone, rather than forming a chelate.

Quenching Methods for Iron, Copper, and Cobalt. These ions, having unfilled d sublevels, do not usually form fluorescent chelates but can quench the fluorescence of chelates of aluminum, etc., and be measured indirectly. Thus the quenching of aluminum 8-hydroxyquinoline can be used to measure clinically important ions such iron(III), copper(II), and cobalt(II).

Iron(III) can also be measured by reducing it to iron(II) and taking advantage of the more stable chelates formed by the iron(II) ion. One method (21) reported the reduction of iron(III) with hydroxylamine and the addition of fluorescent 2,2',2"-terpyridyl to form a chelate. The quenching of the fluorescence of 2,2',2"-terpyridyl could be used for parts per billion of iron(III) or, course, iron(II).

Cobalt as cobalt(II) was measured by Zamochnick and Rechnitz (22) using quenching. They determined 1–20 ppb of cobalt(II) by quenching the fluorescence of the chelate of aluminum(III) with superchrome blue black extra.

8.2.2. Methods for Other Inorganic Species

We include in this section methods for all other forms of inorganic species, including those chelates that occur naturally in the body.

Zinc Protoporphyrin. Zinc is only of clinical interest with regard to lead poisoning. The presence of lead(II) ion facilitates the formation of fluorescent zinc(II) protoporphyrin, as discussed above in Section 8.2.1 as regards the determination of lead.

Ammonia in Blood. Ammonia in blood may be determined by conversion of ammonia to glutamic acid using the enzyme glutamic dehydrogenase (23).

Reduced nicotinamide adenine dinucleotide (NADH) is used as the cofactor, and the reduction in the fluorescence of NADH is measured.

Borate Anion. Borate anion is not normally fluorescent itself but can react with certain organic molecules to form fluorescent derivatives. The classic example is its reaction with benzoin to form a 1:1 addition compound as described by White and Hoffman (24). A five-membered ring involving boron, two oxygens, and the two aliphatic carbons of benzoin is probably formed. This product is more highly conjugated than benzoin and thus fluoresces in the visible region near 480 nm. Dibenzoylmethane, among others, has also been used to form addition compounds with borate (25).

Halide Anions. Obviously none of the halide ions fluoresce themselves, but each is a potential quencher of fluorescence and should be measurable when the quenching effect is linear. Fluoride has been determined fluorometrically by the quenching of many alminum(III) chelates, such as the aluminum(III) chelate of Pontachrome BBR (26). In contrast, chloride has been determined fluorometrically by forming a fluorescent compound with the 6-methoxy-1-(3-sulfonatopropyl)quinolinium cation (27). Iodide has been determined by the same authors using fluorescent compound formation with the 10-(3-sulfonatopropyl)acridium cation.

8.3. DETERMINATION OF ORGANIC SUBSTANCES

It is assumed that the reader can review the fundamentals of organic luminescence using an earlier chapter in Part 1 of this series on the fluorescence of organic natural products by Wolfbeis (28), as well the fluorescence portion of the chapter by Baeyens on the fluorescence and phosphorescence of pharmaceuticals (29). The present chapter will cover only the luminescence of organic substances of special importance in body fluids.

Before starting, it will be useful to recall that only molecules containing a minimum of one benzene ring exhibit fluorescence and/or phosphorescence. Thus molecules like glucose have to be treated with reagents that will produce fluorescence.

8.3.1. Methods for Fluorescent Amino Acids and Proteins

We will focus mainly on *fluorescence* methods for organic molecules since these are the most convenient for clinical analysis. Although the proteins and the aromatic amino acids are the most common fluorescers in body fluids, we will focus on the determination of the aromatic amino acids in this subsection.

Aromatic Amino Acids. Phenylalanine (Phe), tyrosine (Tyr), and tryptophan (Trp) are the only three amino acids exhibiting significant fluorescence. The fluorescence data for these are summarized in the accompanying tabulation.

Phe	$\lambda_{ex} = 260$ nm	$\lambda_f \sim 282$ nm	$\phi_f = 0.04$
Tyr	$\lambda_{ex} = 275$ nm	$\lambda_f = 303$ nm	$\phi_f = 0.2$
Trp	$\lambda_{ex} = 287$ nm	$\lambda_f = 348$ nm	$\phi_f = 0.2$

The native fluorescence of phenylalanine is weak, as seen from its low quantum efficiency, and it is determined mainly using fluorogenic reagents such as ninhydrin (30). In contrast, the native fluorescence of the related molecule tyrosine is much stronger, and it can be measured directly without the use of fluorogenic reagents. Tyrosine may be determined fluorometrically at the picomolar level in hydrolyzates of proteins following high-performance liquid chromatographic (HPLC) separation (31). As emphasized by Wolfbeis, its natural fluorescence is influenced by many factors, so that reagents such as 1-nitroso-2-naphthol have been used to form fluorescent products for analysis (32). The native fluorescence of trytophan is also much stronger than that of phenylalanine, and it too can be measured directly without the use of fluorogenic reagents. Wolfbeis has again pointed out that this fluorescence is affected by a large number of "external factors," such as blue-shifting in organic solvents. Nevertheless, it has been determined after HPLC separation using its native fluorescence (33).

Fluorescent Proteins. Phenylalanine fluorescence in proteins is very weak and usually observed when tyrosine and trytophan are absent (34). The fluorescence emission at the 302 nm maximum of tyrosine can be used to study proteins, especially when tryptophan is absent (class A proteins), or present in small amounts (35). The fluorescence emission of tryptophan in proteins varies from 308 to 350 nm (36) and is a sensitive indicator of the polarity of the groups around tyrptophan.

8.3.2. Methods for Other Selected Organic Substances

Instead of presenting the complete survey of *fluorescent* molecules that can be found in specialized monographs or chapters such as that by Wolfbeis (28), we will present methods mainly for important selected *nonfluorescent* molecules found in the body.

Glucose. Glucose is a nonfluorescent molecule whose measurement is important enough to include in our discussion. Glucose is determined by reaction with adenosine triphosphate (ATP) in a coupled enzyme procedure to form ultimately fluorescent nicotinamide adenine dinucleotide phosphate (NADPH). The NADPH is excited at 340 nm and emits at 460 nm (37). The method should be amenable to the analysis of glucose in blood serum.

Cholesterol. Like glucose, cholesterol is not fluorescent but is of such importance that a fluorescence measurement is useful. Cholesterol has been mainly measured spectrophotometrically using the Lieberman–Burchard reaction, but the product of this reaction is also fluorescent. The analysis of cholesterol in blood serum involves the usual reaction with acetic anhydride and sulfuric acid and yields a precise assay (38).

Uric Acid. Uric acid, $C_5H_4N_4O_3$, like glucose and chlolesterol, is not aromatic and does not fluoresce, but its analysis is important. Uric acid reacts with uricase and homovanillic acid. Addition of hydrogen peroxide and peroxidase results in oxidation of homovanillic acid to a highly fluorescent product that can be used for analysis of uric acid in blood serum (39).

Serum Albumin. Serum albumin, a protein, can be measured fluorometrically after reaction with various reagents. The reagent 8-anilinonaphthalene-1-sulfonic acid (ANSA) gives lower blanks than other reagents and has been studied in more detail (40).

Coenzymes, Vitamins, Nucleic Acids, Alkaloids, Porphyrins, etc. The fluorescences of all of these substances have been reported by Wolfbeis earlier in Part 1 of this series (28) and will not be discussed here.

8.3.3. Methods for Selected Pharmaceuticals

Instead of presenting the complete survey that can be found in specialized monographs or in the earlier chapter by Baeyens in Part 1 of this series (29), we will present methods for a few important selected molecules that may be applied to clinical samples (these methods have not been developed in a clinical laboratory).

Acetylsalicylic Acid and Salicylic Acid. These two acids are of course related in that acetylsalicylic acid (ASA) breaks down to salicylic acid in the body. For a long time ASA was not thought to be fluorescent and was measured by ultraviolet spectrophotometry or by conversion to salicylic acid. However, Miles and Schenk (41) and Schenk et al. (42) found that ASA in aspirin tablets could easily be measured fluorometrically. The fluorescence of ASA was observed at 335 nm, away from the 450-nm band of salicyclic acid, in 1% acetic acid–chloroform solvent. Salicylic acid could also be measured by this method but can be measured in aqueous solution as well (43).

Acetaminophen and *p*-Aminophenol. Acetaminophen is not intensely fluorescent, but Street and Schenk were able to measure the amount of *p*-aminophenol impurity in acetaminophen by fluorescence (44). The acetaminophen exerted an inner filter effect that had to be corrected for, to achieve accurate results. Other workers do report a weak fluorescence for acetaminophen (45).

Phenylethylamines and Barbiturates. Miles and Schenk have reported that many phenylethylamines and barbiturates can be measured fluorometrically (46). Since the phenylethylamines are simply substituted benzenes, they exhibit the typical benzene fluorescence. In sodium hydroxide solution, the barbiturates assume a fluorescent conjugated structure that can be used for analysis.

Antihistamines. Antihistamines can also be measured fluorometrically. Antihistamines with a diphenylmethane chromophore exhibit typical benzene fluorescence near 300 nm (47). Antihistamines with a 2-aminopyridine chromophore exhibit fluorescence typical of that chromophore (48).

Other Antibiotics, Analgesics, Alkaloids, etc. See the earlier chapter by Baeyens in Part 1 of this series (29).

8.4. SPECIALIZED TYPES OF LUMINESCENCE

The aim of this section will be to indicate briefly some recent advances in techniques that the reader might wish to evaluate for a special situation or for research. We have already indicated some reviews of interest at the beginning of this chapter. In addition to the biennial reviews in *Analytical Chemistry*, two other pertinent reviews by Bright (49) and Yguerabide and Yguerabide (50) should be consulted by the reader.

The topics presented below include fiber-optic sensors, fluoroimmunoassays, total and low-temperature luminescence, and luminescence in organized media. For the benefit of the interested reader, the aforementioned reviews also include topics not covered here. For example, Yguerabide and Yguerabide (50) discuss polarized fluorescence spectroscopy and excitation energy transfer. Bright (49) discusses fluorometric methods of analysis for DNA and RNA and protein–ligand interactions. McGown and Warner's 1990 biennial review includes topics such as solid-surface luminescence, fluorescence polarization, and chemiluminescence.

8.4.1. Fiber-Optic Sensors

General reviews by Bright (49) and McGown and Warner (3) have covered some aspects of the field of fiber-optic sensors. Among the excellent specialized reviews on this subject are those by Seitz (51) and Angel (52).

An obvious advantage of fiber-optic sensors for fluorescence is the flexibility of location of the spectrofluorometer. By conveying fluorescence emission from the site of the sample to the spectrofluorometer, very convenient and potentially accurate measurements can be achieved. Thus fiber-optic sensors can be used to improve traditional measurements of glucose, oxygen (pO_2), pH, pCO_2, bilirubin, glucose, metal ions, and even temperature and humidity (3).

Fiber optics can be used to devise so-called fluorescence optrodes. These consist of a reagent immobilized on a fiber-optic probe. The reagent reacts with the analyte, and the resulting fluorescence is propagated through the probe to a detector. Such optrodes are useful for the continuous determination of such species as O_2 and CO_2 (53). Even the sodium ion can be measured; it has been determined using a reagent phase that responds to sodium ion selectively (54).

Fiber optics would seem to have special advantages for the analysis of gases in the body. Indeed, a fiber-optic coupled fluorescence sensor has been used in connection with a gas analyzer to measure O_2, CO_2, and pH in the extracorporeal circuit during mitral valve replacement (55).

Bright (49) also mentions a fiber-optic system employing fluorescence lifetime selectivity. His system involves a fiber-optic-based phase modulation fluorometer. He reports successful resolution of single-, double-, and triple-resolved exponential decays of fluorescence with a bifurcated fiber-optic probe up to 100 m in length.

8.4.2. Fluoroimmunoassays

General reviews by Bright (49) and McGown and Warner (3) have covered some aspects of the field of fluoroimmunoassays. Among the excellent specialized reviews on this subject are those by Hemmila, whose overview of both homogeneous and heterogeneous fluoroimmunoassays in the field of clinical analysis can serve as a starting point for clinical workers needing a survey of the field (56). Guilbault has discussed this assay for the determination of pharmaceutical compounds (57).

Since many drugs are not fluorescent, both homogeneous and heterogeneous fluoroimmunoassays have been employed for their measurement. The literature in this area is very extensive, and only a brief mention of some recent developments is possible.

Because this technique is subject to many potential interferences, several parameters have been used to alleviate errors. Long-lived rare earth chelates of europium(III) and terbium(III) have been used for this purpose (58). A new europium(III) chelate has also been used for the determination of human choroiogonadotropin (59).

In its most fundamental form, the fluoroimmunoassay is based on a binding competition between a highly selective antibody (Ab), an antigen (Ag), and fluorescent-labeled form of the antigen (Ag*):

$$Ag + Ag^* + Ab \longleftrightarrow AgAb + Ag^*Ab$$

The fluoroimmunoassay measures a difference in the luminescence of Ag* and Ag*Ab to measure Ag indirectly. The fluoroimmunoassay can be either *homogeneous* (no physical separation) or *heterogeneous* (physical separation of Ag* and Ag*Ab).

A promising recent development in homogeneous fluoroimmunoassay is the approach of Bright and co-workers based on FAST (the fluorescence anisotropy selectivity technique) (60). FAST allows analysts to resolve directly the individual components in complex mixtures on the basis of differences in the rotational diffusion rates of the individual components in a mixture. Unlike steady-state polarization-based approaches, FAST does not require prior knowledge of the number of components. Even if the fluorescence lifetimes and emission spectra are very similar, the individual emission spectra of the same fluorophore in two different microenvironments can be simultaneously determined. These workers reported results for the simultaneous resolution of dansylamide bound to β-cyclodextrin and human serum albumin.

Another approach to extending fluoroimmunoassay involves the use of energy transfer. Excitation energy transfer has been used for the determination of both haptens and antigenic analytes. For example, Nithipatikom and McGown (61) have combined the effects of excitation energy transfer and direct quenching to measure human lactoferrin. Both fluorescein-labeled antibody and tetramethylrhodamine-labeled antibody bind to multivalent lactoferrin antigen. Excitation energy transfer causes quenching of the fluorescein fluorescence, which is also quenched by the antigen–antibody binding. The combined fluorescence intensity and lifetime changes are incorporated into a single phase-resolved intensity measurement by phase-resolved fluorometry.

8.4.3. Total and Low-Temperature Luminescence

Total luminescence spectra of course offer more information than either fluorescence or phosphorescence spectra alone. In general, total luminescence and low-temperature luminescence measurements ought to have fewer interferences from collisional quenching and the like and ought to offer increased selectivity. The general biennial reviews by McGown and Warner (3) cover many aspects of these fields as they pertain to all types of systems. For specific biochemical applications, the review of Leiner and Wolfbeis (62) is very useful.

The total luminescence spectrometric approach can, of course, be used to study any system in general, but it also can be used to study systems that exhibit excitation-dependent emission. Of special note are the studies by Wolfbeis and co-workers of the total luminescence spectra of human urine (63) and of human blood serum (64).

The low-temperature spectrometric approach offers increased selectivity but so far seems to have been applied mainly to aromatic hydrocarbons. Wehry and Mamantov have written a general review of matrix isolation in molecular spectrometry that should be applicable to clinical fluorometry (65).

Finally, Thornberg and Maple have reported on recent low-temperature phosphorescence measurements in pharmaceutical analysis (66).

8.4.4. Luminescence in Organized Media

An alternate approach to low-temperature luminescence for eliminating interferences such as quenching has been to utilize an organized medium at room temperature. Micelles, reversed micelles, and microemulsions have been studied for this purpose. Cyclodextrins have been used extensively for this purpose.

Warner and co-workers have discussed the use of cyclodextrins in general for enhancing detection in the analytical spectra (67).

Recent biochemical methods have focused on the determination of molecules like glucose. In one method, glucose or glucose oxidase were determined in an inverted (reversed) micellar system involving hexadecyltrimethylammonium chloride. The determination was based on the reaction of glucose with glucose oxidase to generate hydrogen peroxide. The resulting peroxide reacts with chemiluminescent luminol to produce light whose intensity is proportional to the concentration of glucose or glucose oxidase. The presence of the inverted (reversed) micellar medium permits the measurement of both the enzymatic and chemiluminescent reaction simultaneously without added catalyst or cooxidant (68).

Because the chloride ion limited the sensitivity of the aforementioned method, the same authors (69) determined glucose and amino acids using hexadecyltrimethylammonium bromide as the reversed micellar system. This system also permitted both the enzymatic and chemiluminescent reactions to occur simultaneously without added catalyst or cooxidant. Substitution of the bromide anion for chloride lowered the detection limit of glucose to

5.5×10^{-6} M, an 18-fold improvement over the previous method. The detection limit for phenylalanine was 2.1×10^{-6} M.

A good example of the use of cyclodextrins is the fluorescent enhancement of several hallucinogenic drugs by 10^{-5} M solutions of cyclodextrins (70). The limits of detection were 6–13 µg/L for ibogaine and N,N-dimethyltryptamine. Fluorescence intensity enhancements relative to aqueous solution ranged from 1.2- to 4.0-fold enhancement. Differences were also observed between α-cyclodextrin and β-cyclodextrin solutions for the various drugs.

REFERENCES

1. S. G. Schulman, in S. G. Schulman, Ed., *Molecular Luminescence Spectroscopy: Methods and Applicatons*, Part 1, Wiley-Interscience, New York, 1985, Chap. 1.
2. W. R. G. Baeyens, in S. G. Schulman, Ed., *Molecular Luminescence Spectroscopy: Methods and Applications*, Part 1, Wiley-Interscience, New York, 1985, Chap. 2.
3. L. B. McGown and I. M. Warner, *Anal. Chem.*, **62**, 255R (1990).
4. K. E. Stinshoff, W. Stein, W. G. Wood, and P. F. Laska, *Anal. Chem.*, **59**, 337R (1987).
5. D. L. Taylor, A. S. Waggoner, R. F. Murphy, F. Lanni, and R. R. Birge, Eds., *Applications of Fluorescence in the Biomedical Sciences*, Alan R. Liss, New York, 1986.
6. H. Schneckenburger, B. W. Reuter, and S. M. Schoberth, *Trends Biotechnol.*, **3**, 257 (1985).
7. A. Fernández-Gutiérrez and A. M. de la Peña, in S. G. Schulman, Ed., *Molecular Luminescence Spectroscopy: Methods and Applications*, Part 1, Wiley-Interscience, New York, 1985, Chap. 4.
8. H. G. Brittain, in S. G. Schulman, Ed., *Molecular Luminescence Spectroscopy: Methods and Applications*, Part 1, Wiley-Interscience, New York, 1985, Chap. 5.
9. D. Schachter, *J. Lab. Clin. Med.*, **54**, 763 (1959).
10. I. Clark and G. Howe, *Anal. Biochem.*, **19**, 14 (1967).
11. B. L. Kepner and D. M. Hercules, *Anal. Chem.*, **35**, 1238 (1963).
12. J. Rogers, T. R. Kesketh, G. A. Smith, M. A. Beaven, J. C. Metcalfe, P. Johnson, and P. B. Garland, *FEBS Lett.*, **161**, 21 (1983).
13. W. R. Seitz, *Anal. Chem.*, **56**, 16A (1984).
14. G. Smith, R. Jenkins, and J. Gough, *J. Histochem. Cytochem.*, **17**, 749 (1969).
15. K. West and R. T. Pflaum, *Talanta*, **33**, 807 (1986).
16. P. Fernández. C. Perez-Conde, A. M. Gutiérrez, and C. I. Camara, *J. Mol. Struct.*, **143**, 549 (1986).
17. E. Ryan, A. E. Pitts, and R. M. Cassidy, *Anal. Chim., Acta*, **34**, 491 (1966).
18. W. E. Blumberg, J. Eisinger, A. A. Lamola, and D. M. Zuckerman, *J. Lab. Clin. Med.*, **89**, 712 (1977).

19. P. J. Parson, J. R. Meola, and D. G. Mitchell, *Clin. Chem.* (*Winston-Salem, N.C.*), **34**, 1062 (1988).
20. K. Ritchie and J. Harris, *Anal. Chem.*, **41**, 163 (1969).
21. D. Fink, F. Pivnichny, and W. Ohnsorge, *Anal. Chem.*, **41**, 833 (1969).
22. S. B. Zamochnick and G. A. Rechnitz, *Z. Anal. Chem.*, **199**, 424 (1964).
23. B. L. Nazar and A. C. Schoolwerth, *Anal. Biochem.*, **95**, 507 (1979).
24. C. E. White and D. E. Hoffman, *Anal. Chem.*, **29**, 1105 (1957).
25. J. Aznares, A. Bonilla, and J. C. Vidal, *Analyst*, **108**, 368 (1983).
26. J. Block and J. Morgan, *Anal. Chem.*, **34**, 1647 (1962).
27. O. S. Wolfbeis and E. Urbano, *Fresenius' Z. Anal. Chem.*, **314**, 577 (1983).
28. O. S. Wolfbeis, in S. G. Schulman, Ed., *Molecular Luminescence Spectroscopy: Methods and Applications*, Part 1, Wiley-Interscience, New York, 1985, Chap. 3.
29. W. G. Baeyens, in S. G. Schulman, Ed., *Molecular Luminescence Spectroscopy: Methods and Applications*, Part 1, Wiley-Interscience, New York, 1985, pp. 34–122.
30. M. W. McCaman and E. Robins. *J. Lab. Clin. Med.*, **59**, 885 (1962).
31. S. W. Bailey and J. E. Aylin, *Anal. Biochem.*, **107**, 156 (1980).
32. T. P. Waalkes and S. Udenfriend, *J. Lab. Clin. Med.*, **50**, 733 (1957).
33. A. D. Jones, C. H. Hitchcock, and G. H. Jones, *Analyst*, **106**, 968 (1981).
34. E. A. Permyakov, V. V. Yarmolenko, L. P. Kalinichenko, and E. A. Burstein, *Bioorg. Khim.*, **7**, 1660 (1981).
35. A. Rehaeg and F. X. Schmidt, *Biochemistry*, **21**, 1499 (1982).
36. W. C. Galley, in R. F. Chen and H. Edelhoch, Eds., *Biochemical Fluorescence: Concepts*, Vol. II, Marcel Dekker, New York, 1976.
37. H. V. Bergmeyer, *Methods of Enzymatic Analysis*, Academic Press, New York, 1963, 117.
38. D. B. McDougal and H. S. Farmer, *J. Lab. Clin. Med.*, **50**, 485 (1957).
39. G. G. Guilbault, D. N. Kramer, and E. Hackley, *Anal. Chem.*, **39**, 271 (1967).
40. V. H. Rees, J. E. Fides, and D. J. R. Laurence, *J. Clin. Pathol.*, **7**, 336 (1954).
41. C.I. Miles and G. H. Schenk, *Anal. Chem.*, **42**, 656 (1970).
42. G. H. Schenk, F. H. Boyer, C. I. Miles, and D. R. Wirz, *Anal. Chem.*, **44**, 1593 (1972).
43. W. E. Lange and S. A. Bell, *J. Pharm. Sci.*, **55**, 386 (1966).
44. K. W. Street and G. H. Schenk, *J. Pharm. Sci.*, **68**, 1306 (1979).
45. K. J. Child, C. Bedford, and E. G. Tomich, *J. Pharm. Pharmacol.*, **14**, 374 (1962).
46. C. I. Miles and G. H. Schenk, *Anal. Chem.*, **45**, 130 (1973).
47. D. R. Wirz, D. L. Wilson, and G. H. Schenk, *Anal. Chem.*, **46**, 896 (1974).
48. D. L. Wilson, D. R. Wirz, and G. H. Schenk, *Anal. Chem.*, **45**, 1447 (1973).
49. F. V. Bright, *Anal. Chem.*, **60**, 1013A (1988).
50. J. Yguerabide and E. E. Yguerabide, *Radiat. Phys. Chem.*, **32**, 457 (1988).

51. W. R. Seitz, *CRC Crit. Rev. Anal. Chem.*, **19**, 135 (1988).
52. S. M. Angel, *Spectroscopy*, **2**, 38 (1987).
53. D. W. Luebbers and N. Optiz, *Sens. Actuators*, **4**, 641 (1983).
54. Z. Zhujun, J. L. Mullin, and W. R. Seitz, *Anal. Chim. Acta*, **184**, 251 (1988).
55. D. W. Luebbers, J. Gehrich, and N. Öptiz, *Life Support Syst.*, **4**, Suppl. 1, 94 (1986).
56. I. Hemmila, *Clin. Chem. (Winston-Salem, N.C.)*, **31**, 359 (1985).
57. G. G. Guilbault, *J. Pharm. Biomed. Anal.*, **4**, 771 (1986).
58. K. Petersson, H. Siitari, I. Hemmila, E. Soini, T. Lovgren, V. Hanninen, P. Tanner, and U. Stenman, *Clin. Chem. (Winston-Salem, N.C.)*, **29**, 60 (1983).
59. M. J. Khosravi and E. P. Diamandis, *Clin. Chem. (Winston-Salem, N.C.)*, **33**, 1994 (1987).
60. F. V. Bright, *Appl. Spectrosc.*, **43**, 1245 (1988).
61. K. Nithipatikom and L. B. McGown, *Anal. Chem.*, **59**, 423 (1987).
62. M. J. P. Leiner and O. S. Wolfbeis, *Proc. SPIE—Int. Soc. Opt. Eng.*, **909**, 134 (1988).
63. M. J. P. Leiner, M. R. Hubmann, and O. S. Wolfbeis, *Anal. Chim. Acta*, **198**, 13 (1987).
64. M. J. P. Leiner, M. R. Hubmann, and O. S. Wolfbeis, *Anal. Chim. Acta*, **167**, 203 (1985).
65. E. L. Wehry and G. Mamantov, *Prog. Anal. Spectrosc.*, **10**, 507 (1987).
66. S. M. Thornberg and J. R. Maple, *Anal. Chem.*, **57**, 4536 (1985).
67. I. M. Warner, G. Nelson, G. Patonay, L. Blyshak, and S. L. Neal, *J. Res. Natl. Bur. Stand.*, **93**, 438 (1988).
68. S. Igarashi and W. L. Hinze, *Anal. Chem.*, **60**, 446 (1988).
69. S. Igarashi and W. L. Hinze, *Anal. Chim. Acta*, **225**, 147 (1989).
70. O. Jules, S. Scypinski, and L. J. Love Cline, *Anal. Chim. Acta*, **169**, 355 (1985).

CHAPTER

9

LASER-EXCITED MOLECULAR FLUORESCENCE IN ANALYTICAL SCIENCES

J. W. HOFSTRAAT

*Department of Analytical and Environmental Chemistry
Akzo Research Laboratories Arnhem
Arnhem, The Netherlands*

C. GOOIJER AND N. H. VELTHORST

*Department of General and Analytical Chemistry
Free University
Amsterdam, The Netherlands*

9.1. Introduction
    9.1.1. General
    9.1.2. Outline
    9.1.3. Fluorescence Spectroscopy
9.2. Lasers as Excitation Sources
    9.2.1. General
    9.2.2. Characteristics
        9.2.2.1. Wavelength
        9.2.2.2. Directionality
        9.2.2.3. Time
    9.2.3. Developments in Instrumentation
        9.2.3.1. Gas-Ion Lasers: The Neon Laser
        9.2.3.2. Solid-State Lasers: The Titanium:Sapphire Laser
        9.2.3.3. Ultrafast Laser Systems
        9.2.3.4. Semiconductor Diode Lasers
9.3. Laser-Induced Fluorescence as the Detection Method in Separation Techniques
    9.3.1. Introduction
    9.3.2. Instrumental Aspects
        9.3.2.1. Detector Cell Volumes
        9.3.2.2. Detection Limits
    9.3.3. Chemical Aspects

Molecular Luminescence Spectroscopy, Part 3, Edited by Stephen G. Schulman. Chemical Analysis Series, Vol. 77.
ISBN 0-471-51580-9 © 1993 John Wiley & Sons, Inc.

9.3.4. Developments
    9.3.4.1. Indirect Fluorescence Detection
    9.3.4.2. Analytes with Native Fluorescence
    9.3.4.3. Analytes Requiring Chemical Derivatization
    9.3.4.4. Direct LIF Detection in Capillary Electrophoresis
    9.3.4.5. Diode-LIF Detection
9.3.5. Concluding Remarks

9.4. Low-Temperature Spectroscopy
9.4.1. Introduction
9.4.2. Instrumental Aspects
    9.4.2.1. Excitation
    9.4.2.2. Sample
    9.4.2.3. Detection
9.4.3. Shpol'skii Spectroscopy
    9.4.3.1. Principles
    9.4.3.2. Applications of Shpol'skii Spectroscopy
9.4.4. Fluorescence Line-Narrowing Spectroscopy
    9.4.4.1. Principles
    9.4.4.2. Excitation Wavelength Dependence
    9.4.4.3. Analytical Applications
9.4.5. Concluding Remarks

9.5. Flow Cytometry
9.5.1. Introduction
9.5.2. Principles
9.5.3. Instrumental Aspects
    9.5.3.1. The Flow Chamber and Hydrodynamics
    9.5.3.2. Excitation
    9.5.3.3. Detection
    9.5.3.4. Electronics
    9.5.3.5. Data Acquisition and Analysis
    9.5.3.6. Cell Sorters
9.5.4. Data Analysis
9.5.5. Calibration Aspects
9.5.6. Applications
    9.5.6.1. Survey
    9.5.6.2. Classification of Phytoplankton
    9.5.6.3. Phytoplankton Viability Analysis
    9.5.6.4. Phytoplankton Cell Cycle Analysis
9.5.7. Concluding Remarks

9.6. Future Trends
References

9.1. INTRODUCTION

9.1.1. General

The increased availability of lasers and important improvements in their ruggedness have led to a boom in the use of these unique light sources in recent years. The most striking laser type in this respect is most probably the solid-state diode laser. To date diode lasers, with extremely long lifetimes and good power stability, are available from 660 nm to far into the near-IR at prices that are affordable for any laboratory. Many instruments for chemical analysis are already equipped with lasers, for excitation, illumination, or calibration purposes. Well-known, by now almost classical examples are the optical particle size analyzers with simple He–Ne lasers as standard excitation sources.

The field of laser spectroscopy is a rapidly changing and developing one with many challenging opportunities. The years to come will show still important advances in available laser sources and applications. It is our intention to present in this chapter a perspective of the present state of the art and of the potential of laser-excited fluorescence in the field of analytical sciences.

9.1.2. Outline

In this chapter we intend to focus on applications of laser light sources in the analytical sciences, in particular environmental and bioanalysis, limiting ourselves to fluorescence measurements. Even then we are confronted with a vast and rapidly growing range of applications. Therefore we will not try to give an exhaustive review of everything that has been achieved so far, but rather want to present a survey of a number of areas in which laser excitation offers important advantages over conventional light sources. To define the framework, first some general aspects of fluorescence and the crucial properties of lasers in relation to fluorescence measurements will be presented. Next, the application of lasers in the fluid phase, in particular in combination with chromatographic techniques, will be discussed. In this section also some aspects of the application of fluorescent probes will be critically assessed. Furthermore, attention will be paid to the use of laser excitation for the measurement of molecules in the solid state. Here laser excitation offers unique possibilities for very specific and selective detection in ultratrace analysis (down to the "single molecule" level). Finally, another area where the laser is having a major impact will be reported upon: flow cytometry, a technique that provides rapid measurement of properties of single particles. In this application particularly the use of fluorescent probes has great potential.

9.1.3. Fluorescence Spectroscopy

In this subsections a detailed survey on aspects of fluorescence spectroscopy will not be given, as the principles of this technique have been covered in several excellent monographs (1–3). However, a summary will be given of the main aspects of electronic spectroscopy, and in particular fluorescence, as a basis for the remainder of this chapter.

The principle of fluorescence is best illustrated by means of the energy diagram shown in Fig. 9.1. Dubbed the Jablonski diagram after its inventor, this schematic shows the different pathways that may be followed in the interaction between a molecule and an external electromagnetic field. Via absorption of a suitable quantum of light the molecular system may leave the electronic ground state S_0 and be excited to an electronically excited state, S_1, S_2, etc. For most organic molecules in the ground state all electrons are paired; hence it is referred to as a singlet state. Also, at ambient or low temperatures all molecules are in the vibrational ground state, conforming to the Boltzmann distribution. The lowest vibrational states of most molecules have energies of hundreds of cm^{-1}. Upon excitation, however, the molecule may reach a vibronically excited state, i.e., a state in which both electronic and vibrational quanta are excited at the same time. The absorption process is momentanous: the interaction of the molecule and the electromagnetic field is so fast that the molecular structure does not have the time to adjust to the changed electronic structure (the Franck–Condon principle). In Fig. 9.1 this is symbolized by the vertical arrows. In principle, all transitions from the ground state to the vibronically excited states are possible. However,

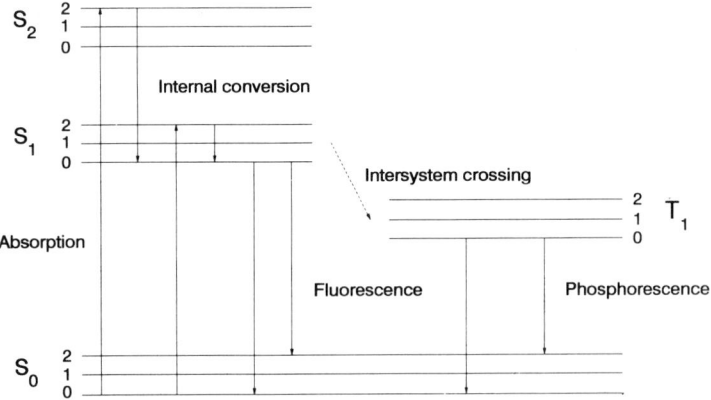

Fig. 9.1. Typical energy diagram of a polyatomic molecule, showing the possible excitation and deexcitation pathways.

INTRODUCTION

the intensity of the transitions is governed by the transition moment that couples the two states involved in the excitation or emission process. In the simplest case, the Born–Oppenheimer approximation, the electronic part and the vibrational part of the vibronic states involved can be treated separately and the transition moment is given by

$$\mathbf{M}_{i,j} = \langle \phi_i \boldsymbol{\mu} \phi_j \rangle \langle \chi_i \chi_j \rangle \tag{9.1}$$

where ϕ_i and ϕ_j are the electronic wave functions and χ_i and χ_j the vibrational wave functions of the vibronic states, and $\boldsymbol{\mu}$ is the electric dipole operator (that only works on the electronic wave functions). The overall intensity of the transition is governed by the first (electronic) term in Eq. (9.1); the second term, the Franck–Condon overlap integral, determines the intensity distribution over fundamentals and overtones.

From the initially excited state the molecules rapidly (time scale: 10^{-12} s) dissipate their excess energy until they end in the S_1 state. This process, in which heat is generated, is called internal conversion. The molecular system stays in the singlet manifold. Another pathway is intersystem crossing: here the singlet excited system crosses over to the triplet manifold. Once the molecular system is in the triplet manifold, again fast internal conversion to the T_1 state occurs and subsequent phosphorescence may be observed when the excited molecule decays radiatively to the ground state. The phosphorescence involves a transition from an excited triplet state to a singlet ground state; this is a forbidden transition, characterized by a very long lifetime (10^{-3}–10^{+1} s). Phosphorescence is rarely observed in organic molecules unless one has taken special precautions, such as working in the solid state or careful deaeration of the sample (4).

Finally, part of the excited molecules arrives in the vibrational ground state of the S_1 electronically excited state. From there radiative decay—or "fluorescence"—may occur, or radiationless deactivation. The fluorescence lifetime in general is on the order of 10^{-9}–10^{-7} ns, which is much slower than the internal conversion processes but much faster than the rate of phosphorescence. Fluorescence can be observed for many organic molecules. Fluorescence transitions are found from the UV (e.g., around 260 nm for benzene) into the near-IR (e.g., around 900 nm for carbocyanine dyes with extended π-electronic systems). The fluorescence decay can both proceed to the vibrational ground state and to vibronically excited states of the S_0 state. Hence, the fluorescence spectrum gives information on the vibrational mode spectrum of the ground state. The absorption spectrum gives similar information on the vibrational structure of the excited states. As the vibrational modes are not strongly affected by the change in electronic structure that is the result of the excitation process, the fluorescence emission spectrum and the absorption spectrum often show mirror symmetry. This is in particular the case for room-temperature spectra of molecules in liquid solutions. Under these circumstances the vibronic fine structure cannot be observed as the individual vibronic lines are blurred owing to short lifetimes and dynamic interactions with the solvent molecules. In low-temperature matrices the vibronic fine structure can be observed, as will be discussed in Section 9.4. Then deviations from mirror symmetry, mainly originating from interactions of vibronic transitions with higher excited electronic states (made allowed via "vibronic coupling effects"), may become clear (5). Spectral bandwidths and vibronic fine structure of fluorescence spectral will be discussed further in Section 9.4.

An important parameter of fluorescence spectroscopy is the intensity of the emission. This parameter is defined by

$$I_f = k\phi\alpha\Phi \tag{9.2}$$

where k represents an instrument-dependent factor (incorporating the efficiency of light collection, sensitivity of detectors, etc.); ϕ is the radiant flux of excitation; α is the spectral absorptance of the sample ($\alpha = 1 - 10^{-A}$, where $A = \varepsilon l c$ is the absorbance of the sample, ε is the extinction coefficient at the wavelength of excitation, l the path length, and c the concentration). The parameter Φ symbolizes the photon yield (or quantum yield) of fluorescence; this factor is—other than in some exceptional cases—independent of the excitation wavelength. The dependence on α implies that the emission intensity will vary as a function of excitation wavelength, as the excitation source scans the absorption spectrum of the molecule. Therefore, it is possible to obtain the absorption spectrum from a fluorescence excitation scan (i.e., a scan of the wavelength of the excitation source, while simultaneously the intensity of the sample emission is monitored), provided that the wavelength dependence of the excitation source is adequately taken into account (6). The quantum yield Φ is defined as the ratio of the number of fluorescent photons emitted to the number of photons absorbed. If we examine Fig. 9.1 it is clear that the quantum yield is given by the relative rates of decay from the excited state:

$$\Phi = \frac{k_{fl}}{k_{fl} + k_{isc} + k_{nr}} \tag{9.3}$$

where k_{fl} is the rate constant of radiative fluorescence decay from the S_1 state; k_{isc}, the rate of intersystem crossing to the triplet manifold; and k_{nr}, the rate constant of nonradiative internal conversion to the ground state.

The fluorescence quantum yield can vary from close to 1, when the rate of fluorescence is much larger than that of radiationless processes (e.g., for many molecules in low-temperature solid solutions, or for extended conjugated systems with a rigid structure, like perylene), to 0 (e.g., for molecules that possess many rotational degrees of freedom).

Equation (9.2) indicates that the fluorescence intensity varies linearly with the radiant flux of excitation. This implies that extremely low detection limits can in principle be reached when high-power lasers are used as excitation sources. At very high excitation intensities sample blanks in general determine the limit of detection; sample blanks can be caused by fluorescent impurities or by background Raman scatter of the solvent. Approaches have been described to eliminate or strongly reduce these blank signals so that shot noise limited detection can be attained. The sensitivity of laser-excited fluorescence measurements will be discussed in more detail in the next paragraph. The inherent high sensitivity of fluorescence spectroscopy, which is caused by the absolute character of the emission measurement, is in contrast with that of relative measurements like absorption spectroscopy. In relative measurements an increase of the source flux does not lead to a concomitant increase in sensitivity as the ratio of the incident and transmitted radiation is recorded. Hence, in relative measurements the stability of the source flux is a much more important determinant of sensitivity. A disadvantage of absolute measurements, however, is that the spectral data are strongly determined, or rather, distorted, by the equipment used to acquire the spectra. Elaborate procedures to correct for instrumental disturbances of the spectra have appeared in the literature (6–8).

Another characteristic of fluorescence is its lifetime. The normally observed fluorescence lifetime is determined by the average time that the molecule spends in the excited state. Thus, it is given by [cf. Eq. 9.3]

$$\tau = \frac{1}{k_{fl} + k_{isc} + k_{nr}} \qquad (9.4)$$

The decay from the excited state is a random process that in most cases follows a single exponential behavior; in this case the fluorescence lifetime τ is defined as the time in which the population of the excited state is reduced to $1/e$ of the initial population. When no nonradiative processes occur (e.g., at very low temperatures, when intersystem crossing is negligible) the intrinsic lifetime of the fluorophore is observed, which is defined as

$$\tau_o = \frac{1}{k_{fl}} \qquad (9.5)$$

The observed lifetime and the intrinsic lifetime of fluorescence are obviously related to the quantum yield, via the relation

$$\Phi = \tau/\tau_o \qquad (9.6)$$

Lifetime measurements give very useful information on molecular structures and in particular on the interactions of molecules and their environment. Also, time-resolved measurement of fluorescent molecules can be used to sense structural and dynamic properties of their environments. Typical fluorescence lifetimes are on the order of 1–100 ns. Processes on shorter time scales can be probed via fluorescence polarization measurements or via indirect techniques (e.g., optical hole-burning or very fast energy transfer measurements). Lasers offer unique opportunities to study such very short-lived phenomena. Presently, femtosecond lasers are commercially available that allow the measurement of events that occur on the time scale of collisions or reactions in solutions. Thus far, time-resolved measurements have not had much impact on analytical applications. The application of lasers for time-resolved measurements will be further discussed in Section 9.3.

Although many naturally occurring substances are fluorescent, important examples of such fluorophores are polynuclear aromatic hydrocarbons, porphyrins (e.g., chlorophyll), as well as a number of compounds of pharmacological interest (e.g., doxorubicin or quinine), the majority of compounds does not fluoresce. An approach that can be used to measure nonfluorescent compounds is to label them selectively with an intensely fluorescent probe molecule. The potential of fluorescent probes is impressive: not only do they allow for sensitive determination of selectively labeled analytes, they also provide the possibility to do functional measurements. The latter application is mainly utilized in biomedical sciences, where fluorescent probes may be employed to measure a variety of properties of cells. Examples are intracellular pH, the activity of a number of specific enzymes, and the rate of DNA synthesis. In particular in Sections 9.3 and 9.5 a number of applications of fluorescent probes will be discussed.

Whenever fluorescent probe molecules are applied to attain a specifically labeled compound, one has to be aware of possible aspecific staining. In particular for trace analysis, where very low concentrations of molecules have to be labeled, interference of unreacted label, fluorescent (side-)reaction products, and/or labeled impurities will limit attainable detection limits. For applications in biomedical science, where in general one is interested in a labeled particle (e.g., a cell), interference of unspecifically labeled substances is more easily reduced than in homogeneous reaction media. The particles can be cleaned more easily than the dissolved analytes by application of centrifugation or filtration techniques, in combination with washing steps. Interference can be further reduced by application of detection techniques that only focus on the labeled *particles*, so that the dissolved fluorescence of

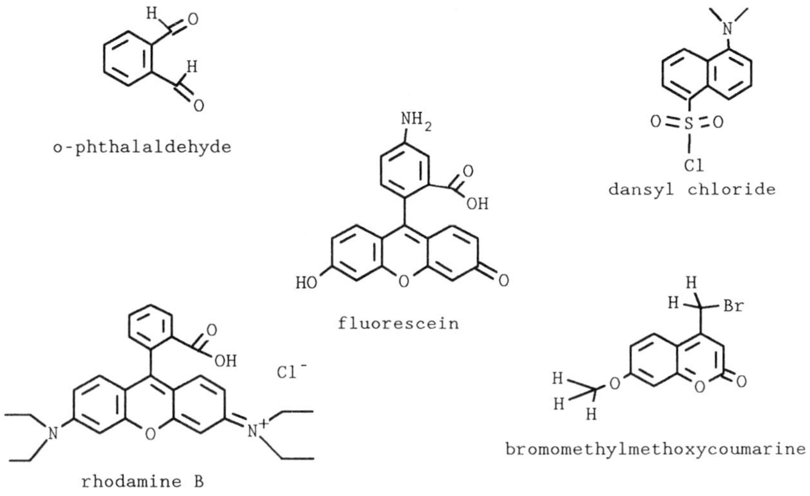

INTRODUCTION

Fig. 9.2. Some examples of fluorescent probes.

unreacted label or fluorescent reaction products does not interfere at all with the determination. This can be easily done in flow cytometry, were all particles are measured individually (see Section 9.5).

An increasing number of highly fluorescent, functionalized, compounds is commercially available, suitable for reaction with a large variety of compounds and functionalities (9–11). The compounds available for fluorescent labeling span the whole wavelength range from UV to, more recently, near-IR. In particular the recent introduction of near-IR fluorescent labels is interesting, in view of the availability of cheap diode lasers that can be used for very efficient excitation in this region. In addition, in this wavelength region there is much less interference of scatter; scatter increases faster than linearly with the wavenumber (with a power of 2–4 depending on the relative dimensions of particle size and wavelength of light). A number of frequently applied fluorescent labels is depicted in Fig. 9.2. In particular, labels such as o-phthalaldehyde (OPT), which are nonfluorescent prior to their selective reaction with primary and secondary amino groups, are very useful. Unreacted OPT does not interfere strongly with the labeled compounds. One of the most frequently applied labels is fluorescein; an impressive number of functionalized fluoresceins are available. Fluorescein is a highly fluorescent compound that can be efficiently excited by one of the main lines of the Ar-ion laser, at 488 nm. Numerous other fluorescent labels have been reported; we refer the reader to the literature for further details (9,10).

9.2. LASERS AS EXCITATION SOURCES

9.2.1. General

In analytical sciences the possibilities afforded by molecular fluorescence techniques have been strongly extended by exchanging classical light sources, such as xenon or mercury lamps, for lasers (12). In some fields the availability of lasers is even crucial. This applies to remote sensing, which requires a reasonable sample irradiance at locations far away from the light source (13); for fluorescence detection in micro–liquid chromatography (micro-LC) and capillary zone electrophoresis (CZE), where extremely small sample volumes must be analyzed (14); for flow cytometry, where particles of small size must be counted (15); and for fluorescence line-narrowing spectroscopy, a high-resolution, cryogenic technique based on the monochromaticity and thus the spectral radiance provided by the light source (16,17). The aforementioned examples illustrate that two inherent features of lasers are of extreme importance: the directionality of the radiation (which enables the beam to be collimated down to spot sizes of 1 μm) and its monochromaticity. It should be emphasized that the monochromaticity of the laser radiation, though very important or even essential in high-resolution molecular fluorescence spectroscopy, is not useful in any other analytical application. The reason is that the excitation spectra of fluid or rigid solutions are normally broad-banded so that only for a small fraction of analyte molecules does adequate spectral overlap with the wavelength profile of the laser line occur.

In addition to the techniques just mentioned, in which the laser plays an essential role, there are various subjects in analytical fluorescence spectroscopy in which improvements can be realized by using a laser instead of a lamp. Of course in this context the positive effects should balance the increment of costs and possible decrease in ruggedness and ease of operation of the analytical technique under concern. An example is Shpol'skii spectroscopy: in this mode of high-resolution molecular fluorescence spectroscopy the observation of narrow fluorescence lines is due to matrix-analyte interactions, in contrast with the fluorescence line narrowing (FLN) technique. Therefore, conventional light sources are adequate to produce the highly resolved Shpol'skii spectra (16,17). Nevertheless strong improvements in sensitivity and selectivity are obtained by using a tunable laser instead of a lamp since not only the fluorescence emission but also the excitation spectra are composed of narrow lines (see Section 9.4.3).

In conventional-size liquid chromatography, detector cell volumes are usually in the 5–10 μL range so that focusing of radiation is not a serious problem and lasers are not strictly needed to apply molecular fluorescence detection. Nevertheless, lasers are involved to improve concentration detection limits, based on the fact that the fluorescence signal is proportional to the irradiance of the sample.

Finally we note that, due to the high spectral irradiances provided by lasers, alternative excitation processes become accessible, such as multiphoton ionization (18) and two-photon-excited fluorescence spectroscopy (19). Within the context of the present chapter only the latter example is of relevance. If classical light sources or continuous lasers are applied, two-photon excitation is extremely inefficient. It has been shown, however, that impressive sensitivities can be reached if appropriate pulsed lasers are utilized, which is readily conceived since the fluorescence signal is proportional to the incident radiation power squared. As a result it is possible to excite UV-absorbing analytes by means of visible light.

The following items, essential in the application of laser-induced molecular fluorescence, will be discussed consecutively: freedom in wavelength selection, advantages and disadvantages of monochromaticity, limits of sample irradiance, and potential of pulsed versus continuous lasers. Furthermore, some recent developments in laser instrumentation will be discussed, with particular attention to gas ion lasers, Ti:sapphire lasers, ultrafast laser systems, and semiconductor diode lasers.

9.2.2. Characteristics

9.2.2.1. Wavelength

From an analytical point of view in the ideal situation there is not any restriction in wavelength choice, so that optimum excitation conditions for the analyte under concern are attainable, corresponding to a maximum signal-to-noise or signal-to-background ratio. Unfortunately, in general lasers provide only a small number of discrete lines, as is shown in Tables 9.1 and 9.2, where the data for the lasers commonly used in molecular fluorescence are collected. Of course dye lasers are wavelength tunable, but it should be realized that the range of tunability per dye is relatively small (typically 50 nm, depending on the particular dye) and that tunability in the UV is only achieved by nonlinear techniques such as frequency doubling of the dye laser output. In the latter approach, the tunable range in the UV is reduced by a factor of 2 in comparison to the tunable range of the dye laser (i.e., to about 25 nm) and the power of the UV laser lines is only 10–20% of the power of the dye laser lines. To summarize, the limited freedom in wavelength choice is one of the main disadvantages in laser-induced molecular fluorescence. Within this context the development of tunable lasers, such as the Ti:sapphire laser, is very interesting, as will be discussed below.

Another aspect related to wavelength is the line width of the excitation light; whereas for conventional sources generally widths of 5 nm or more are utilized since otherwise the irradiance of the sample becoms too low, laser lines are extremely monochromatic, i.e., ranging from typically 0.1 nm for a nitrogen laser down to 0.001 nm for an argon-ion laser. Even much lower values can be achieved under interferometric constraints (etalons). Note, however, that in many applications of laser-induced fluorescence such an extreme monochromaticity can be qualified as overkill. Only in high-resolution molecular fluorescence line widths down to about 0.1 nm are appropriate, but even in this mode of fluorescence higher monochromaticity is not useful. In fluid and frozen solutions excitation bands are usually at least 10 nm wide. Here, the only positive aspect of laser excitation is that the scattered laser light (i.e., Rayleigh and Raman scatter) is also narrow banded, so that it can more easily be discriminated from the broad-banded analyte fluorescence emission.

9.2.2.2. *Directionality*

Laser radiation has a high directionality, the most obvious exception being the recently developed diode lasers, which contrary to conventional lasers reveal a divergent beam (divergences of 30° are not uncommon) of an elliptical shape. The features of diode lasers will be discussed separately.

It is appropriate here to describe the directionality more exactly. The parameter of relevance in laser-induced molecular fluorescence is the number of photons per second that reaches the sample and thus can be used for excitation. This implies that the quality of the source is not determined by the power of the radiation [J(joule)/s(second) = W(watt)] but by the power per unit area of the source (m^{-2}) per unit solid angle [sr(steradian)$^{-1}$], a parameter denoted as the radiance of the source. The most obvious difference between lasers and conventional lamps is that for lasers the emittive surface is very small and furthermore that a divergence of less than 0.5 mrad can be achieved (corresponding to a spot size of 0.5-mm diameter at a distance of 1 m from the laser) whereas classical sources in principle emit equally in all directions.

In fact, laser beams of good quality, such as provided by the argon-ion laser, can be focused easily to spots of 10 μm and by means of microscope lenses even down to about 1 μm (which approaches the limiting value in view of the wavelength of the radiation).

Usually the number of photons per second arriving at the sample is given by the irradiance (expressed in watts per unit area of the sample). It should be realized however that especially in focused laser beams the irradiance is not constant in the direction of propagation, so that it is better to consider the photon density in the irradiated sample volume (14).

In principle the fluorescence signal of the analyte increases with the sample irradiance. In practice, however, the application of too high irradiances causes a lot of trouble. First of all, detector cell damage due to plasma formation has to be circumvented; depending on the materials used, these effects are encountered at irradiances higher than $10^8 \text{ W} \cdot \text{cm}^{-2}$. Furthermore, also in molecular fluorescence, although much less pronounced than in atomic fluorescence, saturation effects are encountered (at a molar absorptivity of $10^5 \text{ M}^{-1} \cdot \text{cm}^{-1}$, typically at irradiances of about 10^8–$10^{10} \text{ W} \cdot \text{cm}^{-2}$). In many cases, however, especially in fluid samples, the maximum signal that can be achieved is not determined by saturation but by photochemical decomposition of the fluorophores. Of course the efficiency of this process depends on the chemical structure of the particular analyte, the composition of the solvent, and the flow rate applied. Noticeable effects at irradiances of as low as $100 \text{ W} \cdot \text{cm}^{-2}$ in flowing systems have been reported in the literature (20,21).

9.2.2.3. Time

Lasers provide either continuous or pulsed radiation. Some of them can be operated both in the continuous and in the pulsed mode, as for instance the argon-ion laser: if it is mode mocked, short pulses are produced at a high repetition frequency.

From an analytical point of view pulsed lasers provide some additional potential compared to continuous lasers. In principle, use can be made of time resolution as a selectivity parameter since different fluorophores generally have unequal fluorescence lifetimes (in practice ranging from a few nanoseconds up to about 500 ns), or the signal-to-background ratio can be improved, as background scattering is momentarily on the fluorescence time scale. If the pulse duration is short enough even completely time-resolved fluorescence spectra can be recorded, i.e., spectra reflecting the complete temporal behavior of the analyte after being excited to the Franck–Condon excited state, thus including the relaxation to the S_1 state (22). Such relaxation phenomena occur on the picosecond time scale. Finally, pulsed lasers with a high pulse energy and a not too low duty factor are far more efficient than continuous lasers in two-photon-excited fluorescence (23–25).

The above considerations imply that various parameters are needed to qualify the performance of a pulsed laser:

- t, the time duration of the pulse (s)
- f, the repetition frequency, i.e., the number of pulses per second (s^{-1})
- The product ft, a dimensionless parameter denoted as the duty factor, representing the fraction of time that photons are produced by the laser

- P, the average power (W), i.e., the amount of energy produced in a 1-s period
- The pulse energy P/f (J) and the pulse power P/ft (W)

In two-photon-excited fluorescence the signal is proportional to the incident power squared. Thus, for a continuous laser with a power P it is proportional to P^2. For a pulsed laser with an average power P the proportionality factor is the pulse power squared $(P/ft)^2$, multiplied by the duty factor ft, giving P^2/ft, which is much higher than P^2 if ft is much smaller than 1. Duty factors for some relevant pulsed laser systems are presented in Table 9.2 (see Section 9.2.3, below).

Apart from a comparison between pulsed and continuous lasers, output fluctuations in time should be considered. These fluctuations are extremely important in analytical chemistry if it is the object of the experiment to perform measurements near the detection limit. It should be realized that lasers are relatively noisy compared to conventional light sources, so that

Table 9.1. Typical Characteristics of CW Lasers Applied in Analytical Molecular Fluorescence Spectroscopy[a]

Type	Stability[b] (%)	Wavelength (nm)	Power (mW)
He:Cd	1–2	325	1–10
		442	2–40
Argon-ion			
Air cooled		488, 514	5–150
Water cooled, small frame	0.5–1 (0.05)	Several lines, 458–530	2000–4000 (514)
Water cooled,[c] large frame	1–2 (0.05)	275 (optional), 334, 351, 364	1000 / 1000–2000 (351)
	0.5–1 (0.05)	Several lines, 458–530	10,000–20,000 (514)
Diode			
Ga:Al:As	0.01	750–860	3–40
In:Ga:Al:P (visible)	0.01	670	10
	0.01	635	3

[a] For details on commercial systems see *Laser Focus World, Buyers Guide*, **26** (1991).
[b] Values within parentheses are achieved if feedback control stabilization is applied.
[c] Can also be operated mode-locked providing pulse, with 150-ps duration at a 76-MHz repetition frequency.

flicker noise in laser-induced molecular fluorescence frequently plays an important role, especially if the analyte signal is recorded simultaneously with a high luminescence background.

For continuous lasers typical stabilities are 1–2% for the He:Cd laser, 0.5–1% for the argon-ion laser, 0.5–1% for the He:Ne laser, and only 0.01% for a stabilized diode laser. Recently, laser stabilizers have become commercially available that, for instance, in combination with the argon-ion laser improve the stability to 0.05% (26).

In pulsed lasers the peak-to-peak fluctuations usually determine the noise of the signal and the background. Furthermore one has to deal with jitter, i.e., fluctuations in the number of pulses provided per second. Peak-to-peak fluctuations can be as large as 10%. Of course these effects are the most serious at a low repetition frequency, whereas they are small if during the response time many signals are averaged (the noise is reduced by a factor of $f^{1/2}$). Besides it is noted that not only do fluctuations in peak power play a role in many experiments, but also the associated spatial fluctuations in the position of the laser spot. This aspect can be of importance if (extremely) small sample volumes are analyzed.

Finally, it should be realized that in experiments where the signal is proportional to the radiation power squared, the effect of laser instability and/or peak-to-peak fluctuations is doubled. Such an increase is observed if for instance the 514 nm argon-ion laser line is frequency-doubled to 257 nm: the instability of the latter line is typically 2%. Similar effects are observed in two-photon-excited fluorescence where the fluorescence signal is proportional to the power squared.

9.2.3. Developments in Instrumentation

The lasers that till now have been most commonly used in analytical sciences are listed in Tables 9.1 and 9.2. Although hardly applied in analytical work as yet, some interesting new laser types and/or developments in laser instrumentation are worthy of consideration in the present context. Consecutively, in the following subsections, progress in the fields of ion lasers, solid-state lasers, ultrafast laser systems, and semiconductor lasers will be discussed with the emphasis on their potential in analytical chemistry.

9.2.3.1. Gas-Ion Lasers: The Neon Laser

The ion laser, with its strongest representative the argon-ion laser, is without doubt one of the most important laser types in research and development laboratories. The basic disadvantage of this laser type is its inherently low efficiency, i.e., typically less than 0.1%. This implies that for every watt of

Table 9.2. Typical Characteristics of Pulsed Lasers Applied in Analytical Molecular Fluorescence Spectroscopy[a]

Type	Wavelength (nm)	Average Power (W)	Peak Power (MW)	Pulse Energy (mJ)	Pulse Duration (ns)	Repetition Frequency (Hz)	Duty Factor
Nitrogen	337	0.1–0.3	0.1–1	1–10	5	10–100	$5 \times 10^{-8} - 5 \times 10^{-7}$
Copper vapor	511, 578	5–50	0.03–0.3	1–10	30	5000	1.5×10^{-4}
Excimer Xe:Cl	308	20	10	200	20	100	2×10^{-6}
Nd:YAG	1064[b]	10	100–200	1000	5–10	10	10^{-8}

[a] For details on commercial systems see *Laser Focus World, Buyers Guide*, **26** (1991).
[b] Can be doubled, tripled, and quadrupled to 532, 355, and 266 nm, respectively.

output power several kilowatts of energy must be dissipated. Furthermore, for the larger lasers the power supply must be water-cooled.

The argon-ion lasers can be grouped into three broad classes (27):

1. Small-frame air-cooled lasers; 488 and 514 nm, 5–150 mW
2. Small-frame water-cooled lasers; 8 wavelengths between 458 and 514 nm; typically 4 W
3. Large-frame water-cooled lasers; typically providing 25 W in the visible, 7 W in the near-UV, and around 1 W in the deep-UV (i.e., from 275–305 nm)

The deep-UV output of the large-frame air-cooled argon-ion laser has recently been achieved and as far as we know not yet applied in analytical chemistry. Evidently, important manufacturer parameters are tube lifetime, efficiency, reliability, and the development of computer-controlled options for operation.

As regards the near-UV, the commercial introduction of a new noble gas ion laser, i.e., the neon-ion laser, seems to be very interesting (27). It provides at least 1-W power output at 332.4 nm [continuous wave (CW)], which is very close to that of the He:Cd laser line at 325 nm, typically 10 mW. The lines of the neon-ion laser are located at

332.4 nm	∼ 1.0 W
334.5 nm	∼ 0.1 W
337.8 nm	∼ 0.4 W
339.3 nm	∼ 0.6 W
371.3 nm	∼ 0.1 W

It should be emphasized tht the neon-ion laser is basically equivalent to a small-frame water-cooled ion laser, but provides UV-output that would require a large-frame argon-ion laser to match. The reason is that the neon ions involved are singly ionized whereas comparable argon UV output involves doubly ionized species. A potential application of the neon-ion laser is pumping dye lasers for "dyes" with an absorption band in the 330–340 nm region. For the other noble gas ion lasers the small-frame versions are not appropriate for dye laser pumping.

9.2.3.2. Solid-State Lasers: The Titanium:Sapphire Laser

In 1982 the lasing of titanium-doped sapphire was demonstrated for the first time. Since then a lot of effort has been devoted to the development of this laser type, and for a few years now the Ti:sapphire laser has been commercially available (28,29). The absorption of titanium-doped sapphire is in the

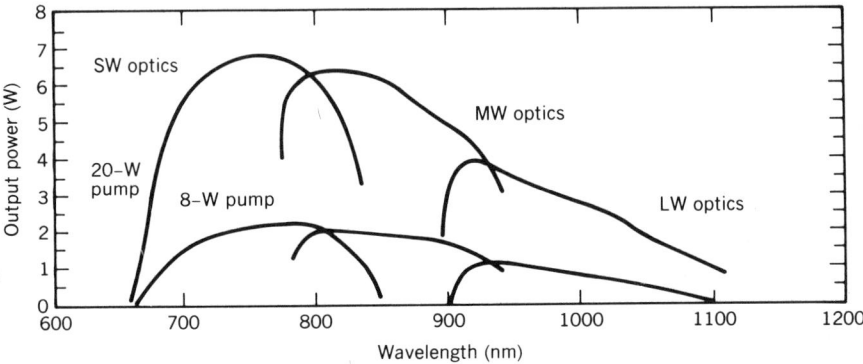

Fig. 9.3. Tuning curves of the Model 890 titanium:sapphire laser (Coherent, Palo Alto, CA), pumped by argon-ion lasers of 8 and 20 W; SW, MW, and LW: short-, medium-, and long-wavelength optics.

blue-green part of the spectrum so that it can be efficiently pumped by the argon-ion laser. The most remarkable feature of this new solid-state laser is that it provides a continuous output in the red and near-IR from 690 to 1100 nm. This wavelength region is "scannable": only three mirror sets are needed. Furthermore, the power output is quite impressive: the curves obtained for 8 and 20 W argon-ion laser pumping are depicted in Fig. 9.3. Thus it will be obvious that this laser is a very good alternative to the dye lasers commonly used to cover this spectral range. This is quite interesting because dye lasers have some inherent disadvantages: nasty organic dyes and solvents, little output power, and much downtime for maintenance.

Also pulsed Ti:sapphire lasers are entering the market, pumped by flashlamp or by a frequency-doubled Nd:YAG or Nd:YLF laser[1] (which in turn may be pumped by a high-power diode laser). Of course, frequency doubling and/or mixing into the visible part of the electromagnetic spectrum will be very interesting from an analytical chemistry point of view. Furthermore, in the coming years sub-picosecond Ti:sapphire lasers will become commercially available, so that time-resolved spectroscopy will be possible in the near-IR part of the spectrum.

9.2.3.3. *Ultrafast Laser Systems*

Ultrafast laser systems are directed to the generation of ultrashort pulses, i.e., pulses of sub-picosecond duration. Picosecond measurements date back to the mid-1960s, when the first mode-locked solid-state laser was made

---

[1] YAG = yttrium aluminum garnet; YLF = yttrium lithium fluoride.

operational. As early as 1970 it was discovered that on passing intense picosecond pulses of a mode-locked laser through (for instance) a beaker of water a broad continuum of light was generated covering typically 10.000 cm^{-1} in pulses on the order of 100 fs. The light produced by this method is usually referred to as an ultrafast supercontinuum laser source.

In the early 1980s, all commercial ultrafast systems were based on mode locking of argon-ion lasers and synchronous pumping of single-jet dye lasers (30). Such a system is capable of generating pulse widths of about 5 ps over a limited range of wavelengths and power levels. Since then, the ion lasers have been replaced by mode-locked solid-state lasers, the most popular representative being the frequency-doubled (by means of a KTP[2] crystal) Nd:YAG laser. More recently a variety of new lasing materials have started to become available, for instance, the Nd:YLF laser in which the crystal yttrium lithium fluoride is the host material for neodymium (31). Two essentially different new technologies have been introduced to reduce the pulse width, i.e., hybrid mode locking and pulse compression. A discussion of these techniques is beyond the scope of this chapter.

Typical systems now have pulse width capabilities of 100 fs, while outputs of 30–50 fs have already been utilized. The wavelength coverage has been extended through nonlinear conversion processes. Furthermore, amplification techniques have become relatively routine in the research. Without amplification the 100-fs pulses provide up to 1-nJ energy. They can be amplified as much as 6 orders of magnitude so that 1-mJ energy can be reached. The latter amount of energy corresponds to a peak power of 1 mJ/100 fs = 10^{10} W!

Three points should be noted. First of all, owing to the Heisenberg time-frequency uncertainty principle $\delta n \, \delta t \geq 1/2\pi$, pulse compression down to the picosecond region has to be at the expense of monochromaticity. Secondly, regardless of the short-pulse generation technique used, femtosecond pulses are not routine. The shorter the pulse width required by the experiment, the more complex and sensitive is the source. For example, at 100 fs a synchronous pumped dye laser must be stable in length to about 100 nm. Finally, in various experiments the fundamental limit of capability is not the pulse width but the form and the stability of the output pulses. A lot of effort is devoted to this aspect. In view of its noise characteristics, the Nd:YLF laser seems to have a promising future. Unfortunately YLF as a host material is much more difficult to handle than YAG.

In ultrafast spectroscopy, i.e., spectroscopy on a time scale from 10 fs (the time span in which light travels only 3 μm!) to 100 fs, photomultipliers cannot be used for detection purposes since they are too slow (22). Instead multiple

---

[2] KTP = potassium titanyl phosphate.

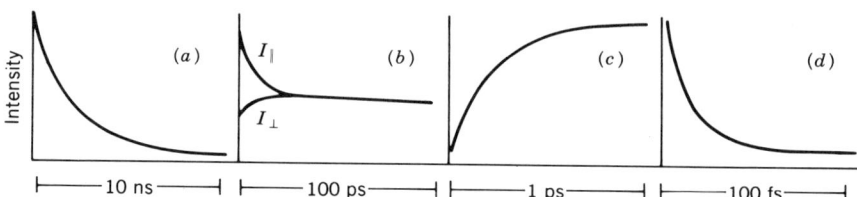

Fig. 9.4. Typical temporal behavior of fluorescence decay (a), fluorescence anisotropy (b), increase of fluorescence (c), and resonance Raman scatter (d).

beam techniques are required. In these techniques, commonly denoted as pump/probe spectroscopy, two laser pulses are used: in fluorescence, a pump pulse to excite the sample and a probe pulse to induce stimulated emission. The stimulated emission intensity is proportional to the fluorescence intensity as long as the probe is not too intense. Crucial here is that the arrival time of the probe pulse is controlled by the optical path length of the probe beam, which can be readily varied.

To date the role of ultrafast spectroscopy in analytical chemistry is rather limited, as has been outlined by Wirth and Blanchard in 1986 (30) and by Wirth in 1990 (22). The most interesting applications are in resonance Raman spectroscopy (RRS) and in time-resolved fluorescence spectroscopy. In RRS rather intense Raman spectra are obtained owing to the resonance effect (i.e., excitation occurs to a real excited state of the analyte and not to a virtual state as in conventional Raman spectroscopy). However in general RRS is severely hindered by a strong fluorescence background. In ultrafast Raman spectroscopy this background is effectively rejected because fluorescence is extremely slow compared to Raman scatter, as clarified in Fig. 9.4.

In time-resolved fluorescence spectroscopy the analytes are monitored immediately after having absorbed a photon: the molecule and its direct surroundings reorient themselves in a process that is too fast to be observed on the nanosecond time scale. Thus this mode of spectroscopy provides detailed information on the short-range structure of a liquid. Fluorescence depolarization measurements (the time scale is indicated in Fig. 9.4) are used in biochemistry to obtain structural information on macromolecules.

9.2.3.4. Semiconductor Diode Lasers

Diode lasers currently represent the most rapidly growing area of the laser market. The best known is the GaAlAs laser emitting at 780 nm. Its 5-mW version can be considered as the workhorse of the compact disc industry and is manufactured in the tens of millions annually (32,33).

Diode lasers are solid-state lasers like the Nd:YAG laser, but their working principle is fundamentally different. In the Nd:YAG solid-state laser, the relevant energy levels are associated with dopant atoms in low concentrations in a host lattice that are essentially isolated from each other. In a diode laser the valence band and the conduction band of the semiconductor play the crucial role. Use is made of a junction between *p*-type and *n*-type semiconductors. The former has impurity atoms in its lattice that possess fewer valence electrons than do the atoms they replace, thus creating positive holes; in the latter type the substitute atoms possess more valence electrons than do the atoms they replace. By applying an electric potential across a junction between a *p*- and *n*-type crystal, electrons crossing the boundary drop from a conduction to a valence band, emitting radiation, so that a light-emitting diode (LED) is obtained. If the *n*-type and *p*-type semiconductor materials are sufficiently well doped and a high enough electric current is applied, then a population inversion of electrons and holes can be induced in the junction region. By enclosing this junction in an optical cavity, a semiconductor diode laser can be produced.

The most familiar semiconductor materials are the Class IV elements silicon and germanium. However also binary compounds between Class III and Class V elements such as GaAs (gallium arsenide) exhibit similar behavior. In the latter semiconductor material in the *p*-type some of the arsenic atoms are replaced by gallium, whereas in the *n*-type the converse is true. By now, further refinements of this approach have been realized and roughly four laser diode categories can be distinguished:

- The Ga:Al:As diodes (metioned above), emitting in the 750–860 nm range
- The In:Ga:As:P diodes, which lase from 1300 to 1600 nm, with applications primarily in fiber-optic communications and potential in fiber sensors
- The In:Ga:As diodes, operating from 910 to 990 nm
- The In:Ga:Al:P diodes, frequently denoted as visible laser diodes, emitting in the 630–680 nm range

Four important developments in diode laser device technology should be mentioned. Continuous advances are being made toward higher powers, lower wavelengths, greatly increased longevity (50,000 h), and lower operating thresholds. Considering the last point, thresholds currents below 10 mA have already been realized, which is very important in view of battery applications.

High-power semiconductor lasers are appropriate pumps for the Nd:YAG laser. The attention is mainly focused on the Ga:Al:As diode. In this approach laser diode arrays are used, providing as much as 10-W power (CW). Pulsed versions have also been developed, and frequency doubling can be invoked to generate visible laser lines.

Maybe more important is the development of the visible small-sized, low-power laser diode, which became commercially available in 1988. The most reliable of this type by now is the 670 nm version with up to 10-mW power. It is expected that the 635 nm version, the lowest wavelength successfully demonstrated thus far, will become commercially available soon. Recently, a Cd:Zn:Se diode laser was reported that produced light at 490 nm.

Diode lasers have the advantage of a very high energy conversion efficiency (typically 20%), and—even more important—their size is extremely small. Currently, the smallest lasers, a diode laser mounted complete with heat sink, photodiode detector, and protective window can measure as little as 50 mm^3, yet deliver as much as 100 mW of continuous power. The collimation of diode lasers is poor: typically the junction of a GaAs laser has an effective thickness of only 0.1 μm, but the active region may be as much as 5 μm wide. Consequently, the beam divergence is considerable (10° is not unusual) and also asymmetric. However, corrective optics are usually included in commercial instruments. In general, diode lasers are temperature tunable from typically -10 to 40 °C, which provides a tuning range of about 10 nm. The modes in a diode laser are typically separated 1–2 cm^{-1}, and the individual modes (utilized in single mode operation) have generally narrow line widths of about 10^{-3} cm^{-1}.

9.3. LASER-INDUCED FLUORESCENCE AS THE DETECTION METHOD IN SEPARATION TECHNIQUES

9.3.1. Introduction

Since the first publication on laser-induced fluorescence (LIF) detection in liquid chromatography, written by Diebold and Zare in 1977 (34), reporting detection limits of 2.5 fmol for aflatoxins, extensive attention has been paid to this subject in the literature. Various reviews have been presented [for instance, van den Beld and Lingeman covering the literature up to 1988 (35)]. For this reason we do not intend to present here a complete treatment of the results achieved up to the very moment. Within the context of this chapter, which is directed toward molecular laser fluorescence in general, it is appropriate to discuss some trends in LIF detection in separation techniques and to focus on important features not always discussed explicitly in the literature.

First, instrumental aspects are considered, i.e., maximum detector cell volumes allowed from a separation point of view and achievable detection limits. As an introduction of the latter aspect, various definitions used in the literature are presented and subsequently the influences of laser power (irradiance), background signals, and noise terms are discussed. Only minor attention is paid to detector cell concepts and geometrics since in this field no recent improvements have been realized. In capillary electrophoresis (CE) the sheath-flow cell has been reported to be very appropriate (36–38).

Secondly, the role of chemistry in LIF detection in chromatography is indicated. Usually this aspect receives only minor attention, in spite of the fact that in real sample analysis it is the chemistry that limits the achievable detection limits.

Thirdly, some important relatively recent achievements are discussed. Short wavelength excitation has been applied by frequency doubling the 514 nm line of the argon-ion laser, making possible 257 nm LIF detection (39,40). Thus, the applicability of LIF to natively fluorescent analytes is significantly extended. Two-photon excitation has been compared with conventional single-photon excitation (25). The analytical potential of indirect laser induced fluorescence detection has been clearly demonstrated, especially in CE (41). For precolumn labeled analytes both in capillary zone and in capillary gel electrophoresis, extremely low detection limits have been published (36–38).

Finally, the very rapid progress in diode laser technology provides the possibility of developing diode LIF detection (33). In this field there is a real opportunity for analytical chemists to generate the appropriate red or near-IR labels.

9.3.2. Instrumental Aspects

9.3.2.1. Detector Cell Volumes

In discussing the potential and performance of LIF as a detection method in combination with separation techniques it is appropriate to make a distinction between conventional liquid chromatography (LC), on the one hand, and micro-LC and capillary electrophoresis (CE), on the other. In our terminology micro-LC includes small-bore LC, packed-capillary LC, and open-tubular LC. In general it can be stated that in conventional LC conventional fluorescence detectors can be applied. In this field the introduction of LIF is only relevant if the laser excitation provides improvements in achievable detection limits or excitation pathways not possible by means of lamp excitation. In contrast, in micro-LC and CE conventional fluorescence detection has a limited potential. It is hardly possible to perform efficient excitation of analytes in detector cell volumes

Table 9.3. Acceptable Flow Cell Volumes for Various Separation Systems (Utilizing Different Column Dimensions)[a]

Separation Systems	Internal Diameter of Column	Acceptable Flow Cell Volume
Conventional LC	3.0 mm	2 µL
	4.6 mm	7 µL
Small-bore LC	1.0 mm	0.5 µL
Packed-capillary LC	0.3 mm	60 nL
Open-tubular LC and CE	10 µm	0.6 nL

[a] Typical values depending on the analyte under concern and various separation parameters (20,35).

smaller than 0.1 µL. Here lasers must be invoked: owing to the directionality and coherence of the laser beams their radiation can be focused very efficiently so that extremely low detection volumes down to 0.1 nL can be handled (41).

In Table 9.3 some typical detector cell volumes are presented, related to the internal diameter of the separation system at hand. These data represent the maximum acceptable cell volumes from a chromatographic point of view.[3] Larger volumes would cause additional band broadening of the eluting chromatographic peaks and therefore loss of chromatographic resolution and detectability of the eluting analytes.

For the highly promising CE techniques capillary zone electrophoresis and capillary gel electrophoresis, the detection cell volume requirements are similar to those of open-tubular LC. As calculated by Dovichi's group (37), for a typical 1-m capillary of 50-µm internal diameter (external diameter, 190 µm) providing 10^6 theoretical plates, the maximum injection volume (allowing a 5% broadening caused by the injection) is as small as 450 pL. Of course the detection cell also should be smaller than 1 nL to avoid band broadening.

To achieve these extremely small volumes, generally on-column optical detection is applied in the direct or in the indirect mode; a window is constructed by removing a portion of the polyimide coating of the capillary. The obvious disadvantage of this setup in (direct) LIF detection is the high background due to the light scatter caused by the walls of the capillary. Recently, significant improvements in achievable detection limits have been reported by applying a postcolumn sheath-flow cuvette for detection (36–38); see Fig. 9.5. In this approach the capillary is inserted into the center of a

---

[3] Although incorrect, strictly speaking, in the following discussion CE also is considered as a chromatographic technique.

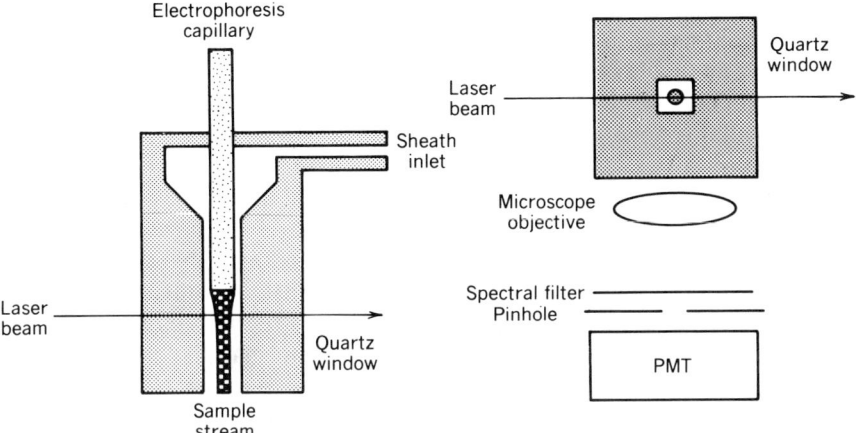

Fig. 9.5. Sheath-flow cuvette for laser-induced fluorescence detection in capillary electrophoresis. The fused-silica capillary, 50 μm i.d. and 190 μm o.d., is inserted into the center of a 200-μm quartz flow chamber. The sample is ensheathed by a flowing buffer stream and forms a thin stream in the center of the flow chamber. For excitation and argon-ion laser operating at 488 nm is used. Fluorescence is collected at right angles with a high-numerical-aperture microscope objective, imaged onto a pinhole that acts as a mask to block scattered laser light, filtered with a band-pass filter, and detected with a photomultiplier tube (PMT). The left part of the figure shows the side view of the flow chamber; the right part depicts the top view of the optical train (37).

200-μm × 200-μm (high-quality) quartz flow chamber resulting in a sample stream with a typical diameter of 10 μm, ensheathed by a buffer stream with a flow rate of typically 0.5 mL/h. The crucial point is that the sample stream and the sheath flow have (almost) identical compositions, so that the light scatter at the interface (which is strongly dependent on the difference in refractive indices) is very weak.

9.3.2.2. Detection Limits

Various Definitions. In considering the performance of a detection technique in chromatography[3] the concentration limit of detection (CLOD) and the mass limit of detection (MLOD), both associated with the injected sample, are of interest. These detection limits are related by

$$\text{MLOD} = \text{CLOD} \times \text{injection volume} \tag{9.7}$$

and correspond to a chromatographic peak with an amplitude of three times the noise mean square.

We note that in micro-LC and CE the extremely low MLODs reported in the literature (42,43) are associated with rather "normal" CLODs; here the injection volume plays the crucial role. For example a 10-nL injection of an analyte solution with a CLOD of 10^{-10} M reveals a MLOD of 10^{-18} mol, i.e., 1 amol (attomole).

Another point is worthwhile emphasizing here. In the literature, instead of reporting the CLOD, in some cases the minimal detectable concentration *at the detector cell* is presented (denoted here as CDET). Since this quantity is independent from the performance of the chromatographic system in use, it is appropriate to make an objective comparison between different detection techniques. It should be realized, however, that owing to the dilution of the injected plug during the elution process, it is not unusual that CLOD is 5–10 times less favorable than CDET.

Finally, we note that the performance of LIF detection can be described by calculating, instead of the MLOD, the corresponding number of moles in the laser-illuminated volume, which can be as small as 1 pL, i.e., much smaller than the volume of the detector cell. In fact, for a CDET of 10^{-10} M and an intersection volume of 10 pL, 10^{-21} mol are detected, i.e., 600 analyte molecules. However, to achieve this extremely low quantity in a chromatographic system utilizing 10-nL injection volume (and assuming dilution by a factor of 10 in the chromatographic system, so that CLOD is 10^{-9} M), 10^{-17} mol or 6×10^6 molecules of fluorescent analyte must be injected.

Influence of Laser Power. Obviously the fluorescence signal observed is determined by F, the number of photons emitted per second by the irradiated analyte solution; F is related to the number of photons absorbed per second, which in turn is proportional to the irradiance I of the sample (W/m^2). When laser excitation is utilized, the irradiance is 10^3–10^6 higher than for conventional light sources. The absorption rate constant k_A, describing the fraction of molecules raised per second to the excited state, is related to the molar absorptivity ε (cm$^{-1} \cdot$M^{-1}) by (44)

$$k_A = \frac{0.23 \varepsilon I}{N_A h v_0} \tag{9.8}$$

In Eq. (9.8) v_0 is the frequency of the excitation light, h is Planck's constant, and N_A Avogadro's number.

For too high irradiances F no longer increases linearly with I, owing to ground state depletion (saturation) and/or to photobleaching (photodecomposition) effects. Saturation effects are observed only at extremely high irradiances (typically $10^{10}\ \text{W}\cdot\text{m}^{-2}$, an irradiance corresponding to a 1 W laser line well focused to a spot of about 10 μm) where k_A approaches k_f, the observed fluorescence decay rate (20,21).

Depending on the fluorophores under concern, photobleaching can be important at much lower irradiances, i.e., $10^6\ \text{W}\cdot\text{m}^{-2}$. The crucial parameter here is the photobleaching efficiency $\phi = k_d/k_f$, wherein k_d is the photodestruction first-order rate constant. In a flow system the overall effect of bleaching depends on the absorption rate constant k_A, the photobleaching efficiency ϕ, and the mean residence time t of the fluorescent molecule in the illuminated part of the detector cell.

Recently, Mathies et al. have presented a thorough study, both theoretical and experimental, devoted to the optimization of high-sensitivity fluorescence detection in continuous flow systems (21). Their general conclusion is that only for not too high irradiances and not too long residence times F increases linearly with I. In order to optimize the conditions, obviously the background signal also should be taken into account. In contrast to the fluorescence signal, the background scattering is always proportional to I. Therefore, the signal-to-noise ratio observed in a flow system is also a function of α, the ratio of background scattering and fluorescence. Mathies et al. conclude that for low values of α, where the signal-to-noise ratio can be approximated as $F/F^{1/2}$ (assuming that shot noise dominates), it is profitable to pump hard (high I) and to use long transit times to extract as many photons from the fluorophore as possible. In contrast, for high values of α the noise on the background dominates, so that the optimal transit time is shorter and the optimal I is lower.

Mathies et al. tested their theoretical expressions for the fluorophore B-phycoerythrin (PE), a compound with a high ε at 514.5 nm ($1.6 \times 10^6\ \text{cm}^{-1}\cdot\text{M}^{-1}$), a t_f of 2.5 ns, and a photodestruction lifetime of 700 μs. In line with their theory, the optimal signal-to-noise ratio was obtained for I equals $3 \times 10^8\ \text{W}\cdot\text{m}^{-2}$ (a relatively low value as a result of the extremely high ε) and a transit time of about 700 μs, close to the photodestruction lifetime.

Background. In the previous subsection we touched upon the important role of the background in optimizing LIF detection. In fact, earlier (in Section 9.2.2.3) it was implicitly assumed that the noise on the background is of the shot-noise type, i.e., equal to $n_b^{1/2}$ if n_b is the number of scattered photons per second. It should be realized, however, that in the practice of LIF detection in liquid chromatography, especially if precolumn labeling has to be applied, the fluorescence background contribution is frequently the most important one and can be so high that flicker noise, being proportional to n_b, dominates the noise (this aspect will be discussed extensively in Section 9.3.4.3). In this situation the signal-to-noise ratio cannot be improved by increasing I even if saturation and photodecomposition effects are prevented (45). Thus, a very important concern in optimization of LIF detection in practical applications is the need to reduce the background as much as possible.

It is appropriate to distinguish the following contributions to the background signal:

1. *Refracted and reflected laser radiation (mainly) at the cell wall boundaries*—To reduce this contribution, in the literature various flow cell geometries and constructions have been evaluated (20,35,46). Of course this problem can be tackled much more easily in conventional LC than in micro-LC. As noted above, in CE the sheath-flow approach has proved to be very successful (36–38).

2. *Rayleigh and Raman scattering from the eluant and maybe from the cell wall material*—The main problem is the Raman contribution, which may interfere with the analyte fluorescence; the Rayleigh light has the same wavelength as the excitation light (20,35,40,47).

3. *Fluorescent background from the flow cell in general (including cell wall material, flow cell holder material, filters and optical fibers), from fluorescent impurities in the eluant, from chemical compounds produced during the derivatization of the analyte, from the derivatization reagent, from photodegradation products formed during laser illumination, and last but not least from coeluting compounds*—Especially in the first contributions not only direct excitation but also excitation by scattered laser light should be considered (39).

In this chapter we will not discuss detector cell construction, as extensive attention has been paid to this subject in the literature (20,35,46). Here our attention is focused on Raman scattering and fluorescence background, signals that may interfere with the fluorescence signal to be detected (20). The relative importance of the Raman term can be delineated as follows:

- Its spatial distribution is not isotropic, in contrast to that of the fluorescence emission.
- Raman scattering is polarized, while fluorescence is not.
- Like Rayleigh scatter, the Raman scatter intensity is proportional to λ^{-4}. This implies that when excitation is compared at 250, 500, and 750 nm the relative Raman intensity decreases as 1000:62.5:12.5. It should also be realized that the fluorescence background generally strongly increases with decreasing excitation wavelength. At short wavelengths many compounds have a nonzero extinction coefficient

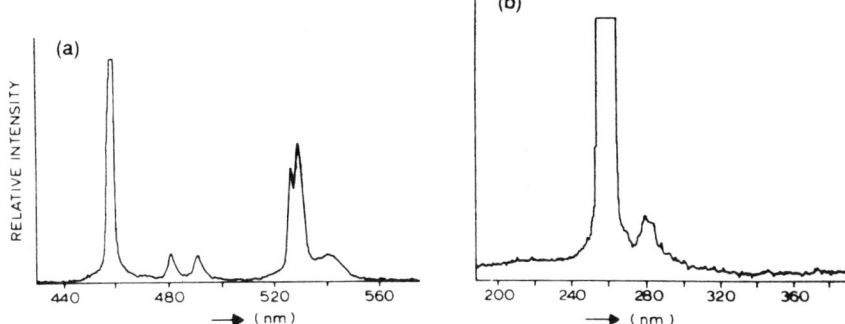

Fig. 9.6. Raman spectra of methanol/water (9:1, v/v) with (a) 458-nm and (b) 257-nm laser excitation (39).

and thus produce a luminescence background even if their quantum yield is extremely low (provided that their concentration is high enough).
- The wavelength range over which the Raman spectrum extends is small for short-wavelength excitation and large for long-wavelength excitation (because the energy range of the spectrum is constant). As a result, by applying 257 nm excitation (the frequency-doubled 514 nm line of the argon-ion laser), fluorescence detection can be done without interfering Raman bands coming from methanol/water eluants, because the longest wavelength Raman peak is found at 285 nm (39); see Fig. 9.6.

Until now LIF detection in LC and CE has been performed almost exclusively using the He:Cd or the argon-ion laser, although recently a low-cost system, i.e., a He:Ne laser operating in the green region (534.5 nm), has been successfully utilized (38). Since only a small number of (natively fluorescent) analytes can be excited by the wavelengths offered by these laser systems, chemical derivatization must commonly be invoked—with its inherent disadvantages, as will be outlined in Section 9.3.3. In view of the foregoing discussion two interesting developments should be noted: first, the introduction of LIF at short-wavelength excitation (e.g., 257 nm), so that the applicability to natively fluorescent analyte is strongly enhanced and Raman can be excluded, with the main disadvantage being the fluorescent background (39,40); secondly, the development of diode laser–induced fluorescence, where long-wavelength excitation is applied (33). Under these excitation conditions the fluorescence originating from impurities is very weak and the Raman interference is the most important one. The crucial problem, however, is a chemical one: how to find an appropriate chemical derivatization reaction that yields tagged analytes excitable by diode lasers and emitting strongly in the red or near-IR, without interference due to the reagent.

Noise. As noted in the preceding subsection, the signal-to-noise ratio in LIF detection increases with I, the irradiance at the sample, as long as shot noise dominates (and as long as saturation and photobleaching can be neglected). In this case the signal is proportional to I and the noise to $I^{1/2}$.

It should be realized, however, that for high laser intensities and high background levels it is not easy to reach the shot-noise limit (45). In many situations flicker noise is the most important noise contribution. Like the signal intensity, this noise is proportional to I, so that under these circumstances the signal-to-noise ratio cannot be improved by increasing the irradiance.

The main contribution to the flicker noise in LIF detection comes from the instability of the laser. It should be realized that lasers are generally noisy in comparison to conventional light sources: 2–5% is not unusual for a He:Cd laser; for an argon-ion laser 0.5% is a typical value, for a He:Ne laser 1%, but for a diode laser only 0.01% [stabilized, even 0.003% (26,33)]. Recently in the literature more attention has been paid to the noise of laser sources: laser stabilizing systems are commercially available which have been applied in combination with He:Cd and argon-ion lasers providing a stability of 0.05% (26). The importance of the reduction of laser noise is exemplified in Table 9.4, where for different background signal intensities, monitored with a photon counter, the shot noise and the flicker noise are compared for lasers with stabilities of 1% and 0.01%. It is obvious from Table 9.2 that for a stability of 1% the contribution of flicker noise can only be neglected at detection levels below 10^4 counts/s whereas for a stability of 0.01% it is negligible up to levels as high as 10^8 counts/s. From that point of view the introduction of the (stabilized) diode laser in LIF detection is very promising [41].

Table 9.4. Relative Importance of Flicker and Shot Noise, for Lasers with Stabilities of 1% and 0.01%

Counts/s	Shot	Flicker (1%)	Flicker (0.01%)
10^8	10^4	10^6	10^4
10^6	10^3	10^4	10^2
10^4	10^2	10^2	1
10^2	10	1	—

9.3.3. Chemical Aspects

Since the introduction of LIF detection in column liquid chromatography and capillary zone electrophoresis, chemical derivatization of analytes has played an important role (35,42,43,46). Nonfluorescent analytes are provided with a fluorescent tag, either in a precolumn or in a postcolumn reaction, in order to enable the application of the extremely sensitive LIF detection technique. Commonly, use is made of existing derivatization procedures, although the availability of only a limited number of laser lines for excitation causes some additional requirements to be met: the maximum in the excitation spectrum of the label should be not too far from the available laser line. To illustrate this point: instead of the well-known OPT (o-phthalaldehyde) reaction for primary amines, in LIF applying a He:Cd laser (wavelengths 325 nm and 442 nm) the analogous NDA (naphthalene-2,3-dialdehyde) derivatization is far more appropriate, since 442 nm is close to the excitation maximum of the fluorescent NDA reaction product (48).

Of course, an important reason to make use of LIF instead of conventional fluorescence detection in LC is to improve concentration detection limits. It should be realized, however, that derivatization of analytes at concentrations as low as 10^{-9}–10^{-10} M gives rise to problems normally not encountered at 10^{-7}–10^{-8} M, as usually applied in conventional fluorescence detection. The same problems are reported in chemiluminescence detection (49–51). Besides we should note that in the literature in some cases derivatizations are carried out at relatively high analyte concentrations and subsequently the product is diluted before injection. However, in everyday practice the derivatization has to be performed with low concentrations of analytes in real samples.

In general the following problems may be encountered. First of all, a surfeit of luminescent label has to be used (a thousandfold to ten-thousandfold excess is not unusual), which possibly interferes in the chromatogram. For instance, labels containing an amine or hydrazine functional group usually provide strongly tailing bands on a reversed-phase column. In such a case, the excess of reagent must be removed prior to HPLC. The second problem is the formation of side products during derivatization and the role of impurities of the label reagent. An example is the NDA derivatization reaction mentioned above, which requires the cyanide ion to produce the fluorescent cyanobenzisoindole structure. However, CN^- also promotes the formation of other fluorescent products (20,50,51). Moreover, in the well-known FITC (fluorescein isothiocyanate) reaction fluorescent side products can be formed (52,53). Finally we note that low surpluses of reagent (let's say a fivefold excess) are not readily applicable in practice: especially at extremely low analyte conditions there is a real chance that the derivatization reaction will be very slow and/or the conversion not complete.

To conclude this subsection it is emphasized that analytes provided with a labeled tag under favorable conditions can be detected by LIF at extremely low levels. Concentration detection limits in conventional size LC as low as 10^{-14} M have been reported (54). However, in everyday practice derivatization reactions at these concentration levels are impossible. Even for derivatization at the 10^{-9}–10^{-10} M level a lot of problems are encountered. Detection at this level can only be carried out if side products and excess of reagent are removed from the derivatization mixture prior to analysis (20,55).

9.3.4. Developments

In this section some recent developments in LIF detection in separation techniques are discussed. First, indirect fluorescence detection is the more generally applicable approach: the basic requirement is that the eluting analytes be able to displace a fluorescing ion present as a solute in the eluant. In this approach, spectroscopic properties of the analyte do not play any role. Substantial progress has been made in combining this technique with miniaturized separation systems (41). Secondly, our attention is focused on analytes that in principle can be detected by conventional fluorescence because of their nonzero fluorescence quantum yield. For these compounds there is no need to utilize chemical derivatization. The constraint is the laser excitation: only a few excitation wavelengths are provided by the laser. Two approaches to solve this problem are the generation of UV laser lines by frequency doubling (39,40) and the application of two-photon excitation (23,25). Thirdly, recent results obtained for LIF detection of peptides are discussed with emphasis on the role of chemical derivatization. It is shown that for a specific dodecapeptide, postcolumn derivatization using the fluorogenic agent OPT is much more favorable than precolumn derivatization using the FITC or NDA label, in spite of the fact that the OPT reaction has some disadvantages from a spectroscopic point of view (20). Subsequently, some recent CE results are presented, i.e., capillary zone electrophoresis of labeled amino acids and capillary gel electrophoresis of labeled DNA fragments, with emphasis on achievable detection limits (36–38). Finally, special attention is focused on LIF detection using diode lasers, a promising field of research that is as yet hardly exploited (33).

9.3.4.1. Indirect Fluorescence Detection

The concept of indirect fluorometry for detection purposes was first developed for conventional size ion chromatography (56). A fluorescing coeluting ion is used to produce a constant background signal. Since electroneutrality is maintained, the analyte ions cause the displacement of an equivalent number of fluorescent ions. Thus negative signals are observed when the analytes elute from the columns.

For the detection limits achievable by this technique various parameters play a role. First of all is the dynamic reserve (DR) of the detection system, i.e., the ability to detect a small change on top of a large fluorescence background signal, denoted as S. Of course, the noise on the background signal, denoted as ΔS, plays an essential role. By definition DR is given by

$$DR = S/\Delta S \tag{9.9}$$

To cite an example: for a laser-induced fluorescence background applying a laser with a stability of 1% (which is not unusual), DR is calculated to be 10^2. This implies that a signal decrease of less than 1% will not be observable because of the noisy background. This DR is very low compared to other detection techniques: for indirect UV/visible light (vis) absorption, DR is 5×10^3; for refractive index detection, DR is 10^6; and for polarimetry, DR is even higher than 10^7.

The second parameter determining the CLOD (concentration detection limit) is the concentration of the eluting fluorescent ion, C_M. The lower this concentration, the lower the CLOD. If C_M is 10^{-3} M and one analyte ion is able to displace one fluorescent ion (i.e., the transfer ratio TR is 1), an analyte concentration of 10^{-6} M produces a signal change of 0.1%. For a C_M of 10^{-5} M the same change of 0.1% is produced by an analyte concentration of 10^{-8} M. Thus, the following equation holds:

$$CLOD = C_M/(TR \times DR) \tag{9.10}$$

In order to improve their detection limits Mho and Yeung developed a double-beam configuration for their indirect LIF detection in ion chromatography (56). Thus they were able to reduce the flicker noise contribution significantly and to realize a DR of 10^4. Nevertheless, the end result is that the CLOD achievable in conventional-size LC for laser-induced indirect fluorescence detection is not better than that achieved for indirect photometry or conductivity detection (41).

However, in miniaturized separation systems indirect fluorometry is without doubt the detection method of choice. In these systems C_M values of 10^{-4} M or even lower are appropriate. The sensitivity of laser-induced fluorescence detection guarantees that even in 10–20 μm capillaries such concentrations can be handled, while (commercially available) external stabilization of the laser beam provides a DR of 10^3 under these conditions, where double-beam correction is not appropriate because of mechanical instabilities. Thus, in principle, CLODs in the 10^{-7} M range are attainable.

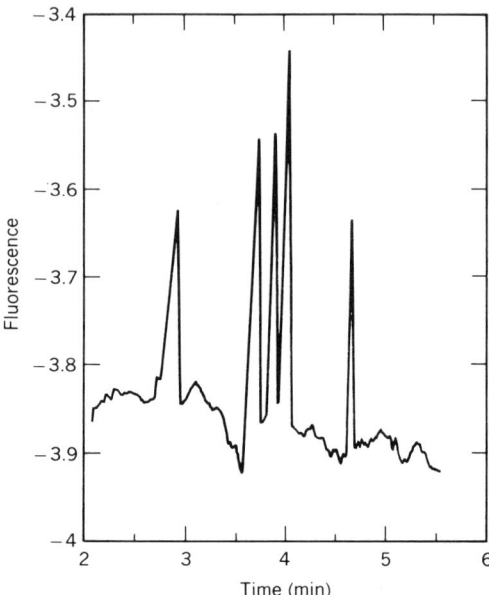

Fig. 9.7. A mixture of alkali and alkaline-earth metal ions separated by capillary zone electrophoresis, utilizing indirect fluorescence detection (the fluorophore is quinine, excited by a stabilized argon-ion laser). Five peaks are observed: K^+ (at 172 s), Ca^{2+} (221 s), Na^+ (233 s), Mg^{2+} (240 s), and Li^+ (279 s); the injected amounts range from 2 to 10 fmol. From Gross and Yeung (59).

Applications for miniaturized separation systems as microbore LC or capillary LC and, more importantly, capillary zone electrophoresis (CZE) have been reported. Capillary LC and CZE provide the highest separation efficiencies available [up to millions of theoretical plates (57,58)]. CZE is an instrumental form of zone electrophoresis without a supporting gel where chemical species are separated on the basis of their electrophoretic mobility. It is used not only for ions [see Fig. 9.7 (59)] but also for amino acids, peptides, and nucleotides and even for sugars. The sugars are electrically neutral but can be slightly ionized at pH 11.5 (60).

In CZE an ionic buffer (which comprises the fluorescing ion to be displaced by the analyte) is present in the capillary tube to provide electrical conductivity so that a constant electric field can be maintained along the length of the column. Since buffer concentrations as low as 10^{-4} M can be used, CLOD is in the 10^{-7} M range (assuming DR is 10^3). Some practical aspects still have to be dealt with, which are in fact related to the separation process. Nevertheless interesting developments are anticipated. Maybe in the near future single cells can be studied by direct injection in CZE so that pharmacokinetics on a single cell basis can be monitored (41).

9.3.4.2. Analytes with Native Fluorescence

At first glance, for analytes with native fluorescence there would seem to be no need to perform chemical derivatization in order to apply LIF detection. It should be realized, however, that for LIF detection in LC most generally the argon-ion laser and/or the He:Cd laser are used, providing only a limited number of excitation lines. Hence, in many examples derivatization is required prior to laser excitation. Analytes with a native fluorescence that can be excited at relatively long wavelengths such as the visible lines of the argon-ion laser are quite exceptional. In this section, first such an exceptional application is presented and subsequently two alternative approaches to excitation are discussed, i.e., LIF detection at short-wavelength excitation (257 nm) and laser-induced, two-photon-excited fluorescence detection. Both approaches are directed to analytes with absorptivities in the UV part of the spectrum.

An interesting example of LIF detection of a natively fluorescent analyte is the bioanalysis of the cytostatic drug doxorubicin (Adriamycin) in human plasma; see Fig. 9.8 (20). This compound (its structure is given in Fig. 9.8) can be appropriately excited with the 488 nm line of the argon-ion laser; its

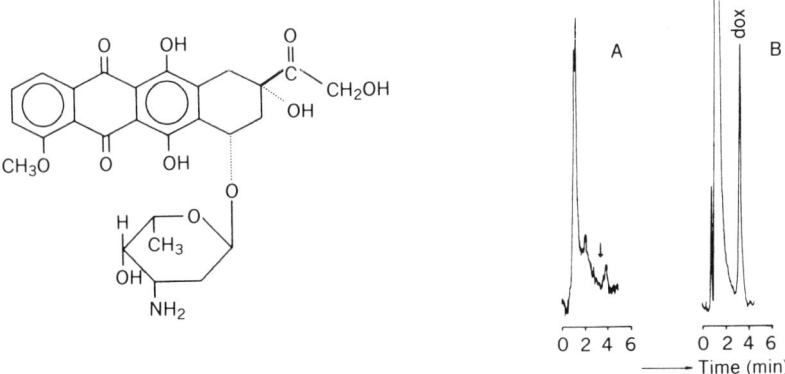

Fig. 9.8. Analysis of plasma samples pretreated by one-step protein precipitation. LIF detection was performed with an air-cooled argon-ion laser (488 nm, 8 mW). The chromatograms were obtained using a conventional-size RP-18 column; the mobile phase was a mixture of aqueous orthophosphoric acid solution (0.01 M) and acetonitrile (3:1, v/v): (A) blank plasma sample; (B) plasma sample spiked with 4 ng/mL doxorubicin (20). The chemical structure shown is that of doxorubicin (Adriamycin).

emission maximum is at 595 nm. For sample cleanup, only a one-step protein precipitation was needed. Compared to conventional fluorescence detection the determination limit (200 pg/mL) was improved more than 1 order of magnitude even by using an air-cooled argon-ion laser providing 8 mW at 488 nm. Presumably an additional gain of at least 1 decade will be obtained by applying a high-power laser giving about 1 W, but the low-power system suffices for the problem at hand.

257-nm-Induced LIF. In our laboratory we have tried to extend the potential of LIF detection by frequency doubling the argon-ion laser 514 nm line and achieved a 257 nm line of about 5 mW power (39, 40). At such a short wavelength a large number of analytes can be excited directly, without prior chemical modification. The most obvious disadvantage of short-wavelength LIF detection is the inherently high fluorescence background, since by far the most fluorescent impurities are also excited. As a positive point we note that in many cases Raman interferences can be simply excluded since the wavelengths of the Stokes–Raman lines are close to the excitation line. This is exemplified in Fig. 9.6, where the Raman spectra obtained for 257 and 459 nm excitation are compared. In the former case the fluorescence at wavelengths longer than about 295 nm is not contaminated with Raman scatter at all.

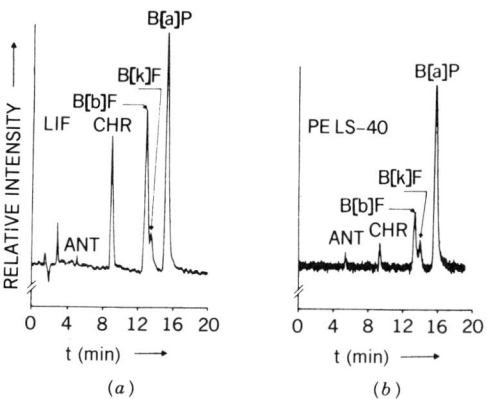

Fig. 9.9. Chromatograms of a mixture of 5×10^{-10} M of anthracene (ANT), 5×10^{-9} M of chrysene (CHR), 1×10^{-8} M of benzo[*b*]fluoranthene (B[*b*]F), 1×10^{-9} M of benzo[*k*]fluoranthene (B[*k*]F), and 1×10^{-8} M of benzo[*a*]pyrene (B[*a*]P), measured with 257-nm excitation (a) with the LIF system with 400-nm emission detection and (b) with the Perkin-Elmer LS-40 fluorometer with 410-nm emission detection. Conventional size RP-18 columns were used; the eluant was methanol/water (9:1, v/v) (39).

Chromatograms for some model compounds, i.e., a mixture of polynuclear aromatic hydrocarbons, detected by 257 nm LIF and conventional fluorescence, are depicted in Fig. 9.9. By application of 257 nm excitation the chromatographic detection limits obtained by LIF are 4–30 times more favorable. It is remarkable that for anthracene the gain is only by a factor of 4. This is due to the extremely high monochromaticity of the laser line compared to the bandwidth of 10 nm applied in conventional fluorescence detection: the absorptivity of anthracene at 257 nm is a factor of 20 lower than at 252 nm.

It should be realized that, in contrast with derivatized analytes, for natively fluorescent compounds the fluorescence spectrum, in the ideal case, shows characteristic features enabling qualitative analysis to be performed or even identification of the eluting analytes. Since on-the-fly recording of complete spectra is rather complicated, a semi-on-line approach has been followed, utilizing eluant deposition on a moving thin-layer chromatography (TLC) plate and subsequent recording of complete conventional fluorescence emission and excitation spectra (61). An on-the-fly technique has been developed by Winefordner and co-workers (62). They designed a multiwavelength detector system based on four photomultiplier combinations, interfaced with the flow cell by fiber-optic bundles. Linear diode array (LDA) detectors, frequently applied in UV/vis absorbance detection, are not appropriate in fluorescence since their sensitivity is poor. However, this sensitivity can be improved considerably by means of image-intensifying techniques. As such the combination of laser excitation and intensified LDA detection (ILDA) seems promising, although of course this combination of instruments is rather expensive (40). We conclude that laser excitation, as expected, compensates for the lower sensitivity of the ILDA compared to the photomultiplier tube (PMT): the detection limits obtained by the LIF–ILDA combination and the Xe lamp–PMT system of Winefordner's group are similar.

Besides it is noted that the coupling of LC and TLC mentioned above also opens the possibility of invoking cryogenic high-resolution molecular fluorescence spectroscopy for identification purposes. Exploratory experiments have shown that TLC plates are suitable matrices for fluorescence line narrowing (FLN) spectroscopy. This will be outlined in Section 9.4.4.3.

Two-Photon-Excited Fluorescence. Instead of generating UV laser lines to extend the applicability of LIF detection, laser-induced two-photon excitation can be utilized: visible laser light is used for excitation of UV-absorbing analytes so that the fluorescence spectrum is on the short-wavelength side of the excitation. As a result, the analyte fluorescence is not perturbed by the Stokes–Raman lines from the eluant. The anti-Stokes Raman lines do not cause any problems since they are very weak; especially the excited states of the high-frequency normal vibrations are hardly occupied at room temperature.

Unfortunately, two-photon excitation (TPE) is extremely inprobable (19). As early as 1977 Sepaniak and Yeung published the first paper on laser TPE fluorescence detection in LC, utilizing a 4-W (CW) argon-ion laser operating at 514 nm (63). In this paper, chromatograms are shown for oxadiazoles in the concentration range 10^{-5}–10^{-6} M, which illustrates the poor sensitivity of the technique. An interesting point noted by the authors is that 10^{-5} M injections of the highly fluorescent compounds phenol, anthracene, and chrysene did not show up in the chromatogram. This indicates the inherent selectivity of TPE.

Since then, strong improvements in detectability have been realized by using pulsed instead of CW lasers. This is readily conceived since the efficiency in TPE is proportional to the laser power squared, as has been outlined in Section 9.2.2.3. For the same test solute as described in the preceding paragraph, Pfeffer and Yeung were able to realize a concentration detection limit of 9×10^{-10} M for the oxadiazole PBD (see Fig. 9.10) in a micro-LC system by utilizing a Cu-vapor laser (510/578 nm), with an average power of 3 W and a peak power of 20 kW (23). Wirth and Fatunmbi have reported a detection limit as low as 2.3×10^{-10} M for bis(methylstyrylbenzene) in batch using a Nd:YAG synchronously pumped dye laser at 600 nm (average power 245 mW; peak power 600 W) (24).

Recently a direct comparison was made between TPE and OPE (one-photon-excited) fluorescence detection in liquid chromatography in order to acquire some insight into the inherent selectivity of TPE (25). This was done for some model compounds, using an excimer–dye laser combination. Three excitation wavelengths, 514, 586, and 650 nm, were used, and furthermore a direct comparison was made between the chromatograms detected by 586-nm-excited and 293-nm-excited fluorescence (obtained by frequency doubling), without changing the chromatographic setup.

It is obvious from Fig. 9.10 that the chromatograms are significantly different; for example, the DPS (4,4'-diphenylstilbene) peak recorded in the TPE fluorescence mode is the most intense one, whereas in the OPE detection mode it is only weak. The results obtained illustrate that the efficiency of TPE as a function of λ and the efficiency of OPE as a function of $\lambda/2$ are basically different. Whereas, under the experimental conditions at hand, for PBD the TPE detection limit is 50 times worse than in OPE, for DPS it is 2.5 times more favorable (25). At this stage of research no general conclusions can be drawn. In particular, information on the fluorescence background encountered in TPE fluorescence applied to real samples is lacking.

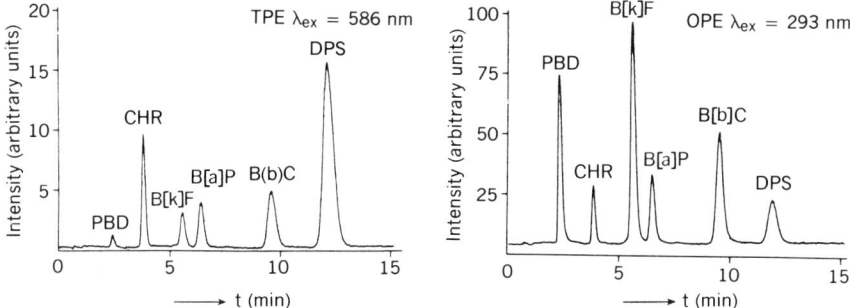

Fig. 9.10. LC chromatograms of a mixture of 2-phenyl-5-(4-biphenylyl)-1,3,4-oxadiazole (PBD), chrysene (CHR), benzo[k]fluoranthene (B[k]F), benzo[a]pyrene (B[a]P), benzo[b]chrysene (B[b]F), and 4,4'-diphenylstilbene (DPS), excited with 586 nm (TPE) and 293 nm (OPE). The fluorescence detection window was between 360 and 480 nm. The concentrations in the left-hand figure were 5×10^{-7} M for chrysene and 5×10^{-8} M for the other analytes; in the right-hand figure the concentrations for all analytes were 2.5-fold lower. Note the different scales on the y-axes; zero intensity corresponds to the dark current.

9.3.4.3. Analytes Requiring Chemical Derivatization

Various important classes of analytes are very bad fluorophores, so that chemical derivatization has to be invoked in order to apply fluorescence detection in combination with high-performance liquid chromatography (HPLC). Chemical aspects of derivatization related to LIF detection were discussed in Section 9.3.3. The subject of LIF detection in LC has been extensively reviewed in the literature (14,35). For this reason we do not intend to give an exhaustive overview here. Instead, some general features of derivatization in relation to LIF detection are considered and our attention is focused on the analysis of peptides, i.e., the dodecapeptide desenkephalin-γ-endorphin, denoted as DEγE, a nonfluorescent neuropeptide, in human plasma (20).

Of course, derivatization reactions originally applied for conventional fluorescence detection have also been utilized for LIF detection. However, in view of the limited number of available laser lines, obviously the fluorophores produced in such reactions cannot always be efficiently excited. A lot of effort has been devoted to the improvement of this aspect. An example is the well-known OPT derivatization, a rapid reaction that has found wide application in the field of amino acid and peptide analysis. The products formed are isoindoles, rather unstable compounds that can only be excited by the UV multiline (350–360 nm) of the argon-ion laser. In order to shift the excitation spectrum to longer wavelengths a lot of effort has been put into the development of the naphthalene analogue of OPT, i.e., NDA (naphthalene-2,3-dicarboxylaldehyde). Compared to OPT, this reagent (in presence of cyanide) provides stable derivatives, i.e., 1-cyanobenz[f]isoindoles (CBI), compounds with high fluorescence quantum yields compatible with the 442 nm laser line of the He:Cd laser and the 458 nm line of the argon-ion laser. Recently Novotny's group developed new variants of the NDA label (64–66). However, both for NDA and the NDA analogues the derivatization is slow, so that it can only be applied in the precolumn mode (20).

Like the NDA derivatization reaction, the fluorophoric agent fluorescein isothiocyanate (FITC) seems to give very favorable results: it shows good selectivity, and the products formed can be excited at 488 nm (an argon-ion laser line). However, this reaction can also only be utilized in the precolumn mode (54).

Recently, Van den Beld has tried to apply LIF detection to the bioanalysis of the neuropeptide DEγE (20). The problems requiring attention are not only the stability of the derivatization products but also the formation of reaction side products. It was already known that the NDA/CN agent mentioned above forms a multitude of interfering fluorescent reaction side products if peptides are labeled precolumn (67,68). Similar problems were encountered for the FITC labeling reaction. Van den Beld developed a chromatographic technique, based on ion pair formation, in which DEγE labeled by reaction with FITC could be measured selectively in the presence of its main metabolites β-endorphin (βE) 7–17 and βE 8–17. However, the determination limit of DEγE in human plasma was as high as 1 μg/mL.

Far more favorable results were obtained by invoking the OPT reaction. OPT is a fluorogenic label, i.e., it has no native fluorescence itself but generates fluorescent derivatives. Furthermore, it shows fast reaction kinetics in aqueous media so that it is applicable in the postcolumn mode. In this mode, side products possibly formed during the reaction do not interfere with the analysis of the derivatives (46). For DEγE note that the isoindole derivatives formed by the ε-amino function of the lysine residue have favorable fluorescence properties. A complete bioanalytical method has been developed, consisting of selective sample pretreatment based on gel permeation solid-phase isolation, and reversed-phase LC, coupled off-line to an ion-pair LC separation system: LIF detection was performed using the UV output of the argon-ion laser (100 mW) and recording the fluorescence emission at 450 nm. The authors claim a determination limit as low as 0.4 ng/mL, an improvement in signal-to-noise ratio by a factor of 30 compared to conventional fluorescence detection (utilizing a Perkin-Elmer LS-4 detector). Figure 9.11 shows the analysis of a human plasma sample containing 9.3 ng/ml DEγE, 7 ng/mL βE 7–17, and 6.5 ng/mL βE 8-17.

The above discussion clearly illustrates the crucial role of the chemical

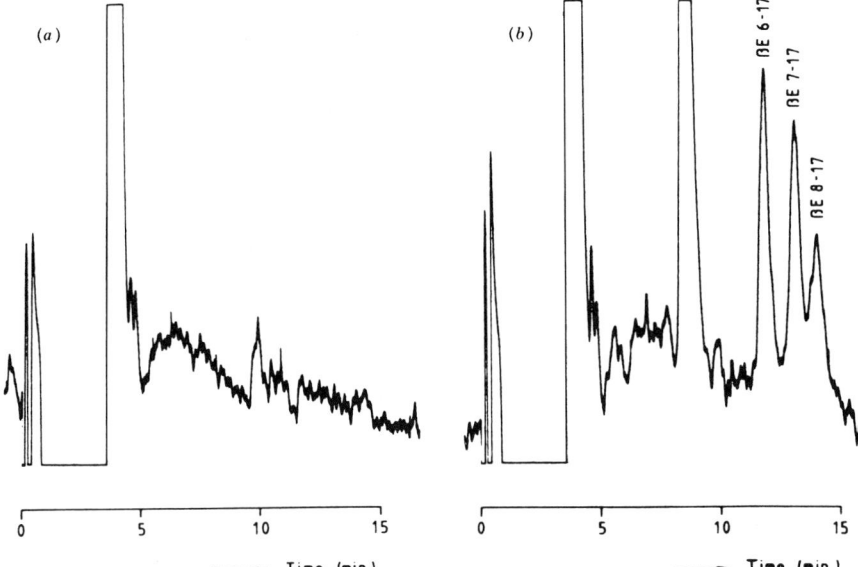

Fig. 9.11. (a) Chromatogram obtained after the analysis of a processed plasma sample (processing consisted of, consecutively, gel permeation, solid-phase isolation, and reversed-phase LC) using ion-pair reversed-phase LC and post-column OPT reaction/LIF detection. (b) Chromatogram obtained after a similar analysis of a human plasma sample, containing 9.3 ng/mL DEγE, 7 ng/mL βE 7–17, and 6.5 n/mL βE 8–17 (20).

aspects of derivatization in real sample analysis, applying LIF detection. Although not the most appropriate from a spectroscopic point of view, the OPT reaction in the above example is the most successful because OPT is nonfluorescent itself and the derivatization is rapid and thus applicable in the postcolumn mode. In general, precolumn derivatization will only be successful if stable derivatives are produced in 100% yield that can be easily separated on the chromatography columns and, secondly, if formation of side products can be prevented, even if the analyte concentration is extremely low. These requirements are not readily met in real sample analyses. In practice, usually additional precolumn purification steps are required to reduce the influence of interferences.

9.3.4.4. *Direct LIF Detection in Capillary Electrophoresis*

In Section 9.3.4.1 recent developments in capillary zone electrophoresis utilizing indirect laser-induced fluorescence detection were discussed. This detection technique is rather universal, as it is based on the displacement of fluorescent buffer ions. For stabilized laser sources, for buffer concentrations as low as 10^{-4} M, the CLOD is in the 10^{-7} M range, which corresponds to typically 5×10^{-17} mol injected on a 10-μm column (41).

Of course for direct LIF detection of labeled fluorescent analytes the achievable detection limits are far more favorable. As an example we refer to the FITC-labeled amino acids detected by means of a 0.05-W argon-ion laser (at 488 nm) using the sheath-flow cuvette flow chamber (described in Section 9.3.2.1) in a CZE setup (1 m $\times$ 50 μm i.d. $\times$ 180 μm o.d. fused silica) that revealed MLODs ranging from 1.7×10^{-21} mol (CLOD 1.3×10^{-12} M) to 6×10^{-21} mol (CLOD 5.6×10^{-12} M), as depicted in Fig. 9.12 (36). It is noted by the authors, however, in line with our discussion in Section 9.3.4.3, that labeling at the 10^{-12} M amino acid concentration level is not straightforward: the purpose of their paper was to illustrate the sensitivity of their LIF detection system in CZE. The FITC-labeled amino acids were prepared at a 10^{-6}–10^{-7} M level and subsequently were diluted. Instead of FITC, fluorescein thiohydantoin has been utilized for the labeling of amino acids (37).

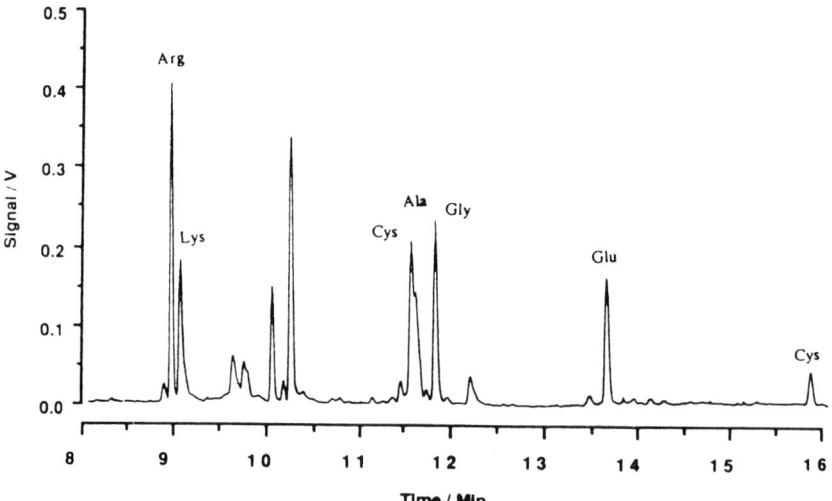

Fig. 9.12. Capillary zone electrophoresis of FITC-labeled amino acids, detected by a 0.05-W argon-ion laser (488 nm) in a sheath-flow cuvette. The concentrations of arginine, lysine, alanine, glycine, and glutamic acid were 1.14×10^{-9} M; that of cysteine was 3.42×10^{-9} M. Presumably cysteine produces two peaks because of multiple labeling of this compound. Reagent peaks elute between 9.5 and 10.5 min (36).

Interesting further improvements have been realized by applying tetramethylrhodamine isothiocyanate (TRITC) as a label (38). TRITC can be efficiently excited by means of a low-cost laser, i.e., the green He:Ne laser (0.75 mW at 534.5 nm). The price of this laser is on the order of $1000. The detection limit for the dye in CZE is as low as 500 ymol (1 yoctomole = 10^{-24} mol), or 300 analyte molecules. Such extremely low detection limits open perspectives for DNA sequencing by capillary gel electrophoresis (utilizing TRITC for labeling) where the sample loading is on the order of 1–10 amol and detection limits in the zeptomole range (1 zeptomole = 10^{-21} mol = 600 molecules) are necessary, with a detection volume of less than 10^{-10} L. Although some separation problems still must be overcome, the results obtained are hopeful: for labeled primer the detection limit reported is 2 zmol (38).

9.3.4.5. Diode-LIF Detection

Until recently, only minor attention has been paid to LIF detection in the red and IR part of the spectrum, despite the inherent advantages of using long-wavelength excitation: negligible background fluorescence, little photodecomposition of the sample, and relatively weak Raman scatter. The most obvious reason for this seeming lack of interest is that only very few fluorophores can be directly excited in this wavelength region and furthermore that appropriate labels and labeling procedures are not available. In addition, we note that conventional photomultiplier tubes have little sensitivity at wavelengths longer than about 700 nm, so that detection of fluorescence in the near-IR is not straightforward. This instrumental problem has been solved by the introduction of GaAs PMTs, which have a good response up to about 900 nm.

The introduction of the diode lasers has obviously changed the situation. This is readily conceivable in view of the extremely favorable properties of this laser type (price, size, stability, longevity) and the recently realized shifts to shorter laser wavelengths and increase of power. As early as 1986, Ishibashi's group published a paper on diode-LIF detection in combination with liquid chromatography (69). They separated several polymethine dyes and achieved detection limits around 0.3 pg applying a 15-mW 780-nm diode laser (modulated at 100 Hz in combination with a lock-in-amplifier detector) and detecting the emission at 830 nm. In batch experiments for laser dyes in a cuvette, Winefordner's group realized concentration detection limits as low as 6×10^{-14} M (70).

From an analytical point of view these model compounds are of little interest: it is essential to develop labeling procedures applicable in the red or near-IR. Imasaka and Ishibashi have shown an interesting example, making use of noncovalent binding of fluorophores (i.e., nonspecific binding, based on adsorption or on electrostatic forces) to very large compounds as proteins (33). Covalent labeling, however, has hardly yet been applied as this chapter goes to press. One example we can mention is the labeling of albumin with oxazine 750, a dye with an amine group that can be utilized for derivatization. The detection limit obtained was no better than 0.13 pmol, which has to be attributed to a low labeling efficiency (71). Recently, Patonay and Antoine have considered the possibility of using derivatives of carbocyanines for covalent labeling (72). The interesting feature of such dyes is that they can be wavelength tuned by chemically modifying their structure, so that in principle the excitation spectrum of the label and the wavelength of the available diode laser can be matched. As an example, striking effects are induced by changing the length of the polymethine chain: addition of an ethylene group leads to a red shift of about 100 nm. However, analytical applications still have to be realized.

In summary we conclude that the main constraint to applying diode-LIF detection in chromatography is not an instrumental one. The available 10-mW 670-nm diode laser is an appropriate device; in the near future a 635-nm diode laser will be available too, and also the use of higher power, if needed, will be possible. The problem to be solved urgently is of a chemical nature: the development of appropriate labeling procedures.

9.3.5. Concluding Remarks

The separation techniques considered in this section are column liquid chromatography and capillary electrophoresis (CE). LIF detection has hardly yet been combined with thin-layer chromatography, but the applicability of high-resolution molecular fluorescence to samples on TLC plates opens interesting possibilities for identification purposes, especially in the bio-analytical field. This will be outlined in Section 9.4.4.

By far the most separations in LC are performed utilizing conventional-size columns, whereas open tubular liquid chromatography thus far has received little attention. Hence, in general LIF detection in LC is only relevant if it provides significant improvements in detectability of analytes in real samples, in comparison to conventional fluorescence detection. During the last few years, people have become strongly aware of the problems inherent in the chemical derivatization of real samples when the objective is to improve detection limits. One trend is to avoid chemistry as much as possible, by making use of short-wavelength laser excitation and/or two-photon excitation. Of course this line can only be followed for analytes with native fluorescence. For nonfluorescent analytes, chemical derivatization must be applied and additional separation steps have to be invoked to realize optimum detection limits. Fluorogenic reagents applicable in the postcolumn mode seem to be the most promising. In this respect, there is a trend toward developing new labeling reactions for use with the cheap, small-sized, and stable visible diode lasers.

At the moment CE, receiving a lot of attention at scientific meetings and in the literature, is considered to be a very promising separation technique. Owing to the dimensions of the detector cell, LIF detection plays a crucial role and conventional fluorescence is not appropriate. Of course, also for this technique chemical derivatization, generally applied in the precolumn mode, has some inherent disadvantages that must be overcome. The results in terms of detection limits for tagged analytes reported thus far are really astonishing, even if cheap, low-power lasers are utilized. In fact, high-power lasers are not appropriate since, owing to the long residence time of the analytes in the irradiated volume of the detector cell, photobleaching effects can be dominant. Another trend is the development of indirect laser-induced fluorescence detection for CE, a method that is widely applicable (also for nonfluorophores) while chemical derivatization is not necessary. The detection limits published recently, although much less favorable than for direct LIF detection of labeled analytes, are promising.

9.4. LOW-TEMPERATURE SPECTROSCOPY

9.4.1. Introduction

So far the application of laser excitation has been considered for molecules in liquid solutions. The main reasons for the application of laser-induced fluorescence techniques in such matrices in general are related to the high intensity and/or the good collimation properties of laser beams. The advantage, obviously, is the possibility of reaching low detection limits and of probing very small sample volumes. In low-temperature spectroscopy laser excitation has an additional advantage: the high "monochromaticity" of the laser light can be used to generate highly specific spectra. In contrast with many approaches, the increase in selectivity in low-temperature spectroscopy is *not* necessarily accompanied by a decrease in sensitivity. In laser-excited Shpol'skii spectroscopy the sharpening of the spectral features, on the contrary, even results in an increase in sensitivity (see Section 9.4.3).

With laser-induced fluorescence at low temperatures, high-resolution fluorescence spectra can be obtained. Spectral bandwidths can be realized that vary from 0.1 to 0.01 nm in most applications, but with suitable equipment bandwidths may be obtained that are orders of magnitude narrower. Such widths are amply sufficient to reveal the vibrational fine structure of fluorescence spectra. The sensitivity that may be realized is impressive; recently even single molecule detection has been reported for laser-induced fluorescence measurement of fluorophores in a cryogenic crystalline matrix. Recent reviews on the analytical potential of high-resolution molecular fluorescence spectroscopy have been published by Wehry (16) and by Hofstraat et al. (17). Particularly in the latter review fundamental aspects of a number of high-resolution techniques are discussed. In this section only brief attention will be paid to fundamental aspects, focusing on two solid-state methods, Shpol'skii spectroscopy and fluorescence line-narrowing spectroscopy. Some instrumental considerations will be presented. The major part of this section, however, will be devoted to a description of applications of the two techniques.

Fluorescence spectroscopy is one of the most sensitive detection techniques for many compounds of environmental interest [e.g., polycyclic aromatic hydrocarbons (PAHs), some of which are among the most potent mutagenic and carcinogenic substances] (17). Unfortunately, fluorescence spectra of molecules in the liquid phase are not very specific. In general such spectra are composed of bands that can be tens of nanometres wide at half maximum. These broad bands are due to two processes that cause blurring of the electronic energy levels: first, the strongly fluctuating interactions of dissolved molecules with the cage of solvent molecules surrounding them and, second, the occurrence of many inelastic collisions of dissolved molecules and solvent molecules, leading to considerable shortening of the excited state lifetime. In addition, for polyatomic molecules there can be numerous vibronic fluorescence transitions, separated by relatively small energetic distances, that will merge into one or a few featureless bands because they are all subjected to the band-broadening processes mentioned above.

In high-resolution techniques samples are cooled to very low temperatures. Many low-temperature solids are so rigid that the solvent cage surrounding the guest molecules remains virtually fixed on the fluorescence time scale. Of course, collision-mediated energy transfer cannot occur in the solid matrix. Although some resolution may be gained in low-temperature solids, in most cases cooling is not a sufficient prerequisite to bring about narrow lines in electronic absorption and emission spectra. The reason is that the immediate environment of the guest molecules may be fixed in time and is not necessarily the same for all guest molecules. Thus, in fluorescence one observes a large ensemble of molecules having different solvent cages, resulting in small differences in excited and ground state energy. As a result a statistical distribution of transition energies is found for the spectral bands, which are hence called inhomogeneously broadened. The extent of this broadening depends on the compatibilty of guest and host molecules, but bandwidths of the same order of magnitude as measured in fluid media are generally observed. In order to obtain sufficiently selective fluorescence spectra the inhomogeneous broadening of the spectral bands must be strongly reduced.

Basically there are two ways to realize this reduction. First, one can strive for homogenization of the solvent cages surrounding the dissolved molecules, in an effort to approach a perfect crystal where all molecules in the ensemble have equal interactions with the matrix (or a limited number of interactions, leading to site structure in the spectrum). This approach is chosen in so-called Shpol'skii spectroscopy (16,17,73–75). Secondly, use can be made of selective—preferably laser—excitation of the guest molecules. Since the interactions in the solid matrix remain fixed on the time scale of fluorescence, the excited state population thus created remains conserved and will also produce a narrow emission signal. This technique is referred to as fluorescence line-narrowing (FLN) spectroscopy (16,17,75,76). In Sections 9.4.3 and 9.4.4 the principles of the two techniques and their analytical applications will be discussed in more detail.

9.4.2. Instrumental Aspects

A short general description of experimental aspects that are crucial for the achievement of highly resolved fluorescence spectra will now be given. More detailed information can be found in our chapter in Part 2 of this series (17). Successively, the excitation system, the sample compartment, and the detection system will be discussed.

9.4.2.1. Excitation

For excitation conventional lamps (e.g., xenon, mercury, or deuterium lamps) can be used in Shpol'skii spectroscopy, where the line-narrowing effect is matrix induced. Such lamps, in combination with a high-throughput monochromator or a filter to select a convenient excitation wavelength region, provide light over the whole wavelength region of interest. If selective excitation is required, which is crucial in FLN spectroscopy but is also advantageous in Shpol'skii spectroscopy, a high-power lamp in combination with a monochromator with sufficient resolving power may be used. Generally xenon lamps are utilized since they provide continuous wavelength output at relatively high powers (for a 2.5-kW xenon lamp powers on the order of 0.1–1 mW can be obtained in the visible part of the spectrum with an excitation bandwidth of 1 nm).

However, for sensitive, selective excitation experiments the laser is the excitation source to use. As outlined in Section 9.2, lasers can deliver very high photon densities, both spectrally and spatially. Thus, effective excitation powers in the microwatt-to-watt range (see Table 9.1) can be realized at a spectral bandwidth of 0.1 nm or much less (even bandwidths of several kilohertz can be attained, at the cost of power). To make optimal use of selective excitation a tunable laser (Ti:sapphire laser: see the discussion of dye lasers in Section 9.2.3.2) should be employed, to apply resonant excitation of the analyte of interest. Most recent work on trace analysis of various compounds, e.g., PAH metabolites, has been done by making use of laser excitation. In so-called laser-excited Shpol'skii spectroscopy (LESS), extremely high sensitivities can be achieved when the laser excitation wavelength is tuned to a narrow absorption transition. In FLN spectroscopy, useful highly resolved spectra of compounds in dilute solutions can only be obtained when excitation is done with a narrow bandwidth laser source.

9.4.2.2. Sample

For the high-resolution techniques discussed here, samples must be kept at low temperatures (in general temperatures below 20–30 K suffice). There are several ways to achieve such temperatures. The most convenient cooling apparatus is the closed cycle helium refrigerator, which is based on repeated compression and expansion (near the sample) of gaseous helium. The closed cycle cooler only requires line power to operate. The helium flow-through cryostat uses liquid helium that is pumped from a container and sprayed at the bottom of the cold tip onto which the sample is mounted. Both types of cryocoolers are based on conductance cooling via the cold tip (although some helium exchange gas may be admitted into the sample chamber), so that care must be taken to optimize thermal contacts. Another disadvantage is that they take a long time to cool down and heat up. The lowest temperatures that can be reached are on the order of 3.5 K. An alternative is the helium bath cryostat, where the sample is cooled by immersion in the liquid refrigerant. In that case thermal contacts are optimal and cooling may be instantaneous. On the negative side the bath cryostat consumes liquid helium (like the helium flow-through cryostat) and, in addition, the presence of boiling helium in the optical pathways interferes with the experiment. When laser excitation is used one has to be aware of possible local heating of the sample, in particular when the thermal contact between the sample and the cryogenic medium is not optimal. This can easily occur when cooling via thermal conductance is applied. Local heating effects can result in a strong reduction of the intensity of the narrow lines, particularly in FLN spectroscopy (77).

The mode of preparation of the sample is critical, especially when the nature of the low-temperature matrix causes the line-narrowing effect (78). In that case experimental parameters such as rate of cooling and freezing and possible application of annealing treatments may profoundly influence the high-resolution spectra (see Section 9.4.3). To remove dissolved gases, especially oxygen, from solutions, application of freeze-pump-thaw techniques may be required. Instrumental aspects of low-temperature spectroscopy are thoroughly discussed in the comprehensive book by Meyer (79).

9.4.2.3. Detection

For dispersion of the light, generally a high-resolution monochromator or spectrograph (optical pathlength 0.5 m or more) is employed. As the low-temperature solid samples are often opaque, one is usually confronted with scatter light problems. Especially when laser excitation is employed, the intensity of the scattered light may be very high. Because of the high degree of monochromaticity of the laser light, this scattered light is efficiently removed by using a double monochromator (i.e., two monochromators used in tandem) or a triple monochromator. The latter type of monochromator is relatively new and is eminently suited for measurement of optical high-resolution spectra at low light intensities. It consists of a high-throughput double monochromator with short focal length to remove the stray light, followed by a single high-resolution monochromator with long focal length to provide sufficient separation efficiency to detect the narrow spectral bands. If a pulsed laser source is used, the scatter background (including Raman scattering, e.g., from the solvent) can be removed by application of time-delayed detection of the fluorescence. It should be realized that the fluorescence lifetime is generally longer in the low-temperature matrix than in liquid solutions.

The mode of detection depends on whether a continuous or a pulsed excitation source is used. When continuous excitation is applied, both single channel detection (using a photomultiplier tube followed by preferably a photon-counting apparatus), and multichannel detection can be employed. The same applies to a pulsed source, but then gated detection is required because of the low duty cycle of the experiment (see Section 9.2.2.3). In combination with pulsed excitation for single-channel detection the boxcar integrator is the instrument to use; for multichannel detection a gateable (intensified) array detector is optimal. In particular in FLN spectroscopy the application of multichannel detection—with a photodiode array, SIT vidicon, or charge-coupled device—is advantageous, as time-dependent effects can be removed from the spectra (80).

9.4.3. Shpol'skii Spectroscopy

9.4.3.1. Principles

In Shpol'skii spectroscopy a suitable host is selected to minimize inhomogeneous spectral broadening (5,6,73–75). In principle, all materials that have a crystalline form at cryogenic temperatures may be employed to yield matrix-induced narrow-line spectra. For instance, use can be made of the mixed-crystal technique, which is based on the preparation of a solid solution of the analyte in a single crystal of host molecules with approximately the same dimensions (81). In this case the guest molecules occupy well-defined substitutional sites in the solid matrix, replacing one host molecule. Obviously the direct environment of the guests is the same for all molecules, so that all inhomogeneous broadening is removed. Examples are naphthalene in durene or isotopically mixed crystals (e.g., deuterated naphthalene in naphthalene). Although the difference in spectral transitions is very small, about 1 nm in the fluorescence spectrum, the resolution of high-resolution spectroscopy is amply sufficient to distinguish isotopes. Analytical applications of the mixed-crystal technique are limited, as the compatibility of the molecular geometries of guest and host is critical and the preparation of a good quality crystal is tedious.

In 1952, however, the Russian scientist E. V. Shpol'skii found that many aromatic molecules dissolved in n-alkanes show well-resolved spectra at low temperatures (82). Important features that make n-alkanes suitable for analytical applications are the ease of preparation of the samples, the fact that they are optically inert in the applied wavelength regions, and the fact that the demands on the compatibility of guest and host geometry are less stringent than in mixed crystals. By now Shpol'skii spectra have been reported for numerous homocyclic and heterocyclic aromatic compounds and derivatives (e.g., methyl-, amino-, or halogeno-substituted compounds, radicals and ions), polyenes, and porphyrin-like molecules (6,83). Even for bulky, nonplanar molecules like hexahydrohexahelicene, narrow-line spectra have been reported (84). Typical bandwidths for Shpol'skii spectra are on the order of 0.01–0.1 nm at temperatures below 20 K (85). This is still considerably broader than the lifetime limited homogeneous linewidth, which is on the order of 10^{-4}–10^{-5} nm (for excited state lifetimes of 1–10 ns). The residual inhomogeneous broadening is because the polycrystalline matrices that are formed when normal sample preparation is applied (i.e., fast cooling) are far from perfect.

The way in which the analyte molecules are incorporatd in the n-alkane host lattice is crucial for the reduction of inhomogeneous broadening. This implies that the composition and mode of preparation of the matrix will influence the appearance of the Shpol'skii spectra (78). There is strong spectral evidence that the guest molecules occupy substitutional sites in the n-alkane lattice, replacing several host molecules (86,87). Therefore some demands—though much less strict than in mixed-crystal techniques—are set by the choice of solvent. In general, the best resoluton is observed if there is a good match between the longest dimension of the guest and the length of the n-alkane ("key-and-hole rule") (88,89). Some, especially larger, compounds may show high-resolution spectra in a relatively broad range of n-alkanes (90).

Often Shpol'skii spectra show several identical spectra that are displaced with respect to each other over a slight energetic distance, mostly not more than a few nanometres. The multiplet structure, which is strongly dependent on the matrix, is attributed to subsets of guest molecules occupying different substitutional sites, each subset being perturbed by a slightly different crystal field (91). The occurrence of multiplets, on the one hand, is disadvantageous as it contributes to the complexity of the spectrum and reduces the sensitivity of the method. On the other hand, the multiplet structure adds to the specificity of the technique, because the spectral positions and relative intensities of the components of the multiplets are reproducible when the same experimental conditions are applied, particularly in relation to the preparation of the sample.

The preparation (rate of cooling and of freezing, annealing treatments) of the solid Shpol'skii matrix may profoundly influence the appearance of the spectra. There are two factors that play a role. First, the thermal history of the solid determines to a great extent the quality of the crystal lattice that is formed. Second, the preparation of the sample involves at least a binary system, so that segregation of sample constituents is likely to occur upon freezing. To prevent segregation effects, which may lead to broad bands in Shpol'skii spectra of analytes frozen out of the n-alkane crystal, fast cooling (or rather *freezing*) of the sample is required (78,92). However, very fast freezing of the sample, as in matrix-isolation sample preparation, where the sample is frozen momentaneously when sprayed on a cooled surface, is detrimental to the quality of the spectra. Significant improvement of the spectral resolution can be realized by application of an annealing treatment following the deposition (93,94). Fortunately, many—especially large— planar molecules yield Shpol'skii spectra that are not very sensitive to thermal history effects. Many PAH priority pollutants, e.g., benzo[a]pyrene, benzo[g,h,i]perylene, and perylene, belong to this class of compounds.

Numerous experiments have been done at a temperature of 77 K, i.e., in liquid nitrogen, which is a readily available cryogenic substance. Although at this temperature resolution is indeed gained in comparison with conventional fluorometry, an impressive resolution and sensitivity increase is realized

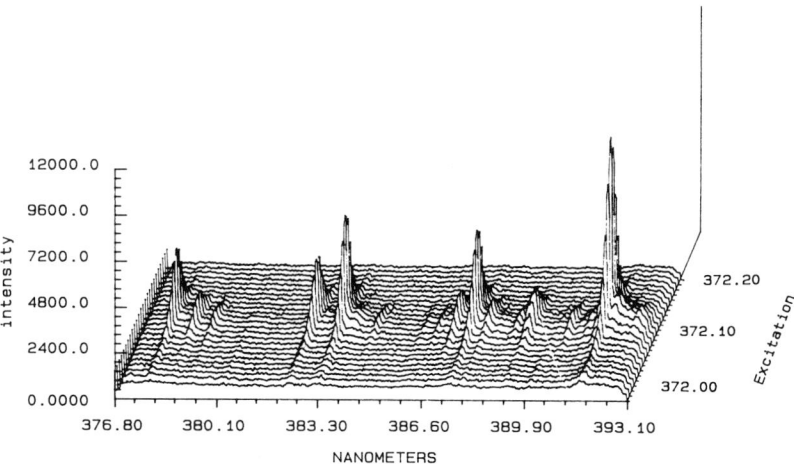

Fig. 9.13. Excitation–emission matrix of pyrene in *n*-octane measured at 28 K. The emission spectrum is derived from a cross section in the horizontal direction. The excitation spectrum is obtained when a vertical cross section is made. The selectivity of LESS is clear when the width of the absorption line is considered: less than 0.05 nm (full width at half maximum). From (106).

by further temperature reduction. For tetracene in *n*-nonane under non-selective excitation bandwidths are reduced by a factor of 2–3 when going from 75 to 30 K; the gain in peak height at the same time is a factor 4–8. Further reduction of the temperature does not produce a significant improvement, owing to the residual inhomogeneous broadening of the spectral bands (85). When selective laser excitation is employed the remaining inhomogeneous broadening can be deminished, so that in this case lowering the temperature has more advantages (85).

As the Shpol'skii effect is a matrix-induced phenomenon, not only the electronic emission spectrum but also the absorption spectrum is considerably narrowed. The emission-excitation matrix depicted in Fig. 9.13 nicely demonstrates the previous statement. The three-dimensional (3D) spectrum has been obtained by taking consecutive excitation scans for increasing emission wavelengths. The high resolution of the excitation spectrum can be used to get a tremendous enhancement of selectivity, by employing resonant laser excitation in the narrow-banded adsorption region of selected analytes. Laser-excited Shpol'skii spectroscopy (LESS) has been reviewed by D'Silva and Fassel (95). Some applications of LESS will be discussed in detail next in Section 9.4.3.2.

9.4.3.2. Applications of Shpol'skii Spectroscopy

General. Surveys of analytical applications of Shpol'skii spectroscopy have been given by Hofstraat et al. (17) and, very extensively, by de Lima (74). To date hundreds of papers dealing with the application of this technique in chemical analysis have appeared. In particular, PAHs have been determined in a variety of mostly environmental matrices (air; car exhaust; harbor, river, and marine sediments; coal and oils).

In the majority of applications only *qualitative* analysis has been performed. Owing to the vibrational resolution of the spectra, in combination with the characteristic wavelength of the electronic origin [which is also temperature dependent! (85)], Shpol'skii spectroscopy is extremely suitable for identification purposes. Even closely structurally related methyl isomers of PAHs can be unequivocally identified. This has been demonstrated by lamp-excited Shpol'skii spectroscopy, using both fluorescence and phosphorescence measurements (96,97), and by LESS (98). Especially the latter technique, with its capabilities of selective excitation, is extremely selective. An extra discriminating feature that can be employed is the fluorescence lifetime; when a pulsed laser is used for excitation, time-resolved detection on the nanosecond time scale, which is relevant for discrimination of fluorescent species, can be applied (99). Even the most advanced chromatographic techniques, e.g., capillary column gas chromatography with mass spectrometric detection, the most popular and effective method for analysis of PAHs in complex environmental samples, cannot provide full characterization of isomers when conventional ionization techniques are used (100,101). However, distinction among isomers is important, as structurally alike compounds may have strongly differing toxicities: for example, 5-methylchrysene is a much more potent carcinogen than the other methylchrysenes and their mother compound (102).

Fewer reports have appeared on *quantitative* analysis with Shpol'skii spectrometry, the reason being that direct quantification of analytes seemed impossible. Owing to irreproducibilities in sample preparation and optical setup, quantification can only be realized with sufficient accuracy if an internal standard, or even a standard addition procedure in combination with an internal standard, is used. Particularly, deuterated analogues of the analytes are suitable internal standards, as they react in the same way to the sample preparation procedure and can be easily discerned spectroscopically from their analogues (the spectra of deuterated compounds are shifted to the blue over about 1 nm) (103,104). With these methods PAHs have been quantified in several types of environmental samples, with acceptable relative standard deviations: in general reproducibilities better than 10% can be realized (17). Quantitative analysis of standard reference materials with Shpol'skii spectroscopy has also proven the accuracy of the low-temperature technique (105).

In LESS, quantitative analysis requires more precautions than in conventional Shpol'skii spectroscopy. In order to make optimum use of the selectivity of laser excitation, the laser wavelength should be tuned to match a narrow absorption line of the analyte of interest. The best correction, as discussed above, can be done by application of deuterated PAHs. However, as the analyte and the deuterated compound are separated by about 1–2 nm and both compounds have narrow absorption bands (0.01–0.1 nm), the laser wavelength has to be scanned in order to excite the reference compound. The selection of the laser wavelength must be done very accurately in view of the absorption line widths. Hence, the use of deuterated analogues as internal standards is extremely time consuming and may add to the uncertainty of the quantification. As for internal standards in LESS, we have had good results with deuterated compounds that are excited in a higher excited state at the wavelength that is applied for resonant excitation of the analyte. Higher excited states are almost without exception very short lived and therefore strongly homogeneously broadened. Under these conditions only the emission wavelength has to be varied to enable one to measure the intensity of the fluorescence of the reference compound. When a photodiode array is used for detection, both the analyte and the reference signal can be measured simultaneously (106).

Detection limits for academic solutions and "clean" samples, i.e., samples that are free from interfering background light emission, have been reported for a number of compounds (74). For instance for the highly fluorescent PAH benzo[a]pyrene in n-octane at 10 K using lamp excitation, detection limits, defined as 3 times the peak-to-peak intensity of the background noise, on the order of 5×10^{-10} M (corresponding to about 100 ng/L, or 100 ppt) have been published (104, 107). When laser excitation is employed the detection limit can be improved to 3×10^{-11} M (about 8 ng/L, or 8 ppt) (108). The linear dynamic range for benzo[a]pyrene may be 5 decades, i.e., from around 10^{-6} M down to the detection limit (17).

Detection limits of analytes in real samples are generally restricted by broad background luminescence of other sample constituents. In principle, if sufficient attention is paid to cleanup of the extract, low detection limits are attainable. Recently, even the detection of single molecules in crystalline matrices has been reported (109–111). By application of a very narrow bandwidth laser (bandwidth on the order of 1 MHz) focused to a very small spot size, single molecules of pentacene could be observed in a p-terphenyl crystal. The pentacene emission was selectively detected by application of high spectral and spatial resolution. As the pentacene emission appears in the red part of the spectrum and selective laser excitation is used (the width of the pentacene absorption lines in the crystalline matrix is relatively narrow, about 0.01 nm, or 200 GHz), interference of impurity emission can be neglected. Upon excitation with the narrow bandwidth laser line the inhomogeneous broadening of the pentacene emission bands is removed. The homogeneous (lifetime-limited) bandwidth observed for the 0–0 fluorescence bands is about 8 MHz, more than 4 orders of magnitude narrower than the original, inhomogeneously broadened width. In fact, the experiment described above illustrates the application of fluorescence line-narrowing spectroscopy on molecules in a well-organized crystalline matrix. The term fluorescence line-narrowing spectroscopy is mostly used in connection with impurity molecules in amorphous low-temperature matrices. In Section 9.4.4 principles and applications of this technique will be discussed in more detail.

An interesting consequence of the high selectivity of Shpol'skii spectroscopy is that sample cleanup procedures can be minimized. Useful narrow-line fluorescence spectra have been obtained for PAHs in a number of environmental matrices without any or with minimal sample pretreatment. For instance, liquid fossil-fuel samples have been analyzed as received; only dilution with the appropriate n-alkane was applied (108). PAHs could be quantitatively measured in harbor sediment by simply dissolving the raw extract in n-octane, without any cleanup (104). Even in very complex matrices, such as biota, useful Shpol'skii spectra can be obtained with minimal cleanup, in particular when laser excitation is employed (105,112). In the next subsection this aspect will be considered in more detail.

PAHs in Biota. To analyze biotic samples by chromatographic methods, extensive cleanup of the crude extract is required to remove the large amount of fatty components that will hamper the determination. For instance, terns contain about 30% of fatty constituents on a dry weight basis (112). The potential of LESS for determination of PAHs in crude biotic extracts will be demonstrated via a discussion of the determination of the PAH pyrene in tern meat (105, 112).

In Fig. 9.14 Shpol'skii fluorescence spectra of the 0–0 region of pyrene in n-octane tern extract are depicted, as obtained with conventional xenon-lamp excitation, both with (Fig. 9.14a) and without (Fig. 9.14b) prior chromatographic cleanup. It is clear that the cleaned extract produces a Shpol'skii spectrum with a much better signal-to-noise ratio. The main reason for the reduced quality of the spectrum shown in Fig. 9.14b is the presence of a large amount of apolar compounds, mainly of fatty origin, in the extract. These compounds have two effects. First, they strongly absorb in the UV and blue part of the spectrum, resulting in strong inner filter effects. Second, they disturb the crystalline structure of the n-octane matrix, which is crucial for the Shpol'skii effect. To eliminate these effects that preclude

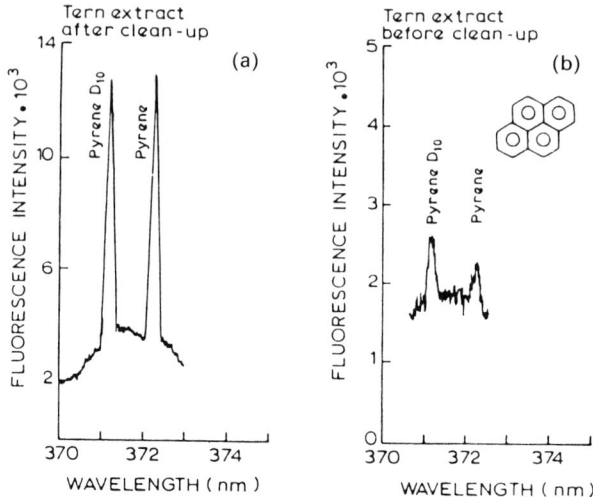

Fig. 9.14. Shpol'skii spectra obtained for pyrene in cleaned (a) and crude (b) tern extract under conventional lamp excitation. Excitation was done at 335 nm. The concentration of the internal standard, pyrene-d_{10}, was 1×10^{-7} M in (a) and 3×10^{-8} M in (b). The Shpol'skii solvent was n-octane and the temperature of measurement about 26 K. From Hofstraat et al. (105).

straightforward quantification, two approaches may be used: either a rather elaborate standard addition procedure can be followed, or the crude extract has to be diluted far enough to remove the influence of the interfering compounds. The latter approach is the easiest but eventually also leads to a reduction in signal of the analytes, as is seen in Fig. 9.14b. The spectrum shown in this figure has been measured for 40 times diluted crude extract, resulting in a solution containing about 5×10^{-9} M of pyrene. Quantification has been done using pyrene-d_{10} as the internal standard. The spectrum has only one site, and the pyrene-h_{10} and -d_{10} peaks are clearly separated. For the crude extract and the cleaned extract similar concentrations of pyrene have been found: 155 and 143 ng/g, respectively, on a dry weight basis. The repeatability is on the order of 10–15% for the crude extract and 5–10% for the cleaned extract. The determination of PAHs in biota using conventional Shpol'skii fluorometry has been discussed in detail by Ariese et al. (112).

When selective laser excitation is employed, the spectrum is much improved compared to that obtained under lamp excitation (see Fig. 9.15). This is easily understood if one realizes that, the Shpol'skii effect being matrix induced, the excitation transitions are also energetically well defined. Hence, it is possible to selectively excite analytes of interest by tuning the laser

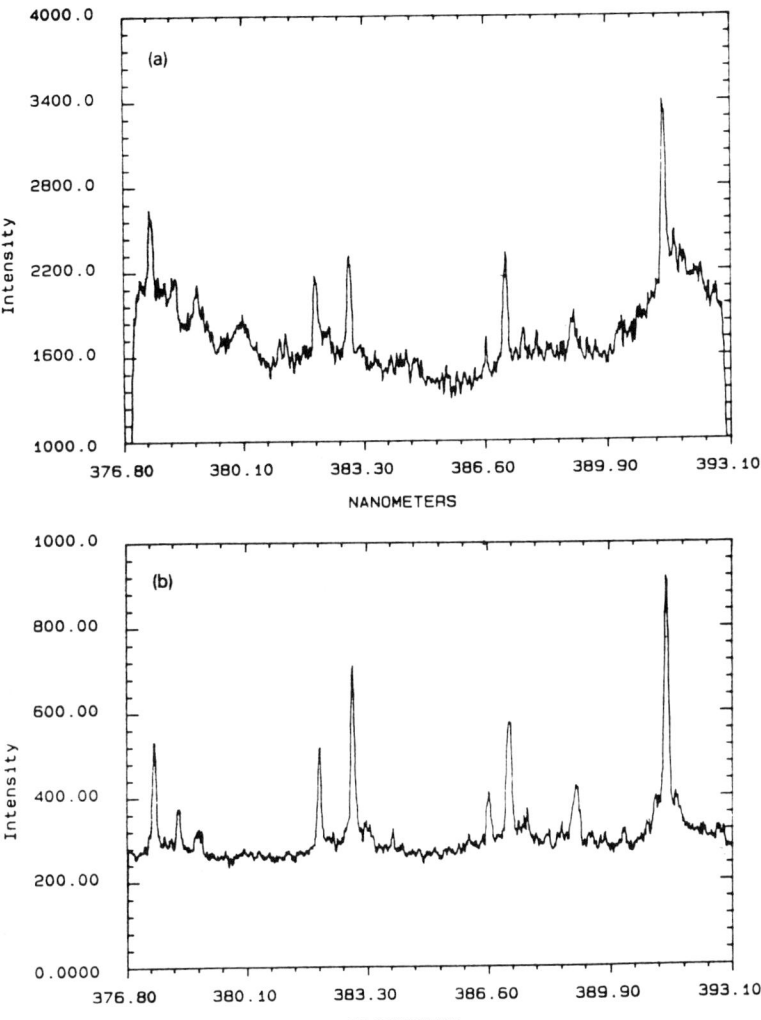

Fig. 9.15. Laser-excited Shpol'skii spectra obtained for pyrene in a crude tern extract. Laser excitation wavelength was 372.1 nm, in the 0–0 transition region. Spectrum (a) was recorded without time resolution; spectrum (b) with a delay of 50 ns. The strong reduction in fluorescent background when time resolution is employed is noteworthy. The spectra were recorded at about 26 K with n-octane as solvent. From Hofstraat et al. (105).

excitation energy to a narrow-line absorption transition. In Fig. 9.15 spectra are given for pyrene in crude tern extract under laser excitation in the 0–0 transition region. The spectral region shown is the vibronic emission region, ranging from 377 to 393 nm. The most prominent band in Fig. 9.15 has an intensity of about 30% of that of the 0–0 emission transition, i.e., the transition depicted in Fig. 9.14b. The latter spectrum was obtained using nonselective lamp excitation. Under these circumstances the vibronic part of the spectrum, which is clearly observed in Fig. 9.15, cannot be discerned from the noise. In Fig. 9.15 even for the vibronic region the signal-to-noise ratio is better than for the 0–0 region shown in Fig. 9.15b!

A further improvement can be made by employing time-resolved detection. This can easily be done by using a pulsed laser for excitation. Time-resolved detection is particularly useful for compounds with a relatively long fluorescence lifetime, like pyrene [about 500 ns (113)]. Therefore a relatively long delay can be used for detection ($t_d = 50$ ns; gate width = 100 ns). As most of the background luminescence has a significantly shorter radiative lifetime than pyrene, a further important improvement of the Shpol'skii spectrum is obtained, as is clear from a comparison of Figs. 9.15b and 9.15a. In the time-resolved spectrum the background signal is only 200 counts/s, in contrast with the 2000 in the steady-state fluorescence spectrum! Furthermore, some peaks that were visible in the non-time-resolved spectrum (mainly around 380 nm) have disappeared in the time-resolved spectrum; apparently these peaks are not due to pyrene. In the time-resolved spectrum even the weakest vibronic transitions of pyrene are clearly visible. The signal-to-noise ratio of the spectrum shown in Fig. 9.15b is about 2 orders of magnitude better than the spectrum shown in Fig. 9.14b, for lamp excitation. This implies that by employing time-resolved laser-excited Shpol'skii spectroscopy levels of pyrene on the order of 1 ng/g on a dry weight basis can be determined in complex biotic extracts without application of any cleanup.

Benzo[*a*]pyrene Metabolites in Fish Bile. PAHs have been widely recognized as high-priority toxic pollutants. Experiments have demonstrated that exposure of animals to PAHs leads to carcinogenic and mutagenic effects. An important part of the carcinogenicity of coal tar and derived products can be attributed to one single compound, benzo[*a*]pyrene (114). Toxicological studies have shown that benzo[*a*]pyrene itself is not toxic. After uptake by higher organisms it is readily transformed into a number of mono- and polyhydroxylated derivatives (115, 116). These metabolites can be bound by polar endogenous substances (e.g., glucuronic acid or glatathione); the conjugates are readily excreted. However, some of the metabolites that are formed are very reactive and may bind covalently to DNA (117). The changes in the DNA structure may finally lead to the formation of malignant tumors (118). Hence, chemical analysis of PAH metabolites is important for toxicological studies. Many aspects of carcinogenic activity of PAHs have yet to be elucidated. Most studies so far have been based on exposure of animals to extremely high doses of PAHs, in particular benzo[a]pyrene, as most analytical approaches are not sufficiently sensitive and lack selectivity. There are some indications that exposure to high levels of PAHs may induce different metabolic reactions of the organism (119). It is therefore considered of the utmost importance to develop methods that are sufficiently sensitive to assess the toxicological effects of PAH exposure at a realistic, low exposure level.

It appears that with Shpol'skii spectroscopy one is able to measure metabolites of benzo[a]pyrene in the bile of fish caught in relatively clean coastal areas in the Netherlands (120). The measurement procedure is relatively simple. Bile is taken from the gallbladder with a syringe, diluted with water and hydrolyzed enzymatically with a glucuronidase/arylsulfatase solution. The monohydroxy metabolites can then be extracted with n-octane and detected directly at low temperatures. With laser-excited Shpol'skii spectroscopy (LESS), using time-resolved detection, metabolites can be measured for laboratory studies in which fish were injected with 0.8–4 mg/kg body weight of benzo[a]pyrene. The relatively polar hydroxy compounds, however, appeared highly incompatible with the n-alkane matrix: the main part of the metabolites was frozen out of the matrix, hence reducing the sensitivity of the detection and impairing quantitative determination of the compounds.

An important improvement was obtained by methylation of the hydroxy metabolites (121). Methoxybenzo[a]pyrene compounds appear to be sufficiently compatible with the Shpol'skii solvent, yielding a significant gain in sensitivity. The methylation involves a rapid and quantitative one-step reaction that can be done in the aqueous hydrolyzed solution (120). After derivatization all methoxy metabolites can be extracted with the n-alkane solvent and subsequently be measured. Even in the bile of fish caught in the Wadden Zee, where the sediment load of benzo[a]pyrene amounts to about 100 μg/kg, i.e., relatively low for coastal areas, metabolites can be measured and quantitated by application of this approach. Figure 9.16 shows the Shpol'skii fluorescence spectrum obtained for 3-methoxybenzo[a]pyrene, isolated from a fish from the Wadden Zee. The spectrum was measured with an intensified linear photodiode array using selective excitation in a narrow vibronic absorption band, 300 cm^{-1} above the 0–0 transition. In addition time-resolved detection was applied. The 3-hydroxybenzo[a]pyrene content of the bile, as inferred from this measurement, is about 1 ng/mL. The detection limit that can be obtained is at least 1 order of magnitude lower. Direct quantitation can be done by comparison of the signals from the analyte with

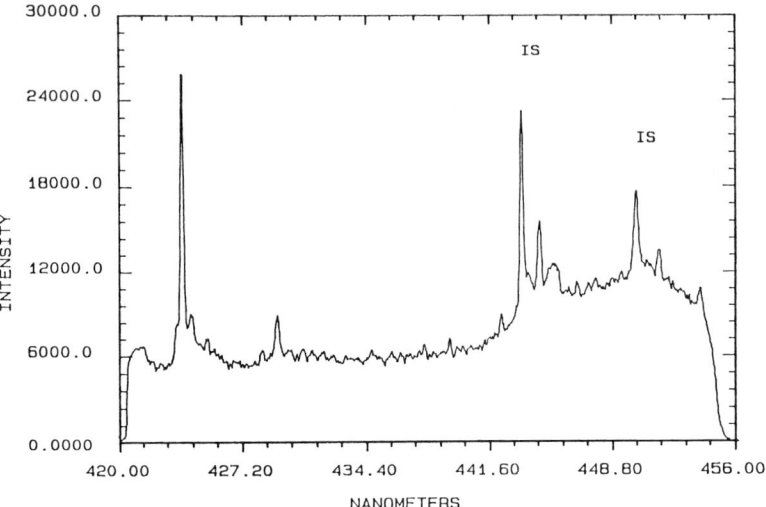

Fig. 9.16. Laser-excited Shpol'skii spectrum of 3-methoxybenzo[a]pyrene formed by methylation of 3-hydroxybenzo[a]pyrene isolated from bile of a flounder caught in the Dutch Wadden Zee. Excitation was done at 418.36 nm. The spectrum was recorded at about 20 K in n-octane. The internal standard (IS) is perylene-d_{12}.

those of the internal standard, perdeuterated perylene, that can be determined simultaneously with multichannel detection. Some quantitative results are summarized in Table 9.5.

All monohydroxybenzo[a]pyrene metabolites can be selectively determined with the procedure described above. Interestingly, contrary to observations in other (laboratory) studies, no 9-hydroxybenzo[a]pyrene could be detected in the bile of the flounders studied thus far. The availability of a technique that combines good selectivity with extremely high sensitivity may provide new insight into toxicological effects of PAHs for environmentally relevant situations. Dihydroxy metabolites are more difficult to determine. Rough methylation yields a complex mixture of products that cannot be straightforwardly quantitated. Only with careful methylation can the dimethoxy compounds be formed. The limits of detection for these compounds, however, are 1-2 orders of magnitude higher than for the monomethoxy derivatives.

Recently, Ariese was able to determine 3-hydroxybenzo[a]pyrene in urine of coke oven workers by means of the derivatization procedure and Shpol'skii fluorescence detection (122). Hence, this technique also has potential for the evaluation of occupational health risks.

Table 9.5. Concentration of 3-Hydroxybenzo[a]pyrene in Fish Bile

Situation[a]	[3-Hydroxybenzo[a]pyrene] (ng/mL)
Harbor sediment (direct contact)	76 ± 24 ($n = 3$)
Harbor sediment (no direct contact)	7.7 ± 2.4 ($n = 4$)
Sand bottom	1.2 ± 0.1 ($n = 3$)
Wadden Zee (pooled sample)	1.6

[a] Data are from a mesocosm exposure experiment. Fish were exposed for a prolonged period to three situations: they were kept in a basin filled with dredging class 3 harbor sediment in direct contact, in a basin with clean sediment but filled with water that had been equilibrated with the harbor sediment (no direct contact with sediment), and in a basin filled with Wadden Zee sediment. For comparison a result is presented for a field measurement. More information is given in Ariese et al. (120).

9.4.4. Fluorescence Line-Narrowing Spectroscopy

9.4.4.1. Principles

In Shpol'skii spectroscopy the inhomogeneous broadening of spectra is reduced by choosing suitable matrix materials. The main disadvantages of this approach are that one needs to employ several matrices to determine all compounds of interest and, more importantly, that for many compounds there may not be suitable matrix materials to obtain narrow-line spectra. As was discussed in Section 9.4.3.2, polar hydroxy-substituted benzo[a]pyrene metabolites had to be permethoxylated before reliable quantitation could be realized. In FLN spectroscopy line-narrowing is induced by employing selective excitation (16, 17,76). By making use of an excitation source with a considerably narrower spectral bandwidth than the inhomogeneous bandwidth of the molecules that are probed, a well-defined subset of excited molecules is created with energy differences between ground and excited states, exactly matching the energy of the excitation source. Hence, the matrix is less crucial for the procurement of highly resolved spectra: an important requirement is that the well-defined excited state population created by the excitation process remains conserved in the emission process. The final bandwidth observed is determined by the convolution of the excitation bandwidth, which influences the excited state population, and the homogeneous bandwidth of the probed and the observed transitions. The homogeneous bandwidth depends on the lifetimes of the states involved in the transition, via the Heisenberg uncertainty principle. In Section 9.4.3.2 attention was paid to very narrow line widths measured for selective excitation of a compound in a crystalline matrix. In general, the term fluorescence line-narrowing spectroscopy is used for molecules in amorphous matrices.

On the negative side we note that only part of the molecules are brought into the excited state by the selective excitation process. In a typical FLN experiment a laser with a bandwidth of $0.1\,cm^{-1}$ probes a $100\,cm^{-1}$ broad inhomogeneous band, so that only 1 out of 10^4 analyte molecules are probed. Therefore FLN spectroscopy is clearly less sensitive than methods that use matrix-induced line-narrowing.

In order to achieve highly resolved FLN spectra a number of conditions must be fulfilled (16,17,76):

1. The emission bandwidth is determined by the cross section of the excitation source and excited levels. Therefore it is necessary to excite to relatively long living states (with lifetimes in the nanosecond-to-picosecond range), as otherwise the lifetime-limited homogeneous bandwidths of the probed levels become too broad. Excitation to higher electronic levels in general does not produce FLN spectra.

2. It is crucial that, if excitation is employed to another level than that which emits the light, the excited state distribution does not lose its well-defined character before emission takes place. This condition is in most cases not fulfilled if another electronic state is involved in the transfer process, e.g., excitation in the singlet manifold does *not* produce narrow lines in the phosphorescence spectrum. Excitation to higher vibronic levels of the S_1 state in general results in narrow emission lines, although they are somewhat broadened with respect to resonantly excited states (123). However, excitation into vibronically excited states may pose a problem. If the energy is increased in the S_1 vibronic envelope, the density of vibrational states grows and hence also the chance that several—inhomogeneously broadened—vibronic bands are probed simultaneously. Molecules probed in several vibronic states will proceed to their corresponding purely electronic states, thus creating several well-defined subsets of excited molecules (see Fig. 9.17). In the FLN spectrum this process results in the appearance of multiplets, which can, on the one hand, make the spectrum difficult to assign, particularly when inhomogeneous broadening is large; on the other hand, the multiplets provide vibrational energies of the analyte in the S_1 state, which can be utilized for identification purposes, an aspect that will be discussed in Section 9.4.4.2. Since the density

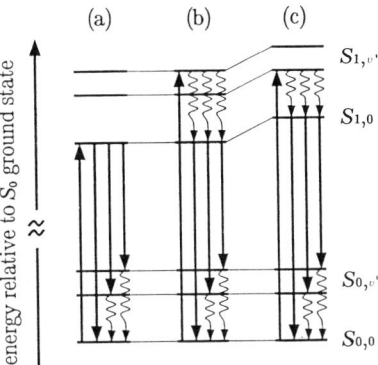

Fig. 9.17. Energy diagram illustrating the influence of the mode of excitation on the appearance of the FLN spectrum: (a) and (b) show a molecule at the red end of the inhomogeneous absorption band. Excitation with low energy leads to a single-site spectrum (situation a). Higher energy excitation into vibrationally excited S_1 states leads to a multisite spectrum if the inhomogeneous broadening is large with respect to the difference in vibrational energy between adjacent bands (situations b and c: two subsets of excited states are created by excitation with the same energy). From Hofstraat et al. (126).

of vibronic states strongly increases at energies above the purely electronic state in excess of 1500–2000 cm^{-1}, excitation into this area generally does not lead to usable FLN spectra: many overlapping bands will appear in the multiplet, so that much of the line narrowing effect is lost. The excitation wavelength dependence will be discussed in more detail below.

3. The selectively excited population must remain conserved during the lifetime of the excited state. This means that no alterations may occur in the environment of the molecules on this time scale—precluding the realization of FLN spectra in liquids where solvent reorientation is fast—and that no guest-to-guest energy transfer may take place.

4. There should be no strong interaction between the electronic transitions of the guest molecules and host crystal modes, as otherwise the FLN spectrum displays apart from the narrow lines also broad "phonon" sidebands. In Shpol'skii matrices these broad bands may be present, but as the interaction is weak they will not impede the measurement. In FLN they play a more important role because they may be created in the emission process, as in Shpol'skii spectroscopy, but will also be probed in the excitation process (123). Since the intensity of the phonon sidebands relative to the narrow lines increases with temperature, this implies that for FLN lower temperatures are required than for Shpol'skii spectroscopy. In a number of cases temperatures

should be so low that they cannot be conveniently reached with a closed cycle refrigerator. Furthermore, a number of matrix materials will be unsuitable for inducing FLN, owing to strong interactions with molecules of interest.

5. Attention should be paid to the occurrence of "hole-burning" effects in FLN spectra (124,125). These effects are due to photochemical reactions and/or to reorientations of guest molecules with respect to their solvent cage while they are in the excited state (124). They lead to a decrease in the ground state population that can be selectively excited by the laser at the wavelength used and, hence, to a reduction of the narrow-line intensities in the spectrum under prolonged irradiation. In the absorption spectrum, hole-burning leads to a narrow gap in the inhomogeneously broadened bands. In FLN spectra hole-burning effects are mostly unwanted and can be reduced in several ways (123). However, sometimes they can also be used to improve FLN spectra, e.g., by diminishing the contribution of broad phonon sidebands, which are much less affected by hole-burning phenomena. Approaches to removal of hole-burning effects are the application of multichannel photodiode array detection (80) or of scanned laser techniques (126,127). This aspect will be considered in more detail below.

9.4.4.2. Excitation Wavelength Dependence

The principles of FLN spectroscopy and the excitation wavelength dependence of the spectra that are obtained can be elegantly demonstrated in an excitation–emission matrix (EEM), as has been discussed by Hofstraat et al. (126). Figure 9.18 shows the EEM obtained for oxazine-4-perchlorate in poly(vinyl butyral) (PVB). The 2D spectrum (or rather 3D spectrum as the intensity can be considered as a third dimension, apart from the excitation and emission wavelengths) provides the emission intensity as a function of excitation and emission wavelength. Single emission and excitation spectra, at fixed excitation and emission wavelengths, respectively, can easily be obtained by taking appropriate cross sections from the EEM. The excitation wavelength dependence of FLN spectra can easily be derived from the 2D plot. The most striking features are the very narrow ellipsoids with a slope of 1 in the emission–excitation plane. These ellipsoids are the result of the energy selection process: as the excitation process determines the molecules brought to the excited state, the emission wavelength shifts linearly with the excitation wavelength, thus yielding a slope of 1. A change in excitation wavelength, therefore, yields a concomitant change in emission wavelengths of the 0–0 transition and of the corresponding vibronic manifold.

Now let us consider in more detail the excitation wavelength dependence of the FLN spectrum. In the upper part of Fig. 9.18, starting at an excitation wavenumber of about 15600 cm^{-1}, 0–0 excitation is predominant, corre-

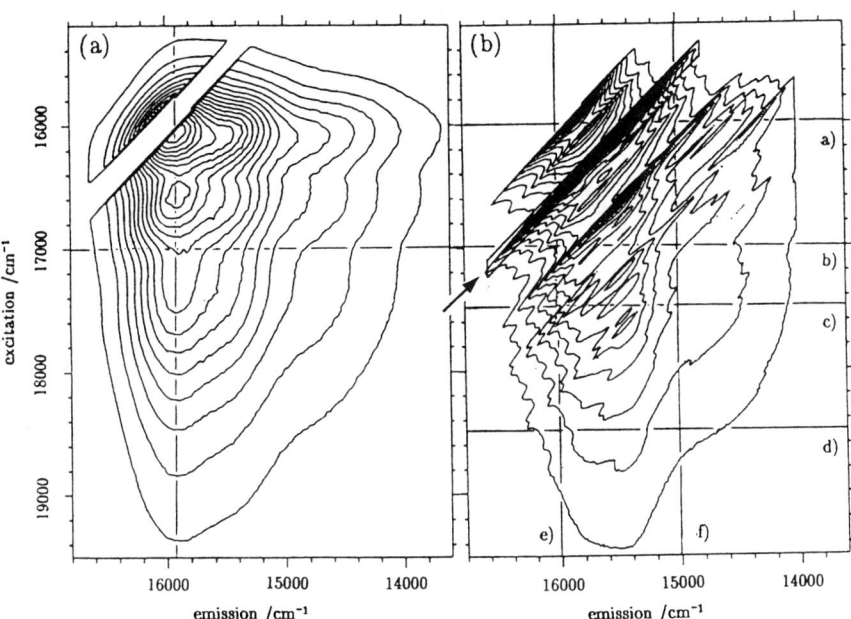

Fig. 9.18. Contour plot of the EEM of oxazine-4-perchlorate in PVB, obtained at (a) room temperature and (b) 8 K. The straight lines in the spectrum indicate the positions of the cuts shown in Fig. 9.19. From Hofstraat et al. (126).

sponding to situation (a) in the diagram presented in Fig. 9.17. In this case the excitation wavenumber coincides with the 0–0 emission wavenumber. Owing to the intense Rayleigh scattering of the excitation light, this emission cannot be measured; the vibronic part of the spectrum is clearly visible (cf. the emission cross section shown in Fig. 9.19a). Increasing the energy within the 0–0 inhomogeneously broadened envelope results in a shift of the complete spectrum: both the 0–0 emission wavenumber, which coincides with the excitation wavenumber, and the vibronic emissions shift with the 0–0 wavenumber.

When the excitation energy is increased further a complex picture emerges, diagrammatically shown in Fig. 9.17b–c. Several subsets of excited molecules are created by simultaneous probing of 0–0 and vibronically excited levels or, at higher excitation energies, simultaneous probing of more than one vibronically excited level. The energy differences of both types of excitation processes may coincide with the laser energy as a result of inhomogeneous broadening. Vibronically excited FLN spectra show a 0–0 emission band situated at an

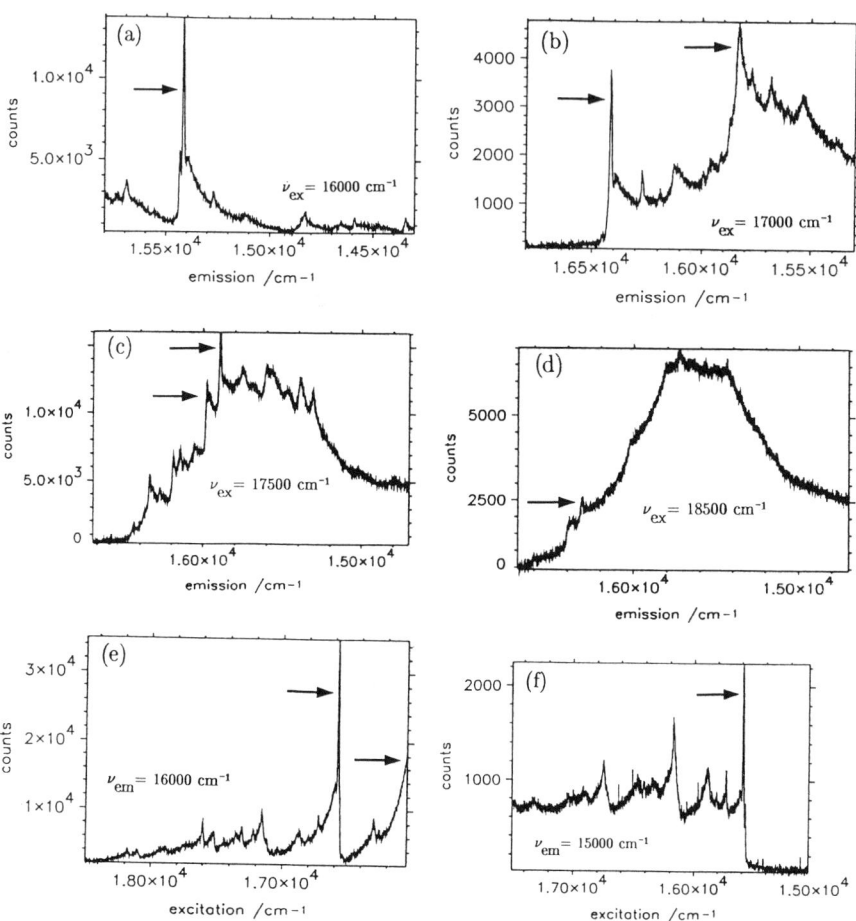

Fig. 9.19. One-dimensional FLN emission (a–d) and excitation (e,f) spectra of oxazine-4-perchlorate in PVB. The positions of the six 1D scans are presented in Fig. 9.18 by straight lines. From Hofstraat et al. (126).

energy difference from the excitation wavenumber, exactly corresponding with the probed excited state vibrational quantum. In this case the complete spectrum, including the 0–0 transition, can be measured. Figure 9.19b shows an example of multiplet structure as expected for higher energy excitation. The FLN spectrum becomes more and more complex as excitation is done at higher excess energies, where more and more vibronic states are probed simultaneously (see Fig. 9.19c). Finally, at excitation wavenumbers even further away from the 0–0 transition area (>2000–$3000\,cm^{-1}$) the energy selection effect diminishes (see Fig. 9.19d). For excitation above $18700\,cm^{-1}$ no narrow lines are observed. Here a region is probed where the density of vibronic states is high, so that many closely lying states with their corresponding vibrational ground states are probed simultaneously. Thus the transition energy selection effect is obliterated by spectral overlap in the emission step.

It is obvious from Fig. 9.19 that the excitation spectrum also shows vibrational resolution, like the emission spectrum that is commonly recorded in FLN spectroscopy. Figures 9.19e and 9.19f show cuts from the EEM taken in the excitation direction. The vibrational resolution obtained implies that excitation FLN spectra are equally useful for analytical applications as emission FLN spectra. An important advantage of excitation FLN spectra over emission FLN spectra is that the former are recorded by scanning the excitation wavelength, while a narrow emission wavelength region is being monitored. Since the excitation wavelength varies while the spectrum is scanned, hole-burning effects will be strongly reduced or absent in the spectrum.

Although it seems that there are a lot of problems in realizing a good FLN spectrum, in practice narrow-line spectra have been reported for a variety of molecules in a variety of matrices. FLN spectra have been measured for highly polar and even ionic species (17) that would never yield highly resolved spectra in Shpol'skii matrices. As matrix materials—apart from the most frequently employed glassy solvents—polymers, membranes, TLC plates, and vapor-deposited rare-gas matrices could be used (17).

9.4.4.3. Analytical Applications

General. As FLN is a newer technique than Shpol'skii spectroscopy and as more sophisticated equipment is needed than for the matrix-induced techniques, to date only few reports on the analytical application of this technique have appeared. The potential of FLN for analysis of environmental samples has been demonstrated for the quantitative determination of a number of PAHs in different samples: cigarette tar (128), soot (128), solvent-refined coal (129), and gasoline (130). Analytes have been identified in the low-ppb range, while no cleanup of the samples has been carried out. Small's group have reported that direct quantitation without application of internal standards is feasible in FLN, in contrast with Shpol'skii spectroscopy (129). The line-narrowing induced by laser excitation does not depend strongly on the fit of the guest molecules in the solid matrix, so that sample preparation effects are small. Information on detection limits is scarce. Although one would expect, on the basis of the selective excitation process used, that FLN is much less sensitive than Shpol'skii spectroscopy, very low detection limits have been published [e.g., Bykovskaya et al. have reported a detection limit of 4×10^{-11} M, i.e.,

10 ng/L, or 10 ppt, for perylene (130)]. However, one should realize that in fluorometric analyses detection limits in general are determined by the background luminescence of solvents, by scattered light, etc., so that the sensitivity differences between the two techniques may be obscured.

The most promising field of application for FLN will be in the determination of polar compounds that do not produce highly resolved spectra via the matrix-induced techniques. Nice examples of the application of FLN in this context have been reported by Small's group (128,129). They were able to determine metabolites of PAHs and their adducts to DNA by simply dissolving these compounds in glassy solutions and cooling. Some examples of the application of FLN spectroscopy in this field will be discussed below.

An interesting aspect of FLN spectroscopy is that it may be applied in combination with chromatographic techniques, thus providing an extremely selective detection method. Of course FLN, being a solid-state technique, can never be applied in direct on-line combination with dynamic chromatographic separations. However, there appear to be a number of promising options. Hofstraat et al. have shown the applicability of FLN in combination with TLC, also for quantitative purposes (131). The semi-on-line detection of compounds separated in LC with FLN appears possible by making use of a TLC plate to immobilize the effluent (132). Finally, it has been demonstrated that FLN spectra of PAHs can be obtained on substances used as packing material in LC concentration precolumns and in LC separation columns, indicating the feasibility of "on-column" detection (133).

PAH Metabolite Adducts to DNA. Carcinogenic effects are generally preceded by metabolic activation of previously harmless compounds. The activated compounds, or metabolites, can form covalent bonds with DNA, which may eventually lead to uncontrolled cell growth and tumor formation (117,118). The main events that cause the carcinogenic effects for many compounds are not well understood. Even for benzo[*a*]pyrene, the most widely investigated PAH in this respect, numerous questions remain. A complicating factor is that at present no technique is available that is sufficiently specific to determine the nature of the DNA adducts formed as a result of exposure of biota to pollutants and that at the same time is sufficiently sensitive to provide the identification for exposures at realistic levels. By means of radioactive detection of modified DNA nucleotides labeled with ^{32}P (in so-called ^{32}P-postlabeling analysis), adducts can be determined for biota exposed to low levels of carcinogens (134). The identification is based on comparison with a standard submitted to similar separation by 2D thin-layer chromatography. However, no structural evidence is obtained, because conventional chromatographic and spectrometric techniques lack sensitivity: they can only be applied for determination of

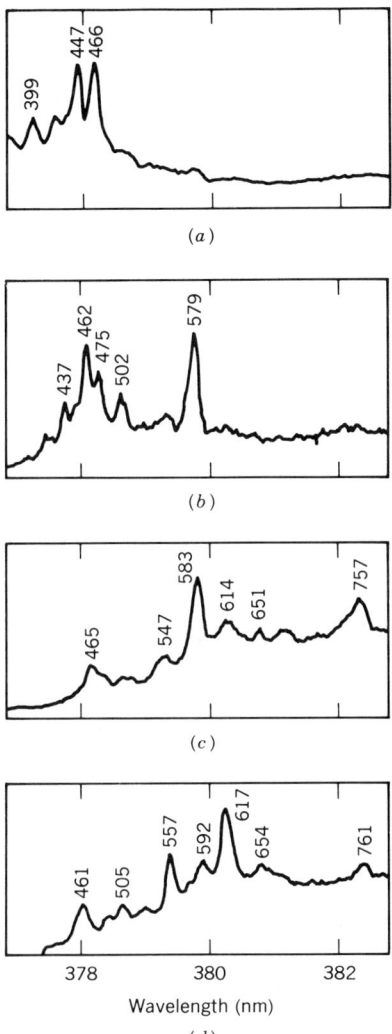

Fig. 9.20. FLN spectra of a mixture of (a) benzo[a]pyrene–tetrol and DNA and of three DNA adducts: (b) (+)-*anti*-BPDE–DNA, (c) (−)-*anti*-BPDE–DNA, and (d) *syn*-BPDE–DNA. The peaks are labeled with their corresponding vibrational frequencies. The spectra were recorded in glycerol/water glass at 4.2 K with an excitation wavelength of 371.6 nm. From Jankowiak and Small (119).

metabolites and adducts in laboratory studies, where biota are exposed to artificially high levels.

Recently, Small's group has shown that FLN spectroscopy can be successfully applied to the determination of metabolites of benzo[a]pyrene bonded to DNA. For the adduct of benzo[a]pyrene-7,8-diol-9,10-epoxide (BPDE), the "ultimate" carcinogen, and DNA a detection limit of about 3 modified bases in 10^8 has been established (135). This value approaches the sensitivity that is needed for analysis of adducts in real samples. FLN has an important advantage over the conventionally applied techniques, in that it provides specific spectral information. Figure 9.20 shows FLN spectra obtained for several BPDE–DNA adducts, in a glycerol/water glass, cooled to 4.2 K (119). Figures 20b–d display spectra measured for adducts of several stereoisomers of BPDE, namely, (+)-anti, (−)-anti, and syn. For analysis, in vivo identification of the actual metabolites that form covalent bonds to DNA is very important. Even minor stereochemical differences may give significant changes in the reactivity and in the mode of attack of the metabolites (136,137). Although the adducts are strongly similar, it appears that they can be readily distinguished on the basis of their FLN spectra. Moreover, the adducts can be easily discerned from the unreacted metabolite benzo[a]pyrene–tetrol, as can be inferred from the spectrum shown in Fig. 9.20a.

In Fig. 9.21 the FLN spectrum obtained for DNA adducts in fish liver, extracted from fish exposed to 100 mg/kg benzo[a]pyrene is depicted. Also shown are reference spectra of *syn*- and (+)-*anti*-BPDE–DNA adducts measured under the same experimental conditions (which are different from the ones used for the spectra depicted in Fig. 9.20). The adducts in fish DNA are clearly identified as the *syn*-BPDE–DNA adduct. This is surprising, as *anti*-BPDE in general has a greater activity than *syn*-BPDE (138). It appears that the exposure level determines the relative activities of both metabolites (119).

Analyses of DNA adducts in fish liver by means of ^{32}P-postlabeling have revealed that, both in laboratory exposure studies to benzo[a]pyrene and in field studies, many spots are observed on the TLC plates that cannot be attributed to adducts for which standards are available (139). The specificity of FLN spectroscopy, which can also be applied to compounds on TLC plates (131,132), might afford important new information in this respect. Knowledge about the identity of the adducts present in the fish liver will provide more insight into the mechanism and effect of PAHs, after uptake by biota.

Combination of FLN Spectroscopy with Thin-Layer Chromatography. In FLN spectroscopy use is made of laser excitation of analyte molecules in a low-temperature solid solution; a temperature lower than 30 K is generally required. As the line-narrowing·effect is solely caused by the selective laser excitation, FLN spectra can be obtained for molecules in a variety of matrix materials

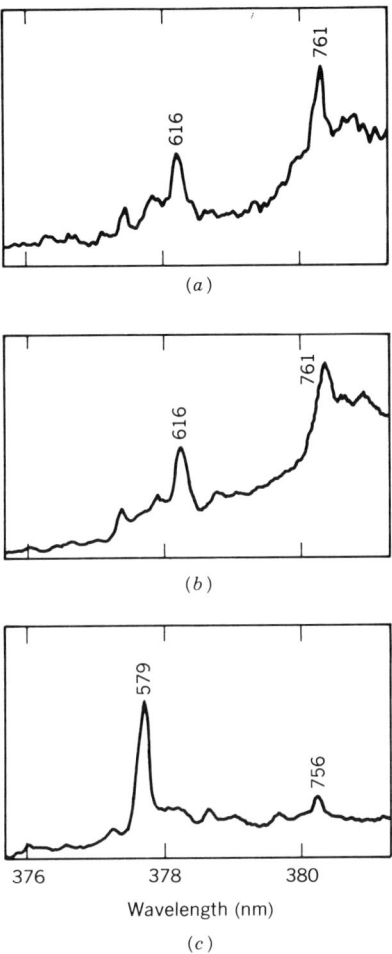

Fig. 9.21. FLN spectra of (a) DNA extracted from liver of fish exposed to benzo[a]pyrene, and of the standards (b) *syn*-BPDE–DNA and (c) (+)-*anti*-BPDE–DNA. Spectra were measured in glycerol/water glass at 4.2 K with an excitation wavelength of 369.6 nm. From Jankowiak and Small (119).

provided that there is no strong interaction between the analyte molecules and the matrix. Hofstraat et al. have demonstrated the applicability of FLN for detection of analytes on various types of TLC materials (131). So far silica, alkyl- and phenyl-modified silica, alumina, and cellulose have been successfully employed. Impurities present on the surface of the plates can be removed by overnight elution, for instance, with methanol; this can lead to an improvement of the detection limit by a factor of 10 (61). A serious problem is the irregular surface of TLC plates, causing serious scatter-light problems.

An unwanted effect in analytical application of FLN spectroscopy is the occurrence of hole-burning effects (see Section 9.4.1). In the literature attention has been paid to the time evolution of the FLN spectra under prolonged laser irradiation as well to the matrix, laser intensity, temperature, and excitation wavelength dependence of physical hole-burning effects (124). A study of tetracene in various matrices leads to the conclusion that spectral hole-burning effects are strongly reduced (by about a factor of 10) for TLC plates in comparison with the glassy ethanol/methanol and 3-methylpentane matrices.

Recently FLN spectroscopy has been used to study an impure 1-chloropyrene sample, on an RP-18 TLC plate (131,140). A comparison of the fluorescence line-narrowed spectrum, obtained by excitation with the 364 nm line of the excimer (Xe:Cl) dye laser, with the highly resolved Shpol'skii spectra reported for this compound by Colmsjö et al. (141) have revealed the 1-chloropyrene fluorescence bands from pyrene and at least a third compound. After elution of the 1-chloropyrene spot on a silica gel TLC plate with acetonitrile as the mobile phase, four separate spots appeared. In view of the method used for the synthesis of 1-chloropyrene it was supposed that higher substituted chloropyrenes were present in the sample. On the basis of the R_f values and the FLN spectra of some relevant chloropyrenes, the highly resolved and vibrational fine structure of pyrene, 1-chloropyrene, 1,6- and 1,8-dichloropyrene, and trichloropyrene could be assigned. On the basis of the pure dichloropyrene samples it became clear that the dichloropyrene spot contains both structural isomers. From this study it is obvious that FLN spectroscopy combined with TLC plates is a powerful tool for identification.

Quantification of FLN spectra is also possible, as has been reported by Brown et al. for glassy matrices (129) and by Hofstraat et al. for thin-layer matrices (131).

The combination of FLN spectroscopy and TLC plates has several advantages. Many samples can be examined in one batch, thus ensuring the same experimental condition and saving time. When a complex sample is studied, the (partial) applied spot can be followed by selective identification and quantification with the highly resolved fluorescence technique. A disadvantage is the strong scattering of the laser excitation light by utilizing expensive apparatus.

Hofstraat et al. have demonstrated that the potential of LC can be extended by combination with TLC (132). If the effluent from the LC column is deposited on a TLC plate moving with a constant speed, an "immobilized" chromatogram is obtained. Such a coupling of LC to TLC can be useful for various purposes. First, separation efficiency may be increased by combining two different separations, e.g., reversed-phase LC followed by thin-layer separation. Secondly, detection modes that are incompatible with LC can be applied, e.g., FLN spectroscopy, phosphorescence, and surface-enhanced spectroscopy, methods generally requiring immobilized analytes. Thirdly, storage of the chromatogram may be useful in improving the feasibility of relatively slow detection techniques like IR spectroscopy and 2D fluorescence and absorption spectroscopy. The essential aim in this coupling is the deposition of the effluent without significant loss of chromatographic information. It has been shown with conventional fluorescence spectroscopy that the room-temperature fluorescence emission and excitation spectra of, for example, benzo[k]fluoranthene on a silica-on-aluminum plate and in a methanol solution are identical (132). This similarity demonstrates the excellent capability of the TLC material to serve as an inexpensive means of increasing the detection potential in LC. The fluorescence emission spectra were also used for quantitative purposes. For benzo[k]fluoranthene the calibration curve was linear over 3 decades (from about 5 ng to the detection limit of 3 pg), with a correlation coefficient of 0.993 ($n = 6$) and relative standard deviation of 1% ($n = 6$) for 76 pg.

FLN spectroscopy can only be applied at low temperatures; thus it cannot be used as an on-line detection method. The feasibility of applying this technique for off-line detection in LC was investigated for immobilization of the column effluent on a TLC plate, with tetracene chosen as the model compound. In Fig. 9.22 part is shown of the spectrum of tetracene (13 ng) deposited on a silica plate, obtained with 5-mW 457.9-nm argon-ion laser excitation at 9.3 K. The 0–0 transition at 474.1 nm and a number of vibronic emission bands can be discerned. The background emission of the TLC plate, however, was much stronger with laser excitation than with a conventional lamp source, because even compounds with very low fluorescence quantum

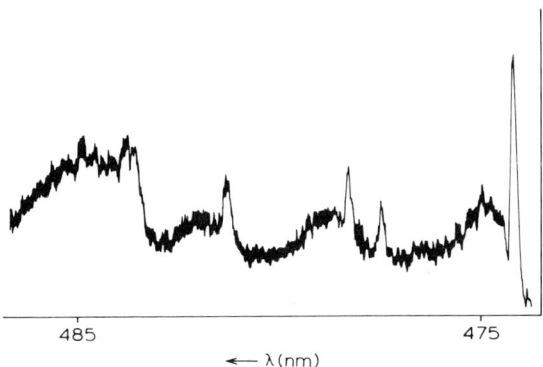

Fig. 9.22. Fluorescence line-narrowing spectra for 13 ng of tetracene on a silica-on-aluminum plate (excitation at 457.9 nm with 5 mW of laser light; $T = 9.3$ K). From Hofstraat et al. (132).

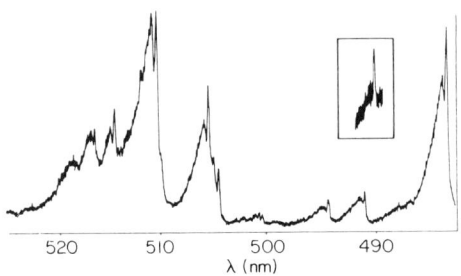

Fig. 9.23. Fluorescence line-narrowing spectra of 3.1 ng of tetracene on a KC18 TLC plate (excitation at 476.5 nm with 1 mW of laser light; $T = 9.3$ K). Inset shows the 483.8 nm emission line recorded for 31 pg of tetracene. From Hofstraat et al. (132).

yields are observed. The luminescence background has about the same intensity as the strongest peak in the spectrum but tends to be broad and structureless. Figure 9.23 displays the spectrum obtained for tetracene on a Whatman KC18 plate with excitation at 476.5 nm. The first part of the spectrum from 476 to 480 nm cannot be recorded, because of Rayleigh scattering of the intense laser light, unless time-resolved detection is used. The first strong line observed in the spectrum, at 483.8 nm, relates to a vibronic emission band at 314 cm^{-1} from the origin. This spectrum was obtained for 3.1 ng of tetracene at a laser power of 1 mW. Clearly, the 476.5 nm excitation provides a better signal-to-noise ratio. The emission line obtained for 31 pg of tetracene on a KC18-silica plate is shown as an inset in Fig. 9.23.

The next step will be identification and quantification of a complex mixture of aromatic hydrocarbons, deposited from an LC column on a TLC plate, with FLN spectroscopy detection. It is expected that the compounds benzo[*a*]anthracene and chrysene, benzo[*e*]pyrene, and benzo[*b*]fluoranthene, which could not be distinguished with conventional fluorescence (142), will be discriminated using FLN detection.

9.4.5. Concluding Remarks

Laser-excited high-resolution molecular fluorescence spectroscopy in low-temperature matrices affords both selective and sensitivie analyses. The spectroscopic features can be used to identify structurally closely related compounds, such as isomers. It also appears possible to selectively determine compounds in complex environmental samples with a minimum of cleanup and in most cases without the need to separate the analytes. Quantification can be done by using perdeuterated analogues of analytes as internal standards.

Such compounds, which are absent in real samples, are spectrally easily distinguished. Detection limits are very low: for "clean" samples, single molecule detection has been realized; in environmental samples, sub-ppt detection limits have been achieved.

On the negative side it must be mentioned that optimization of experimental circumstances (host matrix, temperature of measurement, cooling procedure, excitation and emission wavelengths that are most useful) for a particular application may take some time. Furthermore, rather specialized equipment is needed, like cryogenic apparatuses and laser systems not common in analytical laboratories. In addition, at present no complete setups are commercially available. Finally, it will not be easy to automatize the analysis owing to its static character.

Two techniques have been described: Shpol'skii spectroscopy and FLN spectroscopy. Shpol'skii spectroscopy offers the highest sensitivity due to its matrix-induced nature. Also, when combined with selective excitation, it is the most specific method of the two. A disadvantage of the Shpol'skii technique is that in particular polar compounds are difficult to measure because they do not dissolve in the apolar n-alkanes that are mostly employed. Furthermore, the way samples are prepared may influence the spectra that are obtained. FLN spectroscopy is less sensitive than Shpol'skii spectroscopy but has as its strong point that it can be applied to a wide variety of solvents and matrices. For polar and even for ionic species FLN spectra have been observed. In particular, the suitability of the FLN technique for the analysis of PAH–DNA adducts has resulted in a unique and fascinating application. Finally, the way samples are prepared does not have such a strong influence on the FLN spectrum as it may have on the Shpol'skii spectrum. In conclusion, both cryogenic fluorescence techniques supplement each other quite well, the—preferably laser-excited—Shpol'skii technique being the first method of choice, with the FLN method offering an alternative approach for special problems and affording exciting applications.

9.5. FLOW CYTOMETRY

9.5.1. Introduction

Flow cytometry is a very powerful analytical technique that recently has become much more accessible and has received wide application in particular in biomedical sciences. The availability of high-power lasers—in addition to the further development of electronics and of computer hardware—has made an important contribution to the development of flow cytometric analysis. The unique feature of this technique is that intrinsic or extrinsic properties of particles (i.e., cells, but also algae, bacteria, larvae, polymer beads, etc.) can be quantitatively measured on an individual basis at extremely high speed.

Flow cytometry combines many advantages of microscopy and biochemical analysis techniques. As in microscopy, properties or characteristics of individual cells are measured. However, in flow cytometry objective, quantitative information is obtained via straightforward optical analysis. Also, these data are acquired at very high speed. Hence, hundreds of particles can be analyzed per second, so that meaningful statistical information becomes available. Microscopic analysis obviously gives the most detailed morphological information, but it is very slow. Quantitative information, e.g., on the number of membrane surface receptors, on DNA content, or on intracellular pH, cannot be obtained. Biochemical analysis in general supplies objective quantitative data but lacks the information on individual particles. It provides an average of a cell population; rare cell subpopulations, which may be very important (e.g., toxic phytoplankton species or tumor cells), in general are not observed. Finally, flow cytometry has two unique advantages over any competitive technique: simultaneous multiparameter measurements can be done on individual particles; and on the basis of rapid aquisition and real-time treatment of data, selected particles can be physically separated from the sample [e.g., in the FACS (fluorescence-activated cell sorter) machines].

Modern flow cytometric instrumentation originated in the 1950s, when Coulter introduced the first instruments that measured the volume of individual cells (143). The basis for this measurement was the change of electrical resistance as the cells passed through a narrow orifice suspended in an electrolyte solution. The first flow cytometric instruments that used optical measurement of particles were developed in the 1960s (144). Kamentsky's instrument used a mercury source to measure the absorbance of cells that passed through a flow chamber. Very soon, however, the laser was discovered as the ideal light source for flow cytometric applications (145). The narrow bandwidths of lasers, in combination with their high spectral radiant power and good spatial and temporal stability, made these light sources very attractive (see Section 9.2). Stray light can easily be removed by spectral blocking filters. The laser light can be transported easily over large distances and can be focused down to illuminate very small areas at high radiant density. Presently, laser sources are available that cover the full spectral range. Hence, sensitive and selective analysis of even very small particles (e.g., bacteria) can be realized.

In this section no attempt will be made to afford full coverage of developments in flow cytometry. Several excellent books and monographs are available that provide in-depth information on this topic (15,146–149; see also references therein). In particular, the comprehensive book of Shapiro, which can also be used as a manual for those who want to build their own flow cytometer, is worthwhile. The specialized journal *Cytometry* covers all important developments in instrument design and in interpretation software. Here, only the essential principles and characteristics of flow cytometry will be explained. Furthermore, a number of applications will be discussed to illustrate the potential of laser flow cytometry. The applications will not only be taken from the biomedical field, where the flow cytometer already has become a workhorse for routine cell analysis, but also from environmental studies, where this technique holds great promise.

9.5.2. Principles

In flow cytometry measurements are done on single particles. A first condition for a flow cytometer, therefore, is to produce a well-defined flow of isolated particles that pass through an optically accessible region. Particles, therefore should be present in single-cell suspension, which makes flow cytometry directly applicable for the analysis of e.g., phytoplankton and bacteria in aqueous samples or cells in blood and semen. In solid tissues cells must be brought into suspension via mechanical or chemical means. That often causes damage of a large fraction of the cells.

A schematic representation of the central part of a flow cytometer is shown in Fig. 9.24. The organized flow of particles in general is realized by application of hydrodynamic focusing: the sample is introduced via a capillary at a relatively small flow rate into a conically shaped flow chamber where a much larger "sheath" flow is added. The sheath flow, with different specific density, does not mix with the sample flow but focuses it into the cuvette. A similar approach has recently been successfully applied for LIF detection in capillary

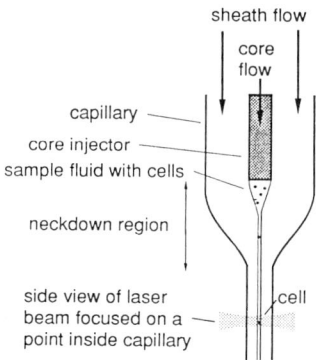

Fig. 9.24. The "heart" of a flow cytometer: hydrodynamic focusing of cells in the sample stream by the sheath flow and interaction with the laser beam in the free-flowing region. From Shapiro (146).

electrophoresis, as discussed in Section 9.3.2.1 and 9.3.4.4. The cells in the sample flow thus are forced into a well-defined part of the cuvette, where they can be probed by the laser beams used for excitation. The relative volumes of sample flow and sheath flow can be controlled, so that single passage of cells is guaranteed—and problems with coinciding cells are avoided.

For excitation one or more lasers are applied that illuminate the cuvette. In the latter case, the cuvette is illuminated at different positions, so that optical signals can be acquired corresponding to the separate laser sources. This obviously puts strong demands on the optical design of excitation and detection optics. By application of advanced optical filtering techniques, the light induced by one laser source can be detected in different spectral regions simultaneously. Of course, the high flow rates and the need for parallel data processing (of more than one detection channel) necessitate the use of dedicated and advanced electronics and data acquisition hardware. For the large amount of data generated (in an average experiment on the order of 5 parameters are measured for 10,000 particles, yielding some 600,000 bits of data), the availability of good, fast, and versatile data analysis software is required.

Several optical parameters can be determined per particle. The main parameters measured are light scatter and fluorescence. Also, the length of the particle can be determined via a time-of-flight measurement.

The intensity of the scattered laser light may be measured in different directions: mainly forward scatter and perpendicular scatter are observed, as these parameters require only simple optical arrangements. Light scatter is affected by cell size, shape, density, content of light-absorbing compounds, and granularity. At angles smaller than $\Theta = 1.22\, \lambda/d$ rad, in the so-called small-angle or forward light scatter (FLS) mode, where λ is the wavelength of the laser light and d the cell diameter, the scatter light is independent of the refractive index of the cell content (150). The diffraction intensity is determined by the silhouette of the particle presented to the illuminating laser beam. It is dispersed over a series of lobes that decrease in intensity with increasing angle, as measured from the propagation direction of the laser light. About 85% of the scattered light is contained by the angle Θ as defined above. Hence, the amount of light scattered in the forward direction can be used to estimate the particle size (151–153). It should be emphasized, however, that since the intensity of the forward scatter also depends on other parameters, e.g., the presence of absorbing cell components, only relative sizes of similar particles can be obtained via this approach. In order to determine absolute particle sizes, well-characterized and representative standards have to be applied. In general, the FLS signal is measured over an angle of $0.5°-2.5°$ from the direction of propagation of the laser beam. It is critical to remove the laser light that is transmitted by the particles: normally, the transmitted light is much stronger in intensity than the FLS signal.

The intensity of the light scattered over a larger angle, e.g., over an angle of 90° in so-called wide-angle or perpendicular light scatter (PLS), is strongly dependent on refractive effects and on reflection. Refraction is determined by internal structures in the cell, as well as differences in refractive index of the particles and the suspending medium. Reflection properties are obviously determined by the shape of the particles. An example is the distinction of blood lymphocytes, monocytes, and granulocytes on the basis of perpendicular scatter (154). The latter cells contain granular structures that cause strong scatter, and hence a highly intense PLS signal. Another example is the discrimination of cyanobacteria that contain gas vacuoles, characterized by strong PLS, from cyanobacteria that lack such internal structures (155). Also, morphological information can be derived from PLS measurements, in particular when they are combined with FLS data, e.g., by taking the ratio of the two parameters. For instance Olson et al. were able to distinguish pennate diatoms and centric diatoms, with typical elongated, ellipsoidal, and disk-like shapes, respectively, on the basis of their scatter signals (156).

When scatter is measured under an angle of about 30°, Hodkinson showed that the intensity of the scattered light depends mainly on absorption processes within the particle (157). Via absorption measurements strongly absorbing pigments (e.g., hemoglobin, measured in the Soret band at 415 nm) or cells that have been stained with a strongly absorbing dye (e.g., trypan blue for cell viability analysis, also observed in the blue part of the spectrum) can be quantitated. Another approach is to register the decrease in intensity of the laser beam as it is propagating through the cell (158). For measurement of the transmitted laser light, or axial light loss, care should be taken to avoid detection of a significant amount of scattered light. Hence, very small detectors or specially constructed optics must be employed (see Section 9.5.3, below).

Obviously, scatter and absorption (or axial light loss) can be measured for any particle. In addition to these parameters, the intensity of fluorescent light can be registered in different wavelength regions. Measurement of fluorescence of cells that either are autofluorescent or contain autofluorescent constituents or are stained with fluorescent dyes presently must be regarded as the most fruitful and promising side of flow cytometry. Fluorescence normally is only observed for very few types of particles. Autofluorescent cells are, for instance, phytoplankters or cyanobacteria. Many cells, however, contain autofluorescent components, e.g., NADH or proteins (tryptophan, for instance, fluoresces when excited in the UV). The most promising approach in flow cytometry, however, is the use of selective fluorescent stains. As fluorescence provides very sensitive detection, particularly when it is coupled with laser excitation, even relatively small amounts of label can be detected. That is obviously necessary when one has to work on a unicellular level. Selective staining can be done by making use of fluorescent dyes with specific functional groups or by application of immunochemical labels that contain fluorescent material. By application of several labels simultaneously, with different excitation or emission properties, more than one property of the cells can be determined at the same time. As multiparameter approaches can be used, problems that are well known for the use of fluorescent stains in homogeneous solutions, e.g., interference of excess stain with the measurement, are much less significant in flow cytometry. Here, one is only interested in the *particles* that have been stained; the particles can be unequivocally recognized via their scatter signals. A problem that can occur in heterogeneous staining, though, is the presence of unspecific binding of the fluorescent dyes. In Section 9.5.6 a number of applications of fluorescence measurements in flow cytometry will be discussed in greater detail.

The evolution of the signals in time can also be followed and registered as a separate parameter. This principle is used in slit-scanning measurements. Here, the laser beam is focused on a spot or line that is smaller in the direction of the flow than the diameter of the particles to be measured. In the simplest application the length of the particles can be obtained via pulse width or time-of-flight (TOF) measurements (159,160). Such measurements can be done for the time evolution of all optical parameters. Of course, when a fluorescence parameter is used, only the size of the fluorescent part of the particle will be measured. An interesting application is the simultaneous measurement of cell (or cytoplasm) length and nucleus length of cells by detecting the TOF for two fluorescence parameters: red fluorescence of the nucleus, stained with propidium iodide; and green fluorescence from the cytoplasm, stained with FITC (161). A more recent development is the accurate classification of chromosomes by means of high-resolution slit-scan analysis. Images of the laser light in the order of $0.25\ \mu m$ were obtained in the object plane, and the fluorescence signal waveform of stained chromosomes was recorded with a high-speed analyzer. In this way the morphology of the chromosomes, with its typical indentations, could be used as an additional classification criterion (162,163).

A final—and most exciting—possibility is to use data of measurements in real time to activate a sorting device. In this way specific particles can in fact be physically separated from the sample and become available for closer investigation or positive identification. Sorting also can be done at very high speeds.

9.5.3. Instrumental Aspects

As flow cytometers are by no means standard laboratory instruments, a rough description will be given of the layout of a typical instrument. As a basic example a flow cytometer will be taken that was especially constructed for the measurement of phytoplankton. In contrast with most commercially available instruments that are primarily built for biomedical applications, this instrument has a number of special options. In particular, this flow cytometer is extremely versatile: it can be used for the measurement of particles in the size range of 1–1000 μm; it has a modular construction that enables simple changes in excitation and emission wavelengths and in the parameters that are applied, and it offers the possibility of simultaneously using three lasers for excitation and eight parameters (including TOF) in detection. The construction of the "Optical Plankton Analyser" (OPA) will be sketched below in brief; more detailed descriptions of the instrument and of its potential for measurement of phytoplankton are given in publications by Dubelaar et al. (164) and Hofstraat et al. (165).

9.5.3.1. The Flow Chamber and Hydrodynamics

The heart of the flow cytometer is the flow chamber and the means of introduction of sample and sheath fluids. It is crucial that the cells pass the interrogation regions one by one and in a spatially well-defined location. Furthermore, the layout of the cuvette should be so as to minimize unwanted scatter signals that could interfere with the measurement.

The sample is brought into the flow cuvette by means of a volumetric, pump-driven, injection technique. This ensures reproducible determination of particle concentrations—by dividing the count rate and the sample flow rate—and gives stable flow conditions even under the low pressures in the flow chamber of the OPA (the internal diameters of the flow system are very wide to enable the analysis of large particles). Sample flow rates can be varied between 0.02 μL/s and 0.55 mL/s in eight steps by changing the transmission of a synchronous motor that drives a 10-mL syringe. Precautions are taken to prevent settling of heavy particles in the flow system and to make sure that the particles remain homogeneously distributed in the suspension.

The sample flow is directed and hydrodynamically focused in the cuvette by combination of a sheath fluid of much larger volume and a specially constructed funnel-shaped cuvette. The hydrodynamic focusing of the sample flow—and the particles therein—is essential to produce the well-defined orientation in the cuvette. The mechanism that is operative in hydrodynamic focusing is via injection of a sample or core fluid into the central part of a flowing stream of sheath fluid. The sheath flow should be absolutely cell free. In our instrument we apply reverse-osmosis water that has been filtered over a 0.45-μm glass fiber filter; for more critical measurements, e.g., of very small particles also special—but expensive—particle-free water can be bought. In the sheath flow supply in-line a 2-μm prefilter and a 0.2-μm postfilter are present. An important requirement for the sheath flow is that it should be sufficiently larger in volume than the sample flow to ensure stationary, laminar flow throughout the cuvette. For introduction of the sheath fluid a volumetric technique also is used: a gear pump provides a discharge that is adjustable between 15 and 250 mL/min. The fastest discharge gives a flow at the point of interrogation of about 5 m/s. For every sample injection flow a sufficiently large sheath flow rate must be chosen to ensure laminar flow and sufficient confinement of the core stream.

The core stream is focused further down the cuvette by using a funnel-shaped introduction of sheath and core flows. As the water mass flow remains constant down the cuvette, reduction of the cuvette diameter results in an increase of the velocity of the flow. Finally, at the point of interrogation we are trying to obtain a core of sufficiently small diameter so that the cells pass one at a time and in laminar flow (thus, also reproducible positioning of the particles is secured). In the OPA the funnel-shaped cuvette is so made as to minimize sharp changes in fluid velocity. By keeping velocity gradients low in the flow cuvette, shear forces on the particles in the core stream are minimized. In the cuvette hence even relatively fragile filamental algae remain intact.

The inlet through which the sample is introduced into the sheath fluid stream has a diameter of 1.2 mm, so that particles with sizes of up to 1000 μm can be measured without problems. Seawater samples, even from eutrophic coastal regions where large, colony-forming algae occur, can be introduced directly into the flow cuvette without any prefiltration. The large range of sample flows that can be realized also enables the direct analysis of water samples, even in the winter period, when phytoplankton concentrations are low (about 10^5 per liter, as compared to 10^7 per liter in the spring bloom period). The end of the sample introduction tube is beveled, as suggested by Pinkel and Stovel (166); the special shape of the tube end results in a ribbon-shaped core. Therefore all particles are oriented according to their two-dimensional shape: the longest dimension is oriented in the direction of the flow, the second longest dimension by the forces induced by the special form of the core stream. The ribbon is placed parallel to the laser beam for optimum resolution, but perpendicular to the laser beam for accurate TOF measurements (then the particles pass through the narrowest point in the laser focus, which is exactly in the middle of the cuvette).

The optical part of the cuvette is a 1-cm × 1-cm quartz cuvette with a 1-mm-wide flow channel. Excitation is done perpendicular to the face of the cuvette, so that undesired scatter is minimized.

9.5.3.2. *Excitation*

Mostly lasers are applied as excitation light sources. Advantages of lasers for application in flow cytometry are their high power, excellent beam pointing possibilities, and monochromaticity. High powers are required to measure weakly autofluorescent cells or small cells that contain little fluorophores. One should realize that fluorescence is only recorded during the few microseconds that the particle passes through the region illuminated by the laser beam. Therefore, strong demands are made on the sensitivity of the instrument. Low divergence and good beam pointing stability is necessary for FLS measurements, where the transmitted beam has to be removed and at the same time forward scatter has to be measured under a relatively small angle. For high-resolution slit scanning or TOF measurements the exciting light should be focused down to a very narrow width in the direction of the flow. The fact that the laser beam is already spatially well defined and has low divergence permits the use of relatively simple and small optics. Hence, multiwavelength excitation can be applied, using several lasers focused at a small distance along the cuvette. The good spectral quality affords efficient removal of scattered light, so that low fluorescence signals can be measured.

Lasers that are commonly applied are of the argon-ion, He:Ne, and He:Cd type (see also Section 9.2). The Ar-ion lasers give a number of laser lines in the blue and green. Equipped with a different set of mirrors they also emit UV laser light. High-power Ar-ion lasers are costly and large. They can also be applied in combination with dye lasers. Small, air-cooled Ar-ion lasers produce tens of milliwatts, only in the visible part of the spectrum. He:Ne lasers emit at 633 nm; they are air cooled and also give tens of milliwatts of power. He:Cd lasers are also air cooled. Two types, with different optics, are available with laser emission in the blue at 442 nm or in the UV at 325 nm. Powers are in the same range as that of the He:Ne lasers. All these laser types are gas lasers. Such lasers in general are bulky, as long paths are required to get sufficient laser output in a diluted gas. Presently, solid state lasers, such as Nd:YAG or diode lasers, are becoming more and more available. In the OPA a frequency-doubled Nd:YAG laser is used that gives almost 100 mW of power at 532 nm. Solid-state lasers are compact and give very stable outputs. A survey of the lasers used in the OPA is given in Table 9.6.

The optical layout of the OPA is sketched in Fig. 9.25. Three colors, generally green, blue, and red, are used for consecutive excitation of the particles. The laser beams are focused at positions in the cuvette with an approximate distance of 700 μm. The beam diameters are matched by correction optics and directed by beam-steering prisms, which have been mounted on optical rails. It is therefore very easy to change the order of the laser foci on the cuvette or to exchange lasers. The three laser beams are focused simultaneously on the cuvette by one exchangeable spherical achromatic lens in combination with one exchangeable cylindrical lens. The available combinations of lenses allow for ellipsoidal foci on the cuvette with dimensions varying from $5 \times 75\,\mu\text{m}$ to $17 \times 2700\,\mu\text{m}$ (height $\times$ width; $1/e^2$ intensity).

Table 9.6. Lasers Used in the OPA

Laser Type	Wavelength (nm)	Power (mW)
Ar-ion	334/351/364	200
	458	350
	477	600
	488	1,500
	497	600
	502	400
	515	2,000
	529	350
He:Ne	633	40
He:Cd	442	50
Nd:YAG (2nd harm)[a]	532	100

[a] Frequency doubled output of 1064 nm Nd:YAG laser.

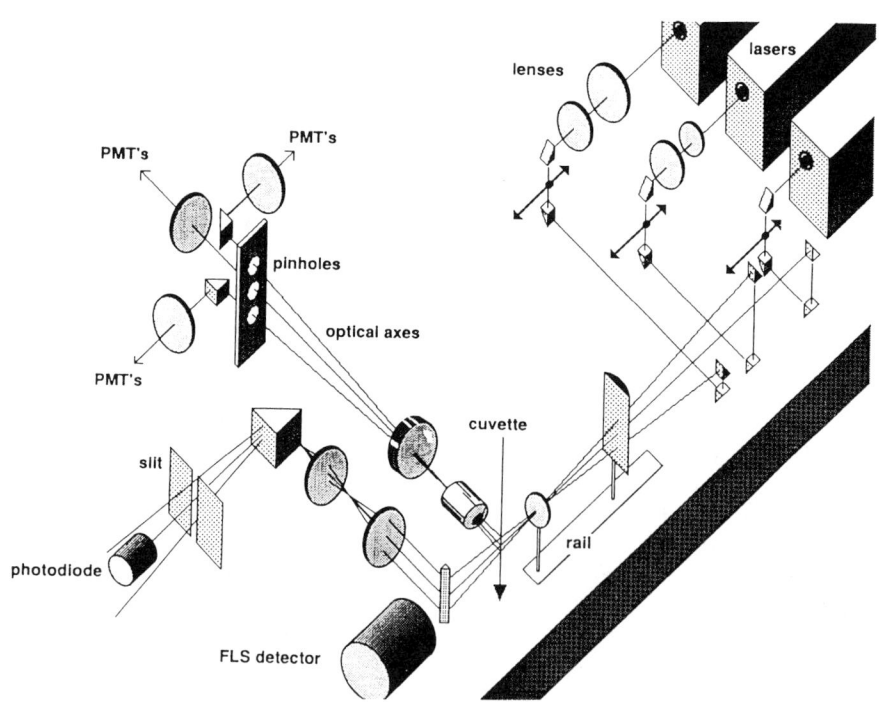

Fig. 9.25. Optical layout of the OPA. From Dubelaar et al. (164).

9.5.3.3. Detection

Detection is done in the forward and the perpendicular direction. In the forward direction FLS and axial light loss (ALL) is detected. The FLS is collected by a lens with a numerical aperture (N.A.) of 0.53 that collects light scattered between 1° and 25° relative to one of the laser beams. It is detected by a Si photodiode. The transmitted laser light is stopped by a vertically placed $1 \times 1 \times 10$ mm right-angle prism. The reflected laser light is imaged by a pair of lenses on a vertical slit in front of another Si photodiode. This photodiode can be used to measure beam attenuation or ALL.

In the perpendicular direction the PLS and fluorescence light is collected by a 0.50 N.A. $40\times$ microscope objective. Light originating from between 68° and 112° relative to the laser beams is imaged by the objective in combination with an achromatic field lens on three pinholes. In this way spatial filtering is achieved, so that signals induced by the three lasers can be measured independently. The field of view can be chosen to be 0.1 or 0.5 mm by exchanging the pinholes. Through the pinholes the light enters a detector box, in which prisms, lenses, and dichroic mirrors transfer the light to photomultipliers that convert the light into electrical signals. In the detector box the fluorescence and scattered light are separated very efficiently by application of dichroics. For

Table 9.7. Standard Parameters Used for Phytoplankton Measurement

Parameter	Excitation Wavelength	Emission Wavelength	Property
Detection at 90°			
PLSG	529 or 532	529 or 532	Particle shape, internal structure
FGR	529 or 532	>665	Chlorophyll *a* content, accessory pigments
FGO	529 or 532	550–650	Phycobiliprotein content
FBR	442 or 458	>665	Chlorophyll *a* content
FBO	442 or 458	550–650	Fluorescent stain content
FBG	442 or 458	470–530	Fluorescent stain content
FRR	633	>665	Chlorophyll *a* content, phycocyanin content
TOF	529 or 532	>665	Particle length
Detection at 0°			
FLSG	529 or 532	529 or 532	Particle size
ALLG	529 or 532	529 or 532	Light absorption, absorptive stain content

every laser beam several wavelength regions can be selected. The spectral separation optics is contained in exchangeable filter blocks, so that wavelength regions can be changed easily. A survey of the standard parameters used in the measurement of phytoplankton is given in Table 9.7.

9.5.3.4. Electronics

The high speed of the measurements (up to 1000 particles per second, 8 parameters per particle), the multilaser excitation at different positions in the cuvette, and the necessity of having a sufficiently large dynamic range pose very strong demands on the quality of the electronics. The two most important aspects are pulse correlation and pulse quantitation.

The pulse detection in the flow cytometer is governed by a central timing unit. The measurement is started when a signal from one of the preamplified detectors exceeds a preset electronic threshold. In general, a fluorescence signal is used to "trigger" the measurement, i.e., to discriminate the phytoplankton from nonfluorescent aqueous particles. As long as the trigger is above the threshold an electronic gate is opened, during which the signal is accumulated. The TOF is also measured for the trigger parameter (see below). The other parameters that are associated with the first laser are acquired during the same period of time as the trigger. For multilaser measurements a correlation must be made between the parameters associated with the three lasers, which are measured at different, consecutive times but pertain to the same particle. To achieve the correlation, the original trigger gate is supplied to two digital shift registers, which can be adjusted to generate time delays of up to $1684\,\mu s$. The delayed gates are manually positioned over the pulses induced by the second and third laser, which can be inferred from an oscilloscope. The oscilloscope projects traces of the pulses generated for all parameters. From the time delays, in combination with the distance between the consecutive laser beams (about $700\,\mu m$ each), the fluid velocity in the cuvette can be calculated. In combination with the TOF measurement this provides a direct measure of the length of the particles.

Another important aspect is pulse quantification. In most commercial flow cytometers quantification of the parameter signals is done simply by measurement of pulse height. A drawback of this approach is that the dynamic range of the instrument is limited by the linearity of the electronics. When logarithmic amplifiers are used the dynamic range is on the order of 10^3-10^4, which is also the limiting range for pulse height analysis. For analysis of phytoplankton with the variety of shapes and sizes present in eutrophic water samples, a much larger dynamic range is required. One should realize that the fluorescence signal intensity for a spherical particle varies, as the volume, with the third power of the radius; a dynamic range of 10^3, therefore, corresponds to only a factor of 10 in particle radius. The approach taken in the OPA is to apply fast digital integration of the signal during the time that the particles pass the laser focus. The current pulses from the detectors are converted into voltage pulses by linear preamplifiers, followed by logarithmic amplification. The height of these voltage pulses is determined every 200 ns and digitized. The digitized heights are converted into linear numbers with the help of a look-up table and stored in an accumulator. The maximum capacity of the accumulator is 12-bit (4096 numbers). At the end of the pulse, when the particle has passed, the sum of all numbers in the accumulator is read out into a first-in first-out (FIFO) buffer. The combination of analog-to-digital converters, log-to-linear conversion table, accumulator, and FIFO is dubbed the "digital integrator."

When the values have been stored for all parameters, all numbers are combined and transfered to the computer.

From the capacity of the accumulator, the integration is limited to 819 μs,

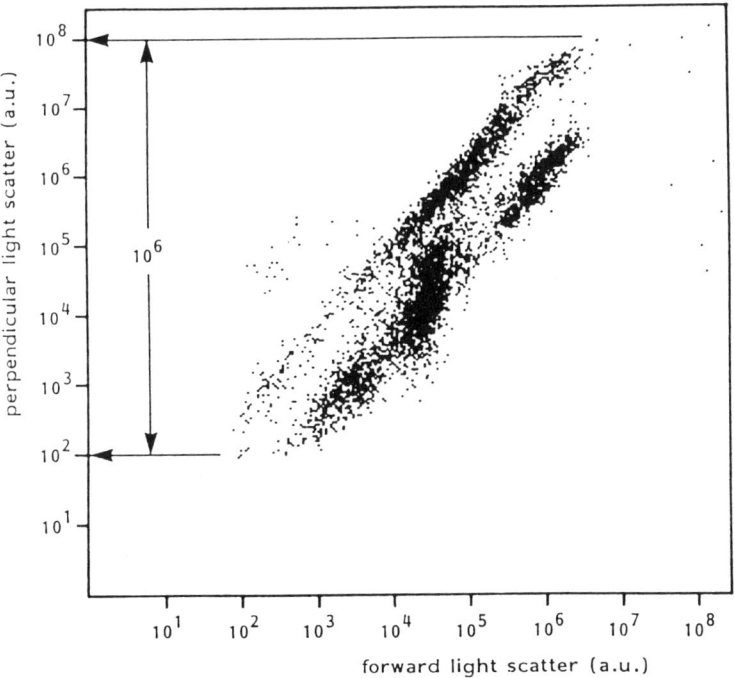

Fig. 9.26. Demonstration of the dynamic range of the OPA for a field sample from Lake Kaag. The sample contained mainly the cyanobacterium *Microcystis aeruginosa*, which may be present as small single cells and in large colonies. The total dynamic range exceeds 10^6. From Dubelaar et al. (164).

corresponding to a particle length of a few millimeters. Many phytoplankton species form filamentous colonies, consisting of series of connected single cells, which may reach such a length. The digital integration approach is a very successful one for such particles. It extends the dynamic range to more than 10^6, which is sufficient to measure water samples without the need for fractionation. In Fig. 9.26 an illustration is given of the large dynamic range of the OPA. Another advantage of the digital integration approach is that the TOF can be obtained directly from the number of pulse height samples that have been taken. A problem associated with digital integration is that the integrated signal will be inversely proportional to the fluid velocity. This makes careful calibration a necessity in all measurements. Calibration of flow cytometers will be discussed separately below.

9.5.3.5. *Data Acquisition and Analysis*

Data from the FIFOs are read under direct memory access into a desktop computer. Data of "intensity" parameters are converted from 28-bit linear data and stored in memory as 12-bit logarithmic words in list mode. The TOF-value is already a 12-bit linear number; it is also log-transformed. List mode means that all parameter values are stored per particle in a single file. The data can be analyzed afterward, either with an especially developed program or using commercially available software. Data analysis methods will be discussed in Section 9.5.4.

9.5.3.6. *Cell Sorters*

In flow cytometers that are also used as cell sorters the flow chamber is constructed so that at the end of the cuvette the water stream breaks up into tiny droplets (166). Normally this is done by attachment of the cuvette assembly to a piezoelectric crystal that vibrates at high frequency (20–40 kHz). The droplets can be charged electrically, either negative or positive, and sorted by a pair of deflection plates. The decision to charge a particle has to be taken very rapidly. In real time the parameter values obtained have to be evaluated to check whether the particle fits in the preselected parameter space that has been stored in hardware as a bitmap; if so, the particle is charged by a short voltage pulse. The time between measurement and sorting with the sample speeds used is not more than a few microseconds. Two types of particles can be selected in this way, by employing negative and positive charges, respectively.

For large volumes of fluids, as in the OPA, the application of droplet sorting is not feasible. In that case fluid-switching techniques have to be applied. Several methods to induce switching of the flow direction after analysis have been published, using electroacoustic or electromechanical techniques (167).

9.5.4. Data Analysis

In a flow cytometric measurement an incredible amount of data is assembled. A typical experiment with the OPA yields eight 12-bit numbers for each of 10,000 particles. It is clear that the presentation and interpretation of such a wealth of information poses a number of interesting challenges to software developers.

The simplest approach to analysis is the presentation in the form of histograms, plots in which the intensity distribution of the separate parameters is given (Fig. 9.27a). From histograms sometimes conclusive information can be derived, for instance, in cell cycle analysis, where one is only interested in the distribution of the DNA content over the cells. Special programs for cell cycle analysis are commercially available. Many programs can be purchased that allow for the quantitative analysis of histograms, e.g., to calculate coefficients of variance (CVs), averages and medians for selected parts of the histogram.

The next step is to represent the data in two-dimensional (2D) space, in so-called bivariate plots, or dot-plots (Fig. 9.27b). In bivariate plots the values of two parameters are shown simultaneously as a dot in the framework spanned by the two parameter intensity axes. Obviously, more information can be derived from such representations. In many programs it is possible to select clusters of particles in the dot-plot in order to calculate statistical data or to take the selected population for further analysis with the other parameters. Other approaches to 2D representation of data exist as well, e.g., by making use of intensity contours, shown as lines, grays, or colors, or with pseudo-3D mountain projections.

The best visual representation is displayed in Fig. 9.27c: here three axes are shown simultaneously in a pseudo-3D plot. One can do even more, as computer programs exist that allow for the rotation of the 3D plot in real time. Again, all selections can be done as mentioned for the bivariate plots. For many applications the interactive analysis that has been sketched above is sufficient. However, the interactive approach has some drawbacks: on the one hand, it is more or less subjective and not automatic; on the other hand, it is impossible to take more than one parameter at a time into account. Obvious approaches to deal with the first drawback are the implementation of statistical analysis and classification methods, which have been well documented (168; and in a series of papers by Bartels, 169). To make the distinction of clusters objective and automatic, for many applications simple linear discriminant function analysis can be applied. In this approach functions are developed, built up from linear combinations of the parameters, which, when evaluated for different particle types, assume distinctly different ranges of values. When the discriminant functions have been calculated, they can even be implemented for real-time analysis and used in combination with a sorter.

A simple approach to data processing that may significantly simplify the data analysis is transformation to parameter ratios (170). Advantages are that spectral information can be taken into account not only of emission but also of excitation properties. For phytoplankton in particular excitation properties may differ owing to the presence of accessory pigments. This will be discussed in more detail below. Additional advantages are that instrumental variations

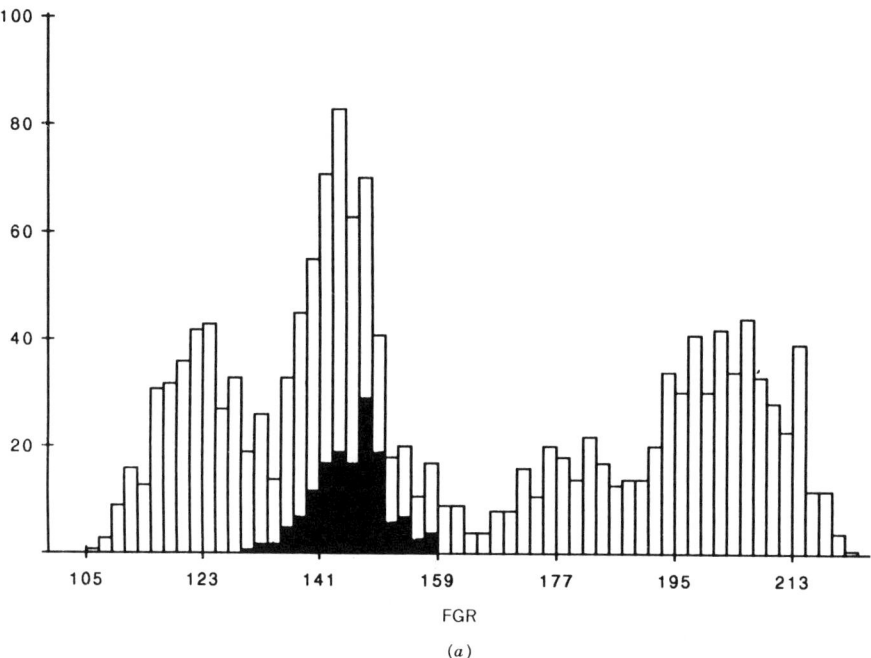

(a)

Fig. 9.27. Representations of flow cytometer data for a mixture of 10 marine phytoplankton species. In all figures logarithmic intensity values are shown. (a) Histogram of FGR; in the histogram the frequency distribution of the red fluorescence intensity is given. The part of the histogram that is drawn in black is due to the cryptophyte *Rhodomonas minuta*. (b) Bivariate plot of FGO and FGR; the points in the plot represent intensities of these two parameters as measured for each individual particle. Similar particles form clusters of points in the plot. The clustering is indicated in the figure for three algal species. *Rhodomonas minuta*, which is clearly separated from the rest; *Dunaliella tertiolecta*; and *Tetraselmis* sp. (c) Trivariate plot of FGO, FGR, and FRR. Again the *Rhodomonas* cluster is clearly separated from the rest. Now also the *Tetraselmis* and *Dunaliella* clusters can be distinguished. These two algal species are efficiently excited in the red.

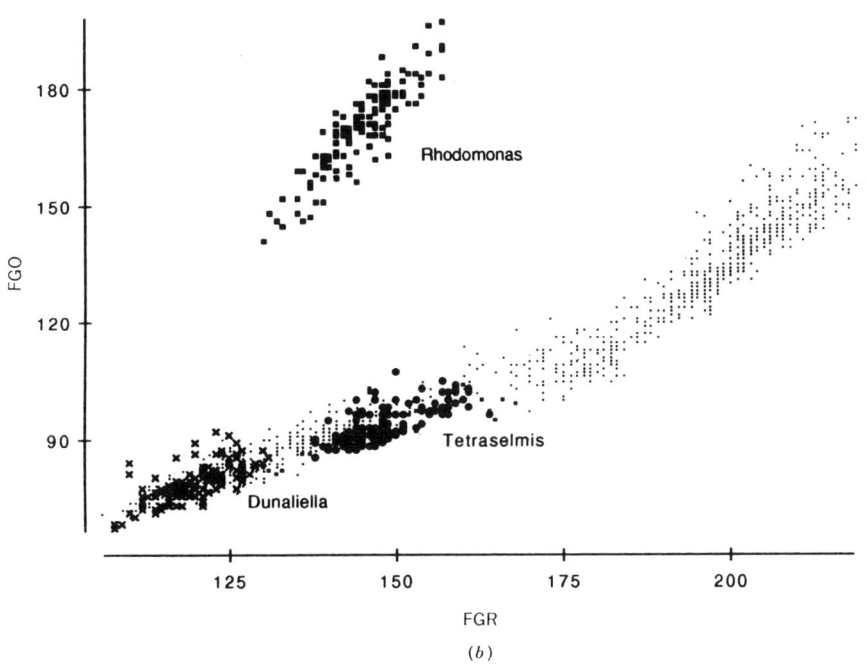

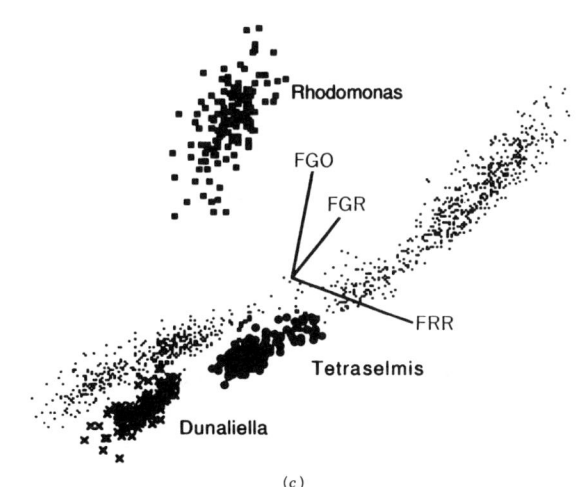

Fig. 9.27. (*Continued*)

will be removed to a great extent when relative intensities can be used for analysis.

More advanced statistical techniques are not always suitable for analysis of flow cytometric data. There are several reasons for the insufficiency of these approaches. The large amount of data generated may lead to prohibitively long computational efforts, for instance, in nonparametric cluster analysis [e.g., k-nearest-neighbor (171)]. In addition, many clusters, particularly in flow cytometry of phytoplankton, have very irregular shapes, with different, and not necessarily normal, distributions in different directions of the parameter space. This precludes the application of faster, parametrized cluster analysis methods that assume a generally normal distribution of the objects in the cluster (171).

Two approaches appear promising for the automated interpretation of complex, multiparameter flow cytometric data: neural network analysis and a density-driven computational technique. In neural network analysis classification of clusters is done via an adaptive approach. In principle, no knowledge is required of the position of the particles in the N-parameter data space. A variety of neural networks is available. The multilayer feedforward network has already shown to be successful for the automated classification of algae, based upon flow cytometric data (172–174). The analysis is simple (see Fig. 9.28): first, the network is trained by iteratively forcing the problem (i.e., flow cytometric data—the input) and the solutions (i.e., the classifications—the

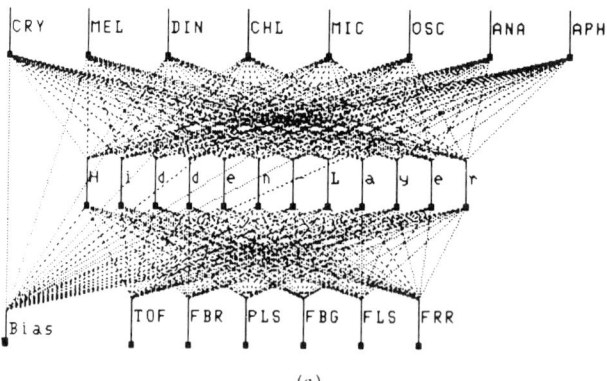

Fig. 9.28. (a) A diagram of the multilayer feedforward network that was used to analyze OPA data. The input pattern (the problem: OPA parameter values for one particle) is fed to the neural network; the signals are propagated through the network via the internal units to the output units (the solution: the identification). Via this network input patterns and output patterns can be associated with each other. (b) Result of the analysis of real mixtures by a neural network. Discrimination is shown between cyanobacteria and other algae. (See p. 415.)

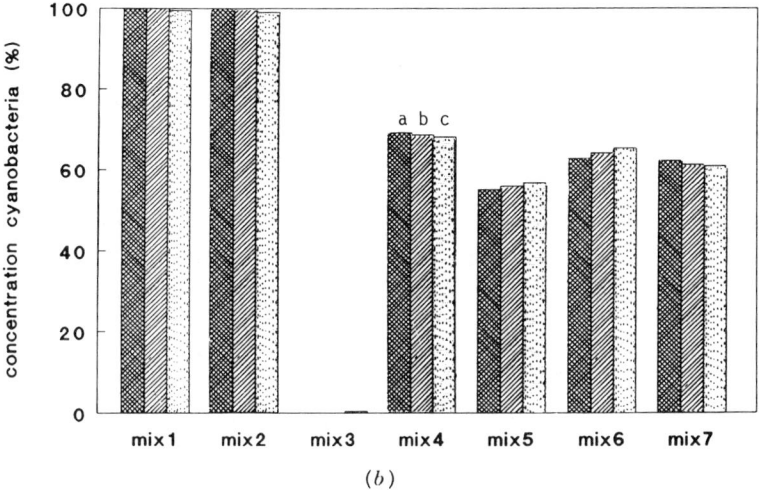

Fig. 9.28. (*Continued*)

output) upon the network. After the learning phase, in which connections are made between the input and the output side of the network, the neural network is optimized for the automated classification of the particles it was trained with. Training of the network may take several hours, but automated analysis with the trained network can be done in real time; if it is built in hardware it is sufficiently fast to steer a cell sorter (175). The limitation of the supervised learning approach, obviously, is that the classification is limited to particles, whose flow cytometric properties are known. For many problems this poses no limitation, as pure suspensions of the particles, which have to be classified, can be isolated and measured. If this knowledge is not available, other approaches have to be followed. One can be by sorting the unknown particle and subsequently feeding the purified sample to the flow cytometer. Another—software—approach may be the use of an unsupervised neural network (171). Our experiences with this approach so far are not positive. An important advantage of neural network analysis is that all parameters can be taken into account for classification of the data.

A recently reported approach is the application of density-driven computation of population boundaries (176). With this method cluster boundaries can be specified that keep automatic track with intensity shifts in the data. The density-driven computational technique offers optimal objective specification of cluster boundaries. On the other hand, the clustering is surveyed by direct viewing of the clusters, which obviously limits the number of parameters that can be considered simultaneously.

9.5.5. Calibration Aspects

Flow cytometers are very complex optical instruments. Alignment of the optical parts is critical and *not* trivial. In most instruments excitation and the main part of the detection optics is on two perpendicular optical axes. Of course there are ways to align these two axes in a relatively simple way, e.g., by using pinholes illuminated from the back by red-light-emitting diodes to realize alignment of the perpendicular detection box, but proper performance checks must be done on a regular basis. In the OPA, in addition to optical alignment, changes in flow velocity also may cause significant changes in signal intensities, due to the application of the digital integration technique. The generally used approach to check the performance of a flow cytometer is to measure standard beads. Fluorescent standard beads are available from a range of suppliers (Polysciences, Flow Cytometry Standards Corporation, Coulter, Pandex) for most applications. They are mostly made from poly-(styrene), a polymer than can be synthesized in a very narrow molecular weight, and thus size, distribution. Available sizes range from 1 to 25 μm. The spheres are labeled during the polymerization process. Labeled spheres can be obtained which fluoresce in the green, yellow, orange, or red part of the spectrum, with different excitation spectra. A problem for multilaser instruments is that not many "broad" spectral beads are available, so that generally more than one bead must be used to calibrate the instrument. The stability of most labeled beads is good, as long as they are kept in the dark.

Three properties should be examined when the performance of flow cytometers has to be quantified: the precision, the sensitivity, and the accuracy. The last property is particularly important when one wants to do quantitative flow cytometry.

The coefficient of variance, or CV, is defined as the ratio of the standard deviation divided by the mean, of a distribution of fluorescence or light scattering intensities. The CV is a measure for the precision of the measurement. Standard beads generally have very low CVs, on the order of 1–3%, so that they are eminently suited to determine the correct alignment of the optics in the flow cytometer. As far as the sensitivity is concerned, obviously a lot of variation occurs, dependent on the construction of the flow cytometers. Most larger instruments can detect around 1000 fluorescein molecules per particle. When the flow is slowed down and laser beams are focused on smaller spots, however, even single fluorescent molecules can be detected. The sensitivity of the instrument should be measured regularly to detect deterioration of the optics, or faulty electronics. Finally, the accuracy of the measurement, i.e., the degree to which the measured values approach the true values of the measured variables, should be considered. A problem that is often encountered in this respect is nonlinearity of the electronics, in particular of the logarithmic amplifiers. When one wants to do quantitative flow cytometry, e.g., to determine concentrations of particular compounds in the cell via immunochemical approaches, careful calibration over the whole intensity range must be done. Beads with different amounts of label are available that enable one to perform this type of calibration.

For instruments that use digital integration techniques it is important to ensure that the signal intensities measured are comparable. Two approaches can be followed to ascertain that the detection efficiency remains the same, which is necessary when a time series must be measured. One way is to vary flow speed and calibrate electronic settings (amplification factors, PMT high voltages, etc.) so that the same values are obtained for all parameters of the standard beads in every measurement. The other is to add standard beads to every sample and use the intensities measured for the beads to correct the intensities obtained for the particles in the sample. When the latter approach is followed, care should be taken to ensure that the measured intensities all are in the linear part of the instrument response.

With the OPA a routine measuring program of coastal phytoplankton samples is maintained throughout the year at a number of locations. Without correction of the intensities and adequate calibration procedures no comparable data are obtained. Mostly chlorophyll-labeled beads are used for calibration purposes. They give signals that are detectable in the red fluorescence channels, induced by blue, green, and red lasers; of course they are also useful for correction of scatter signals and TOF. For correction of green and orange fluorescence, fluorescein beads are employed. The calibration approach applied is a combination of the methods mentioned above. First, the flow rate is determined from the delay between the three laser foci and adjusted so that it is about 1 m/s. Next, the flow cytometer is optimized with the standard beads, so that an optimum CV is obtained (in the OPLA the CV is mostly instrument determined and is on the order of 5%). After these optimizations the intensity of the fluorescence and scatter parameters remains the same to within 25%, even over long periods of time (see Fig. 9.29). In addition, standard beads are added to every sample to perform finer corrections in the software, following analysis. An indication of the result of the correction is given in Table 9.8, where parameters measured for a culture of the phytoplankter *Dunaliella tertiolecta* and of chlorophyll beads with two different green laser powers are presented. For the two laser powers the same ratios between the parameters measured for both types of particles are obtained, which illustrates the suitability of this correction procedure. The difference in signal intensity is much larger for the two experiments shown in Table 9.8 than in normal flow cytometric analysis, as can be inferred from Fig. 9.27. It is also clear from Table 9.8 that the CV of the phytoplankton measurement is much larger than for the standard beads, obviously as a result of biological variations. For most

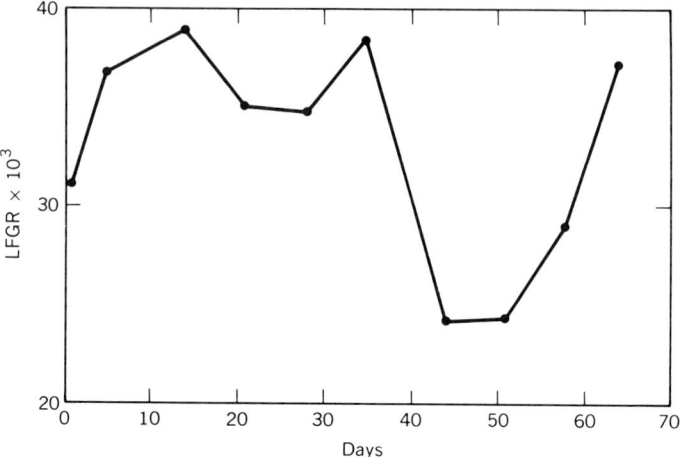

Fig. 9.29. Variations of FGR of chlorophyll standard beads as measured over a period of 70 days. Data were taken from weekly measurements on North Sea samples to which standard beads were added. The variation is on the order of 20%.

Table 9.8. Correction of Flow Cytometric Measurements of *Dunaliella tertiolecta* with Chlorophyll Standard Beads

Parameter[a]	40-mW Excitation (529 nm)		100 mW-Excitation (529 nm)	
	Intensity	CV (%)	Intensity	CV (%)
$LFGR_{Duna}$	56,000	20	103,000	19
$LFGR_{Bead}$	29,000	6	55,000	5
$LFGR_{Duna}/LFGR_{Bead}$	1.9		1.9	
$LPLSG_{Duna}$	203,000	30	440,000	25
$LPLSG_{Bead}$	540,000	10	972,000	10
$LPLSG_{Duna}/LPLSG_{Bead}$	0.38		0.45	
$LFLSG_{Duna}$	126,000	15	298,000	16
$LFLSG_{Bead}$	140,000	4	338,000	4
$LFLSG_{Duna}/LFLSG_{Bead}$	0.90		0.88	

[a] The prescript L indicates that the intensity values have been linearized. The parameters are as described in Table 9.7.

parameters that are measured for particles of biological origin large CVs are observed, so that the relative low resolution of the OPA does not pose a problem. Only for DNA analysis—the content of DNA in the nondividing populations varies by less than 2%—low CVs are critical. DNA analysis will be discussed in the following section.

9.5.6. Applications

The variety of applications, in particular in biomedical science, at present is virtually endless. In many hospitals flow cytometers are employed for automatic, routine analysis of cells in body fluids, mainly blood. Procedures have been developed to count the main constituents of blood, red cells, white cells and, more recently, platelets, automatically (177). Clinical applications have been reported in the fields of hematology, immunology, oncology, microbiology, genetics, and parasitology. In most cases the flow cytometer is used as a fast and objective way to quantify numbers of cells or to identify aberrant, e.g., malignant, cells. A number of comprehensive reviews on clinical flow cytometry have been published (178–180). Recently, however, outside the field of biomedical science, applications of flow cytometry have appeared in the literature, for example, in marine biology, microbiology, food science, and pharmacology and toxicology. In most of these fields the history of flow cytometry is still short, and many applications have yet to be developed. In this section we will try to give an idea of the potential of flow cytometry via a more general survey of possible approaches and a more detailed discussion of a few examples. The examples have been taken from the field of marine biology but can easily be translated into other research areas.

9.5.6.1. Survey

Flow cytometry can be applied for the determination of all kinds of particles and their properties. Most applications thus far deal with *classification* of particles and subsequent quantification of the classified cells. The classification is mainly based on optical properties of the particles themselves, or of particular compounds that have some sort of specific interaction with the particles to be identified. In most cases the classification is based on structural characteristics of the particles. Another application of flow cytometry is in the determination of *functional properties* (i.e., biological properties or activity, or chemical parameters that are strongly dependent on cell physiology) of cells or of parts of cells. For functional studies almost exclusively specific stains are applied. In both areas of application, flow cytometry offers an extremely fast and objective means of analysis.

A second distinction that can be made in a survey of the applications of flow cytometry is based on whether the parameters that are measured are of the particles as such or of specific reagents that have been added to the particles. The two approaches have been called intrinsic and extrinsic, respectively. For the determination of extrinsic parameters several options are available. Mostly applied is the use of fluorescent reagents that have some sort of specific interaction with the particle or constituents of the particle that are to be determined. The specific interaction can either be based on well-defined reactivity or on stereospecific properties, as in the application of fluorescently labeled immunochemical labels. Another approach, mostly taken in functional analysis, is the application of labels that undergo a change in spectral properties or in fluorescence intensity. The change is generally the result of enzymatic activity or, for instance, of pH.

Table 9.9 (181–213) gives a—by no means exhaustive—list of cellular parameters that can be measured via flow cytometry. A distinction has been made in structural and functional parameters, as can be measured for particles as such or with the help of reagents.

Table 9.9. A Summary of Applications of Flow Cytometry

Property	Parameter	Reference
I. INTRINSIC PARAMETERS		
A. *Structural determinations*		
Particle size	Forward scatter	151–153
	Pulse length (TOF)	159,160
Particle shape	Multiangle scatter	181–183
	Scatter ratios	156,184
	Pulse shape	161,162
Internal structure (e.g., granularity)	Perpendicular scatter	154,155
Pigment content (e.g., hemoglobin, photosynthetic pigments, luminescent bacteria)	Fluorescence	170,185–188
Protein content (e.g., tryptophan)	Fluorescence	189
B. *Functional determinations*		
Pigment content	Fluorescence, fluorescence ratios	170,190,191
Redox state	Fluorescence (NADH)	192

Table 9.9. *(Continued)*

Property	Parameter	Reference
II. EXTRINSIC PARAMETERS		
A. *Structural determinations*		
DNA content	Ethidium bromide	193
	Propidium iodide	194
DNA base ratio	DAPI[a] (A-T)	195
	Hoechst 33258 (A-T)	196
	Hoechst 33342 (A-T)	197
	Mithramycin (C-G)	198
	Chromomycin A (C-G)	199
RNA content	Acridine orange	200
	Pyronin Y	201
Protein	FITC[b]	194
	Fluorescamine	202
Sulfhydryl groups	OPT[c]	203
	MBCL[d]	204
Antigens	Immunochemical labels	146,178,205
Toxic phytoplankton	Immunochemical labels	206
B. *Functional properties*		
Viability	FDA[e]	207
Enzyme activity	FDA[e]	207
	4-MUP[f]	208
Endocytosis/phagocytosis	FITC-labeled particles	209
DNA synthesis	DNA dyes (see above)	210
Membrane fluidity	DPH[g]	211
Intracellular $[Ca^{2+}]$	fura-2, indo-1	212
Intracellular pH	COF[h]	213

[a] DAPI = 4′,6-diamidino-2-phenylindole. [b] FITC = fluorescein isothiocyanate.
[c] OPT = *o*-phthalaldehyde. [d] MBCL = monochlorobimane.
[e] FDA = fluorescein diacetate. [f] 4-MUP = 4-methylumbelliferone phosphate.
[g] DPH = diphenylhexatriene. [h] COF = carboxyfluorescein.

In the following subsections more detailed examples will be given of application of flow cytometry for classification purposes and for functional analysis. Both applications have been taken from the area of marine biology. The classification of phytoplankton species will be discussed on several levels: based on intrinsic parameters; with the aid of dyes that are specific for the cell wall of dinoflagellates, an important phytoplankton class; and, most specifically, with immunochemical labels for particular toxic phytoplankton species. Functional analysis will be illustrated with two examples: cell cycle analysis and viability analysis, both for phytoplankton cultures.

9.5.6.2. Classification of Phytoplankton

The high degree of eutrophication of many coastal waters has led to increasing problems with phytoplankton. In particular in the spring and summer periods large abundances of algae occur. Decay of algal blooms may lead to oxygen deficiencies (and hence to mortality of fish and other sea animals) and unwanted formation of slime and foam on the beaches. Trends in presence of phytoplankton along the Dutch coast have been described by Cadée et al. (214). Recently, a number of reports have appeared on blooms of toxic phytoplankton in many European and North American coastal regions (215).

In view of the importance of phytoplankton in relation to the quality of the marine environment, many countries have started a monitoring program in which algal species composition is followed at a number of locations throughout the year. In principle, phytoplankton species abundance is measured by microscopic inspection, which involves a lengthy procedure: counting of a typical sample takes 3–4 h. It is clear that the application of flow cytometry for automatic phytoplankton classification and quantification may yield a tremendous increase in efficiency, and in addition offers a much more objective and statistically more sound set of data. The application of flow cytometry for the measurement of algae was prompted by three almost simultaneous publications (186–188). However, currently a number of groups throughout the world are applying this advanced counting method.

Main Group Analysis. Obviously, the optical analysis that is applied in flow cytometry does not provide the same highly detailed morphological information provided by visual observation of the samples. For many monitoring applications, fortunately, this kind of information is not needed. What flow cytometry can do is to afford fast, objective, and statistically sound information and, by combination with size measurements, the calculation of biovolume and biomass, which are considered to be the main parameters for the assessment of the phytoplankton load. In addition, in the most relevant period of the year, in the spring and in the summer, when algae are most abundant, generally not more than 1–5 algal species are dominant. The numbers of these most abundant species can often be easily determined from flow cytometric measurement.

With optical analysis, distinction among taxa can be made via several approaches. The most straightforward approach is to distinguish phytoplankton main groups via their characteristic pigment composition. All autotrophic phytoplankton contains photosynthetic pigments that supply the algal cells with the energy needed for their biosynthesis, respiration, and reproduction. All phytoplankton contains the red fluorescing pigment chlorophyll *a*. The presence of red fluorescence, hence, is used to distinguish phytoplankton from all other, much more numerous particles that are present in surface water. In addition to chlorophyll *a*, phytoplankton contains other, accessory pigments, the main function of which is to make more efficient use of the incident solar irradiation. Most of these pigments do not fluoresce appreciably, but rather transfer their energy efficiently to chlorophyll *a*. They do, however, change the light absorption (and thus the color) of the phytoplankton. Green algae (chlorophytes), for instance, absorb light in the red and blue part of the spectrum. By measurement of the red fluorescence of chorophyll *a*, as induced by excitation with several laser wavelengths, a discrimination can be made between major phytoplankton classes, which are distinguished by pigment composition. Some phytoplankton classes contain orange fluorescent pigments in addition to chlorophyll *a*. Of course, such algae are measured most easily. Cyanobacteria, which also contain orange and/or orange/red fluorescent pigments, also can be quantified without much difficulty. The major phytoplankton classes and their main characteristics are summarized in Table 9.10. The classification of phytoplankton main groups via flow cytometric analysis of fluorescence properties has been described in detail in Hofstraat et al. (170).

Table 9.10. Optical Characteristics of Major Phytoplankton Classes

Class	Main Pigments	Absorption Maxima (nm)	Fluorescence Maximum (nm)
Chlorophytes	Chlorophyll *a*	420–450	680–690
	Chlorophyll *b*	470–490, 630–650	—
Bacillariophytes	Chlorophyll *a*	420–450	680–690
	Fucoxanthin, β-carotene	450–550	—
Dinophytes	Chlorophyll *a*	420–450	680–690
	Peridinin	450–550	—
Cryptophytes	Chlorophyll *a*	420–450	680–690
	Phycoerythrin	540–570	580–630
Cyanobacteria	Chlorophyll *a* (not always)	420–450	680–690
	Phycoerythrin	540–570	580–630
	Phycocyanin	620–640	640–670

In addition flow cytometry gives information on cell size and length, and on internal structures, which can also be used for classifical purposes. Examples are discrimination of pennate (long, ellipsoid) diatoms from centric (spherical) diatoms by multiangle light scatter (156) and measurement of the gas-vacuole-containing cyanobacterium *Microcystis aeruginosa* via its high perpendicular scatter (155).

In Fig. 9.27 some examples of plots of flow cytometric data have already been shown that illustrate the classification of phytoplankton species on the basis of differences in fluorescence emission and excitation properties. Based on the histogram (Fig. 9.27a) some discrimination in the phytoplankton mixture can be made. However, no clear assessment of the identity of the species is possible, as the histogram does not provide information on accessory pigments that are present. The bivariate plot (Fig. 9.27b) clearly displays the well-separated cluster of the orange fluorescent cryptophyte *Rhodomonas*. The rest of the phytoplankton species is not well separated. The (pseudo-3D plot in Fig. 9.27c also shows separated clusters for the chlorophyte *Dunaliella* and the prasinophyte *Tetraselmis*. These algal species absorb efficiently in the red, owing to presence of chlorophyll b, and can hence be easily identified by their relatively high FRR. The other species in the sample cannot be discerned on the basis of their fluorescence characteristics; they are all bacillariophytes and dinophytes. Further distinction can be made, however, based on length (TOF) and shape (scatter).

A very simple and straightforward approach to determining the main groups in the sample makes use of parameter ratios, which, of course, are obtained per individual particle. Ratios show much less variability than the absolute parameter values, since size effects that give large intensity variations are removed. The parameter ratios also clearly show the pigment composition. Figure 9.30 illustrates two examples of algal classification based on parameter ratios, for real water samples. In Fig. 9.30a cyanobacteria can be clearly separated from other, mainly diatomic algae (bacillariophytes), by taking the ratio of the red- and blue-induced red fluorescence intensities. The cyanobacteria contain phycocyanin, which is efficiently excited in the red region.

Fig. 9.30. Examples of flow cytometric analysis of algae in field samples. (a) Sample from the freshwater Lake Wolderwijd. The cyanobacteria are straightforwardly identified from the bivariate plot by their high FRR values. They are indicated in the plot by bold **X**. They can also be easily classified on the basis of the histogram of the FRR/FBR intensity ratio. The parameters were linearized before the ratio was taken, as indicated by the prefix L. (b) Samples from the North Sea, 10 km off the Dutch coast. Several clusters of orange fluorescing algae are indicated: *Rhodomonas* (●) and *Mesodinium* (**X**). The broad range in the bottom of the figure is due to dinoflagellates and diatoms. *Rhizosolenia*, is easily recognized on the basis of the relatively large size of this alga. From the histogram of the FGO/FGR ratio the orange-fluorescing algae and the chlorophyll standard beads can easily be quantified.

FLOW CYTOMETRY

Lake Wolderwijd

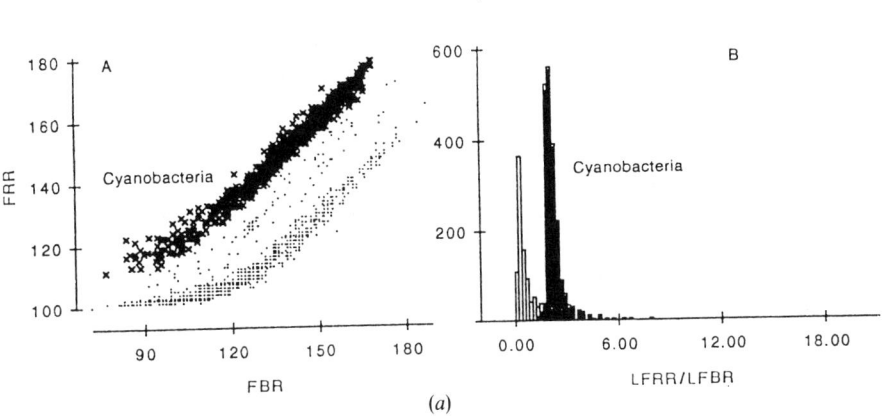

(a)

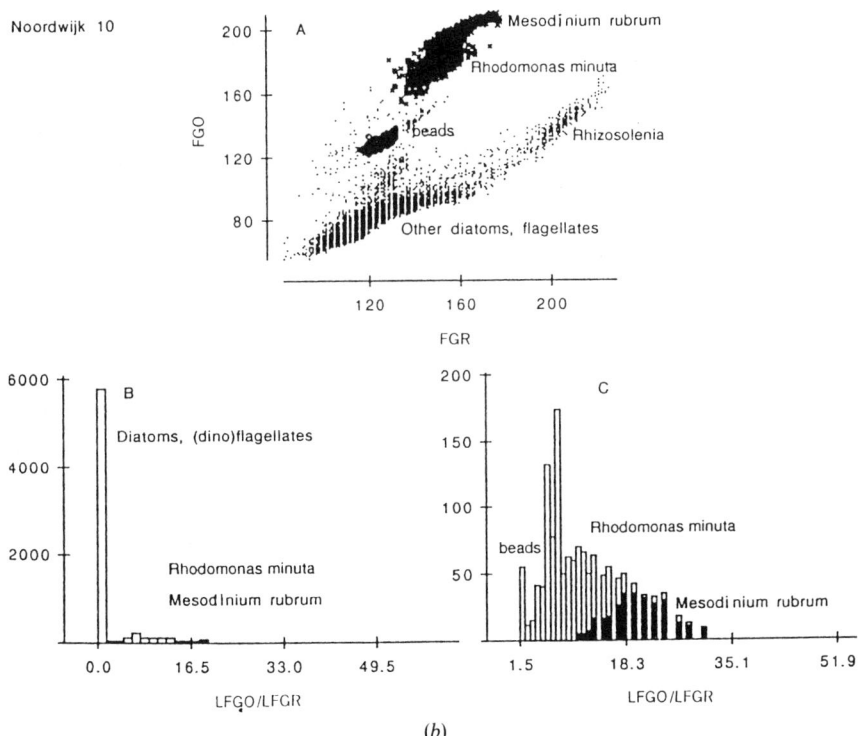

(b)

Cyanobacteria are nuisance algae that frequently cause problems in the summer, when blooms may form a thick greenish layer on the lake surface. Some cyanobacteria, e.g., *Anabaena flos-aquae*, contain toxins. Figure 9.30b shows a bivariate plot obtained for a typical coastal water sample. The cryptophytes *Rhodomonas minuta* and *Mesodinium rubrum* (the latter actually being a heterotrophic ciliate species that lives in symbiosis with a cryptophyte) are easily distinguished from the much more abundant diatoms and dinoflaggelates. One diatomic species, *Rhizosolenia*, is discerned very well owing to its large size.

A fully automated approach to classification that takes into account all eight parameters simultaneously is via the application of neural network analysis. The classification of algae with neural networks is described in Balfoort et al. (173) and Smits et al. (174).

Above we have demonstrated the facile discrimination of many types of phytoplankton classes by means of their fluorescence characteristic. For marine applications, however, there is a serious limitation: diatoms and dinoflagellates, which both are efficiently excited in the green, cannot be discerned. Just these two types of phytoplankton are most numerous in coastal regions. Discrimination of diatoms and dinoflagellates calls for another approach.

Determination of Stained Dinoflagellates. Selective determination of dinoflagellates is very important: many dinoflagellate species contain toxins (see the next subsection). Fortunately, most dinophytes can be selectively stained since they have a chitin (cellulose) cell wall. The diatoms have silicium cell walls, which are not easily stained. A number of stains for cellulose have been reported. Examples are primulin (Direct Yellow 59) and calcofluor white, two dyes that are excited in the UV and show blue fluorescence (216,217). Another fluorescent stain that can be applied is proflavine (3,6-diaminoacridine hemisulfate), a dye that is excited in the blue-green region (with the 442 nm He:Cd line or with one of the Ar-ion laser lines in the 458–488 nm region) and emits green/yellow light (216). The advantage of the latter stain is that the determination of the dinoflagellates can be done simultaneously with that of the other species by measuring the green fluorescence induced by the blue laser as an extra parameter. Application of UV fluorescence requires changing of the mirrors of the Ar-ion laser to transmit the 334–364 nm lines; in this configuration, however, no visible laser lines are obtained. The staining procedure is extremely simple: proflavine is added to the sample to a final concentration of 10^{-5} M; after 5 min all phytoplankton cells are pelleted by centrifugation and redissolved in clean, salty water (the salt is added to retain the osmotic pressure of the seawater). The suspension then can be analyzed by flow cytometry. Stained algal cells can be discriminated from any other

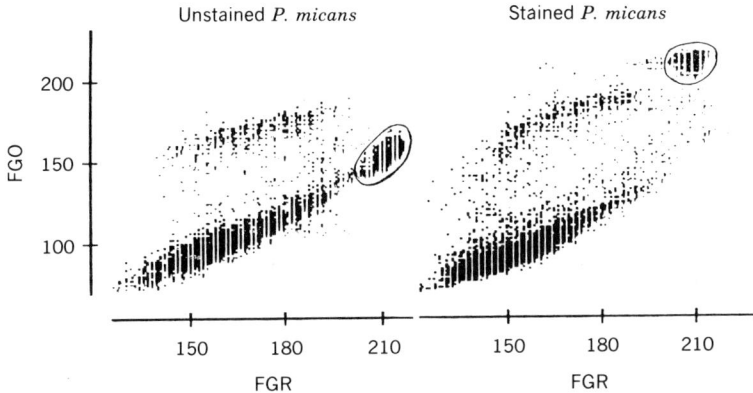

Fig. 9.31. Selective staining of the dinoflagellate *Prorocentrum micans* in a North Sea sample, taken 135 km off the Dutch coast, with the green/orange-emitting dye proflavine. The cluster of the stained algae is clearly separated from the rest by its high FGO intensity.

particles that may have been stained by applying simultaneous detection of green (stain) and red (phytoplankton chlorophyll) fluorescence. In Fig. 9.31 an example is shown of specific staining of dinoflagellates. The dinophyte *Prorocentrum micans* can be clearly discerned after staining of a field sample taken some 135 km off the Dutch coast, near the Dogger Bank.

Obviously, an abundance of selective stains is available to color compounds with all kinds of cell walls. Examples of their use with phytoplankton are the staining of mucus-forming algae (e.g., the nuisance-bloom-causing *Phaeocystis pouchetii*) and of heterotrophous, i.e., non-autofluorescent, flagellates (216). Examples of staining of other particles can be found in the references mentioned in the survey.

Immunochemical Determination of Toxic Phytoplankton Species. In the last decades more and more toxic marine algae have been observed in samples taken along the Western European and North American coast (215). Along the Dutch coast the dinoflagellates *Gymnodinium* spp. and *Dinophysis* spp. are quite common. The also toxic *Alexandrium* cf. *tamarense* and *Gyrodinium* cf. *aureolum* were observed for the first time in 1989, but also seem to be quite common in the Dutch part of the North Sea. The algae mentioned here may cause diarrhetic shellfish poisoning (DSP) and paralytic shellfish poisoning (PSP). Effects are primarily mortality of fish and invertebrates, but the consumption of poisoned fish and shellfish may also cause toxic effects in humans. To be able to detect toxin-producing phytoplankton cells at an early stage, before they can produce any detrimental effects, they have to be measured at very low concentrations: on the order of 100–1000 cells per liter. Measurement of such low concentrations requires extremely selective detection of algae. The approach we have chosen for specific detection of toxic phytoplankton species is immunochemical labeling of the dinoflagellate cell wall and flow cytometric detection of the selective fluorescent label.

In a first study, purified cell wall extract of the dinoflagellate *Prorocentrum micans* was used as an antigen to generate antibodies in rabbits (206). Antibodies were collected after 30 days of immunization. The primary antiserum was added to algal suspensions, which were incubated for about 1.5 h. Next, the samples were washed and FITC-conjugated goat–anti-rabbit IgG (immunoglobulin type G) was added to the suspension. The green fluorescence of FITC was measured in conjunction with the red phytoplankton fluorescence to identify the labeled algae. Confocal scanning laser microscopy confirmed that the attachment of the antibody was on the cell wall of *Prorocentrum*, mainly near the pores and near the haptonema. Cross-reactivity with a number of phytoplankton species was investigated but appeared negligible. Figure 9.32 shows the flow cytometric results obtained for immunochemically labeled *Prorocentrum*. On the basis of the significantly

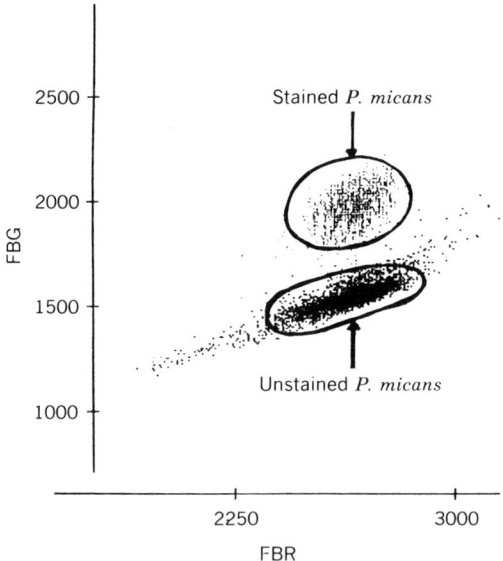

Fig. 9.32. Immunochemical staining of the dinoflagellate *P. micans* with a FITC-labeled antibody directed to the cell wall of the phytoplankton. The stained species can be distinguished on the basis of their high FBG fluorescence intensity.

increased green fluorescence signal after blue excitation, the stained phytoplankton species can be adequately detected. Selective stains for other toxic phytoplankton species presently are being produced.

9.5.6.3. Phytoplankton Viability Analysis

An important parameter for functional investigation of phytoplankton is the determination of the viability of single cells. Viability determination cannot

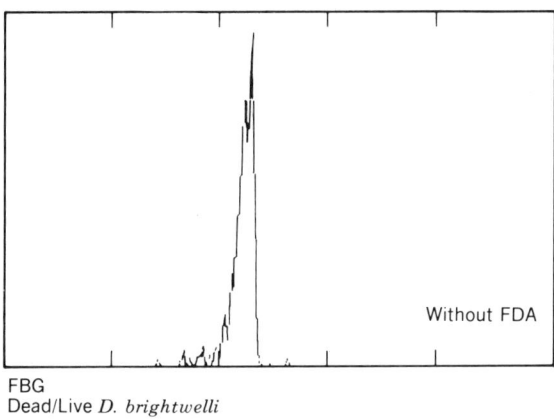

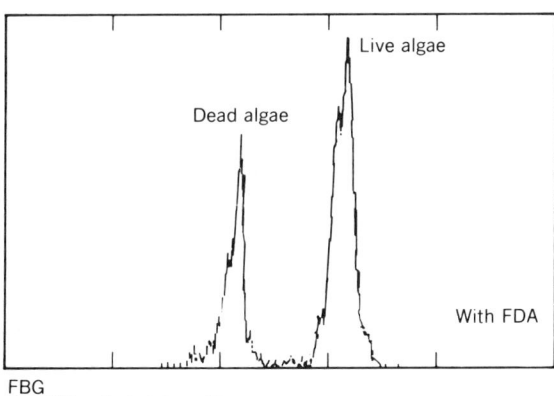

Fig. 9.33. Viability analysis with fluorescein diacetate (FDA), as illustrated for the diatom *Ditylum brightwelli*. In viable cells the green fluoescent fluorescein molecule is formed that cannot pass through the cell wall. Hence, an increased FBG fluorescence intensity is measured for these cells.

be done unequivocally by examination of the morphology of the cell, as many phytoplankton species have firm cell walls. Knowledge of the concentration of viable cells can be used to obtain a better estimate of phytoplankton growth rates and as a more sensitive way to determine the effect of contaminants, e.g., in phytoplankton growth inhibition tests.

An elegant way to establish phytoplankton viability is by measurement of the enzymatic activity of single cells. A simple approach is to add a solution of fluorescein diacetate (FDA) to the cell suspension. This apolar, nonfluorescent dye readily passes the cell membrane; once inside the cell, the acetate moieties are broken down by nonspecific esterases, resulting in strongly fluorescent fluorescein. The fluorescein is much more polar and does not easily leak through a healthy cell membrane. Therefore, viable cells show bright green fluorescence (218). Simultaneously application of FDA in combination with the intercalating red fluorescent dye propidium iodide, which can only pass through the membrane of dead or dying cells (219), yields a reliable cell viability test (220).

Application of the FDA viability assay gives very clear results for many planktonic algae. Figure 9.33 shows the result obtained for stained *Ditylum brightwelli*, a fragile marine diatom. From the histogram the viable algal cells, with increased blue-induced green fluorescence, can be directly quantitated. Conversion of FDA into fluorescein can also be followed with time. The esterase activity gives a steady increase in fluorescent product, which can be followed with the flow cytometer (221). From the time-dependent measurement the metabolic activity of the esterases can be inferred, which is also an important physiological parameter. When other, more selective dyes are applied, the activity of other enzyme systems can be determined as well.

9.5.6.4. Phytoplankton Cell Cycle Analysis

As a last example of functional cell analysis, an approach to quantify the cell cycle of phytoplankters will be discussed. Cell cycle analysis is one of the main types of determination that is done with flow cytometers. Reproduction of cells is done in an ordered sequence of events, during which cellular DNA is replicated prior to cell division. DNA replication occurs in a very well-defined manner, ensuring that the daughter cells will have exactly the same set of genetic information. Also, the DNA sequence (and hence also the DNA content) is well defined, since it is genetically determined as well. The cell cycle has been subjected to intensive study (222–227).

The cell cycle can be described as follows. In the first growth phase, G_1, the organelles and the cytoplasma of the cells grow, in preparation for the next phase, in which DNA is synthesized (S phase). Subsequently, a second

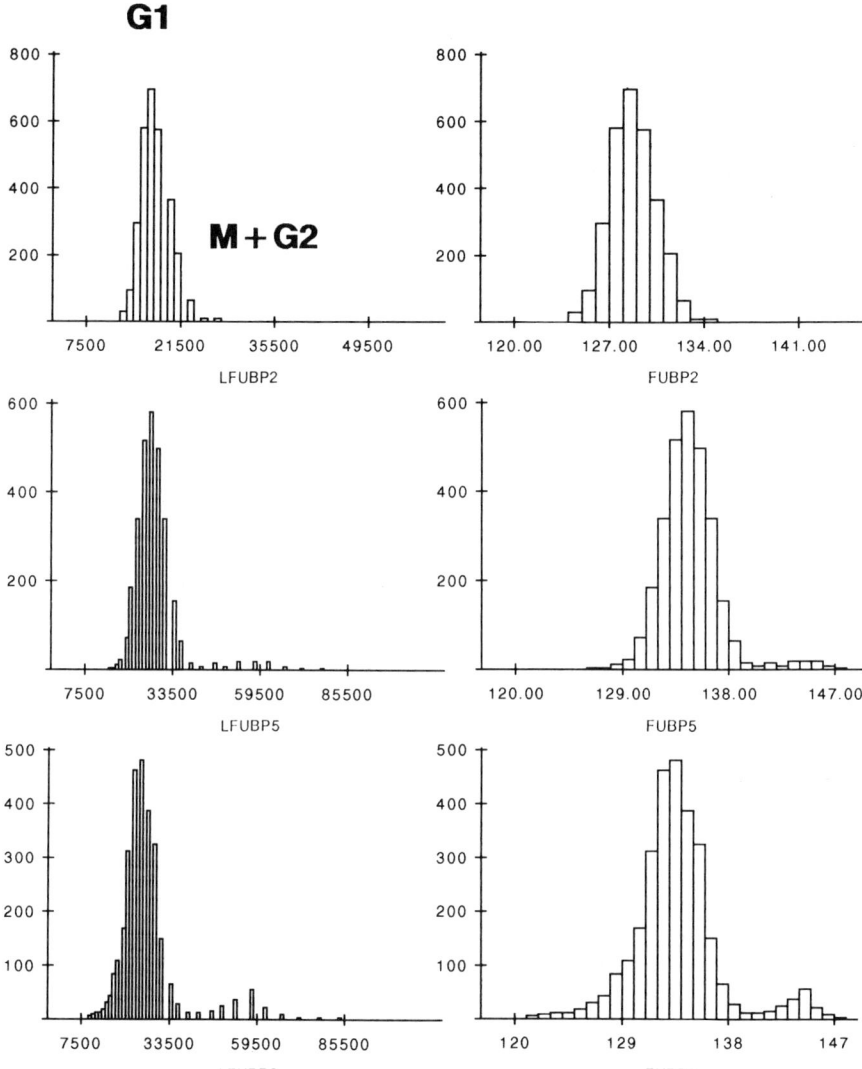

Fig. 9.34. DNA content for several cultures of *Dunaliella*. From the histogram the relative number of cells with the double amount of DNA can directly be inferred. The percentage of such cells is indicative of the pupulation of cells that is in the G_2 stage or M stage, i.e., the population of cells that is involved in division. The histogram is shown for both logarithmic and linear FUB intensity values. The cells were stained with DAPI. The figure shows the limited resolution of the OPA, which is due to the optimization of the design to afford a high dynamic range. As a result the histogram given for the linearized intensity values is not equidistant.

growth phase, G_2, takes place. In G_2 the cells are prepared for mitosis. In the mitosis phase (M phase) DNA, in the form of chromosomes, separates into two identical sets. Finally, the cell divides and the cycle is completed. By application of specific fluorescent stains (see Section 9.5.6.1) DNA can be quantitatively labeled, so that the fraction of cells in each phase can be determined. In the flow cytometric measurement the G_1 phase shows a narrow distribution for the fluorescence parameter, which is related to the well-defined DNA content. Cells in the S phase display a gradual increase of fluorescence. Finally, cells in the G_2 and the M phase have exactly twice the amount of fluorescence (and hence of DNA) than the cells that are in the G_1 phase. Measurement of the distribution of the cells over the four phases provides very useful information on the rate of cell growth. For phytoplankton analysis an estimate of the growth rate can be used to predict the species succession, when measurements are done on algae that can still be discerned on the basis of their pigment composition, i.e., living or adequately conserved cells.

An example of flow cytometric cell cycle analysis for a number of cultures of the chlorophyte *Dunaliella* is shown in Fig. 9.34. The cells were stained with DAPI, a dye that is excited in the UV and emits blue light (hence the parameter FUB that is detected). From the figure a limitation of the OPA can be inferred: the instrument has a tremendous dynamic range, but at the cost of resolution. When linearly amplified signals are detected, CVs for histograms of DAPI fluorescence intensity will be much lower than those shown in the figure. In cell cycle analysis fluorescence signals generally do not differ by more than a factor of 2, so that only a limited dynamic range is needed. From Fig. 9.34 an adequate estimation can be obtained for the *Dunaliella* cell growth. To obtain a complete picture, however, the cell cycle analysis should be carried out throughout the day. Some algal species are known to divide at a particular time of day, e.g., during the night or in the early morning, directly after dawn. Obviously, cell cycle analysis can be done easily and automatically with a flow cytometer. Commercial software packages are available for dedicated data analysis.

9.5.7. Concluding Remarks

Flow cytometry is a technique that has rapidly come of age, in particular due to significant developments in commercially available and affordable laser systems, fast computers, and advanced electronics. In addition, commercial flow cytometers are available, still mainly geared toward biomedical applications, i.e., more suitable for small particles.

Important improvements are still needed in flow cytometer quality control and in the development of reproducible staining and cell preparation techniques, two subjects that have not developed as fast as the instrumental improvements.

In addition to the "classical" biomedical applications more and more applications in other areas are reported. Flow cytometry has a tremendous potential in any field where quantitative data are needed for particles. Many areas can be envisaged, including biological, biochemical, environmental, and polymer analysis, where flow cytometry may have a significant contribution. A research effort that merits particular attention in this context is the development of a technique, based on flow cytometric detection, that can be used for the automated determination of the sequence of bases in DNA fragments (228) (cf. our discussion of similar developments in capillary gel electrophoresis in Section 9.3.4.4). The approach relies on fluorescent labeling of the bases in a single strand of DNA, movement of the DNA strand into a moving stream, and detection of the *individual* specifically fluorescently labeled bases as they are cleaved from the strand by an exonuclease. This method of DNA sequencing will have important advantages over existing approaches. Obviously, the realization of this apparatus will demand the utmost of available instrumental techniques; on the other hand, it will show the incredible potential of flow-cytometry-based methods.

9.6. FUTURE TRENDS

In this chapter a number of areas of molecular fluorescence in analytical sciences have been discussed in which lasers have proved to be very useful spectroscopic tools. Lasers can be used to enhance selectivity, to increase sensitivity, or to enable one to examine small areas or volumes. As more and more cheap and user-friendly lasers are commercially available this field will show an increasing number of applications. A new and promising development in this respect is the optical parametric oscillator, a nonlinear optical element that, when used in combination with a suitable laser source, can provide continuously tunable laser output over an extremely wide spectral range.

For a number of applications that have been discussed in this chapter the use of lasers is of particular interest. For instance, the development of capillary separation techniques—where analytes have to be probed in sub-microliter flow cells—strongly depends on the availability of suitable laser-based detection systems. Application of laser-induced fluorescence detection even allows for the analysis of very small samples, down to the single-cell level. Analysis of constituents of single cells can also be achieved via other approaches, like flow cytometry or a technique that has not been discussed in this chapter, confocal scanning laser microscopy. In the former technique many individual cells can be measured at a high rate, thus providing a statistically sound measurement. In the latter technique highly detailed spatially resolved information is obtained for a single particle.

Another new and promising range of application of lasers is offered by the possibility of performing "single molecule" detection. DNA sequencing and study of microstructural aspects of doped solids are but a few of the areas in which such approaches to ultratrace analysis can be envisaged. Presently, only the first indications of such applications have appeared in the literature.

Another type of application is in high-resolution low-temperature spectroscopy, which enables selective ultratrace analysis. This approach holds great promise for the elucidation of the mechanism of carcinogenicity of (fluorescent) xenobiotics, such as PAHs. Laser excitation is also crucial for an approach of low-temperature spectroscopy that has not been discussed in this chapter, supersonic jet (or molecular beam) spectroscopy. In this technique analyte molecules are seeded into expanding carrier gas. The concentration of the analyte molecules in the gas obviously is very low, so that only by sensitive laser-induced approaches can useful information is gathered. Many important applications in environmental and biomedical analysis can be envisaged for high-resolution spectroscopy.

Finally, we would like to point out the interesting possibilities of laser-induced fluorescence for remote sensing. The first applications, e.g., determination of phytoplankton in surface water or of contaminants in groundwater, have just been published for "real" samples. Remote sensing offers the possibility of directly obtaining synoptic information on physical and chemical parameters. In combination with imaging techniques and time-resolved excitation and detection, even 3D spatially resolved measurements can be performed.

REFERENCES

1. C. A. Parker, *Photoluminescence of Solutions*, Elsevier, Amsterdam, 1968.
2. J. B. Birks, *Photophysics of Aromatic Molecules*, Wiley, New York, 1970.
3. J. R. Lakowicz, *Principles of Fluorescence Spectroscopy*, Plenum Press, New York, 1983.
4. R. J. Hurtubise, *Anal. Chem.*, **55**, 669A (1983).
5. I. B. Bersuker, *The Jahn–Teller Effect and Vibronic Interactions in Modern Chemistry*, Plenum Press, New York, 1984.
6. W. H. Melhuish, *J. Res. Natl. Bur. Stand.*, **76A**, 547 (1972).
7. K. D. Mielenz, Ed., *Optical Radiation Measurements*, Vol. 3, Academic Press, New York, 1982.

8. J. W. Hofstraat, K. Rubelowsky, and S. Slutter, *J. Plankton Res.*, **14**, 625 (1992).
9. Molecular Probes, Eugene, OR, USA, and Lambda Fluoreszenztechnologie, Graz, Austria, are two companies that special in the supply of fluorescent labels.
10. N. Ichinose, G. Schwedt, F. M. Schnepel, and K. Adachi, *Fluorometric Analysis in Biomedical Chemistry. Trends and Techniques including HPLC Applications*, Wiley, New York, 1991.
11. H. Lingeman and W. J. M. Underberg, Eds., *Detection Oriented Derivatization Techniques in Liquid Chromatography*, Marcel Dekker, New York, 1990.
12. See, for instance, E. H. Piepmeier, Ed., *Analytical Applications of Lasers*, Wiley, New York, 1986.
13. R. M. Measures, in E. H. Piepmeier, Ed., *Analytical Applications of Lasers*, Wiley, New York, 1986, Chap. 11.
14. E. S. Yeung, in E. H. Piepmeier, Ed., *Analytical Applications of Lasers*, Wiley, New York, 1986, Chap. 17.
15. S. J. Hein and L. C. Thomas, in E. H. Piepmeier, Ed., *Analytical Applications of Lasers*, Wiley, New York, 1986, Chap. 16.
16. E. L. Wehry, in E. H. Piepmeier, Ed., *Analytical Applications of Lasers*, Wiley, New York, 1986, Chap. 7.
17. J. W. Hofstraat, C. Gooijer, and N. H. Velthorst, in S. G. Schulman, Ed., *Molecular Luminescence Spectroscopy, Methods and Applications: Part 2*, Wiley, New York, 1988, Chap. 4.
18. R. S. Houk, in E. H. Piepmeier, Ed., *Analytical Applications of Lasers*, Wiley, New York, 1986, Chap. 18.
19. A. C. Koskelo and M. J. Wirth, in E. H. Piepmeier, Ed., *Analytical Applications of Lasers*, Wiley, New York, 1986, Chap. 9.
20. C. M. B. van den Beld, Ph.D. Thesis, Leiden University, The Netherlands, 1991.
21. R. A. Mathies, K. Peck, and L. Stryer, *Anal. Chem.*, **62**, 1786 (1990).
22. M. J. Wirth, *Anal. Chem.*, **62**, 270A (1990).
23. W. D. Pfeffer and E. S. Yeung, *Anal. Chem.*, **58**, 2103 (1986).
24. M. J. Wirth and H. O. Fatumnbi, *Anal. Chem.*, **62**, 973 (1990).
25. R. J. van de Nesse, A. J. G. Mank, G. P. Hoornweg, C. Gooijer, U. A. T. Brinkman, and N. H. Velthorst, *Anal. Chem.*, **63**, 2685 (1991).
26. X. Xi and E. S. Yeung, *Anal. Chem.*, **62**, 1580 (1990).
27. H. W. Messenger, *Laser Focus World*, November, pp. 85–87 (1990).
28. B. Dance, *Laser Focus World* (*Eur. Suppl.*), Autumn, pp. 18–20 (1990).
29. C. Miyaka, *Lasers Optron.*, October, pp. 44–50 (1990).
30. M. J. Wirth and G. J. Blanchard, in E. H. Piepmeier, Ed., *Analytical Applications of Lasers*, Wiley, New York, 1986, Chap. 14.
31. K. Ibbs, *Lasers Optron.*, February, pp. 39–50 (1990).
32. T. Mahony, *Photon. Spectra*, January, pp. 103–106 (1991).
33. T. Imasaka and N. Ishibashi, *Anal. Chem.*, **62**, 363A (1990).

34. G. J. Diebold and R. N. Zare, *Science*, **196**, 1439 (1977).
35. C. M. B. van den Beld and H. Lingeman, in W. R. G. Bayens, D. de Keukeleire, and K. Korkidis, Eds., *Luminescence Techniques in Chemical and Biochemical Analysis*, Marcel Dekker, New York, 1991, Chap. 9.
36. S. Wu and N. J. Dovichi, *J. Chromatogr.*, **480**, 141 (1989).
37. J. Z. Zhang, D. Y. Chen, S. Wu, H. R. Harke, and N. J. Dovichi, *Clin. Chem. (Winston-Salem, N.C.)* **37**, 1492 (1991).
38. D. Y. Chen, H. P. Swerdlow, H. R. Harke, J. Z. Zhang, and N. J. Dovichi, *J. Chromatogr.*, **559**, 237 (1991).
39. R. J. van de Nesse, G. P. Hoornweg, C. Gooijer, U. A. T. Brinkman, and N. H. Velthorst, *Anal. Chim. Acta*, **227**, 173 (1989).
40. R. J. van de Nesse, C. Gooijer, G. P. Hoornweg, U. A. T. Brinkman, N. H. Velthorst, and S.-J. van der Bent, *Anal. Lett.* **23**, 1235 (1990).
41. E. S. Yeung and W. G. Kuhr, *Anal. Chem.*, **63**, 275A (1991).
42. Y. F. Cheng and N. J. Dovichi, *Science*, **242**, 562 (1988).
43. H. Drossman, J. A. Luckey, A. J. Kostichka, J. D'Cunha, and L. M. Smith, *Anal. Chem.*, **62**, 900 (1990).
44. O. Svelto, *Principles of Lasers*, Plenum Press, New York, 1989.
45. See for instance, J. D. Ingle and S. R. Crouch, *Spectrochemical Analysis*, Prentice-Hall, Englewood Cliffs, NJ, 1988, Chap. 5.
46. U. A. T. Brinkman, G. J. de Jong, and C. Gooijer, *Pure Appl. Chem.*, **59**, 625 (1987).
47. E. S. Yeung and M. J. Sepaniak, *Anal. Chem.*, **52**, 1465A (1980).
48. M. C. Roach and M. D. Harmony, *Anal. Chem.*, **59**, 411 (1987).
49. P. J. M. Kwakman, Ph.D. Thesis, Free University, Amsterdam, The Netherlands, 1991.
50. L. M. Nicholson, H. B. Patel, F. Kristjansson, S. C. Crowley, Jr., K. Dave, J. F. Stobaugh, and C. M. Riley, *J. Pharm. Biomed. Anal.*, **8**, 805 (1990).
51. P. J. M. Kwakman, H. Koelewijn, I. Kool, U. A. T. Brinkman, and G. J. de Jong, *J. Chromatogr.*, **511**, 155 (1990).
52. A. Vairavamurthy and K. Mopper, *Anal. Chim. Acta*, **237**, 215 (1990).
53. M. Kobayashi and E. Ichishima, *Anal. Biochem.*, **189**, 122 (1990).
54. C. M. B. van den Beld, H. Lingeman, G. J. van Ringen, U. R. Tjaden, and J. van der Greef, *Anal. Chim. Acta*, **205**, 15 (1988).
55. D. S. Stegehuis, U. R. Tjaden, C. M. B. van den Beld, and J. van der Greef, *J. Chromatogr.*, **549**, 185 (1991).
56. S. Mho and E. S. Yeung, *Anal. Chem.*, **57**, 2253 (1985).
57. J. W. Jorgenson and E. J. Guthrie, *J. Chromatogr.*, **255**, 335 (1983).
58. H. H. Laner and D. McManigill, *Anal. Chem.*, **58**, 166 (1986).
59. L. Gross and E. S. Yeung, *Anal. Chem.*, **62**, 427 (1990).
60. T. W. Garner and E. S. Yeung, *J. Chromatogr.*, **515**, 639 (1990).

61. R. J. van de Nesse, G. J. M. Hoogland, J. J. M. de Moel, C. Gooijer, U. A. T. Brinkman, and N. H. Velthorst, *J. Chromatogr.*, **552**, 613 (1991).
62. K. Tanabe, M. Glick, B. Smith, E. Voigtman, and J. D. Winefordner, *Anal. Chem.*, **59**, 1125 (1987).
63. M. J. Sepaniak and E. S. Yeung, *Anal. Chem.*, **49**, 1554 (1977).
64. S. C. Beale, Y. Z. Hsieh, D. Wiesler, and M. Novotny, *J. Chromatogr.*, **499**, 579 (1990).
65. J. Liu, Y. Z. Hsieh, D. Wiesler, and M. Novotny, *Anal. Chem.*, **63**, 408 (1991).
66. J. Liu, O. Shirota and M. Novotny, *Anal. Chem.*, **63**, 413 (1991).
67. L. A. Sternson, J. F. Stobaugh, J. Reid, and P. de Montigny, *J. Pharm. Biomed. Anal.*, **6**, 657 (1988).
68. M. Mifune, D. K. Krehbiel, J. F. Stobaugh, and C. M. Riley, *J. Chromatogr. Biomed. Apl.*, **496**, 55 (1989).
69. K. Sauda, T. Imasaka, and N. Ishibashi, *Anal. Chim. Acta*, **187**, 353 (1986).
70. P. A. Johnson, T. E. Barber, B. W. Smith, and J. D. Winefordner, *Anal. Chem.*, **61**, 861 (1989).
71. T. Imasaka, A. Tsukamoto, and N. Ishibashi, *Anal. Chem.*, **61**, 2285 (1989).
72. G. Patonay and M. D. Antoine, *Anal. Chem.*, **63**, 321A (1991).
73. E. V. Shpol'skii and T. N. Bolotnikova, *Pure Appl. Chem.*, **37**, 183 (1974).
74. C. G. de Lima, *CRC Crit. Rev. Anal. Chem.*, **16**, 177 (1985).
75. J. W. Hofstraat, *High-Resolution Molecular Fluorescence Spectroscopy in Low-Temperature Matrices: Principles and Applications*, Free University Press, Amsterdam, 1988.
76. R. I. Personov, in V. M. Agranovich and R. M. Hochstrasser, Eds., *Spectroscopy and Excitation Dynamics of Condensed Molecular Systems*, North-Holland Publ., Amsterdam, 1983, pp. 555–619.
77. J. W. Hofstraat, A. J. Schenkeveld, C. Gooijer, and N. H. Velthorst, *Anal. Chem.*, **60**, 377 (1988).
78. J. W. Hofstraat, I. L. Freriks, M. E. J. de Vreeze, C. Gooijer, and N. H. Velthorst, *J. Phys. Chem.*, **93**, 184 (1989).
79. B. Meyer, *Low Temperature Spectroscopy*, American Elsevier, New York, 1971.
80. J. W. Hofstraat, M. Engelsma, J. H. de Roo, C. Gooijer, and N. H. Velthorst, *Appl. Spectrosc.*, **41**, 625 (1987).
81. S. M. Thornberg and J. R. Maple, *Anal. Chem.*, **56**, 1542 (1984).
82. E. V. Shpol'skii, A. A. Il'ina, and L. A. Klimova, *Dokl. Akad. Nauk SSSR*, **87**, 935 (1952).
83. R. N. Nurmukhametov, *Russ. Chem. Rev. (Engl. Transl.)* **38**, 180 (1969).
84. K. Palewska and Z. Ruziewicz, *Chem. Phys. Lett.*, **64**, 378 (1979).
85. J. W. Hofstraat, A. J. Schenkeveld, M. Engelsma, C. Gooijer, and N. H. Velthorst, *Spectrochim. Acta*, **44A**, 1019 (1988).

86. A. M. Merle, M. Lamotte, S. Risemberg, C. Hauw, J. Gaultier, and J. P. Grivet, *Chem. Phys.*, **22**, 207 (1977).
87. A. M. Merle, W. M. Pitts, and M. A. El-Sayed, *Chem. Phys. Lett.*, **54**, 211 (1978).
88. E. V. Shpol'skii, *Sov. Phys.—Usp. (Engl. Transl.)* **6**, 411 (1963).
89. C. Pfister, *Chem. Phys.*, **2**, 171 (1973).
90. C. Pfister, *J. Chim. Phys.*, **67**, 418 (1970).
91. T. Vo-Dinh and U. P. Wild, *J. Lumin.*, **6**, 296 (1973).
92. J. Rima, L. A. Nakhimovsky, M. Lamotte, and J. Joussot-Dubien, *J. Phys. Chem.*, **83**, 4302 (1984).
93. P. Tokousbalides, E. L. Wehry, and G. Mamantov, *J. Phys. Chem.*, **81**, 1769 (1977).
94. M. B. Perry, E. L. Wehry, and G. Mamantov, *Anal. Chem.*, **55**, 1893 (1983).
95. A. P. D'Silva and V. A. Fassel, *Anal. Chem.*, **56**, 985A (1984).
96. P. Garrigues, E. Parlanti, M. Radke, J. Bellocq, H. Willsch, and M. Ewald, *J. Chromatogr.*, **395**, 217 (1987).
97. P. Garrigues, M.-P. Marniesse, S. A. Wise, J. Bellocq, and M. Ewald, *Anal. Chem.*, **59**, 1695 (1987).
98. Y. Yang, A. P. D'Silva, and V. A. Fassel, *Anal. Chem.*, **53**, 894 (1981).
99. K. M. Bark and R. K. Forcé, *Talanta*, **38**, 181 (1991).
100. R. J. Crowley, S. Siggia, and P. C. Uden, *Anal. Chem.*, **52**, 1224 (1980).
101. D. W. Later, T. G. Andros, and M. L. Lee, *Anal. Chem.*, **55**, 2126 (1983).
102. S. S. Hecht, M. Loy, and D. Hoffmann, in R. I. Freudenthal and P. W. Jones, Eds., *Carcinogenesis*, Raven Press, New York, 1976, Vol. 1, pp. 325–340.
103. Y. Yang, A. P. D'Silva, and V. A. Fassel, *Anal. Chem.*, **53**, 2107 (1981).
104. J. W. Hofstraat, H. J. M. Jansen, G. P. Hoornweg, C. Gooijer, N. H. Velthorst, and W. P. Cofino, *Int. J. Environ. Anal. Chem.*, **21**, 299 (1985).
105. J. W. Hofstraat, W. J. M. van Zeijl, F. Ariese, J. W. G. Mastenbroek, C. Gooijer, and N. H. Velthorst, *Mar. Chem.*, **33**, 301 (1991).
106. F. Ariese, C. Gooijer, N. H. Velthorst, and J. W. Hofstraat, *Fresenius' Z. Anal. Chem.*, **339**, 722 (1991).
107. L. Paturel, F. Jarosz, C. Fachinger, and J. Suptil, *Anal. Chim. Acta*, **147**, 293 (1983).
108. Y. Yang, A. P. D'Silva, V. A. Fassel, and M. Iles, *Anal. Chem.*, **52**, 1351 (1980).
109. W. E. Moerner and L. Kador, *Anal. Chem.*, **61**, 1217A (1989).
110. M. Orrit and J. Bernard, *Phys. Rev. Lett.*, **65**, 2761 (1990).
111. W. P. Ambrose and W. E. Moerner, *Nature (London)*, **349**, 225 (1991).
112. F. Ariese, C. Gooijer, N. H. Velthorst, and J. W. Hofstraat, *Anal. Chim. Acta*, **232**, 245 (1990).
113. J. L. Kropp, W. R. Dawson, and M. W. Windsor, *J. Phys. Chem.*, **73**, 1747 (1969).
114. H. V. Gelboin, *Physiol. Rev.*, **60**, 1107 (1980).
115. H. V. Gelboin, *Rev. Cancer Biol.*, **31**, 39 (1972).

116. J. K. Selkirk, R. G. Croy, J. P. Whitlock, Jr., and H. V. Gelboin, *Cancer Res.*, **35**, 3651 (1975).
117. A. H. Conney, *Cancer Res.*, **42**, 4875 (1982).
118. O. Pelkonen, K. Vahakangas, and D. W. Nebert, *J. Toxicol. Environ. Health*, **6**, 1009 (1980).
119. R. Jankowiak and G. J. Small, *Anal. Chem.*, **61**, 1023A (1989).
120. F. Ariese, S. J. Kok, C. Gooijer, N. H. Velthorst, and J. W. Hofstraat, *Polynuc. Aromatic Hydrocarbons*, in press.
121. S. Weeks, S. Gilles, R. Dobson, S. Senne, and A. P. D'Silva, *Anal. Chem.*, **62**, 1472 (1990).
122. F. Ariese, unpublished results.
123. J. W. Hofstraat, A. J. Schenkeveld, M. Engelsma, C. Gooijer, and N. H. Velthorst, *Spectrochim. Acta*, **45A**, 139 (1989).
124. J. W. Hofstraat, M. Engelsma, C. Gooijer, and N. H. Velthorst, *Spectrochim. Acta*, **45A**. 491 (1989).
125. G. J. Small, in V. M. Agranovich and R. M. Hochstrasser, Eds., *Spectroscopy and Excitation Dynamics of Condensed Molecular Systems*, North-Holland Publ., Amsterdam, 1983, pp. 515–554.
126. J. W. Hofstraat, R. Locher, and U. P. Wild, *Appl. Spectrosc.*, **44**, 1317 (1990).
127. B. B. Price and J. C. Wright, *Anal. Chem.*, **62**, 1989 (1990).
128. J. C. Brown, J. M. Hayes, J. A. Warren, and G. J. Small, in G. J. Hieftije, J. C. Travis, and F. E. Lytle, Eds., *Lasers in Chemical Analysis*, Humana Press, Clifton, NJ, 1981.
129. J. C. Brown, J. A. Duncanson, Jr., and G. J. Small, *Anal. Chem.*, **52**, 1711 (1980).
130. L. A. Bykovskaya, R. I. Personov, and Yu. V. Romanovskii, *Anal. Chim. Acta*, **125**, 1 (1981).
131. J. W. Hofstraat, H. J. M. Jansen, G. P. Hoornweg, C. Gooijer, and N. H. Velthorst, *Anal. Chim. Acta*, **170**, 61 (1985).
132. J. W. Hofstraat, M. Engelsma, R. J. van de Nesse, U. A. T. Brinkman, C. Gooijer, and N. H. Velthorst, *Anal. Chim. Acta*, **193**, 193 (1987).
133. J. W. Hofstraat, C. Gooijer, and N. H. Velthorst, *Appl. Spectrosc.*, **42**, 614 (1988).
134. K. Randerath, E. Randerath, H. D. Agraval, R. C. Gupta, M. E. Schurdak, and M. V. Reddy, *Environ. Health Perspect.*, **62**, 57 (1985).
135. R. Jankowiak, R. S. Cooper, D. Zamzow, G. J. Small, G. Doskocil, and A. M. Jeffrey, *Chem. Res. Toxicol.*, **1**, 60 (1988).
136. P. Brooks and M. R. Osborne, *Carcinogenesis* (*London*), **3**, 1223 (1982).
137. M. R. Osborne, S. Jacobs, R. G. Harvey, and P. Brooks, *Carcinogenesis* (*London*), **2**, 553 (1981).
138. R. F. Newbold and P. Brooks, *Nature* (*London*), **261**, 53 (1976).
139. J. W. Hofstraat, M. L. Eggens, A. Bergman, F. Ariese, C. Gooijer, N. H. Velthorst,

M.-J. S. T. Steenwinkel, R. A. Baan, and J. P. Boon, *Polynuc. Aromatic Hydrocarbons*, in press.

140. R. J. van de Nesse, I. Vinkenburg, C. Gooijer and N. H. Velthorst, to be published.
141. A. Colmsjö, Y. Zebühr, and C. Ostman, *Chem. Sci.*, **20**, 123 (1982).
142. R. J. van de Nesse, G. J. M. Hoogland, J. J. M. de Moel, C. Gooijer, U. A. T. Brinkman, and N. H. Velthorst, *J. Chromatogr.*, **552**, 612 (1991).
143. W. H. Coulter, *Proc. Natl. Electron. Conf.*, **12**, 1034 (1956).
144. L. A. Kamentsky, M. R. Melamed, and H. Derman, *Science*, **150**, 630 (1965).
145. M. A. van Dilla, T. T. Trujillo, P. F. Mullaney, and J. R. Coulter, *Science*, **163**, 1213 (1969).
146. H. M. Shapiro, *Practical Flow Cytometry*, Alan R. Liss, New York, 1988.
147. M. R. Melamed, P. F. Mullaney, and M. L. Mendelsohn, Eds., *Flow Cytometry and Sorting*, Wiley, New York, 1979.
148. M. A. van Dilla, P. N. Dean, O. D. Laerum, and M. R. Melamed, Eds., *Flow Cytometry: Instrumentation and Data Analysis*, Academic Press, Orlando, FL, 1985.
149. J. A. Steinkamp, *Rev. Sci. Instrum.*, **55**, 1375 (1984).
150. M. Born and E. Wolf, *Principles of Optics*, Pergamon, Oxford, 1959.
151. P. F. Mullaney, M. A. van Dilla, J. R. Coulter, and P. N. Dean, *Rev. Sci. Instrum.*, **40**, 1029 (1969).
152. G. C. Salzman, P. F. Mullaney, and B. J. Price, in M. R. Melamed, P. F. Mullaney, and M. L. Mendelsohn, Eds., *Flow Cytometry and Sorting*, Wiley, New York, 1979, p. 105.
153. M. Kerker, *Cytometry*, **4**, 1 (1983).
154. G. C. Salzman, J. M. Crowell, J. C. Martin, T. T. Trujillo, A. Romero, and P. F. Mullaney, *Acta Cytol.*, **19**, 374 (1975).
155. G. B. J. Dubelaar, J. W. M. Visser, and M. Donze, *Cytometry*, **8**, 405 (1987).
156. R. J. Olson, E. R. Zettler, and O. K. Anderson, *Cytometry*, **10**, 636 (1989).
157. J. R. Hodkinson, in C. N. Davies, Ed., *Aerosol Science*, Advanced, New York, 1966, p. 237.
158. L. A. Kamentsky, *Adv. Biol. Med. Phys.*, **14**, 93 (1973).
159. T. Sharpless, F. Traganos, Z. Darzynkiewicz, and M. R. Melamed, *Acta Cytol.*, **19**, 577 (1975).
160. T. K. Sharpless and M. R. Melamed, *J. Histochem. Cytochem.*, **24**, 257 (1976).
161. H. A. Crissman and J. A. Steinkamp, *J. Cell Biol.*, **59**, 766 (1973).
162. L. S. Cram, D. J. Arndt-Jovin, B. G. Grimwade, and T. M. Jovin, *J. Histochem. Cytochem.*, **27**, 445 (1979).
163. L. S. Cram, D. J. Arndt-Jovin, B. G. Grimwade, and T. M. Jovin, in O. D. Laerum, T. Lindmo, and E. Thorud, Eds., *Flow Cytometry IV*, Universitetsforlaget, Bergen, Norway, 1980, p. 256.
164. G. B. J. Dubelaar A. C. Groenewegen, W. Stokdijk, G. J. van den Engh, and J. W. M. Visser, *Cytometry*, **10**, 529 (1989).

165. J. W. Hofstraat, W. J. M. van Zeijl, J. C. H. Peeters, L. Peperzak, and G. B. J. Dubelaar, *SPIE Proc.*, **1269**, 116 (1990).
166. D. Pinkel and R. Stovel, in M. A. van Dilla, P. N. Dean, O. D. Laerum, and M. R. Melamed, Eds., *Flow Cytometry: Instrumentation and Data Analysis*, Academic Press, Orlando, FL, 1985, p. 78.
167. R. G. Sweet, in M. R. Melamed, P. F. Mullaney, and M. L. Mendelsohn, Eds., *Flow Cytometry and Sorting*, Wiley, New York, 1979, p. 177.
168. T. K. Sharpless, in M. R. Melamed, P. F. Mullaney, and M. L. Mendelsohn, Eds., *Flow Cytometry and Sorting*, Wiley, New York, 1979, p. 359.
169. P. H. Bartels, *Anal. Quant. Cytol.*, **1**, 20, 77, 153 (1979); **2**, 19, 77, 155 (1980); **3**, 1, 83, 167, 251 (1981); **4**, 81, 241 (1982); **5**, 229 (1983).
170. J. W. Hofstraat, M. E. J. de Vreeze, W. J. M. van Zeijl, L. Peperzak, J. C. H. Peeters, and H. W. Balfoort, *J. Fluoresc.*, **1**, 249 (1991).
171. D. L. Massart, B. G. M. Vandeginste, S. N. Denning, Y. Michotte, and L. Kaufman, *Chemometrics: A Textbook*, Elsevier, Amsterdam, 1988.
172. D. S. Frankel, R. J. Olson, S. L. Frankel, and S. W. Chisholm, *Cytometry*, **10**, 540 (1989).
173. H. W. Balfoort, J. Snoek, J. R. M. Smits, L. W. Breedveld, J. W. Hofstraat, and J. Ringelberg, *J. Plankton Res.*, **14**, 575 (1992).
174. J. R. M. Smits, H. W. Balfoort, L. W. Breedveld, J. Snoek, J. W. Hofstraat, M. W. J. Derksen, and G. Kateman, *Anal. Chim. Acta*, **258**, 11 (1992).
175. D. S. Frankel and S. L. Frankel, *Cytometry, Suppl.*, **5**, 63 (1991).
176. P. Bierre, R. Mickaels, and D. Thiel, *Cytometry, Suppl.*, **5**, 64 (1991).
177. J. D. Bessman, *Automated Blood Counts and Differentials: A Practical Guide*, Johns Hopkins University Press, Baltimore, MD, 1986.
178. M. Andreeff, Ed., *Ann. N.Y. Acad. Sci.*, **468**, 1–408 (1986).
179. P. Quirke and J. E. D. Dyson, *J. Pathol.*, **149**, 79 (1986).
180. D. H. Ryan, M. A. Fallon, and P. K. Horan, *Clin. Chim. Acta*, **171**, 125 (1988).
181. M. Kerker, H. Chew, P. J. McNulty, J. P. Kratohvil, D. D. Cooke, M. Sculley, and M. P. Lee, *J. Histochem. Cytochem.*, **27**, 250 (1979).
182. M. R. Loken, R. G. Sweet, and L. A. Herzenberg, *J. Histochem. Cytochem.*, **24**, 284 (1976).
183. P. F. Mullaney, J. M. Crowell, G. C. Salzman, J. C. Martin, R. D. Hiebert, and C. A. Goad, *J. Histochem. Cytochem.*, **24**, 298 (1976).
184. M. R. Loken, D. R. Parks, and L. A. Herzenberg, *J. Histochem. Cytochem.*, **25**, 790 (1977).
185. G. J. Weil and T. M. Chused, *Blood*, **57**, 1099 (1981).
186. B. J. Trask, G. J. van den Engh, and J. H. B. W. Elgershuizen, *Cytometry*, **2**, 258 (1982).
187. R. J. Olson, S. L. Frankel, S. W. Chisholm, and H. M. Shapiro, *J. Exp. Mar. Biol. Ecol.*, **68**, 129 (1983).
188. C. M. Yentsch, P. K. Horan, K. Muirhead, Q. Dortch, E. Haugen, L. Legendre,

L. S. Murphy, M. J. Perry, D. A. Phinney, S. A. Pomponi, R. W. Spinrad, M. Wood, C. S. Yentsch, and B. J. Zahuranec, *Limnol. Oceanogr.*, **28**, 1275 (1983).
189. D. J. Arndt-Jovin and T. M. Jovin, *Annu. Rev. Biophys. Bioeng.*, **7**, 527 (1978).
190. S. Demers, K. Davis, and T. L. Cucci, *Cytometry*, **10**, 644 (1989).
191. E. Sakshaug, S. Demers, and C. M. Yentsch, *Mar. Ecol.: Prog. Ser.*, **41**, 275 (1987).
192. B. Thorell, *Cytometry*, **6**, 745 (1980).
193. J.-B. LePecq, *J. Mol. Biol.*, **27**, 87 (1967).
194. H. A. Crissman and J. A. Steinkamp, *J. Cell Biol.*, **59**, 766 (1973).
195. I. W. Taylor, *J. Histochem. Cytochem.*, **28**, 1021 (1980).
196. R. H. Jensen, R. G. Langlois, and B. H. Mayall, *J. Histochem. Cytochem.*, **25**, 954 (1977)
197. K. J. Puite and W. R. R. ten Broeke, *Plant Sci. Lett.*, **32**, 79 (1983).
198. H. A. Crissman and R. A. Tobey, *Science*, **184**, 1297 (1974).
199. R. V. Lebo, F. Gorin, R. J. Fletterick, F.-T. Kao, M.-C. Cheung, B. D. Bruce, and Y. W. Kan, *Science*, **225**, 57 (1984).
200. Z. Darzynkiewicz, F. Traganos, and M. R. Melamed, *Cytometry*, **1**, 98 (1980).
201. Z. Darzynkiewicz, J. Kapuscinski, F. Traganos, and H. A. Crissman, *Cytometry*, **8**, 138 (1987).
202. S. P. Hawkes and J. C. Bartholomew, *Proc. Natl. Acad. Sci. U.S.A.*, **74**, 1626 (1977).
203. J. Treumer and G. Valet, *Exp. Cell Res.*, **163**, 518 (1986).
204. G. C. Rice, E. A. Bump, D. C. Shrieve, W. Lee, and M. Kovacs, *Cancer Res.*, **46**, 6105 (1986).
205. K. A. Muirhead, P. K. Horan, and G. Poste, *Bio/Technology*, **3**, 337 (1985)
206. E. G. Vrieling, W. W. C. Gieskes, F. Colijn, J. W. Hofstraat, L. Peperzak, and M. Veenhuis, in T. J. Smayda (Ed.), *Proceedings of the 5th International Conference on Toxic Marine Phytoplankton*, Elsevier, Amsterdam, in press.
207. B. Rotman and B. W. Papermaster, *Proc. Natl. Acad. Sci. U.S.A.*, **55**, 134 (1966).
208. F. A. Dolbeare and R. E. Smith, in M. R. Melamed, P. F. Mullaney, and M. L. Mendelsohn, Eds., *Flow Cytometry and Sorting*, Wiley, New York, 1979, p. 317.
209. A. Rolland, G. Merdrignac, J. Gouranton, D. Bourel, R. LeVerge, and B. Genetet, *J. Immunol. Methods*, **96**, 185 (1987).
210. H. A. Crissman and J. A. Steinkamp, *Exp. Cell Res.*, **173**, 256 (1987).
211. T. M. Jovin, in M. R. Melamed, P. F. Mullaney, and M. L. Mendelsohn, Eds., *Flow Cytometry and Sorting*, Wiley, New York, 1979, p. 137.
212. G. Grynkiewicz, M. Poenie, and R. Y. Tsien, *J. Biol. Chem.*, **260**, 3440 (1985).
213. R. Nutticelli and R. Deamer, Eds., *Intracellular pH: Its Measurement, Regulation and Utilization in Cellular Functions*, Alan R. Liss, New York, 1982.
214. G. C. Cadée and J. Hegeman, *Hydrobiol. Bull.*, in press.
215. T. J. Smayda and E. Granéli, Eds., *Toxic Marine Phytoplankton*, Elsevier, Amsterdam, 1990, p. 29.

216. H. Kuosa, *Arch. Hydrobiol. Beih. Ergebn. Limnol.*, **31**, 301 (1988).
217. S. Hara and E. Tanoue, *Deep-Sea Res.*, **36**, 1777 (1989).
218. M. D. Persidsky and G. S. Baillie, *Cryobiology*, **14**, 322 (1977).
219. P. K. Horan and J. W. Kappler, *J. Immunol. Methods*, **18**, 309 (1977).
220. K. H. Jones and J. A. Senft, *J. Histochem. Cytochem.*, **33**, 77 (1985).
221. J. Dorsey, C. M. Yentsch, S. Mayo, and C. McKenna, *Cytometry*, **10**, 622 (1989).
222. H. A. Crissman and R. A. Tobey, *Science*, **184**, 1297 (1974).
223. A. Krishan, *J. Cell Biol.*, **66**, 188 (1975).
224. Z. Darzynkiewicz, F. Traganos, T. Sharpless, and M. R. Melamed, *J. Histochem. Cytochem.*, **25**, 875 (1975).
225. D. W. Galbraith, K. R. Harkins, J. M. Maddox, N. M. Ayres, D. P. Sharma, and E. Firoozabady, *Science*, **220**, 1049 (1983).
226. F. Dolbeare, H. Gratzner, M. G. Pallavicini, and J. W. Gray, *Proc. Natl. Acad. Sci. U.S.A.*, **80**, 5573 (1983).
227. M. Kubbies and G. Pierron, *Exp. Cell Res.*, **149**, 57 (1983).
228. J. H. Jett, R. A. Keller, J. C. Martin, B. L. Marrone, R. K. Moyzis, R. L. Ratliff, N. K. Seitzinger, E. B. Shera, and C. C. Stewart, *J. Biomol. Struct. Dyn.*, **7**, 301 (1989).

INDEX

ABEI. *See* Aminobutylethylisoluminol
Absorption, of fluorescent probes, 32
Absorption spectroscopy:
 application in spectral-hole burning studies, 155–163
 NIR dyes for, 248
Acetaminophen, fluorometric detection of, 315
N-Acetyl-β-D-glucosaminidase, photographic detection of, 8
Acetylsalicylic acid:
 fluorometric detection of, 314–315
 photochemical fluorometry of, 119, 120, 126
Acousto-optic modulators:
 for microspectrofluorometry, 277, 280
 for transient spectral-hole detection, 165, 166
Acridans, chemiluminescence of, 11
Acridine derivatives, chemiluminescence of, 10–12
Acridinium salts, oxidation of, 11, 12
6-Acroloyl-2-(dimethylamino)naphthalene. *See* ACRYLODAN
ACRYLODAN, as polarity probe, 41
Aflatoxins, laser-induced fluorescence detection in liquid chromatography of, 344
Aggregation numbers, measurement by fluorescence correlation spectroscopy, 267
Albumin:
 oxazine 750 derivative of, 366
 semiconductor laser fluorometric detection of, 240
Alcohols, photochemical fluorometry of, 114, 116
Aldehydes, photochemical fluorometry of, 111, 114
Aldicarb, photochemical fluorometry of, 128
Alexandrium cf. *tamarense*, occurrence of, 427
Aliphatic alcohols, photochemical fluorometry of, 111

Alkaline phosphatase, chemiluminescent immunoassay of, 9
Alkaloids, fluorometric detection of, 314, 315
7-Alkoxycoumarin, structure of, 42
8-Alkoxypyrene-1,3,6-trisulfonic acid, trisodium salt of, as fluorescent probe, 31
Alumina powder, quinizarin adsorption onto, spectral hole-burning studies on, 212
Aluminum, fluorescent chelate of, 311
Aluminum oxide plates, in photochemical fluorescence studies, 93
Ambroxol, photochemical fluorometry of, 126
Amines, high-performance liquid chromatography of, 8
Amino acids:
 capillary electrophoresis of, 356, 364
 chemiluminescent determination of, 16
 fluorometric detection of, 313, 318
 isoluminol derivatives of, 7
 laser-excited molecular fluorescence-capillary electrophoresis of, 354
9-Aminoacridine, in ethanol glass, spectral hole-burning of, 153
Aminobutylethylisoluminol:
 isothiocyanate derivatives of, use in immunoassays, 7
 structure of, 7
 use in immunoassays, 7
5-Amino-2,3-dihydro-1,4-phthalazinedione. *See* Luminol
p-Aminophenol, fluorometric detection of, 315
Ammonia, detection in blood, 311–312
Anabaena flos-aquae, toxins in, 426
Analgesics, fluorometric detection of, 315
Anhydro-3,3,3',3'-tetramethyl-1,1'-bis(4-sulfomethyl)-4,5,4',5'-dibenzoindotricarbocyanine, NIH laser fluorometry of, 237, 238
2-Anilino-8-naphthalene, structure of, 42

445

1-Anilinonaphthalene sulfonate:
  as polarity probe, 41–43
  structure of, 41
8-Anilinonaphthalene-1-sulfonic acid, use in fluorometric detection of serum albumin, 314
Anisotropic rotations, of fluorescent probes, 59
Annulenes, absorption in the near-infrared region, 244
Annulenoannulenes, absorption in the near-infrared region, 244
ANO2 (antibody), binding to lipid haptens, 289–290, 297–298
ANS. *See* 1-Anilino-8-naphthalene sulfonate
Anthracene, two-photon excitation fluorescence studies on, 360
Anthraquinones:
  as NIR dyes, 246, 247
  photochemical fluorometry of, 111, 118
Anthroyloxy stearic acids, fluorescent probe studies on, 29
Antibiotics, fluorometric detection of, 315
Antibodies:
  F_c receptor binding to, 290–292
  lipid hapten binding of, 260, 288–290, 296–298
Antihistamines, fluorometric detection of, 315
Antihole structure, in spectral hole-burning spectrum, 177, 182
Antimalarial compounds, photochemical fluorometry of, 119
Argon-ion laser:
  frequency doubling of, 345, 351
  monochromatic lines of, 334
  noise from, 352
  stability of, 337
  use in flow cytometry, 405, 426
  use in fluorescence spectroscopy, 331, 341, 357, 362, 364
Aryloxalates, structures and chemiluminescent reactions of, 15
Aspirin tablets, acetylsalicylic acid determination in, 314
Auramine:
  as molecular rotor, 48
  structure of, 49
Azo dyes, as NIR dyes, 246, 247

Bacteria, flow cytometric studies on, 399

Barban, photochemical fluorometry of, 127
Barbiturates, fluorometric detection of, 315
Barium fluorochlorobromide, samarium-doped, use in spectral hole-burning studies, 194–197, 213, 214, 216
Benfluralin, photochemical fluorometry of, 127
Benzo[*a*]anthracene, fluorescence line-narrowing spectroscopy of, 396
Benzocoumarin, as intermediate in chemiluminescent reaction, 16, 17
1,4-Benzodiazepine, photochemical fluorometry of, 126
Benzofluoranthrenes, fluorescence line-narrowing spectroscopy of, 395, 396
Benzo[*g*]phthalazine-1,4-dione, chemiluminescent reaction of, 6
Benzoic acid:
  phototautomerization of, 184
  thioindigo-doped, use in spectral hole-burning, 183–186
Benzo[*g,h,i*]perylene, Shpol'skii spectroscopy of, 373
Benzo[*a*]pyrene:
  as carcinogen, 390
  metabolites of, in fish bile, 380–382
  Shpol'skii spectroscopy of, 373, 380–382
Benzo[*e*]pyrene, fluorescence line-narrowing spectroscopy of, 396
Benzo[*a*]pyrene-7,8-diol-9,10 epoxide, DNA adduct of, fluorescence line-narrowing spectroscopy of, 39, 393
Benzoxazinone derivative, use in flow cytometry, 33
Bichromophoric molecules, formation of, 54, 55
Bile salts:
  in luminescence analysis, 133–148
    advantages, 137–138
    of complex salts, 140–141
    energy transfer between probes in, 139–140
    fluorescent probes, 142–144
    metal cation enhancement of, 137–138
    mixed micellar phase formation by, 145
    structure and aggregation of, 135–136
    in trope formation, 145–146
    use in enantiomer separation, 144–145
Bilirubin, fiber-optic sensor detection of, 316
Binding constant(s):

of antibody-F_c receptor, 292
of antibody-lipid haptens, 288–289
derived from total internal reflection fluorescence microscopy, 300–301
Biological materials, chemiluminescent trace analysis of, 2
Biological membranes, fluorescent probe studies on, 27
Biological reductants, chemiluminescent determination of, 11
Bioluminescence:
 definition of, 3
 quantum yield of, 16
Biosensors, preparation of, 283
Biota, polynuclear aromatic hydrocarbon detection in, 377–380
Biotin, immunoassay using chemiluminescence, 7
Bivariate plots, in flow cytometric data analysis, 411
Blood:
 ammonia detection in, 311–312
 detection using luminol, 3
 lead detection in, 310–311
Blood serum, low-temperature luminescence spectra of, 318
Borate anion, fluorometric detection of, 312
Born–Oppenheimer approximation, 327
Bovine serum albumin:
 binding on glass, fluorescence measurements of, 260
 conformation studies using TIR/RET, 277
Boxcar integrator, use in low-temperature spectroscopy, 371
Bromomethylmethoxycoumarine, as fluorescent probe, 331
Brownian motion:
 of antibody bound to lipid hapten on membranes, 297
 effects on fluorescence recovery after photobleaching, 261
Butylmalonic acid mono-(N,N'-diphenyl)hydrazide, photochemical fluorometry of, 120
n-Butyl-2-naphthylmethyldithiocarbamate complexes, fluorometric detection of, 109–110, 114, 115
Butyltartronic acid mono-(N,N'-diphenyl)hydrazide, photochemical fluorometry of, 120

Cadmium, fluorometric detection of, 310
Calcein chelates, use in calcium detection, 310
Calcium, fluorometric detection of, 309
Calcium fluoride, holmium-doped, use in spectral hole-burning, 170–172
Calcium ion, photoprotein reaction with, 3
Calcium sulfate:samarium, spectral hole-burning studies on, 215
Calcofluor white, as cellulose stain, 426
Capillary electrophoresis:
 background signal origin in, 350
 detection limits in, 345, 348, 356, 364
 DNA sequencing by, 433
 laser excitation sources in, 332
 laser-excited molecular fluorescence use with, 353, 354, 355, 356, 363–365, 366
 sheath-flow cell use in, 345, 350
N-Caproylhydrazobenzene, photochemical fluorometry of, 120
Carbocyanines, use in derivatization, 366
Carbon dioxide, fiber-optic sensor detection of, 316
Carbon dioxide laser, use in spectral hole-burning, 208
Carboxylic acids, high-performance liquid chromatography of, 8
Cardiac glycosides, photochemical fluorometry of, 111, 122, 123–124
Catecholamine-O-sulfates, photochemical fluorometry of, 109, 125–126
Cd:Zn:Se diode lasers, properties of, 344
Cell cycle analysis, of phytoplankton, 430–432
Cells, fluorescent probe studies on, 28
Cell sorters, flow cytometers as, 410–411
Cerium fluoride, curium-doped, use in spectral hole-burning, 168
Chain melting transition, of lipids on supported planar membranes, 293–295
Chelates, of metals, fluorescence of, 309
Chemical light, peroxyoxalate chemiluminescent system as, 16
Chemically initiated electron exchange luminescence. *See* CIEEL mechanism
Chemiluminescence, 1–23
 definition and types of, 2, 3
 detection of, 353
 history of, 3–4
 intensity of, 10
 phenomenon of, 2, 18
 quantum yields of, 4, 5, 11–12, 16

Chemiluminescent compounds, 3, 4
  reaction mechanisms and applications of, 4–18
Chemiluminescent systems, in trace analysis, 2
Chiral selectivity, use with luminescence analysis, 144–146
Chlorbromuron, photochemical fluorometry of, 127
Chloride, fluorometric detection of, 312
Chlorin:
  adsorption on porous silica surfaces, spectral hole-burning studies on, 212
  n-octane Shpol'skii matrix doped with, use in spectral hole-burning, 170–172
  photoproduct of, 171, 172
Chlorin-poly(vinyl butyral) film, spectral hole-burning in, 159
9-Chlorocarbonyl-10-methylacridinium chloride, structure and chemiluminescent reaction of, 11, 12
Chloromethane, in spectral hole-burning studies, 215
Chlorophenols, photochemical fluorometry of, 109–110, 126–127
Chlorophyll(s):
  as fluorophore, 330
  spectral hole-burning studies on, 210–211
Chlorophytes, optical anlysis of, 423
Chloropropham, photochemical fluorometry of, 127
Chloropyrenes, fluorescence line-narrowing spectroscopy of, 394
Chloroquine, photochemical fluorometry of, 119, 120, 126
Chlorpromazine, photochemical fluorometry of, 118, 120, 125
Cholesterol:
  chemiluminescent determination of, 16
  fluorometric detection of, 314
Cholic acid, three-dimensional model of, 137
Choroiogonadotropin, fluoroimmunoassay of, 317
Chromosomes, high-resolution slit-scan analysis of, 402
Chrysene:
  fluorescence line-narrowing spectroscopy of, 396
  two-photon excitation fluorescence studies on, 360
CIEEL mechanism, 4, 18

Circular dichroism, fluorescence spectroscopy of, 144–145
Clinical applications, of luminescence spectroscopy, 307–341
Clobazam, photochemical fluorometry of, 119, 122
Clomiphene, photochemical fluorometry of, 105, 119, 122, 124
Cluster analysis, application in flow cytometry, 414
Coal liquids, luminescence analysis of, 140–141
Cobalt ion:
  chemiluminescent determination of, 10
  luminescence spectroscopy of, 311
Coefficient of variance, in calibration in flow cytometry, 416–418
Coenzymes, fluorometric detection of, 314
Color center, in diamond, spectral hole-burning studies on, 177–180
Confocal scanning laser microscopy:
  in single-cell analysis, 433–434
  of toxic phytoplankton, 428
Conformer conversion mechanism, in spectral hole-burning, 208
Copper(II), fluorometric detection of, 311
Copper vapor laser, properties of, 338
Cresyl violet, spectral hole-burning studies on materials containing, 173, 175–176, 205–207
Cresyl violet perchlorate, adsorption on porous silica surfaces, spectral hole-burning studies on, 212
Critical micelle concentration, of bile salts, metal cation effects on, 138–139
Cross talk, minimization in spectral hole-burning holography, 217
Crude oils, luminescence analysis of, 140
Crystal violet:
  as molecular rotor, 48–49
  structure of, 49
Curium, cerium fluoride doped with, use in spectral hole-burning, 168
Cyanines:
  as molecular rotors, 49
  as NIR dyes, 246
Cyanobacteria:
  flow cytometric studies on, 401, 426
  optical anlysis of, 423
1-Cyano[f]isoindoles, use in derivatization, 362

Cyclic hydrazides, chemiluminescence of, 6
Cyclodextrins, fluoroimmunoassays using, 317, 318, 319
Cytochrome b_5, incorporation in supported membranes, binding studies on, 298–299

DANCA:
  as polarity probe, 41
  structure of, 42
Dansyl chloride, as fluorescent probe, 331
DAPI, as phytoplankton stain, 432
DDTC:
  structure of, 232
  use in semiconductor laser luminometry, 230–232, 234
n-Decane, porphyrin-doped, use in spectral hole-burning studies, 207
Decomposition, chemiluminescent, 9
Defect sites, spectral hole-burning studies on, 165–176
Demoxepam:
  photochemical fluorometry of, 119, 122
  photocyclization reaction of, 105–106
Density-driven computation technique, application in flow cytometry, 415
Dephasing rate, in spectral hole-burning studies, 212
Desenkephalin-γ-endorphin, laser-excited molecular fluorescence of, 361
Desmethylclobazam, photochemical fluorometry of, 119
Dexter's theory, 66
Diamond, nitrogen-vacancy color center in, spectral hole-burning studies on, 177–180
Diarrhetic shellfish poisoning, from toxic algae, 427
Diatoms, flow cytometric studies on, 401, 424, 426
Diazoquinone, as intermediate in chemiluminescent reaction, 5–6
Dibromoquinolinol, zinc fluorescence with, 310
Dichloropyrenes, fluorescence line-narrowing spectroscopy of, 394
3,3'-Diethyl-2,2'-(4,5,4',5'-dibenzo)thiatricarbocyanine iodide. *See* DDTC
Diethylstilbestrol, photochemical fluorometry of, 105, 122, 123
Diffraction efficiency, in spectral hole-burning, 160
Diffusion, translational and rotational, of lipids, 293–296
1,2-Difluoroethane, krypton-doped, use in spectral hole-burning studies, 207–209
bis(2,6-Difluorophenyl)-oxalate, reactive intermediate of, 16
Diginatin, photochemical fluorometry of, 122, 123
Digoxin, photochemical fluorometry of, 122, 123
3,4-Dihydroxyphenylalanine, photochemical fluorometry of, 120
p-Dimethylaminobenzoate, as molecular rotor, 49
p-Dimethylaminobenzonitrile, as molecular rotor, 49, 51
p-N,N'-Dimethylaminobenzylidenemalononitrile, as molecular rotor, 49, 50
2-p-Dimethylaminolphine, structure and chemiluminescent reaction of, 10
10-(3-Dimethylamino-2-methylpropyl)phenothiazine. *See* Nedaltran
4,2'-(Dimethylamino)-6'-naphthoylcyclohexanecarboxylic acid. *See* DANCA
N,N'-Dimethyl-9,9'-biacridinium dinitrate. *See* Lucigenin
Dimethyl-s-tetrazine, tetramethylbenzene doped with, use in spectral hole-burning, 186–187
N,N-Dimethyltryptamine, fluorometric detection of, 319
Dinitroaniline derivatives, of herbicides, photochemical fluorometry of, 111, 127, 128
Dinoflagellates, flow cytometric detection of, 421, 426–429
Dinophysis spp., occurrence of, 427
Dioctadecylcarbocyanine, in studies of orientation distribution, 299
Diode array detector, use in low-temperature spectroscopy, 371
Diode laser luminescence, NIR dyes for, 248
Diode lasers, 325, 331, 333, 345
  fluorescence detection using, 354, 365–366, 367
  semiconductor type, 342–344
  stability of, 337
  use in liquid chromatography combined

Diode lasers *(Continued)*
    with laser-induced fluorescence, 365
    use in spectral hole-burning, 208
1,2-Dioxetane(s):
    chemiluminescent reaction of, 4, 8
    derivatives of, 9, 11, 12
1,2-Dioxetanediones, as intermediates in chemiluminescent reaction, 4, 15–16
Diphenoylperoxide:
    chemiluminescence of, 4, 16–18
    structure of, 17
Diphenylamine:
    photochemical fluorometry of, 114, 115, 116
    photocyclization reaction of, 107–108
Diphenylanthracene, in chemiluminescent reactions, 4, 16–17
Diphenyl-1,3,5-hexatriene:
    as fluorescent probe, 30, 31
    structure of, 31
4,4′-Diphenylstilbene, two-photon-excited fluorescence detection of, 360, 361
Dipole moment difference:
    in spectral hole-burning, 160
    spectral hole-burning studies on, 173, 175
Dipole moments, of fluorophores, 270
Dipyrenylalkanes, as fluorescent probes, 54, 56
Discriminant function analysis, application in flow cytometry, 411–412
Distance:
    calculation in electron energy transfer, 69–70, 71
    distribution in electron energy transfer, 70–71
Diuron, photochemical fluorometry of, 127
DMAL. *See* 2-*p*-Dimethylaminolphine
DNA:
    benzo[*a*]pyrene metabolite binding to, 380–381
    in cell cycle analysis of phytoplankton, 430–432
    cell distribution of, 411
    flow cytometric analysis of, 419, 421, 433
    fluorometric detection of, 315, 330
    laser-excited molecular fluorescence-capillary electrophoresis of, 354, 365
    polynuclear aromatic hydrocarbon adducts to, fluorescence line-narrowing spectroscopy of, 390–392
    sequencing studies on, 365, 433, 434
DNA hybridization assays, using chemiluminescence, 7
DNA oligonucleotide, photophysical hole-burning in, 211
Doolittle equation, 48, 55
DL-Dopa, photochemical fluorometry of, 118
Dopamine, photochemical fluorometry of, 118, 125
Dot-plots, in flow cytometric data analysis, 411
Doxorubicin:
    as fluorophore, 330, 357–358
    structure of, 357
DPH. *See* Diphenyl-1,3,5-hexatriene
Dunaliella, flow cytometric detection of, 417, 418, 424, 431, 432
Duty factors, for pulsed laser systems, 336
Dye lasers:
    pumping of, 339, 360
    use in low-temperature spectroscopy, 370
    use in spectral hole-burning, 157
    wavelength tuning of, 333
Dyes, with absorption in the near-infrared region, 244–248
Dynamic reserve, of detection systems, 355

Einstein–Simoluchowski expression, 52
Electric dipole moments, spectral hole-burning studies on, 169–172
Electronic energy transfer:
    determination of, 67–69
    mechanisms of, 64–67
    spectroscopic ruler for, 63–71
Electronic interactions, spectral hole-burning studies on, 176–183
Emission dipoles, in polarized fluorescence microscopy, 269
Emission spectra, of fluorescent probes, 32
Enantiomers, bile salts in separation of, 144–145
β-Endorphin, fluorescence detection of, 362
Energy, absorption in luminescence, 2–3
Energy transfer, application in fluoroimmunoassays, 317
Environmental samples:
    chemiluminescent trace analysis of, 2
    photochemical fluorometry of, 126–128
Enzyme-linked immunosorbent assays, of membrane receptors, 300
Enzymes, near-infrared semiconductor laser

fluorometric assays of, 241–244
Epi-fluorescence microscopy, antibody visualization by, 297
Epinephrine sulfates, photochemical fluorometry of, 125
Equilibrium binding constants, of cell membrane receptors, 301
Erbium, yttrium lithium fluoride doped with, use in spectral hole-burning, 182–183
Erythrocyte ghosts, in studies of orientation distribution, 299
Erythrosin derivatives, as fluorescent probes, 30
Ethanol–cresyl violet glass, in spectral hole-burning studies, 202–204
Ethanol:methanol glass, quinizarin-doped, in spectral hole-burning studies, 201–202
Ethers, photochemical fluorometry of, 111, 114
Europium:
 use in fluoroimmunoassays, 317
 yttrium aluminum perovskite doped with, use in spectral hole-burning, 180–182
Evanescent field, spatial extension of, 2897
Evanescent waves, 257, 258–259, 300
Exchange mechanism theory, spectral hole-burning studies and, 190, 192
Excimer formation:
 fluidity determination by, 45, 51–54
 intramolecular, 54–56
 kinetic scheme for, 51
Excimer–laser combination, use in two-photon-excited fluorescence studies, 360
Excimer Xe:Cl laser, properties of, 338
Excitation, of fluorescent probes, 32
Excitation-emission matrices:
 of fluorescence line-narrowing spectroscopy, 386
 in luminescence analysis, 140
Excitation sources, lasers as, 332–337
Excitation spectroscopy, in spectral hole detection, 155–163
Excitation wavelength dependence, of fluorescence line-narrowing spectroscopy, 386–389
Excited state lifetime, of fluorescent probes, 33–35

FAST. See Fluorescence anisotropy selectivity technique
F_c receptor:
 binding to antibody, 290–292
 incorporation into supported planar membranes, in lipid diffusion studies, 294
 from macrophages, binding studies on, 298–299
Fenbendazole, photochemical fluorometry of, 122, 124
Fenergan, photochemical fluorometry of, 119, 122, 123, 125
α-Fetoprotein, chemiluminescent assay of, 12
Fiber-optic bundles, use with laser-induced fluorescence, 359
Fiber-optic sensors, use in luminescence spectroscopy, 315, 316
Fick's second law, in derivation of lateral diffusion, 263
Filter paper, in photochemical fluorescence studies, 93
Flicker noise, in laser-excited molecular fluorescence, 349, 352
Flip-flop processes, in spectral hole-burning studies, 177
Flow cells, for near-infrared semiconductor laser fluorometry, 233–234
Flow cytometry, 397–433
 advantages of, 432–433
 applications of, 419–432
 axial light loss detection in, 407
 calibration aspects of, 416–419
 as cell sorters, 410–411
 classification of particles by, 419
 data acquisition in, 410
 data analysis in, 411–415
 detection in, 407–408
 electronics of, 408–410
 flow chamber and hydrodynamics of, 403–404
 forward light scatter measurements in, 400, 405
 functional property determination by, 419
 instrumentation for, 402–411
 laser excitation sources in, 332, 404–406
 perpendicular light scatter in, 401, 407
 principles of, 399–402
 time-of-flight measurement in, 400, 402, 404, 405, 408, 417
 using fluorescein-labeled ligands and antibodies, 33

Flow injection analysis:
  chemiluminescence use in, 8
  photochemical fluorometry use with, 86, 90, 103–104, 128
Flow rate, in fluorescence correlation spectroscopy, 268
Fluctuation spectroscopy, applications of, 269
Fluidity, fluorescence techniques for, 44–63
Fluometuron, photochemical fluorometry of, 127
Fluorenylium dye ethynologues, as NIR dyes, 246, 247
Fluorescein:
  derivatives of, as fluorescent probes, 30, 33
  flow cytometric detection of, 416
  as fluorescent probe, 331, 353, 416
Fluorescein diacetate, use in phytoplankton viability test, 430
Fluorescein isothiocyanate, as fluorescent probe, 353, 354, 362, 364, 402, 428
Fluorescein thiohydantoin, as amino acid label, 364
Fluorescence, definition of, 3
Fluorescence-activated cell sorter machines, use in flow cytometry, 398
Fluorescence anisotropy, derivation of, 269
Fluorescence anisotropy selectivity technique, use in fluoroimmunoassays, 317
Fluorescence correlation spectroscopy, 267–269
  combined with total internal reflection fluorescence microscopy, 275–276
  early development of, 255
  scanning type, 269
Fluorescence depletion anisotropy, rotational diffusion coefficient measurement by, 271
Fluorescence depolarization measurements, use in biochemical macromolecule studies, 342
Fluorescence-detected circular dichroism spectroscopy, 145
Fluorescence intensity enhancements, in drug detection, 319
Fluorescence lifetime, 327, 329–330, 368
  fiber optics use in selectivity of, 316
  polymethine dye studies on, 235
Fluorescence line-narrowing spectroscopy
  advantages of, 397

  analytical applications of, 389–397
  detection method in, 371
  excitation wavelength dependence of, 386–389
  hole-burning effects in, 386, 389, 394
  inhomogeneous bands in, 384
  laser excitation sources in, 332, 368, 369, 370
  matrix materials for, 389
  multiplets in, 384
  phonon sidebands in, 385
  phosphorescence spectrum in, 384
  photodiode array detection in, 386
  principles of, 383–386, 405
  in spectral hole-burning, 200
  thin-layer chromatography combined with, 359, 392–396
Fluorescence microphotolysis. See Fluorescence recovery after photobleaching (FRAP)
Fluorescence optrodes, fiber optics use in, 316
Fluorescence photobleaching recovery. See Fluorescence recovery after photobleaching (FRAP)
Fluorescence polarization:
  fluidity determination by, 45, 56–63
  measurements using, 330
  principles of, 58
Fluorescence quenching, fluidity determination by, 45, 51–54
Fluorescence recovery after photobleaching (FRAP):
  combined with total internal reflection fluorescence microscopy, 275–276
  fringe pattern photobleaching, 263–264, 264–267
  of membranes, 255
  principles of, 261–267
  spot photobleaching, 261–263
  stripe pattern photobleaching, 263–264
Fluorescence spectroscopy, 133–134, 326–331
Fluorescent probes, 25–84
  absorption of, 32
  analytical applications of, 26, 27, 330
  anisotropic rotations of, 59
  bile salt microenvironments for, 142–144
  characteristics of, 28
  emission spectra of, 32
  environmental effects on, 26
  errors in use of, 72
  examples of, 330–331

INDEX 453

excitation of, 32
excited state lifetime of, 33–35
fluidity determination by, 44–63
free volume effects of, 46, 50
order parameters of, 62, 63
photochemical stability of, 36–37
quantum yields of, 35–36
Stokes shift of, 32–33, 40–41
Fluoride, fluorometric detection of, 312
Fluoroimmunoassays, 28, 34, 315, 316–317
energy transfer applications of, 317
Fluorometers, 103
Fluphenazine, photochemical fluorometry of, 112, 118, 120
Förster's mechanism, 65, 72
Forward light scatter measurements. See under Flow cytometry
Franck–Condon principle, 39, 326, 327, 335
Free volume effects, of fluorescent probes, 46, 50
Freeze-pump-thaw techniques, in low-temperature spectroscopy, 371
Fringe fluorescence recovery after photobleaching, with total internal reflection fluorescence microscopy, 276
Fringe pattern photobleaching, principles of, 264–267
Functional dyes, NIR dyes as, 248

GaAlAs laser, properties of, 342–344
β-D-Galactosidase, chemiluminescent immunoassay of, 9
Gallium arsenic photocathode, use in semiconductor laser luminometry, 230, 231
Gas chromatography, of polynuclear aromatic hydrocarbons, 375
Gas-ion lasers, properties of, 337–339
Gating ratio:
 definition of, 212
 increase in, 215
Gierer–Wirtz microviscosity theory, 50, 53
GK14-1 (antibody), binding to lipid haptens, 288–289, 290
Glasses, spectral diffusion in, 200–205
γ-Globulin, semiconductor laser fluorometric detection of, 240
Glucose:
 chemiluminescent determination of, 6, 16
 fiber-optic sensor detection of, 316

fluorometric detection of, 313–314, 318
Glucose oxidase, fluorometric detection of, 318
Glutamic dehydrogenase, use in ammonia detection, 311–312
Glycerol:ethanol glass, porphyrin-doped, in spectral hole-burning studies, 197–199
Granulocytes, flow cytometric studies on, 401
Gratings, in spectral hole-burning, 157
Groundwater, pesticide detection in, 128
g-values, for Zeeman interactions in spectral hole-burning, 167
Gymnodinium spp., occurrence of, 427
Gyrodinium cf. aureolum, occurrence of, 427

Halide anions, fluorometric detection of, 312
Hallucinogenic drugs, fluorometric detection of, 319
Haptens, antibody interaction with, 296–298
He:Cd lasers:
 noise from, 352
 properties of, 339
 stability of, 337
 use in flow cytometry, 405, 426
 use in liquid chromatography-laser-induced molecular fluorescence, 351, 353, 357, 362
Heisenberg uncertainty principle, 341, 383
Hematofluorometer, use in lead detection, 311
Hemoglobin, flow cytometric studies on, 401
He:Ne lasers, 325
 in flow cytometry, 405
 noise from, 352
 stability of, 337
 use in liquid chromatography-laser-induced molecular fluorescence, 351, 365
Heparin, chemiluminescent determination of, 11
Herbicides, photochemical fluorometry of, 101, 111
Hexahydrohexahelicene, Shpol'skii spectroscopy of, 372
High-performance liquid chromatography (HPLC):
 bile salt–alcohol phases for, 146
 chemiluminescence used with, 7–8, 11, 16
 fluorescence detection with, 361
 photochemical fluorometry use with, 86, 90, 94–103, 105, 111, 117, 118, 119, 123, 125, 126, 128

HPLC (Continued)
  use with near-infrared semiconductor laser fluorometry, 234, 236–241, 244
Histocompatibility antigens:
  incorporation into supported planar membranes, in lipid diffusion studies, 294, 298–299
  Texas red-labeled, support membrane studies on, 300
Hole frequency shift, in spectral hole-burning, 192
Holmium, calcium fluoride doped with, use in spectral hole-burning, 170–172
Holography, use in spectral hole-burning, 157–163, 172, 217–221
Hydrocortisone, photochemical fluorometry of, 111, 122, 123, 124
Hydrogen peroxide:
  chemiluminescent analysis of, 6, 16
  indocyanine green complex with, 242
Hydroperoxyoxalate ester, as reactive chemiluminescent reaction intermediate, 16
9-Hydroxybenzo[a]pyrene, Shpol'skii spectroscopy of, 382
8-Hydroxyquinoline, calcium chelate of, fluorescence, 309–310
8-Hydroxyquinoline-5-sulfonic acid, magnesium chelate of, fluorescence, 310
Hydroxysquarylium, as NIR dye, 246
4-Hydroxytamoxifen, photocyclization reaction of, 105, 106
5-Hydroxytryptamine, photochemical fluorometry of, 118, 120
DL-Hydroxytryptophan, photochemical fluorometry of, 118, 120
3-Hydroxytyramine, photochemical fluorometry of, 120
Hyperfine structure, in spectral hole-burning studies, 180

Ibogaine, fluorometric detection of, 319
Image storage, by spectral hole-burning holography, 217–219
Immunoassays, based on chemiluminescent reactions, 7, 11, 16
Immunochemical determination, of toxic phytoplankton, 427–429
Immunochemical labels, in dyes for fluorescent splectroscopy, 401–402
Immunoglobulin E, bound on supported membranes, hapten binding to, 296
Immunoglobulin G:
  bound on supported membranes, hapten binding to, 296, 298
  total internal reflection fluorescence microscopy studies on, 288
Impurity sites, spectral hole-burning studies on, 165–176, 205–206
Indicators, chemiluminescent, 10
Indocyanine green:
  structure of, 239
  use as label for NIH laser fluorometry, 239–241, 242
Indole, chemiluminescence of, 8, 15
Indole derivatives:
  chemiluminescence of, 13–14
  photochemical fluorometry of, 115
Indophenol–metal complexes, as NIR dyes, 246, 247
Indophenols, as NIR dyes, 246, 247
Infrared spectral hole-burning, 209
Infrared spectroscopy, polarized attenuated total reflection type, in studies of lipids in supported bilayers, 294
Infrared studies, germanium plates for, 283
In:Gas:Al:P diode lasers, properties of, 343
In:Gas:As diode lasers, properties of, 343
In:Gas:As:P diode lasers, properties of, 343
Inorganic compounds:
  chemiluminescent trace analysis of, 2
  fluorometric detection of, 308–312
Integral membrane proteins, incorporation into supported membranes, 298–299
Internal conversion, in fluorescence spectroscopy, 328–329
Intersystem crossing, in fluorescence spectroscopy, 328
Iodide, fluorometric detection of, 312
Ion chromatography, indirect fluorometry use in, 354
Iron(III), fluorometric detection of, 311
Irradiance, of lasers, 345, 348
Isoluminol, derivatives of, 7
Isopropalin, photochemical fluorometry of, 127
Isothiocyanate derivative, of isoluminol, amino acid derivatives of, 7

Jablonski diagram, 326

Julolidinebenzylidenemalononitrile, as molecular rotor, 49

Karbutilate, photochemical fluorometry of, 127
Key-and-hole rule, in Shpol'skii spectroscopy, 373
Krypton, 1,2-difluoroethane-doped, use in spectral hole-burning studies, 207–209
KTP crystal, laser frequency doubling by, 341

Lactoferrin, fluoroimmunoassay of, 317
Laminin, sulfatide binding of, 292–293
Langmuir–Blodgett films:
　spectral hole-burning studies on, 212
　supported bilayer preparation from, 283–284, 286, 293, 294
Lanthanide probes, use in fluoroimmunoassays, 34
Largactil, photochemical fluorometry of, 119, 122, 123
Laser-excited molecular fluorescence, 323–443
　of analytes require chemical derivatization, 361–363
　background signal origin in, 350
　capillary electrophoresis use with. *See* under Capillary electrophoresis, 355
　chemical aspects of, 353–354
　derivatization for, 352, 353, 359, 361–363
　detector cell volumes in, 345–347
　flow cytometry method in, 397–443
　fluorescence line-narrowing spectroscopy in, 332, 368, 383–396
　fluorescence spectroscopy, 326–331
　frequency doubling of lasers for, 358
　future trends in, 433–434
　indirect fluorescence detection in, 354–357
　inhomogeneous broadening of spectral bands in, 358
　instrumental aspects of, 345–352
　liquid chromatography use with. *See* under Liquid chromatography
　low-temperature spectroscopy, 367–383
　recent developments in, 354–366
　Shpol'skii spectroscopy in. *See* Shpol'skii spectroscopy
　signal-to-noise ratios in, 352
　thin-layer chromatograph combined with, 359
　use in separation techniques, 344–367

Laser-excited Shpol'skii spectroscopy, 370, 372–382
Laser fluorometry, of organic compounds, 230, 236
Laser luminometry, of ultratrace species, 229
Lasers:
　characteristics of, 333–337
　continuous or pulsed radiation from, 335
　directionality of, 332, 334–335
　divergence in, 334
　as excitation sources, 332–337
　for flow cytometry, 404–406
　frequency doubling of, 358
　instrumentation for, 337–344
　irradiance in, 334
　for microspectrofluorometry, 279–280
　monochromaticity of, 332, 334, 359, 367
　saturation effects of, 348–349, 352
　semiconductor diode type, 342–344
　signal-to-noise ratio of, 333
　solid-state type, 339–340
　stability of, 353
　stabilizers for, 337
　ultrafast systems of, 340–342
　use in single molecule detection, 434
　wavelength of, 333–334
Lateral diffusion coefficients, determination by fluorescence recovery after photobleaching, 261–262, 264, 265, 276
Lateral diffusion studies:
　of antibodies on supported planar membranes, 296–298
　of cytochrome b_5, 299
　of lipids on supported planar membranes, 293–296, 298
Lead, detection in blood, 310–311
Level anticrossing, spectral hole-burning studies on, 177–180
Levopromazine, photochemical fluorometry of, 119, 122, 123
Ligand-receptor interaction as, microspectrofluorometric studies on, 300
Linear diode array detectors, use with laser-induced fluorescence, 359
Linuron, photochemical fluorometry of, 127
Lipid bilayers, wobble-in-cone model for, 60–61
Lipids:
　antibody binding to, microspectrofluorometric studies on, 288–290
　order parameters of, from polarized at-

Lipids *(Continued)*
  tenuated total reflection infrared spectroscopy, 294
  rotational diffusion coefficients of, by polarized fluorescence microscopy, 271
Lipid vesicles, fusion to hydrophilic surfaces, 285–286
Lipoperoxide assay, using chemiluminescence, 7
α_1-Lipoprotein, semiconductor laser fluorometric detection of, 240
Liquid chromatography:
  background signal origin in, 350
  detection limits in, 347–352, 355
  flow cell volumes for, 346
  laser-induced fluorescence detection use with, 344, 345, 353, 355, 362, 366
  thin-layer chromatography combined with, 394–395, 396
Liquid solutions, photochemical fluorescence studies on, 90–92, 119
Local modes, in spectral hole-burning, 194, 199, 207, 208
Loperamid, photochemical fluorometry of, 126
Lophine:
  as chemiluminescent indicator, 10
  chemoluminescence of, 2, 8, 9–10, 15
  structure of, 10
Lorazepam, photochemical fluorometry of, 108–109, 119, 120
Lorentzian hole profile, in spectral hole-burning, 155
Low-temperature luminescence, applications of, 315, 317–318
Low-temperature spectroscopy:
  detection methods for, 371
  excitation sources for, 369–370
  instrumentation for, 369–370
  laser use in, 367–383, 434
  sample preparation in, 370–371
Lucigenin:
  as analytical reagent, 10, 11
  chemiluminescence of, 3, 10–11
  structure of, 11
Luminescence:
  low-temperature type, 315, 317–318
  specialized types of, 315–319
Luminescence analysis:
  biles salts in, 133–148
  organized media in, 134–135
Luminescence spectroscopy:
  clinical applications of, 307–341
  of inorganic substances, 308–312
  in the near-infrared region, 229–251
Luminol:
  chemiluminescence of, 3, 5–8
  analytical uses of, 7–8
  as chemiluminescent indicator, 10
  derivatives of, 5–8
  structure of, 6
Lymphocytes:
  flow cytometric studies on, 401
  supported planar membranes in studies of, 255–256
Lyotropes, formation by bile salts, 145–146

Macromolecules, binding to supported planar membranes, 286–293
Macrophages, F_c receptor from, binding studies on, 298–299
Magnesium, fluorometric detection of, 309
Magnesium porphyrin, *n*-octane doped with, use in spectral hole-burning studies, 192–194
Magnetic resonance, optical detection of, 170, 180
Matrix effects, fluorescence susceptibility to, 134
Matrix isolation, in molecular spectrometry, 318
Mefloquine, photochemical fluorometry of, 119, 120
Membrane proteins, incorporation into supported membranes, 298–299
Membrane receptors:
  lateral mobility of, 301
  microspectrofluorometric studies on, 300
Membranes:
  integral proteins of, microspectrofluorometry on supported membranes, 298–299
  microspectrofluorometry of, 254
Merocyanines, as NIR dyes, 246
Mesoridazine, photochemical fluorometry of, 122
Metal ions:
  chemiluminescent determination of, 10
  fiber-optic sensor detection of
  fluorometric detection of, 309–311

INDEX 457

luminescence enhancement by, 138–139
Metals, effect on ligand fluorescence, 309
Methamphetamine, high-performance liquid chromatography of, 8
Methoxybenzo[a]pyrene compounds, Shpol'skii spectroscopy of, 381–382
N-Methylacridine, excitation of, 11
N-Methylacridone, chemiluminescence and structure of, 11, 12
5-Methylchrysene, Shpol'skii spectroscopy of, 375
N-Methyl-9-(dicarboalkoxymethyl)-acridans, chemiluminescent reactions and structure of, 12
1-(1-Methyl-3-indolyl)-6-phenyl-2,5,7,8-tetraoxabicyclo[4.2.0]-octane, structure and chemiluminescence of, 14
2-Methyl-3-phytyl-1,4-naphthoquinone. *See* Vitamin K1
bis(Methylstyrylbenzene), two-photon excitation fluorescence studies on, 360
Mianserin, photochemical fluorometry of, 126
Micelles, effects on fluorescence characteristics, 134
Microcystis aeruginosa, flow cytometric detection of, 409, 424
Micro-liquid chromatography:
  detection levels in, 348
  laser excitation sources in, 332, 345
Microscope, for microspectrofluorometry, 277–279
Microscopic fluorescence experiments, resonance energy transfer use in, 275
Microspectrofluorometry:
  combined techniques in, 275–277
  electronic control and data handling in, 281–282
  fluorescence correlation spectroscopy, 267–269
  fluorescence detection in, 281–282
  instrumentation for, 277–282
  laser light sources for, 279–280
  microscope for, 277–279
  polarized fluorescence microscopy, 269–273
  resonance energy transfer in, 273–275
  substrates for, 281–282, 282–283
  on supported planar membranes, 253–305
  techniques of, 255, 256–277
  total internal reflection fluorescence microscopy, 255, 256–277

vesicle fusion techniques in, 285–286
Microviscosity, determination of, 46
Mixed micellar phases, formation by bile salts, 145, 146
Mode-locked semiconductor laser, 236
Molecular computer, image storage by spectral hole-burning, 219–221
Molecular interactions, resonance energy transfer studies on, 299–300
Molecular rotors, fluidity determination by, 45, 48–51
Monalide, photochemical fluorometry of, 127
Monocytes, flow cytometric studies on, 401
Monuron, photochemical fluorometry of, 127
Multiphoton ionization, laser-induced, 333
Myoglobin, photophysical hole-burning in, 211

NADH:
  autofluorescent activity of, 401
  chemiluminescent determination of, 16
  use in fluorometric detection of ammonia, 312
NADPH, use in fluorometric detection of glucose, 314
Naphthalene-2,3-dialdehyde, use in derivatization, 353, 362
Naphthalic acid dichloride, perhydrolysis of, 18
Naphthoquinones:
  as NIR dyes, 246, 247
  photochemical fluorometry of, 118
Naphthalene, Shpol'skii spectroscopy of, 372
Nd:YAG laser:
  properties of, 338, 340, 342, 343
  use in flow cytometry, 405
  use in two-photon excitation studies, 360
Nd:YLF laser, properties of, 340, 341
Near-infrared luminescence spectroscopy, 229–251
  instrument diagram, 231
  pulsed, 235–236
Near-infrared semiconductor laser fluorometry, high-performance liquid chromatography use with, 234, 236–241
Neburon, photochemical fluorometry of, 127, 128
Nedaltran, photochemical fluorometry of, 119, 122, 123
Neon laser, properties of, 337–339

Neural network analysis:
  of algae, 426
  by spectral hole-burning techniques, 221
Neuropeptides, fluorescence detection of, 362
Nile red:
  poly(vinyl butyral) doped with, use in spectral hole-burning, 172–175
  use as impurity for spectral hole-burning studies, 159, 160
NIR dyes, for diode laser absorption spectroscopy, 244–248
Nitrogen laser:
  monochromatic lines of, 334
  properties of, 338
n-Octane Shpol'skii matrix, chlorin-doped, use in spectral hole-burning, 170–172
Noncylic hydrazides, as luminol derivatives, 6
Norepinephrine, photochemical fluorometry of, 118, 125
Norphenylephrine, photochemical fluorometry of, 120
Nuclear magnetic moments, in spectral hole-burning studies, 182, 183
Nuclear quadrupole moments, in spectral hole-burning studies, 181
Nucleic acids, fluorometric detection of, 28, 314
Nucleotides, capillary electrophoresis of, 356
Numerical aperture, in polarized fluorescence microscopy, 269

Octaethylporphyrin, polystyrene-doped, in spectral hole-burning studies, 199–200
n-Octane:
  magnesium porphyrin-doped, use in spectral hole-burning studies, 192–194
  porphyrin-doped, use in spectral hole-burning studies, 191–192
Oils, luminescence analysis of, 140
Oligomerization reactions, in membrane receptor binding, 301
One-photon-excited fluorescence detection, two-photon-excited fluorescence detection compared to, 360
Optical data storage, by spectral hole-burning, 215–217
Optical dephasing, spectral hole-burning studies on, 155, 189–200, 191, 192, 197, 199, 200, 207

Optical dephasing time, 151
Optical parametric oscillator, laser use with, 433
Optical Plankton Analyser (OPA), 403, 404, 432
  calibration of, 416, 417
  operation of, 409, 410
  optical layout of, 405–406
Optical pumping, spectral hole-burning from, 176, 177, 180
Optrodes, fluorescence type, 316
Order parameters, of fluorescent probes, 62, 63
Organic acids, high-performance liquid chromatography of, 11
Organic compounds:
  chemiluminescent trace analysis of, 2
  fluorometric detection of, 312–315
  semiconductor laser luminometry of, 229–251
Organized media, in luminescence analysis, 134–135
Orientation distributions:
  of fluorophores in membranes, 299
  measurement by polarized fluorescence microscopy, 269
  in polarized fluorescence microscopy, 270–271
Orientation selection, in spectral hole-burning, 176
Oryzalin, photochemical fluorometry of, 127
Oxalates, in chemiluminescent systems, 3–4, 15–16
Oxalyl chloride, in chemiluminescent system, 3
Oxalyl esters, in chemiluminescent systems, 3
Oxamides, in chemiluminescent systems, 3, 15–16
Oxazine 750, use in derivatization, 366
Oxazine-4-perchlorate, excitation–emission matrix for, 386, 387, 388
Oxonol dyes, as NIR dyes, 246
Oxygen:
  chemiluminescent analysis of, 6, 13
  fiber-optic sensor detection of, 316
  quenching of triplet states by, 272

Paralytic shellfish poisoning, from toxic algae, 427
Parinaric acids, as fluorescent probes, 31
Pentacene, Shpol'skii spectroscopy of, 376

Peptides:
  capillary electrophoresis of, 356
  fluorescence detection of, 361
Periodic pattern photobleaching, 263–264
Peroxyoxalate chemiluminescent reactions, 3–4, 16
  analytical use of, 16
Perphenazine, photochemical fluorometry of, 114, 118, 120, 125
Perrin's relationship, 59
Perylene:
  in ethanol glass, spectral hole-burning of, 153
  Shpol'skii spectroscopy of, 373
Pesticides, photochemical fluorometry of, 109–110, 126–128
Petrolatums, luminescence analysis of, 140
Phaeocystis pouchetii, flow cytometric detection of, 427
Pharmaceuticals:
  fluorometric detection of, 314–315
  low-temperature phosphorescence measurements of, 318
  photochemical fluorometric analysis of, 118–125
Phase-modulation fluorometry, 34, 35, 58
Phase transition, in bilayers in presence of antibody, 297
Phenmedipham, photochemical fluorometry of, 127
Phenol, two-photon excitation fluorescence studies on, 360
Phenothiazines, photochemical fluorometry of, 104, 112, 118, 119, 122, 123, 125
Phenylalanine, fluorometric detection of, 313
Phenylamides, photochemical fluorometry of, 127, 128
2-Phenyl-5-(4-biphenyl)-1 1,3,4-oxadiazole, two-photon excitation fluorescence studies on, 360, 361
Phenylbutazone, photochemical fluorometry of, 119, 120
Phenylcarbamates, photochemical fluorometry of, 127, 128
Phenylethylamines, fluorometric detection of, 315
9,10-bis(Phenylethynyl)anthracene, as fluorescent compound, 16
Phenylurea herbicides, photochemical fluorometry of, 101, 127, 128

Phonon coupling, in spectral hole-burning, 189
Phonon sideband, spectral hole-burning studies on, 205
Phospholipid bilayers:
  antibody binding studies on, 288–289, 292
  lipid binding studies on, 294–296
Phospholipid chains, melting phase transition of, 294–295
Phospholipids, fluorescent probe studies on, 29, 60
Phosphorescence, triplet state transition in, 327
Phosphorus-32, DNA nucleotide labeling with, 390, 392
Photocathode, red-sensitive, 237
Photochemical fluorescence enhancement factors, 114
Photochemical fluorometry, 85–131
  applications of, 114–128
  in clinical analysis, 125–126
  flow injection analysis use with, 86, 90, 103–104, 128
  high-performance liquid chromatography use with, 86, 90, 94–103, 105, 111, 117, 118, 119, 123, 125, 126, 128
  instrumentation for, 90–104
  for flow analysis studies, 103–104
  for high-performance liquid chromatography, 94–103
  for liquid solution studies, 90–92
  photochemical reactors, 95–102
  in pesticide analysis, 126–128
  in pharmaceutical analysis, 118–125
  photoisomerization reactions in, 108–109
  photolysis reactions in, 109–110
  photooxidation reactions in, 112, 114
  photoreactions in, 104–114
  photoreduction reactions in, 110–112
  theoretical intensity expression for, 88–90
  theory of, 87–90
  thin-layer chromatography use with, 86, 90, 92–93
Photochemical reactors, 95–102
Photochemical stability, of fluorescent probes, 36–37
Photocyclization reactions, in photochemical fluorometry, 105–108
Photodecomposition:
  from diode lasers, 365

Photodecomposition *(Continued)*
  in laser-excited molecular fluorescence, 335, 348, 349, 350, 352, 367
Photodiode detection, in Shpol'skii spectroscopy, 376
Photodissociation, in spectral hole-burning, 186
Photoisomerization reactions, in photochemical fluorometry, 108–109
Photoluminescence, definition of, 2–3
Photomultiplier tubes, use with laser-induced fluorescence, 359, 365, 371
Photon-counting apparatus, use in low-temperature spectroscopy, 371
Photon echo techniques, in spectral hole-burning, 200
Photon-gated spectral hole-burning:
  materials for, 212–215
  storage device, 216
Photooxidation reactions, in photochemical fluorometry, 112, 114
Photoreactions, in photochemical fluorometry, 104–114
Photoreduction fluorescence method, 98, 117–118, 128
Photoreduction reactions, in photochemical fluorometry, 110–112
Photoselection, in polarized fluorescence microscopy, 269–270
O-Phthalaldehyde, as fluorescent probe, 331, 353, 354, 361, 362
2-(3-Phthalhydrazidylazo)-1,3-diketones:
  chemiluminescence of, 6
  structures of, 7
Phthalocyanine, free base, in *n*-octane Shpol'skii matrix, spectral hole-burning of, 153
Phycocyanin, in cyanobacteria, 424, 426
B-Phycoerythrin, laser-excited molecular fluorescence of, 349
Phytoplankton:
  cell cycle analysis of, 430–432
  flow cytometric classification of, 422–429
  flow cytometric detection of, 398, 399, 401, 403, 404, 407, 408–409, 412–413, 414, 417, 421
  optical characteristics of, 423
  remote sensing of, 434
  toxic, immunochemical determination of, 427–429
  viability analysis of, 429–430
Plasmocid, photochemical fluorometry of, 119
Polarity:
  of binding sites, of fluorescent probes, 142
  definition of, 37–39
  estimation by fluorescent probes, 37–44
  estimation by vibronic band changes, 43–44
  estimation from spectral shifts, 39–43
Polarization spectroscopy, application in spectral-hole burning, 163
Polarized attenuated total reflection infrared spectroscopy:
  of lipids in supported bilayers, 294
Polarized fluorescence microscopy:
  fluorescence lifetimes in, 269
  orientation distributions in, 270–271
  principles and applications of, 269–273
  rotational diffusion coefficient measurement by, 269, 270–271
  in studies of orientation distribution, 299
Polarized fluorescence photobleaching recovery:
  rotational diffusion coefficient measurement by, 271, 272
  theory of, 272–273
Pollutants, environmental, photochemical fluorometry of, 127
trans-Poly(acetylene), absorption in the near-infrared region, 244
Polyenes, Shpol'skii spectroscopy of, 372
Poly(ethylene), porphyrin-doped, in spectral hole-burning studies, 197–199, 201
Polymers, fluorescent probe studies on, 27, 29, 54
Polymethine dyes:
  detection by diode laser-induced fluorescence combined with liquid chromatography, 365, 366
  as NIR dyes, 246, 248
  use in semiconductor laser luminometry, 230–231, 235, 236–238
Poly(methyl methacrylate), as co-dopant in spectral hole-burning studies, 215
Polynuclear aromatic hydrocarbons:
  adducts to DNA, fluorescence line-narrowing spectroscopy of, 390–392
  as carcinogens, 434
  exposure and toxicity of, 381, 382
  fluorescence spectroscopic detection of, 368, 370

as fluorophores, 330
gas chromatography of, 375
laser-excited molecular fluorescence detection of, 359
metabolites of, fluorescence line-narrowing spectroscopy of, 390–392
Shpol'skii spectroscopy of, 373, 375, 376
Polystyrene, octaethylporphyrin-doped, in spectral hole-burning studies, 199–200
Poly(tetrafluoroethylene) capillaries, for photochemical fluorometry, 95–99
Poly(vinyl alcohol), cresyl violet-doped, use in spectral hole-burning, 205–207
Poly(vinyl butyral):
   chlorin-doped, use in spectral hole-burning holography, 217
   cresyl violet-doped, use in spectral hole-burning, 175–177
   Nile red-doped, use in spectral hole-burning, 172–175
   use in spectral hole-burning, 159
Porous silica surfaces, chlorin and cresyl violet perchlorate adsorption onto, spectral hole-burning studies on, 212
Porphyrin(s):
   fluorometric detection of, 314
   as fluorophores, 330
   in Langmuir–Blodgett films, spectral hole-burning studies on, 212
   n-decane doped with, use in spectral hole-burning studies, 207
   n-octane doped with, use in spectral hole-burning studies, 191–192
   photoreversible tautomerism in, 170–172
   polyethylene doped with, in spectral hole-burning studies, 197–199, 201
   similarity to chlorophylls, 210
Porphyrin-like molecules, Shpol'skii spectroscopy of, 372
Potassium *tert*-butoxide, in chemiluminescent compound formation, 13
Power broadening, in spectral hole-burning, 189
Praseodymium:
   silicate glass doped with, use in spectral hole-burning, 200
   strontium fluoride doped with, use in spectral hole-burning, 187–189
Pressure, effects on spectral hole-burning, 186
Primaquine, photochemical fluorometry of, 119, 121, 126
Primary charge separation, in spectral hole-burning, 211
Primulin, as cellulose stain, 426
Procentrum micans, flow cytometric detection of, 427, 428–429
PRODAN:
   as polarity probe, 41, 42
   structure of, 43
Proflavine, as cellulose stain, 426
Progesterone, immunoassay using chemiluminescence, 7
Promethazine hydrochloride. See Fenergan
Propachlor, photochemical fluorometry of, 127
Propanil, photochemical fluorometry of, 127
2-Propanol, use in photoreduction reaction, 111
Propham, photochemical fluorometry of, 127
Propidium iodide:
   as nuclear stain, 402
   use in phytoplankton viability test, 430
6-Propionyl-2-(dimethylamino)naphthalene. *See* PRODAN
Propoxur, photochemical fluorometry of, 128
Proteins:
   autofluorescent activity of, 401
   binding studies on, on supported planar membranes, 297
   diode laser-induced fluorometric detection of, 366
   fluorometric detection of, 27, 313
   lipid interactions with, microspectrofluorometric studies on, 300
   in membranes, microspectrofluorometry of, 254
   rotational diffusion coefficients of, by polarized fluorescence microscopy, 271
   semiconductor laser fluorometric detection of, 239–241
   thermochemiluminescent immunoassay of, 8
Protoporphyrin, of zinc, fluorometric detection of, 311
Pseudolocal vibrations, in spectral hole-burning, 199–200
Pseudo-phonon sideband hole, in spectral hole-burning, 205, 206
Pseudoquadrupole interactions, spectral hole-burning studies on, 187–189

Pulsed near-infrared luminescence spectroscopy, 235–236
Pulse fluorometry, 34, 35
Pump-probe procedure, for transient spectral-hole detection, 164
Purpurogallin, photochemical fluorometry of, 108, 114
Pyracarbolid, photochemical fluorometry of, 127
Pyranine:
  as fluorescent probe, 30, 31
  structure of, 31
Pyrene:
  fluorescence spectra of, 43, 377, 378, 379
  as fluorescent probe, 30, 31, 43, 44, 54, 377, 378
  structure of, 31, 42
Pyrene-1-carboxaldehyde, structure of, 42
1-Pyrenedodecanoic acid, as fluorescent probes, 31
Pyridylbis(quinolylhydrazone), zinc fluorescence with, 310

Quadrupole splittings, in spectral hole-burning studies, 181, 182
Quantum efficiency, of fluorescene, 313
Quantum yields:
  of chemiluminescence, 4, 5, 11, 12, 16
  of fluorescence spectroscopy, 328, 330, 354
  of fluorescent probes, 35–36
Quenching fluorescence, of metal ions, 311
Quinine:
  as fluorophore, 330
  photochemical fluorometry of, 118, 121
Quinizarin:
  adsorption onto alumina powder, spectral hole-burning studies on, 212
  ethanol:methanol glass doped with, in spectral hole-burning studies, 201–202
Quinones, photochemical fluorometry of, 110–111, 112, 117, 118

Radioimmunoassays, of membrane receptors, 300
Raman scattering:
  in fluorescence spectroscopy, 329
  by lasers, 334, 359, 365
  in liquid chromatography-laser-induced molecular fluorescence, 350, 358
  in low-temperature spectroscopy, 371
  in spectral hole-burning, 196
Rayleigh scattering:
  in fluorescence line-narrowing spectroscopy, 389
  by lasers, 334
  in liquid chromatography-laser-induced molecular fluorescence, 350
Remote sensing, of phytoplankton, 434
Reserpine, photochemical fluorometry of, 122
Resonance energy transfer:
  combined with total internal reflection fluorescence microscopy, 277
  in microspectrofluorometry, 273–275
  rate equations for, 274
  in studies of molecular interaction in supported planar membranes, 299–300
  in two-dimensional systems, 274
Resonance Raman spectroscopy, ultrafast laser systems in, 342
Resorufin, in Langmuir–Blodgett films, spectral hole-burning studies on, 212
Reversed micellar systems, fluorometric measurements in, 318–319
Rhizosolenia, flow cytometric detection of, 426
Rhodamine and rhodamine derivatives, as fluorescent probes, 30, 33, 331
Rhodomonas, flow cytometric detection of, 424
Rhodomonas minuta, flow cytometric detection of, 426
RNA:
  flow cytometric analysis of, 421
  fluorometric detection of, 315
Rotational correlation time, in fluorescence depletion recovery, 272
Rotational diffusion coefficients:
  of antibodies on supported phospholipid bilayers, 296–298
  of lipids in supported phospholipid bilayers, 296, 298
  measurement by fluorescence recovery after photobleaching, 261
  measurement by polarized fluorescence microscopy, 269, 271–273
Rubrene, in chemiluminescent reactions, 4, 16–17
Ruby, fluorescence line narrowing detection in, 151

Saccharides, photochemical fluorometry of, 111
Salicylic acid, fluorometric detection of, 314–315
Samarium, barium fluorochlorobromide doped with, use in spectral hole-burning studies, 194–197, 213, 214
Saturation hole-burning, 182
Scanning fluorescence correlation spectroscopy, 269
Schiff bases, chemiluminescent reactions of, 14–15
Selenium hydride, centers in germanium–arsenic–selenium glass, in spectral hole-burning studies, 209
Selenoxanthenylium, as NIR dye, 246, 247
Semiconductor diode lasers, properties of, 342–344
Semiconductor laser, mode-locked, 236
Semiconductor laser fluorometry, 252
  flow cells for, 233–34
Semiconductor laser luminometry, in the near-infrared region, 230–232
Serum albumin:
  fluoroimmunoassay of, 317
  fluorometric detection of, 314
Sheath-flow cell:
  use in capillary electrophoresis, 345, 350, 364
  use in flow cytometry, 399
  use in laser-induced fluorescence detection, 346
Shot-noise background, in laser-excited molecular fluorescence, 349
Shpol'skii spectroscopy, 367, 368, 369, 372–382, 385, 389
  applications of, 375–383
  disadvantage of, 397
  emission–excitation matrix of, 374
  inhomogeneous broadening in, 372, 374, 377
  laser-excited, 370, 372–382, 397
  matrix-induced narrow-line spectra in, 372
  multiplet structure in, 373
  principles of, 372
  sample preparation in, 373
  time-resolved detection in, 380
Siduron, photochemical fluorometry of, 127
Signal-to-noise ratios, in laser-excited molecular fluorescence, 349–350
Signal transduction pathways, role of lateral mobility of cell membrane receptors in, 301
Silica gel, in photochemical fluorescence studies, 93
Silicate glass, praseodymium-doped, use in spectral hole-burning, 200
Single-cell fluorometry, of calcium, 310
Single molecule detection, laser use in, 434
Single molecule fluorescence detection technique, in spectral hole-burning studies, 201
Skatole, structure and chemiluminescence of, 13
Slit scanning, excitation sources for, 405
Smoluchowski expression, 52
Sodium dodecyl sulfate, in micelles for luminescence analysis, 139–140
Sodium taurocholate, use in luminescence analysis, 138, 139
Sodium taurodeoxycholate:
  helical aggregates proposed for, 145
  in micelles for luminescence analysis, 139–140, 144
Solid-state lasers, properties of, 339–340
Solid surfaces:
  fluorescent probe studies on, 27
  photochemical fluorescence equipment for, 92–94
  photochemical fluorometric studies on, 90, 119
Solubilization microenvironments, in bile salt media for luminescence analysis, 143–144
Solvation, effects on molecular parameters, spectral hole-burning studies on, 172–175
Solvent relaxation, around a fluorescent probe, 39, 40
Spatial modulation system, use in spectral hole-burning, 163
Spectral diffusion, spectral hole-burning studies on, 155, 200, 201, 202–204
Spectral hole-burning, 149–228
  in biological compound studies, 210–211
  efficiency of, 154, 208, 212, 215
  electronic interaction studies using, 176–183
  experimental techniques in, 155–164
  in fluorescence line-narrowing spectroscopy, 386, 389, 394
  hole bleaching in, 177

Spectral hole-burning *(Continued)*
  hole-burning mechanisms, 154, 165, 168, 177, 181, 184, 186, 188, 202, 207, 208, 210, 211, 215
  hole shift frequency in, 196
  hole width in, 153, 194, 203, 215
  in holography, 215–217
  homogeneously broadened spectral lines, 151, 152, 159, 190, 216
  in impurity and defect site studies, 165–176
  inhomogeneously broadened spectral lines in, 151, 166, 194, 216
  interference effects in, 161, 163, 176, 218, 219
  nonresonant hole-burning in, 165, 205
  optical data storage by, 215–217
  persistent type, 153, 155–163, 207
  in photon-gated spectral materials, 211–212
  saturation type, 153, 182–183
  scientific application of, 164–215
  spatial modulation system in, 163
  in studies of special materials, 209–215
  in studies of spectral diffusion in glasses, 200–205
  technological applications of, 215–221
  in thin-film studies, 211–212
  transient type, 153, 164
  tunneling process studies by, 183–189
  two-level systems in, 190
  in vibronics studies, 204–209
Spectral lines:
  homogeneously broadened. *See* Homogeneously broadened spectral lines
  inhomogeneously broadened. *See* Inhomogeneously broadened spectral lines
Spectrofluorometers, 103
Spectroscopic ruler, in electronic energy transfer, 63–71
Spinach chloroplast, spectral hole-burning studies on, 211
Spin flips, in spectral hole-burning studies, 183
Spin-lattice relaxation, in spectral hole-burning studies, 177
Spin-spin cross-relaxation, spectral hole-burning studies on, 177–180
Spot photobleaching, theory and applications of, 261–262
Squarylium dyes, as NIR dyes, 246
Stark effect, in spectral hole-burning, 160, 161, 163, 169–170, 172, 175, 219
Stereographic projections, from spectral hole-burning studies of electric dipole moments, 171
Stern–Volmer relation, 53
Steroids, high-performance liquid chromatography of, 11
Stokes–Einstein relation, 50, 59
Stokes shift, of fluorescent probes, 32–33, 40–41
Strontium fluoride, praseodymium-doped, use in spectral hole-burning, 187–189
Strontium magnesium fluorochlorobromide: samarium:
  in optical data storage by spectral hole-burning, 216
  in spectral hole-burning studies, 195
Strontium tungstate, uranium-doped, use in spectral hole-burning, 165–168
Strontium tungstate:uranium, spectral hole-burning in, 166, 167
Substitutional ligand disorder, in spectral hole-burning studies, 195
4-(2-Succinimidyloxycarbonylethyl)phenyl-10-methylacridinium-9-carboxylate fluorosulfonate:
  structure of, 12
  use in immunoassay, 12
Sugars, capillary electrophoresis of, 356
Sulfatide, laminin binding of, 292–293
Sulfonamides, photochemical fluorometry of, 122
Sulforidazine, photochemical fluorometry of, 122
Superhyperfine ground structures, spectral hole-burning studies on, 170, 182–183
Supersonic jet spectroscopy, laser use in, 434
Supported phospholipid bilayers:
  antibody binding studies on, 288–289, 292
  lipid binding studies on, 294
  preparation from Langmuir–Blodgett films, 283–285
Supported planar membranes:
  advantages of, 255–256
  lipid diffusion studies on, 293–296
  microspectrofluorometry in, 253–305
  preparation of, 282–286
Supramolecular rotors, examples of, 54
Surface adsorption, kinetic rate constant for, 275
Surface desorption, kinetic rate constant for, 275

INDEX   465

Surface quantitation of fluorescence, in total internal reflection fluorescence microscopy, 257–258
Surfactants:
  fluorescent probe studies on, 27, 29, 53
  trace anlysis by semiconductor laser fluorometry, 232

Tamoxifen, photochemical fluorometry of, 105, 106, 122, 123, 124–125
Tautomerism:
  photoreversible, in spectral hole-burning, 170
  of protons, photochemical hole-burning due to, 183
TDE. See Tetrakis(dimethylamino)ethylene
Terbium(III), use in fluoroimmunoassays, 317
p-Terphenyl, pentacene doped with, in spectral hole-burning studies, 201
2,2′,2″-Terpyridyl, fluorescent iron chelate of, 311
Tetracene:
  fluorescence line-narrowing spectroscopy of, 394, 395–396
  Shpol'skii spectroscopy of, 374
Tetrakis(dimethylamino)ethylene, chemiluminescence and structure of, 12–13
Tetramethylbenzene, dimethyl-s-tetrazine-doped, use in spectral hole-burning, 186–187
3,3,3′,3′-Tetramethyl-1,1′-dimethyl-4,5,4′,5′-dibenzoindotricarbocyanine perchlorate, NIH laser fluorometry of, 237
Tetramethylrhodamine isothiocyanate, use in labeling, 365
$meso$-Tetra-p-tetrabenzoporphyrin, spectral hole-burning studies on, 215
$Tetraselmis$, flow cytometric measurements of, 424
Texas red, histocompatibility antigen labeled with, supported bilayer studies on, 300
Theophylline, photochemical fluorometry of, 119, 121
Thin films:
  spectral hole-burning studies on, 211–212
  total internal reflection fluorescence microscopy on, 257–258
Thin-layer chromatography (TLC):
  fluorescence line-narrowing spectroscopy combined with, 392–396

  of labeled modified DNA, 390, 392
  use with laser-induced fluorescence, 359, 366
  use with liquid chromatography, 394–395, 396
  use with photochemical fluorometry, 86, 90, 92–93
Thiodiphenylamine, photochemical fluorometry of, 122
Thioindigo, benzoic acid doped with, use in spectral hole-burning, 183–186
Thioridazine, photochemical fluorometry of, 118, 121, 122
Thioxanthenylium, as NIR dye, 246
Thiram, photochemical fluorometry of, 128
Thyroxine, immunoassay using chemiluminescence, 7
Time-of-flight measurements. See under Flow cytometry
Time-resolved fluorescence spectroscopy, ultrafast laser systems in, 342
Time-resolved measurement, of fluorescent molecules, 330
Ti-sapphire lasers, 333
  properties of, 339–340
  use in low-temperature spectroscopy, 370
  wavelength tuning of, 333
Titration indicators, based on chemiluminescent reactions, 6–7
T-lymphocytes, supported bilayer studies on, 300
TMA-DPH:
  as fluorescent probe, 30, 31
  structure of, 30
TNS:
  as polarity probe, 42, 43
  structure of, 42
2-p-Toluidinyl-6-naphthalene sulfonate. See TNS
Total internal reflection anisotropy, 276–277
Total internal reflection fluorescence microscopy (TIRFM), 256–261
  antibody binding studies using, 288–289
  combined with fluorescence correlation spectroscopy, 275–276
  combined with fluorescence recovery after photobleaching, 275–276
  combined with resonance energy transfer, 277
  evanescent wave decays in, 257
  excitation in, 256–257

TIRFM *(Continued)*
  fluorescence recovery after photobleaching in, 261–267
  fluorescence quantitation in, 259–261
  fringe fluorescence recovery after photobleaching with, 276
  lack of artifacts in, 300–301
  in studies of antibody binding, 288–290
  in studies of laminin-sulfatide binding, 288–290
  in studies of protein binding, 286–287
  surface quantitation of fluorescence, 257–258
  thin film effects in, 257–258
Trace analysis, using chemiluminescent systems, 2
Tranquilizers, photochemical fluorometry of, 119, 123
Translational diffusion, of lipids, on supported planar membranes, 293–296
Translational diffusion coefficient, calculation from fluorescence correlation spectroscopy, 268
Transparent electrodes, preparation of, 283
Trichloropyrene, fluorescence line-narrowing spectroscopy of, 394
Tricyanoaminopropene, fluorescent copper complex of, 311
Trifluoperazine, photochemical fluorometry of, 112, 118, 121
1-(4-Trimethylammoniumphenyl)-6-phenyl-1,3,5-hexatriene, *p*-toluene salt of. *See* TMA-DPH
3,3,4-Trimethyl-1,2-dioxetane:
  chemiluminescence of, 4, 8
  structure of, 9
Triofluralin, photochemical fluorometry of, 127
Trioxan, structure and chemiluminescent reaction of, 14
2,4,5-Triphenylimidazole. *See* Lophine
Triplet bottleneck, in spectral hole-burning, 153, 203, 210
Triplet state, lifetime of, in fluorescence depletion recovery, 271–272
Trypan, as cell stain for flow cytometry, 401
Tryptamine, photochemical fluorometry of, 118, 121
Tryptophan, fluorometric detection of, 118, 121, 313
Tumor cells, flow cytometric studies on, 398

Tunneling processes, in spectral hole-burning, 183–189, 201, 204, 211
Twisted molecular charge transfer state, in molecular rotors, 51
Two-level systems, in spectral hole-burning, 197, 199, 201, 202, 204
Two-photon-excited fluorescence spectroscopy:
  laser-induced, 333, 336, 345, 354, 357, 359–360, 366
  one-photon-excited fluorescence detection compared to, 360

Ultrafast laser systems, properties of, 340–342
Uranium, strontium tungstate doped with, use in spectral hole-burning, 165–168
Uric acid, fluorometric detection of, 314
Urine, low-temperature luminescence spectra of, 318

Van der Waals solid, organic molecule–doped, use in spectral hole-burning studies, 207–209
Vesicle fusion techniques, in microspectrofluorometry, 285–286, 298–299
Vesicles, fluorescent probe studies on, 27
Vibrational fine structure, of spectra, 367–368
Vibronic bands:
  of fluorescent probes, 142, 143, 144
  polarity estimation from changes in, 43–44
Vibronic fine structure, in fluorescence spectroscopy, 328
Vibronics, spectral hole-burning studies on, 204–209
Vitamin K_1, photochemical fluorometry of, 111–112, 113, 118, 121, 123, 125
Vitamins, fluorometric detection of, 314

Water, pesticide detection in, 127, 128
Wavefunction symmetry, spectral hole-burning studies on, 170
Wavelength, of lasers, 333–334
Williams, Landel, and Ferry relation, 48
Wobble-in-cone model, for characterization of bilayers, 60–61

Xanthenylium, as NIR dye, 246

OHIO UNIVER
Please return

Xanthine, near-infrared semiconductor laser fluorometric assay of, 241–244
Xanthine oxidase, near-infrared semiconductor laser fluorometric assay of, 241–244

Yttrium aluminum perovskite, europium-doped, use in spectral hole-burning, 180–182

Yttrium lithium fluoride, erbium-doped, use in spectral hole-burning, 180–182

Zeeman effect, in spectral hole-burning, 163, 165, 167, 177, 179, 182, 183
Zinc, fluorometric detection of, 310
Zinc protoporphyrin, luminescence spectroscopy of, 311

QD 96 .L85 M65 1985 pt.3

Molecular luminescence
 spectroscopy

(*continued from front*)

Vol. 63. **Applied Electron Spectroscopy for Chemical Analysis.** Edited by Hassan Windawi and Floyd Ho

Vol. 64. **Analytical Aspects of Environmental Chemistry.** Edited by David F. S. Natusch and Philip K. Hopke

Vol. 65. **The Interpretation of Analytical Chemical Data by the Use of Cluster Analysis.** By D. Luc Massart and Leonard Kaufman

Vol. 66. **Solid Phase Biochemistry: Analytical and Synthetic Aspects.** Edited by William H. Scouten

Vol. 67. **An Introduction to Photoelectron Spectroscopy.** By Pradip K. Ghosh

Vol. 68. **Room Temperature Phosphorimetry for Chemical Analysis.** By Tuan Vo-Dinh

Vol. 69. **Potentiometry and Potentiometric Titrations.** By E. P. Serjeant

Vol. 70. **Design and Application of Process Analyzer Systems.** By Paul E. Mix

Vol. 71. **Analysis of Organic and Biological Surfaces.** Edited by Patrick Echlin

Vol. 72. **Small Bore Liquid Chromatography Columns: Their Properties and Uses.** Edited by Raymond P. W. Scott

Vol. 73. **Modern Methods of Particle Size Analysis.** Edited by Howard G. Barth

Vol. 74. **Auger Electron Spectroscopy.** By Michael Thompson, M. D. Baker, Alec Christie, and J. F. Tyson

Vol. 75. **Spot Test Analysis: Clinical, Environmental, Forensic and Geochemical Applications.** By Ervin Jungreis

Vol. 76. **Receptor Modeling in Environmental Chemistry.** By Philip K. Hopke

Vol. 77. **Molecular Luminescence Spectroscopy: Methods and Applications** (*in three parts*). Edited by Stephen G. Schulman

Vol. 78. **Inorganic Chromatographic Analysis.** Edited by John C. MacDonald

Vol. 79. **Analytical Solution Calorimetry.** Edited by J. K. Grime

Vol. 80. **Selected Methods of Trace Metal Analysis: Biological and Environmental Samples.** By Jon C. VanLoon

Vol. 81. **The Analysis of Extraterrestrial Materials.** By Isidore Adler

Vol. 82. **Chemometrics.** By Muhammad A. Sharaf, Deborah L. Illman, and Bruce R. Kowalski

Vol. 83. **Fourier Transform Infrared Spectrometry.** By Peter R. Griffiths and James A. de Haseth

Vol. 84. **Trace Analysis: Spectroscopic Methods for Molecules.** Edited by Gary Christian and James B. Callis

Vol. 85. **Ultratrace Analysis of Pharmaceuticals and Other Compounds of Interest.** Edited by S. Ahuja

Vol. 86. **Secondary Ion Mass Spectrometry: Basic Concepts, Instrumental Aspects, Applications and Trends.** By A. Benninghoven, F. G. Rüdenauer, and H. W. Werner

Vol. 87. **Analytical Applications of Lasers.** Edited by Edward H. Piepmeier

Vol. 88. **Applied Geochemical Analysis.** by C. O. Ingamells and F. F. Pitard

Vol. 89. **Detectors for Liquid Chromatography.** Edited by Edward S. Yeung

Vol. 90. **Inductively Coupled Plasma Emission Spectroscopy: Part I: Methodology, Instrumentation, and Performance; Part II: Applications and Fundamentals.** Edited by J. M. Boumans

Vol. 91. **Applications of New Mass Spectrometry Techniques in Pesticide Chemistry.** Edited by Joseph Rosen